International Mining Geology Conference 2026

21–22 April 2026
Brisbane, Australia

The Australasian Institute of Mining and Metallurgy
Publication Series No 1/2026

AusIMM

Published by:
The Australasian Institute of Mining and Metallurgy
Ground Floor, 204 Lygon Street, Carlton Victoria 3053, Australia

All papers published in this volume were peer reviewed before publication.

ISBN 978-1-922395-55-9

Advisory Committee

Daniel Howe
MAusIMM
Conference Advisory Committee Chair

Fiona Bodycoat
MAusIMM(CP)

Angela Dimond
MAusIMM

Jan Graham
MAusIMM(CP)

Rayleen Hargreaves
MAusIMM(CP)

Niamh McMunn
MAusIMM

Danielle Robinson
FAusIMM(CP)

Dale Sims
FAusIMM(CP)

Krista Sutton
MAusIMM

AusIMM

Julie Allen
General Manager, Events

Michelle Portelli
Senior Event Manager, Events

Sasha Pell
Senior Manager, Conference Growth, Events

Sharon Lim
Program Manager, Events

James Maters
Senior Manager, Events Marketing and Communications, Events

Reviewers

We would like to thank the following people for their contribution towards enhancing the quality of the papers included in this volume:

Mark Berry
James Biggam
Paul Blackney
Geoff Booth
John Carswell
Simon Dominy
Christopher Esterhuizen
Ian Glacken
Sean Helm
James Llorca
Steve Loach
Hamish Pescini
Geoff Phillips
James Pocoe
Bill Shaw
Chris Taylor
Jill Terry
Gregory Zhang

Foreword

On behalf of the Conference Advisory Committee, we are delighted to welcome you to the AusIMM's International Mining Geology Conference 2026.

The conference will highlight the depth, diversity, and innovation of our global mining geology community at a time when the industry is transforming more rapidly than ever before. Across the technical program, you will find new approaches to orebody knowledge, advances in geological modelling, practical solutions that enhance reconciliation and create value in modern mining operations.

Mining geologists continue to sit at the critical intersection of science, technology, and value creation. The committee commends the authors for their contributions and their commitment to advancing our collective understanding. I trust that these proceedings will inspire collaboration, spark new ideas, and support the ongoing growth of our profession. We also acknowledge the work of the many peer reviewers who volunteer their time to guide the authors towards delivering a very high standard of technical papers.

Whether you are a seasoned industry professional, graduate, researcher, or student, we look forward to welcoming you to the AusIMM's International Mining Geology Conference 2026.

Yours faithfully,

Daniel Howe

MAusIMM

International Mining Geology Conference 2026 Advisory Committee Chair

Sponsors

Major Conference Sponsor

BHP

Supporting Partner

Australian Institute of Geoscientists

Platinum Sponsors

DATAMINE
SNOWDEN Optiro

tNavigator®
Rock Flow Dynamics

Gold Sponsors

Evolution MINING

GOLD FIELDS

MAPTEK

micromine

GeologicAI

Networking Function Sponsor

Rio Tinto

Young Professionals Day Sponsor

Rio Tinto

Coffee Cart Sponsors

Capricorn Metals Ltd

Intertek — Total Quality. Assured.

Lanyard Sponsor

entech. — Mineral Resources | Mentoring | Mining

Contents

Geology – New orebody insights

In-mine exploration – Extending mine life

Innovation – Rethinking mining geology

Open to anything – Case studies

Reconciliation – Data to decisions

Value creation – Plan to product

Geology

New orebody insights

Bridging geochemistry and mineralogy – an element-to-mineral model workflow to enhance orebody knowledge

V Andrade[1], C Curtis-Morar[2] and V Rocha[3]

1. Geologist, Glencore, Brisbane Qld 4000. Email: vinicius.andrade@glencore.com.au
2. Senior Geometallurgist, Glencore, Brisbane Qld 4000. Email: catherine.curtis@glencore.com.au
3. Senior Resource Geologist, Glencore, Brisbane Qld 4000.
   Email: vinicius.rocha@glencore.com.au

ABSTRACT

Distinguishing recoverable from non-recoverable metal-bearing mineral phases is critical for defining recoverable resources and forecasting metallurgical performance. However, comprehensive mineralogical characterisation is often constrained by analytical cost and limited sample availability, limiting the ability to map element deportment at the scale required for mine planning, processing strategy selection and revenue estimation.

To address this limitation, a Python-based element-to-mineral model (EMM) workflow was developed to predict modal mineralogy from geochemical assays, using quantitative X-ray diffraction (QXRD) mineralogy and scanning electron microscopy (SEM)-QEMSCAN-derived mineral chemistry as calibration anchors. The workflow was developed for the Black Star Open Cut (BSOC) Feasibility Study in Mount Isa, Queensland, and targets orebodies where metal deportment varies between sulfide and non-sulfide hosts and where chemical signatures overlap across mineral groups, meaning that simple stoichiometric approaches can produce non-unique or unstable mineral estimates.

INTRODUCTION

Methodology overview

The EMM is framed as a linear mass-balance inversion, where modal mineralogy multiplied by an elemental chemistry matrix reproduces geochemical assays. QXRD and SEM-QEMSCAN measurements were collected on subsamples taken from the assay pulp rejects, ensuring consistent sampling support across data types. Mineral calculation is posed as a regularised optimisation that minimises assay misfit while enforcing non-negativity, closure and simple penalty terms to stabilise underdetermined systems in the training data set, so that metal deportment between sulfides and non-sulfides is tracked at block scale.

Workflow and implementation

The implementation in Python is organised into three sequential scripts aligned to uncertainty analysis, calibration and assay-only deployment integrating assays, QXRD and SEM-QEMSCAN mineral chemistry (Figure 1). Geometallurgical domain-specific mineral chemistry and mineral range constraints are defined and then refined through global optimisation to enhance estimation accuracy.

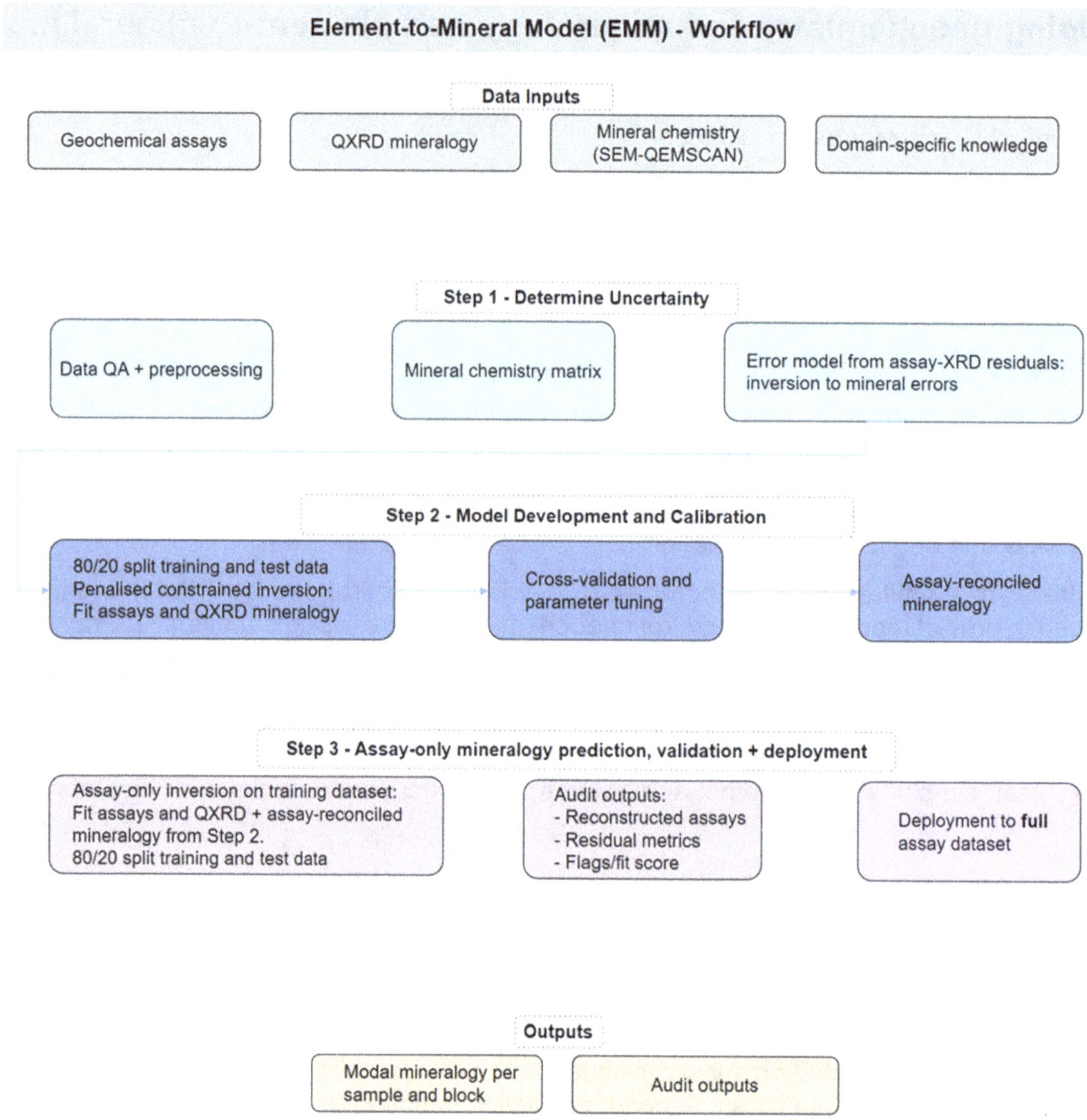

FIG 1 – The EMM workflow follows: [Step 1] Data preparation and uncertainty estimation; [Step 2] Calibration against assay + quantitative mineralogy; and [Step 3] Assay-only mineral prediction with robust optimisation and domain constraints.

Uncertainty and conditioning

Script 1 preprocesses assays, builds a matrix from mineral compositions, diagnoses conditioning, applies small ridge regularisation where near collinearity exists, and empirically estimates uncertainty using residual distributions between reconstructed and observed assays.

Calibration through mineralogy recalculation

Script 2 solves a penalised constrained inversion on training data where both assays and quantitative mineralogy are available, honouring assays while leveraging measured mineralogy as a reference. Regularisation hyperparameters are tuned via cross-validation to improve predictive accuracy without overfitting.

Assay-only deployment

Script 3 predicts mineralogy from assays only, using the calibration-informed constraints derived in earlier steps and validating results against measured or recalculated mineralogy where available. For block-model deployment, only this step is required, as the trained inversion can be applied directly to block estimations.

Case study framework

Contrasting geometallurgical domains are used as a case study, in which the same workflow is applied using domain-specific mineral libraries, mineral chemistry, and constraint ranges. Transitional zones are emphasised because increased variability in mineral species affects metallurgical performance, making block-scale mineralogical calculation critical for forecasting recoverable Zn and Pb. The training data set samples ranged from 5 to 28 SEM-QEMSCAN samples and from 33 to 144 QXRD samples per domain.

The workflow is demonstrated across two primary and one transitional geometallurgical domains. The evaluation data set comprises more than 260 000 ore blocks that meet the economic cut-off. For each block, the EMM predicts economic and gangue modal mineralogy, enabling a mineralogy-informed recovery calculation. For this case study, the authors assumed that only metal deportment drives metallurgical performance to assess the potential impact of using a mineralogy-informed metallurgical recovery model. Block metallurgical recovery predictions were generated using the existing recovery equations re-expressed in terms of sulfide-hosted metal and compared against the existing assay-only implementation.

Analysis of the studied domains showed that the calculated content of sphalerite ranged from near zero to about 25 per cent ($\mu \approx 5$ per cent). This indicates variable alteration and the presence of low sulfide-hosted metal areas within the domains (Figure 2a).

Figure 2b displays the block model coloured by the ratio between sulfide-hosted Zn and total Zn. The results indicate that Zn deportment is predominantly to sphalerite. However, the western zone shows some proportion of Zn hosted in Zn non-sulfides, whereas the eastern zone shows a lower sulfide-hosted Zn ratio, indicating a greater proportion of Zn hosted in non-sulfide minerals.

A small tolerance band was used to assess the potential impact of the EMM on Zn recovery. In general, at the block-model level, the mineralogy-informed metallurgical recovery model shows potential uplift in Zn recovery. In the section, negative deltas cluster where the EMM redirects Zn into non-sulfides, while positive deltas coincide with sulfide-rich ores, illustrating how the model translates metal deportment into the block model (Figure 2c).

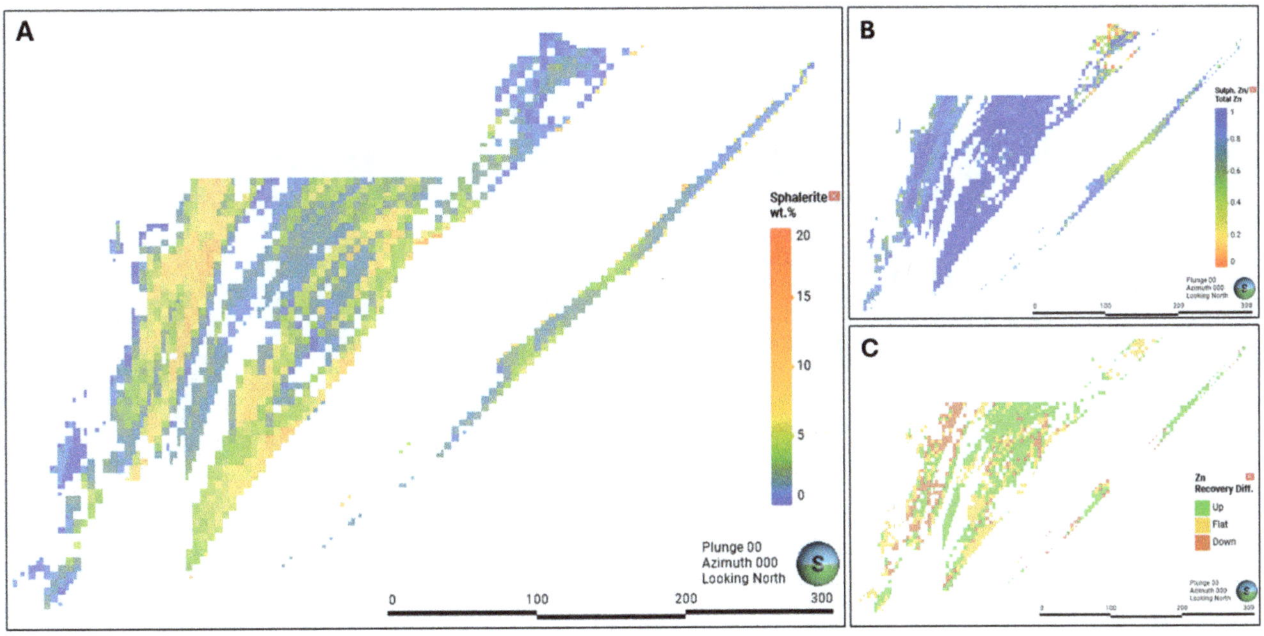

FIG 2 – Representative north-looking section through the combined domains illustrating EMM outputs on ore blocks with trusted mineral prediction. (a) Predicted sulfide modal abundance (wt per cent) from EMM. (b) Ratio between sulfide-hosted Zn and total Zn. (c) Change in block-level metal sulfide recovery between the mineralogy-based and baseline formulations grouped into three classes: Up (increase greater than one percentage point), Down (decrease greater than one percentage point), and Flat (change within range).

Mining geology value and integration pathway

The intended mining geology outcome is a block-model-ready mineralogical variable set derived from assays that can be incorporated into resource model workflows and used to inform geometallurgical domaining, recovery modelling and economic assumptions, with recovery equations re-expressed in terms of EMM-derived recoverable metal in sulfides rather than total grades, while retaining their original structure and constants. When extended across domains, predicted mineralogy supports a shift from generic grade proxies toward mineralogy-informed frameworks and provides a mechanism to back-validate and refine domaining by testing whether predicted mineralogical transitions align with current domain boundaries.

CONCLUSIONS

The EMM provides a reproducible Python-based workflow that links assays, QXRD mineralogy and SEM-QEMSCAN mineral chemistry in a constrained, regularised inversion. When implemented, the proposed model delivers chemically consistent, geologically plausible mineral predictions. The case study shows that these predictions can meaningfully adjust block-level recovery prediction and relative economic response for mine-planning decisions.

Innovations in density modelling at Bingham Canyon – integrating advanced technologies and best practices

J F Ellis[1], G Austin[2], D Bagshaw[3], R Butz[4], G Chapdelaine[5], P Collier[6], J Heim[7] and M Paine[8]

1. Superintendent Mine Geology and Resource, Rio Tinto, Salt Lake City UT 84095, USA. Email: jennifer.ellis@riotinto.com
2. Senior Geologist Resources, Rio Tinto, Salt Lake City UT 84095, USA. Email: gerry.austin@riotinto.com
3. Geologist Underground Technical Studies, Rio Tinto, Salt Lake City UT 84095, USA. Email: don.bagshaw2@riotinto.com
4. Geologist, Rio Tinto, Salt Lake City UT 84095, USA. Email: robin.butz@riotinto.com
5. Geologist, Rio Tinto, Salt Lake City UT 84095, USA. Email: gordon.chapdelaine@riotinto.com
6. Principal Advisor OBK Risk and Assurance, Rio Tinto, Brisbane Qld 4000. Email: perry.collier@riotinto.com
7. Practice Leader Exploration and Evaluation, Rio Tinto, Brisbane Qld 4000. Email: jonathan.heim@riotinto.com
8. Practice Leader OBK Technology, Rio Tinto, Perth WA 6000. Email: mark.paine@riotinto.com

INTRODUCTION

In situ bulk density modelling is fundamental to resource estimation, mine planning, geotechnical design, and reconciliation. Because density directly affects tonnage estimates, minor errors can significantly impact metal or product estimates. Industry standards such as the JORC Code require density to be treated with the same rigour as grade, including transparent documentation of sampling methods, porosity and moisture treatment.

At Rio Tinto Kennecott's Bingham Canyon Mine, recent innovations have improved density model accuracy through advanced measurement technologies and new workflows, including disciplined validation. The key developments include CoreLok® for dry bulk density in porous materials and the first operational deployment of downhole muon tomography in an active open pit. Combined with robust QA/QC and geostatistical modelling, these approaches deliver more representative and reliable models for mine planning and operational decision-making.

BEST-PRACTICE FRAMEWORK

Rio Tinto's Orebody Knowledge (OBK) guidelines require systematic characterisation of wet and dry bulk density across ore and waste, continued data acquisition through to production, and standardised metadata management. The framework emphasises representative sampling across lithologies, alteration, weathering and mineralisation types; appropriate measurement methods with calibration and QA/QC; and integration into block models via domain averages, regression, or geostatistical interpolation (eg kriging/co-kriging). Density data must be treated like assay data, using duplicates, replicates, standards, and cross-checks with independent methods, and reconciliation against production tonnages (wet versus dry). In addition, spatial analysis is used to identify sampling gaps or biases, effectively reducing risks in tonnage estimation and supporting defensible mineral resource classification for public reporting.

CoreLok® – addressing bias in porous and friable materials

Traditional water displacement methods are prone to error in porous or friable materials, where air gaps, water absorption, and sample degradation distort volume measurements. CoreLok® mitigates these issues by vacuum sealing samples in polymer bags before applying Archimedes' principle, enabling rapid and reliable dry bulk density and porosity measurements. In practice, CoreLok® is helping to reduce historical measurement bias, improving representation of low density lithologies, and is now recognised as leading practice at multiple Rio Tinto sites. Our approach aligns with standard QA/QC workflows through routine instrument calibration, duplicate measurements across material types (Figure 1). This layered approach improves confidence in density estimates, particularly for unconsolidated, altered, or highly absorptive units that have traditionally been under-sampled or mischaracterised.

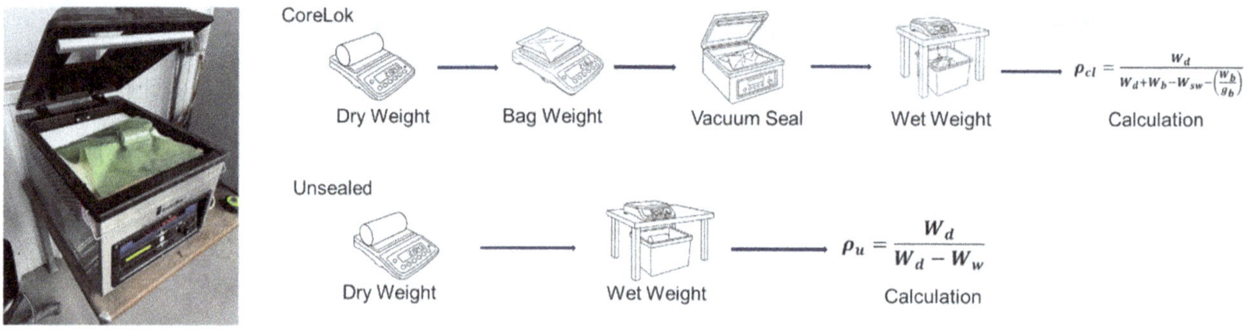

FIG 1 – Overview of CoreLock® system and unsealed traditional water-displacement method.

Muon tomography – volumetric imaging for orebody knowledge

Muon tomography uses cosmic ray muons to image subsurface density contrasts at metre-scale resolution (Figure 2). In 2024, Rio Tinto Kennecott partnered with Ideon Technologies to deploy the world's largest downhole muon system at Bingham Canyon Mine, marking the first application in an active open pit mine. Using existing drill holes in the Revere mining area, billions of muon trajectories were inverted into 3D density estimates, revealing lithological and structural features, as well as alteration halos critical for characterising porphyry systems. Muon tomography provides continuous volumetric imaging, minimising sampling bias and bridging gaps between drill holes, without disrupting operations. As such, it complements data from CoreLok® and conventional logging, providing a bridge between discrete measurements and model scale variability. Bingham Canyon's deployment of muon tomography demonstrates how particle physics imaging can be applied in an active mine to improve orebody knowledge in-line with corporate and industry standards.

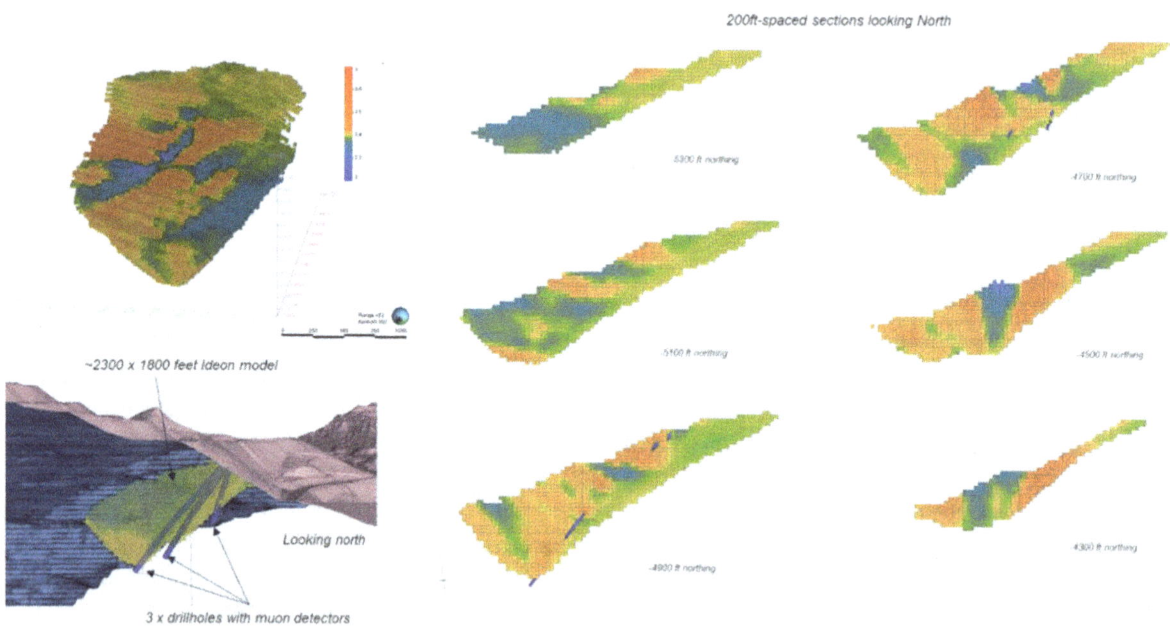

FIG 2 – Muon-derived density model generated from data collected in the Revere area of the Kennecott Bingham Canyon open pit during the project.

By reducing sampling bias and improving volumetric coverage, this combined approach increases confidence in resource classification and supports proactive geotechnical risk management. The success at Bingham Canyon also highlights the scalability of this workflow to other large, complex deposits, paving the way for broader adoption of particle-physics imaging and advanced laboratory techniques within OBK-compliant workflows.

CONCLUSION

An integrated workflow combining OBK best practice, CoreLok® for challenging materials and muon tomography for volumetric imaging has improved density model fidelity and operational utility at

Bingham Canyon. This approach demonstrates how advanced geophysical techniques, combined with rigorous QA/QC, deliver models that enhance planning, reconciliation and geotechnical confidence in an active mining environment.

ACKNOWLEDGEMENTS

We thank Rio Tinto leadership for supporting presentation and work. We extend our sincere thanks to the Ideon Technologies team for their invaluable contributions to this project, including strategic guidance, operational support, technical expertise and data analysis. Their collaboration was instrumental in the successful deployment and interpretation of muon tomography at Bingham Canyon.

Evaluating the reliability of bulk density measurements in nickel laterite deposits using predictive modelling

I Lipton[1], R Carlson[2], R Setyawan[3], F Raseno[4] and L Torres[5]

1. FAusIMM, Principal Geometallurgist, AMC Consultants Pty Ltd, Brisbane Qld 4000. Email: ilipton@amcconsultants.com
2. MAusIMM, Technical Lead Geoscience, AMC Consultants Pty Ltd, Brisbane Qld 4000. Email: rcarlson@amcconsultants.com
3. Mineral Resource Inventory Geologist, PT Vale Indonesia, MGEI, Sorowako, Sulawesi, Indonesia. Email: reza.setyawan@vale.com
4. MAusIMM, Mineral Resource Inventory Geologist, PT Vale Indonesia, Sorowako, Sulawesi, Indonesia. Email: fisco.raseno@vale.com
5. Senior Data Scientist, AMC Consultants Pty Ltd, Melbourne Vic 3000. Email: ltorres@amcconsultants.com

ABSTRACT

Accurate measurement of dry bulk density is essential for Ore Reserve estimates and tracking the mass and grade of ore during mining operations. This is particularly so in young, tropical nickel laterite deposits, which typically show large variations in dry bulk density depending on ore composition. Diamond core drilling is the method of choice for definition of mineral resources in these deposits. If carefully executed, core drilling can provide excellent sample recovery rates and dry bulk density data.

An unusual feature of the core recovery in young tropical nickel laterites is that, when drilling through very soft materials, recoveries of significantly greater than 100 per cent are commonly recorded over multiple contiguous drilled intervals. The cause of this phenomenon has been debated for decades. The critical question is: are these extra-long recovery samples representative of the grade and bulk density of the *in situ* drilled interval?

As part of a study into the measurement and modelling of bulk density within nickel laterite mines at Sorowako in Indonesia, the origin of these extra-long recovery samples was investigated. The methods used for core drilling and density measurement were reviewed. A statistical comparison of the geochemistry of extra-long recovery and normal (recovery less than or equal to 100 per cent) samples was completed. Machine learning was used to predict bulk density from multivariate assay data. Comparison of the measured bulk density and predicted bulk density of the extra-long recovery samples showed that the measured bulk densities are generally reliable. This leads to identification of a physical explanation of the phenomenon of extra-long cores in nickel laterite deposits.

INTRODUCTION

Accurate measurement of bulk density is essential for Mineral Resource and Mineral Reserve/Ore Reserve estimates and tracking the mass and grade of ore and waste as they move along the chain of mining operations. Bulk density is controlled by the mineral assemblage and porosity. In many mineral deposits, the association of dense sulfide minerals and iron oxides with the minerals of economic interest results in wide ranges of density values in the mineralised zones (Parrish, 1993; Lipton, 2000). Consequently, in these deposits, the application of average bulk densities to mineralised domains introduces errors in the Mineral Resource estimates and is unacceptable. Collection of precise and unbiased bulk density data is therefore a prerequisite for robust estimation of Mineral Resources (Dominy, Noppe and Annels, 2002).

Selection of an appropriate method for measuring bulk density requires consideration of the physical characteristics of the rock being measured, such as friability, porosity, presence of vughs, tendency to slake, and moisture content. The most relevant measures of density for mining purposes are dry bulk density, which is used for the estimation of Mineral Resources, and wet bulk density which must be considered when estimating material movements, such as truck and conveyor haulage, and when forecasting the characteristics of mine products delivered to processing plants. There are many valid methods of measuring the bulk density of drill core (Lipton and Horton, 2014) and correct choice of method should allow any scenario to be addressed.

The necessity of sampling continuously along the length of a mineralised drill hole intercept to avoid the risk of sub-sampling errors and biases has long been recognised with respect to assay samples but is commonly ignored when measuring bulk density. Lipton (2000) demonstrated that measuring the density of discontinuous pieces of core much shorter than the assay samples can lead to significant biases. This is a common problem in mineral deposits where alteration and mineralisation are inhomogeneous but is easily rectified by selecting a method that measures continuous core intervals.

This paper considers density data from the Hill X nickel laterite deposit in Central Sulawesi, Indonesia. This deposit forms part of the Sorowako mining operations of PT Vale Indonesia (PTVI). PTVI's method of choice for definition of Mineral Resources is diamond core drilling which provides excellent sample recovery and core that is suitable for bulk density measurement. Bulk density within the Hill X deposit is highly variable as a result of the large variations in chemical composition through the laterite profile. PTVI's approach to the measurement of bulk density is based on measurement of the whole core and avoids the risks associated with subsampling.

An unusual feature of the core recovery at Hill X and many other young tropical nickel laterites is that, when drilling through very soft materials, core recoveries of significantly greater than 100 per cent are commonly recorded over multiple consecutive drilled intervals. The cause of this phenomenon has been debated for decades and various theories have been advanced including caving of the drill hole sidewall, or length-wise expansion of the core after it is cut by the drill bit. Whatever the cause, the critical question has always been: are these extra-long recovery samples representative of the grade and bulk density of the *in situ* drilled interval? This is an important consideration for Mineral Resource estimation and mine production reconciliation.

As part of a study into the measurement and modelling of bulk density for the Hill X deposit, the possibility of predicting bulk density from the assay data was investigated. This, in turn, led to an opportunity to evaluate the reliability of the extra-long recovery samples.

GEOLOGY

The nickel laterite deposits of the Sorowako area are underlain by ultramafic rocks which have been highly serpentinised and structurally modified (highly fractured) by north-west–south-east fault zones. These tectonic features create a rugged landform of ridges, steep hills, and dendritic drainage patterns, which strongly influence lateritisation processes.

Nickel laterite deposits form from ultramafic rocks by the dissolution of magnesium and other elements by vertically percolating groundwater in tropical environments and residual accumulation of iron and aluminium (Golightly, 2010). This geochemical differentiation produces a stratified laterite profile: unaltered bedrock at the base, saprolite zone, an important intermediate/transitional horizon often referred to as the smectite or nontronite zone, and ferruginous zone at the top (Figure 1).

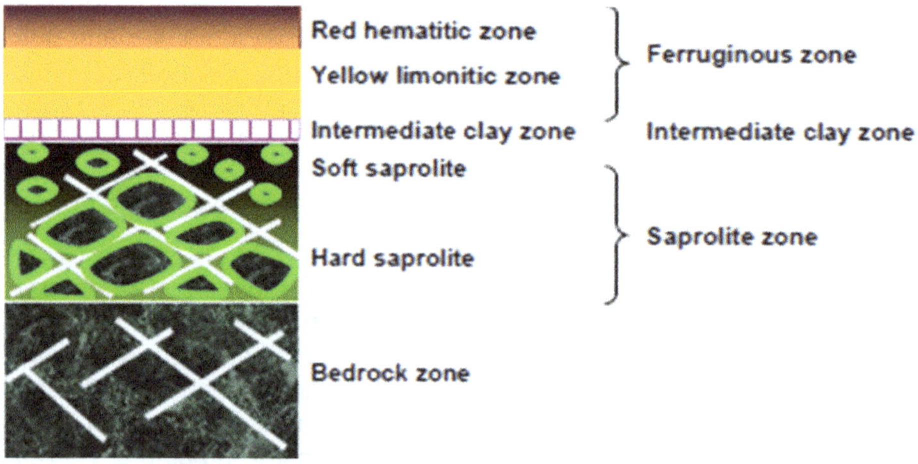

FIG 1 – Schematic section through the nickel laterite profile (Ahmad, 2008).

The fresh bedrock consists of serpentinised harzburgite, lherzolite, and dunite, which typically have dry bulk densities above 2.6 t/m^3.

The saprolite zone represents an intermediate stage of weathering where magnesia, silica, and alkalis are leached rapidly, leaving residual concentrations of iron, aluminium, chromium, and manganese. This zone varies between 6 m and 12 m thick, averages about 10 m thick, and commonly contains minerals such as nontronite at the contact with boulders of bedrock.

The intermediate or transition zone, generally 0.5 m to 3.0 m thick, is characterised by soft goethite supported by a very fine-grained network of hard crystalline quartz, which preserves much of the original rock texture and structure. Drill hole samples from the intermediate zone at Hill X were analysed using high-resolution backscattered electron energy dispersive X-ray mapping with a TIMA-X FEG GMH scanning electron microprobe system (SGS, 2023). The mapping showed that the transitional zone is dominated by goethite and aluminous goethite (Figure 2) which, in the core, have a massive clay-like, earthy appearance. Phyllosilicates such as talc and chlorite and trace nontronite are also present. The development of this horizon depends strongly on climatic conditions: in regions with heavy rainfall year-round, silica and magnesia are completely flushed out, preventing clay formation; whereas in tropical wet-dry climates, limited chemical weathering allows some magnesia and silica to remain, forming smectite/nontronite clays. This zone often exhibits the highest porosity within the weathering profile, with abundant intergranular spaces and larger voids created by leaching of ferromagnesian minerals and underground water channels.

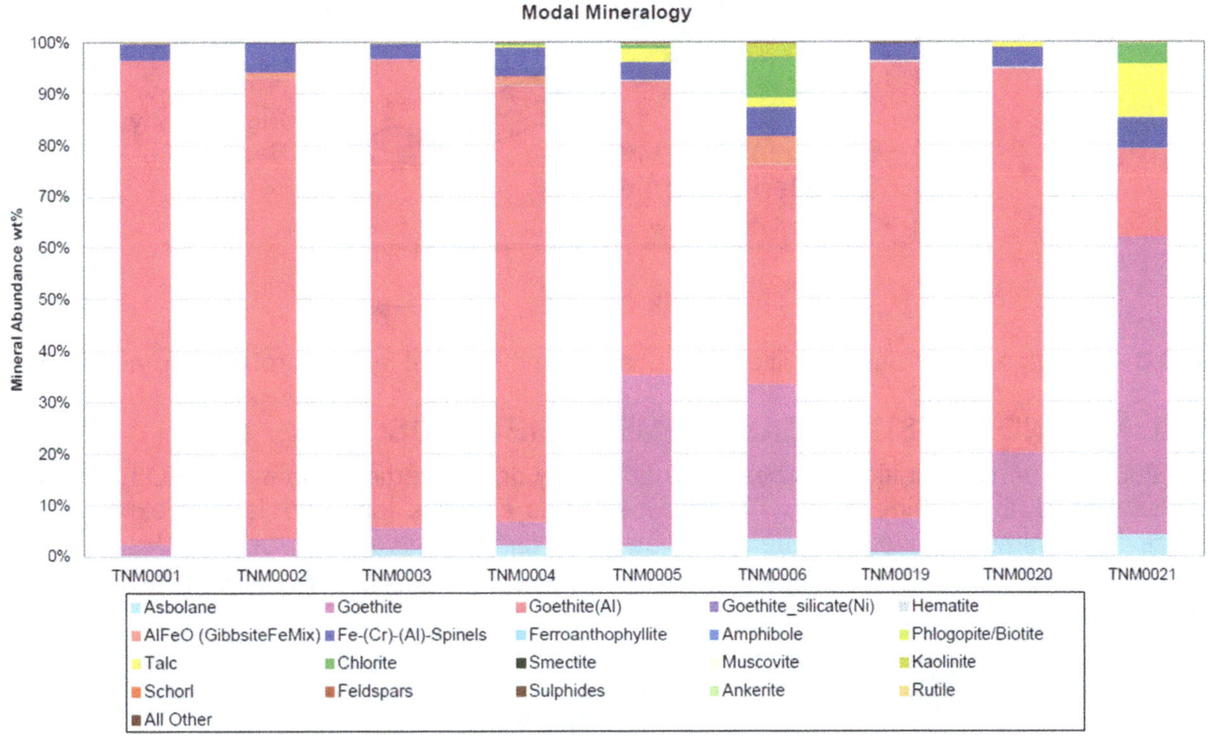

FIG 2 – Modal mineralogy of intermediate/transition zone samples from Hill X.

The ferruginous zone, generally 0.5 m to 2.0 m thick, is the zone where residual concentration of non-mobile elements has reached its maximum value. All soluble components have been leached leaving behind essentially oxides of iron, aluminium and manganese. The upper part of the ferruginous zone is rich in goethite and hematite while the lower part is composed of more hydrated iron oxides generally grouped under 'limonite'. The ferruginous zone represents laterite that has collapsed under its own weight. Due to collapse, the original structure and texture of the rock is completely obliterated and the dry bulk density increases (Ahmad, 2008).

Nickel mineralisation in laterites originates entirely from the leaching of primary minerals such as olivine, pyroxene, and nickeliferous magnetite, with no external source. While residual nickel occurs throughout the profile, its highest concentrations are typically found in the saprolite zone as secondary or supergene enrichment minerals. In the limonite zone, iron hydroxides such as goethite and limonite dominate, often accompanied by asbolite, lithiophorite, and nickeliferous clays like nontronite. In the saprolite zone, nickel occurs in serpentine-group minerals (nepouite, pecoraite), talc (willemsite), kerolite (pimelite), chlorite (nimite), genthite, and various nickeliferous clays. The presence of these clay minerals underscores their role as critical hosts for nickel, especially during the transition from limonite to saprolite, where conditions favour their formation and stability.

Bulk density trends in nickel laterites

There are strong trends in dry bulk density related to development of the nickel laterite profile at Sorowako (Figure 3). As lateritisation advances, soluble elements such as magnesium, silicon, and alkalis leach from the bedrock, creating porosity and reducing dry bulk density — often less than 1.0 t/m^3 in well-leached saprolite. The pores typically remain water-filled, so wet bulk density changes less. With continued leaching, the laterite becomes too weak to support the overlying material, leading to structural collapse and a gradual increase in dry bulk density in the upper parts of the limonite zone (Ahmad, 2008; Kholghifard *et al,* 2005). Eventually, an iron and aluminium-rich cap develops, where induration can increase bulk densities to 2.0–2.4 t/m^3.

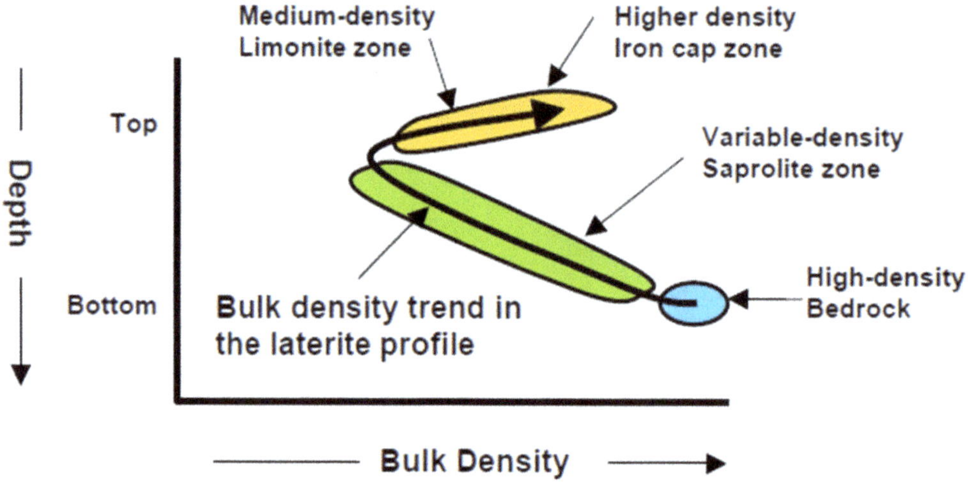

FIG 3 – Relationship of bulk densities with position in the nickel laterite profile (Ahmad, 2008).

DRILLING AND DENSITY MEASUREMENT METHODS

Since 2003, resource definition has been carried out by drilling vertical, triple tube HQ3 core with a Jacro skid-base rig machine (Figure 4). This method is effective for nickel laterite exploration for productivity, minimum mixing between intervals, and can preserve both rocky and soft material without significant disturbance of the core sample. It also produces samples suitable for bulk density measurement.

FIG 4 – Drilling and sampling activity using HQ3 core size.

Core samples are collected using 1 m drill runs (ie over 1 m vertical intervals). Both the length of the core and the length of the drill run are recorded. Core trays are delivered to the sample preparation facility within 24 hrs.

On receipt at the sample preparation facility, each interval is split into up to four size fractions by dry screening at 1 inch, 2 inches, and 6 inches. The screening process is possible because the material is naturally friable as a result of the lateritisation process. The core readily disaggregates by controlled, low energy manual handling upon removal from the core tray, allowing particle size separation by screening without mechanical crushing or addition of water. Clay-rich samples typically form aggregated lumps at natural moisture content that are retained on the screens and do not smear or migrate excessively through the mesh. This approach mimics the particle size distribution that results from mining and screening the ore at production scale.

The size fractions are weighed after screening but before quartering. The combined wet weight of the size fractions is the total wet weight of the sampled interval. The wet bulk density is calculated from this weight and the volume of the sample.

The size-fraction subsamples are then each homogenised and quartered. The subsamples for assaying are then weighed, dried in the oven at 105°C, and weighed again. This allows calculation of the moisture content and the dry weight for each size fraction. The weighted average moisture content of the combined size fractions is used as the estimate of the moisture content of the whole sample interval. This enables the calculation of dry bulk density from wet bulk density of the whole sample.

Wet bulk density and dry bulk density are then calculated using the nominal diameter of the core (61.1 mm), the recovered core length, and the total wet sample weight and total dry sample weight.

EXTRA-LONG CORE RECOVERY

Core drilling in wet tropical laterite deposits commonly results in recovery of lengths of core that are significantly greater than the length of the drilled interval, over multiple consecutive drilled intervals (Figure 5).

FIG 5 – Example of five contiguous core intervals with recovery greater than 100 per cent.

At Hill X, this phenomenon is most evident in soft material with high moisture content across the limonite zone, intermediate/transition horizon, and the upper part of the saprolite zone (Figure 6).

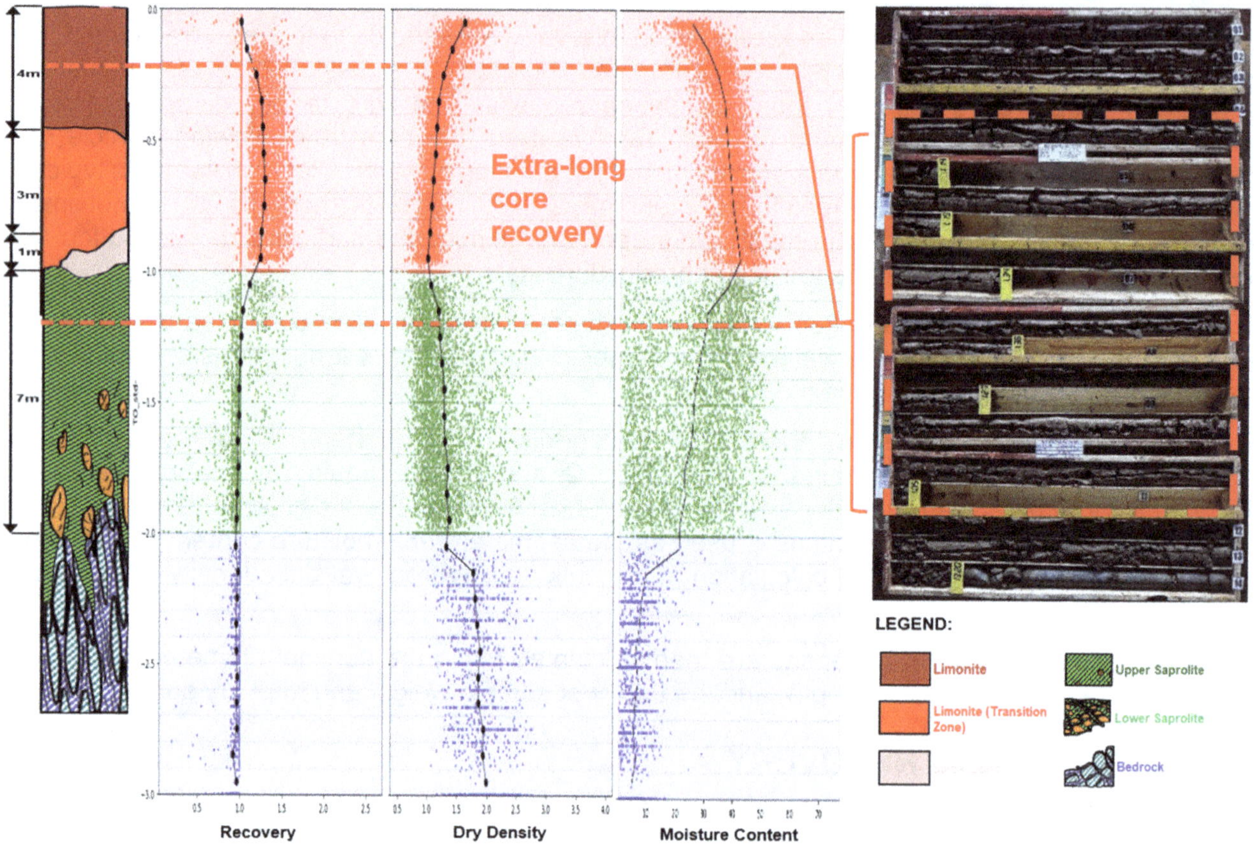

FIG 6 – Schematic diagram of the relationship between extra-long core recovery, dry bulk density, and moisture content in Hill X.

Samples with extra-long recovery form a significant portion of the mineralised samples at Hill X. It is important to determine if the extra material within a sample interval with recovery greater than 100 per cent is representative of the assay grades of the drilled interval. If not, the assays will introduce errors into the Mineral Resource estimates. Similarly, the reliability of the bulk density measurements needs to be confirmed because they directly affect the estimation of Mineral Resource tonnage.

Various theories have been advanced to explain the phenomenon including introduction of material that has caved from the drill hole sidewall, or lengthwise expansion of the core after it is cut by the drill bit. Although commonly described as expansion or swelling, there is no evidence of significant swelling in the diametral direction in the core trays or of fracturing or voids that might be expected if the core had expanded lengthways due to release of overburden pressure. There are no obvious visual changes in the macroscopic appearance of the core, except near the circumference where friction at the contact with the inner face of the bit and the core catcher causes larger-scale shearing and rotation of dislocated slivers of material which are visible to the naked eye.

The phenomenon has not been investigated with borehole acoustic televiewer at Hill X. One of the authors has observed drill holes in a nickel laterite deposit in the Philippines that closed soon after withdrawal of the drill rods due to lateral movement of the limonite into the void. This behaviour would make a televiewer ineffective and risk of loss of the instrument.

At Sorowako, drillers have modified typical HQ3 drilling triple tube procedures to ensure core loss is minimised in all parts of the laterite profile. Drilling of the high moisture content puggy material in the limonite and upper saprolite zones is managed with careful handling of drill bit pressure and water. Too little water and the bit can become clogged, too much and core loss increases.

The authors infer that this careful attention to bit pressure and water usage results in a unique condition where some of the material that is cut away at the bit face and typically washed clear by the drilling fluids is instead retained and forced into the spilt inner tube. Pressure on material inside the drill bit is lower than pressure under the drill bit. This pressure differential appears sufficient to force material under the drill bit into the core barrel, where it deforms, increasing the length of the recovered core.

The volume of material that would be produced by over-cutting of the core in this way can be easily calculated (Table 1).

TABLE 1

Volume comparison between inside and outside core bit diameters.

	HQ3 inside diameter (ID)	HQ3 outside diameter (OD)	Partial over-cutting
Effective coring diameter	61.1 mm	95.6 mm	74.83 mm
Volume of core	2932 mm	7178 mm	4398 mm
Apparent core recovery	100%	245%	150%

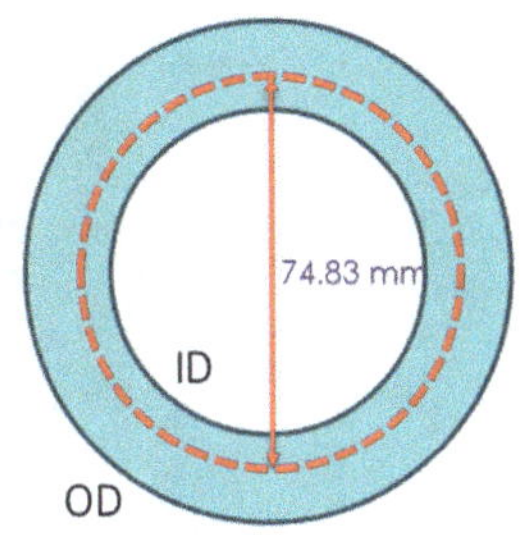

If material is not washed away during drilling, there is a potential volume increase of 245 per cent if all the material under the drill bit is forced into the core barrel and deforms in a plastic manner parallel to the core axis with no change in bulk density. To achieve just 150 per cent recovery (or a 1.5 m of recovered core in a 1 m interval) then the cut radius would only need to be 6.8 mm greater than the nominal HQ3 inside diameter. If, rather than maintaining the *in situ* bulk density, the deformation of the additional material reduced its bulk density, even less over-cutting would be required to account for the extra-long recoveries.

Understanding the bulk density of the extra-long recovery samples therefore would help to explain the behaviour of the extra material when it is forced into the core barrel. Three scenarios are possible:

1. If the bulk density of the extra material squeezed into the core barrel remains the same as the bulk density, the bulk density of the drilled interval can be correctly calculated from the recovered length of the core and the nominal core diameter.
2. If the extra material has lower bulk density because of remobilisation and the creation of new void spaces, the measurement of bulk density based on the recovered length would be biased low.
3. If the extra material has higher bulk density because of collapse of void spaces, the measurement of bulk density based on the recovered length would be biased high.

Only if the bulk density of the extra material is preserved can it be used for Mineral Resource estimation and mine planning.

PREDICTION OF DRY BULK DENSITY

The strong genetic relationship between major element chemistry and porosity on the one hand and dry bulk density on the other indicated that there were good prospects of developing predictive models of dry bulk density from assay data. The porosity of the drill core is not measured at Hill X but moisture content is measured for every sample so that both dry and wet bulk density can be calculated. In saturated core, moisture content is an excellent proxy for porosity. The relationship between porosity and dry bulk density can be inferred from the relationship between moisture content and dry bulk density.

Data analysis was performed on 48 172 samples for which one of the four size fractions formed 100 per cent of the assay sample. In this subset of the data (72 per cent of the total samples), the dry density and moisture content, which are measured for the whole sample interval, are directly correlated to the assay data, which is measured for each fraction.

Figure 7 shows the relationship between the moisture content and dry bulk density of samples by size fraction (-1 inches, -2 inches, -6 inches) and simplified laterite domain. The limonite domain includes the intermediate zone. In strongly leached zones, high porosity develops and, if the laterite has not collapsed under the weight of overlying material, dry bulk densities less than 1.0 t/m^3 are common.

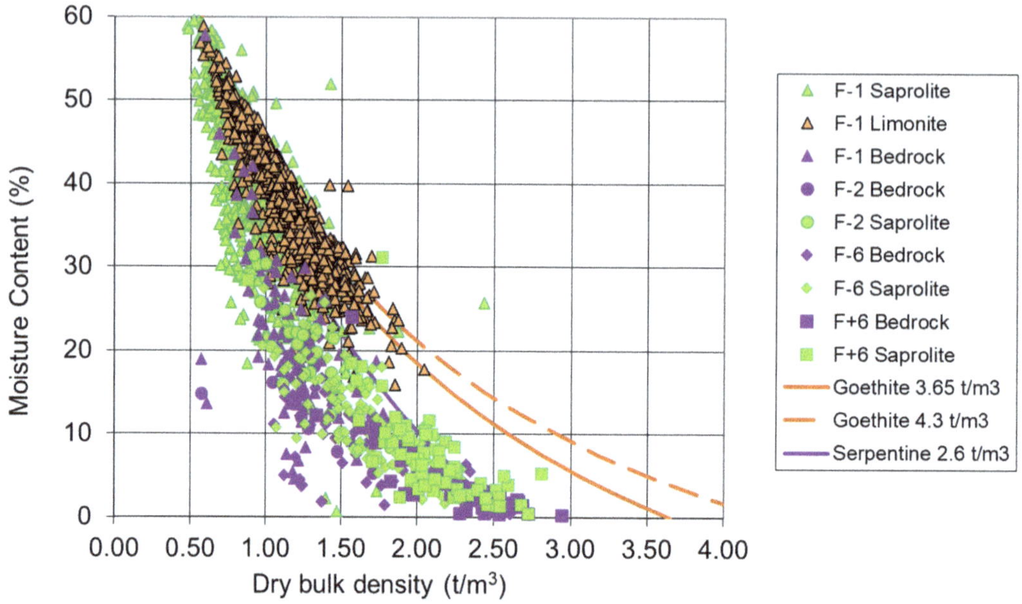

FIG 7 – Scatterplot of dry bulk density versus moisture content.

Curves representing the theoretical dry bulk density and moisture content of rocks containing only goethite and pore water or serpentine and pore water are also shown. The theoretical density of pure crystalline goethite is 4.2 t/m^3 but goethite with porous microstructures may be present and skeletal densities (density of the mineral and occluded pores) of 3.6 t/m^3 or less have been reported (eg Manuel and Clout, 2017). Two curves for goethite are plotted in Figure 7 to illustrate this range.

The figure shows that, broadly speaking, the samples follow the expected trends arising from leaching of the parent ultramafic rocks and the development of secondary porosity. Samples that plot significantly below the theoretical curves are inferred to be unsaturated. The saprolite and bedrock samples, having coarser grain size and less limonite/goethite, drain more quickly than the limonite samples, and commonly arrive at the sampling station in an unsaturated state.

An Index of Lateritisation (IOL) (Babechuk, Widdowson and Kamber, 2014) was calculated using the following formula:

$$IOL = 100 * [(Al_2O_3 + Fe_2O_3)/(SiO_2 + Al_2O_3 + Fe_2O_3)]$$

Although the IOL was developed primarily for the study of weathering in granitic rocks and the formation of bauxite, it provides a useful index of the degree of lateritisation in the ultramafic rocks at Sorowako.

Data validation was completed and a few samples with potentially unreliable assays were removed. Samples with combinations of dry density and moisture content that plotted well outside the expected envelope of data points were also considered unreliable and removed. As the reliability of the data from extra-long recovery samples was uncertain, data analysis and predictive modelling were initially carried out using only the samples with core recovery less than or equal to 100 per cent.

Bivariate correlations between the variables are illustrated in the correlation matrices presented in Figure 8. The side scaling bar indicates the Pearson correlation coefficient ('r'). Key features of the correlations are as follows.

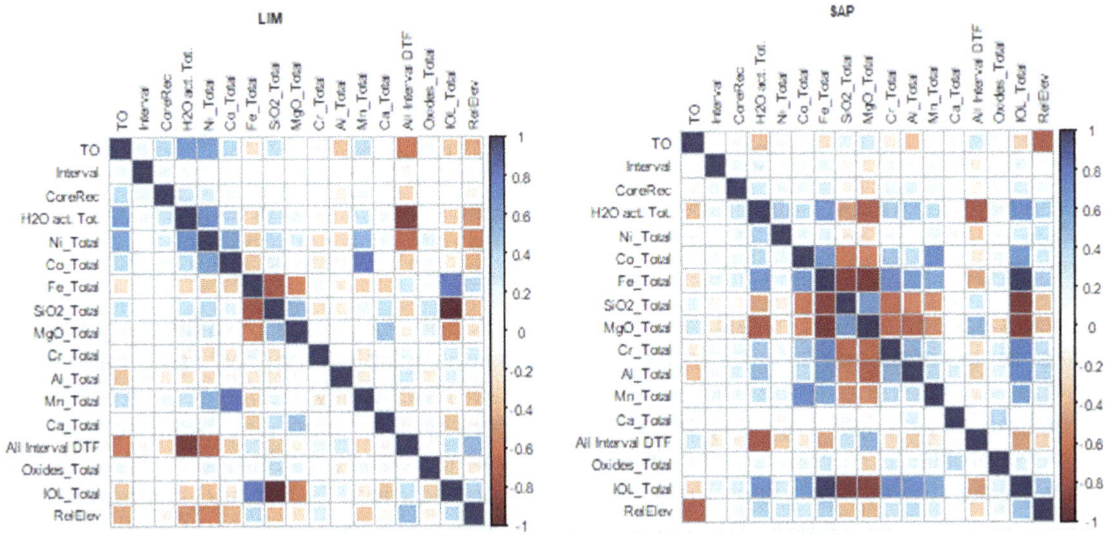

FIG 8 – Correlation matrices for selected variables in limonite and saprolite.

In the limonite:

- Dry bulk density is very strongly negatively correlated with depth_to, moisture content (H2O_act_tot) and nickel.
- Dry bulk density is strongly positively correlated with elevation relative to the limonite-saprolite boundary but this boundary was not intersected in many of the drill holes.
- Dry bulk density is strongly to moderately correlated with iron, silicon, magnesium, aluminium, manganese and the IOL.

In the saprolite and bedrock:

- Dry bulk density is very strongly negatively correlated with moisture content (H2O_act_tot), nickel, iron and IOL.
- Dry bulk density is strongly positively correlated with magnesium.
- Dry bulk density is strongly to moderately negatively correlated with chromium, aluminium, manganese, and elevation relative to the limonite-saprolite boundary.

Cobalt and manganese have strong correlations with nickel and IOL in both limonite and saprolite. They were considered to be redundant for further analysis because it is unlikely that they provide any relevant information that is not already provided by the nickel and IOL values. Saprolite and saprolite + bedrock showed virtually identical correlations.

Supervised machine learning was used for regression modelling. Several regression models were tested. The results for the Cubist model and multiple linear regression are presented here. Multivariate linear regression was used as the initial candidate and benchmark for modelling. In linear regression, the goal is to find the parameter values, or coefficients, that minimise the sum of the squared differences between observed values and the values predicted by the model.

The Cubist algorithm incorporates decision trees with multivariate linear regression along with ensemble learning. This method was chosen as it can deal with non-linearity in the data using decision trees and boosts the accuracy of a single multivariate linear regression equation through additional models (committees). Ensemble learning is a powerful method of training multiple models on different parts of the data set. The approach takes advantage of the wisdom of the crowd where the average of multiple models tends to be more accurate than any single model. The technique also helps to minimise the risk of over-fitting by training multiple models on different parts of the data.

The correlation matrices and scatter plots provided an initial guide to the variables that may be useful as independent variables for predicting dry bulk density. Further tests using the Cubist algorithm were carried out to assess the relative importance of variables for regression modelling. From the correlation analysis and variable importance testing, the following features were selected as the independent variables for prediction of dry bulk density using machine learning:

- Limonite: H2O_act_tot, Ni_total, IOL_total, Depth_to
- Saprolite + bedrock: H2O_act_tot, Ni_total, IOL_total, MgO

Based on their similar multivariate properties, saprolite and bedrock samples were combined. The multiple linear regression and Cubist models were trained using the normal recovery sample data to predict dry bulk density for the limonite and saprolite+bedrock.

Figure 9 shows scatter plots of measured (true) dry bulk density versus predicted dry bulk density for limonite samples and saprolite + bedrock samples, using multiple linear regression and the Cubist model. In the limonite, the linear regression model (lm) generally performed well but showed evidence of bias at very low and very high dry bulk density. The bias was reduced in the Cubist model. In the saprolite, the linear regression model performed poorly for samples with measured dry bulk density greater than 2 t/m³. The Cubist model performed much better and showed significantly less bias and very high R^2 values. The Cubist models were checked using 10-fold cross validation with an 80:20 (training:test) split which confirmed highly repeatable results.

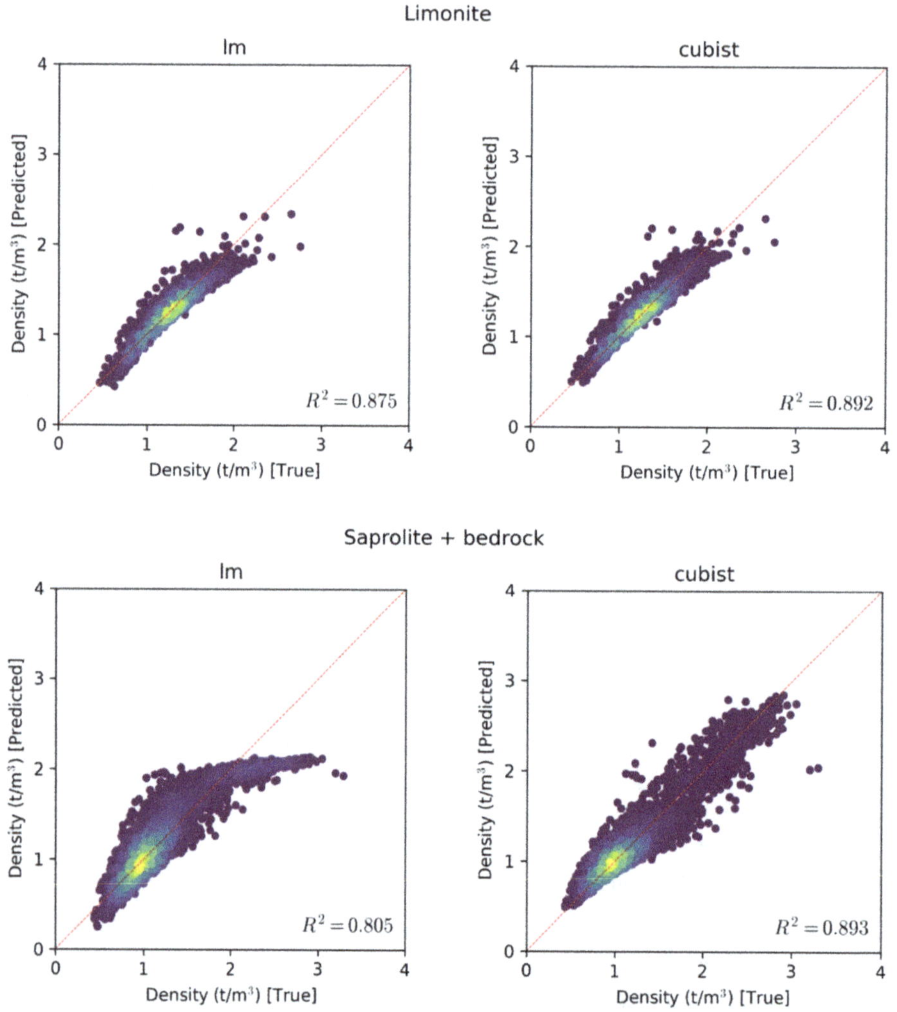

FIG 9 – Scatter plots of measured (true) dry bulk density versus predicted dry bulk density for limonite samples and saprolite+bedrock.

The data analysis demonstrated that machine learning models can be used to predict the dry bulk density of assayed samples for which field measurements of dry bulk density were not possible. This allows gaps in the dry bulk density data to be filled prior to estimating dry bulk density in the Mineral Resource models. Furthermore, there is potential to estimate dry bulk density from the assay database rather than using the field measurements of dry bulk density.

VALIDATION OF DRY BULK DENSITY OF EXTRA-LONG RECOVERY SAMPLES

Having demonstrated that dry bulk density at Hill X can be predicted accurately from the assay database, the question of the reliability of the extra-long recovery samples was addressed. About 20 per cent of the samples in the limonite zone have recovery greater than 100 per cent and some saprolite + bedrock samples also show this characteristic (Figure 10).

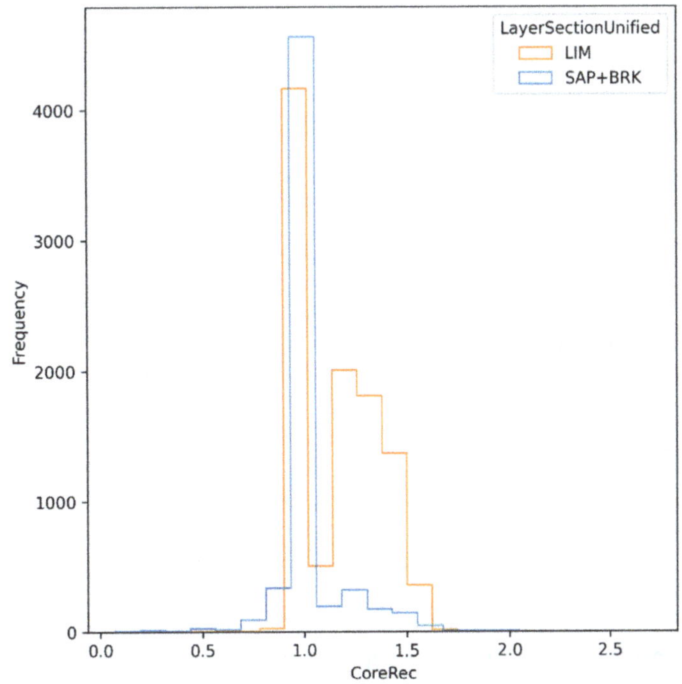

FIG 10 – Histograms of core recovery from the F_Rec-1 = 100 subset, grouped by simplified laterite zone.

Firstly, the chemistry of the extra-long recovery and normal samples was compared. Scatter plots comparing iron, silica and magnesium grades show that the univariate and multivariate distributions of the extra-long recovery and normal samples are very similar in the limonite (Figure 11). The extra-long recovery samples (blue) generally overlap the more iron-enriched part of the distribution of the normal samples (red) in the saprolite.

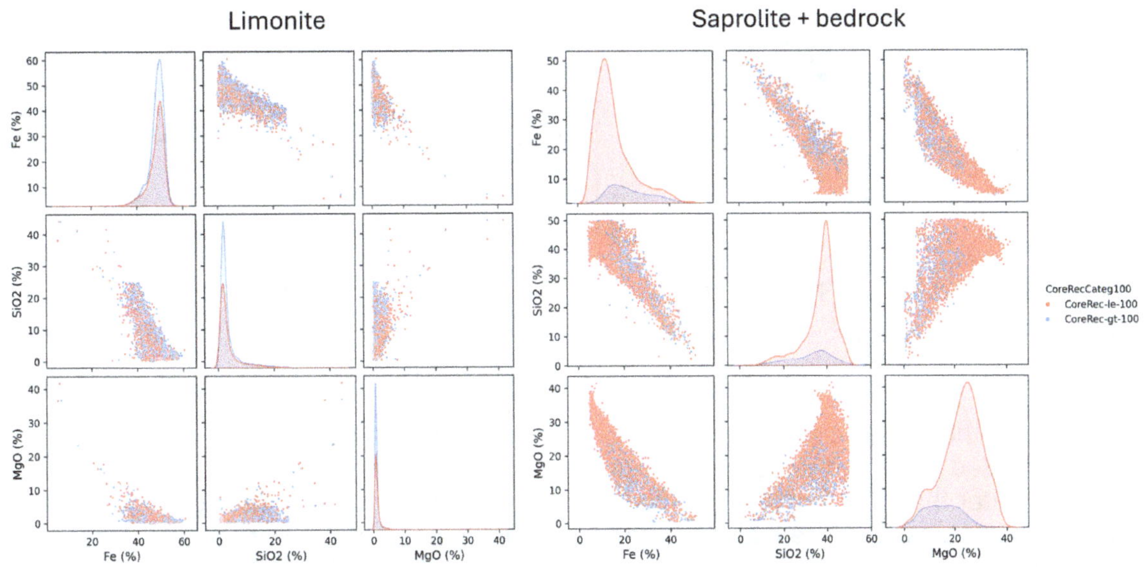

FIG 11 – Scatter plots comparing iron, silicon, and magnesium grades of normal recovery samples and extra-long recovery samples.

The ranges of composition of the two data sets were sufficiently similar that it was reasonable to predict the dry bulk density of the extra-long recovery samples using the Cubist models trained with the normal sample data. This was completed and then the measured and predicted dry bulk densities were compared.

Figure 12 shows scatter plots of the measured and predicted dry bulk densities for limonite and saprolite + bedrock. There are strong correlations with apparently little bias. The R^2 values (R^2 = 0.80 and 0.73) are only slightly lower than obtained when training the Cubist models using the samples with normal core recovery (R^2 = 0.89 and 0.89). This indicates that the Cubist models provide very good predictions of dry bulk density and are not significantly over-fitted.

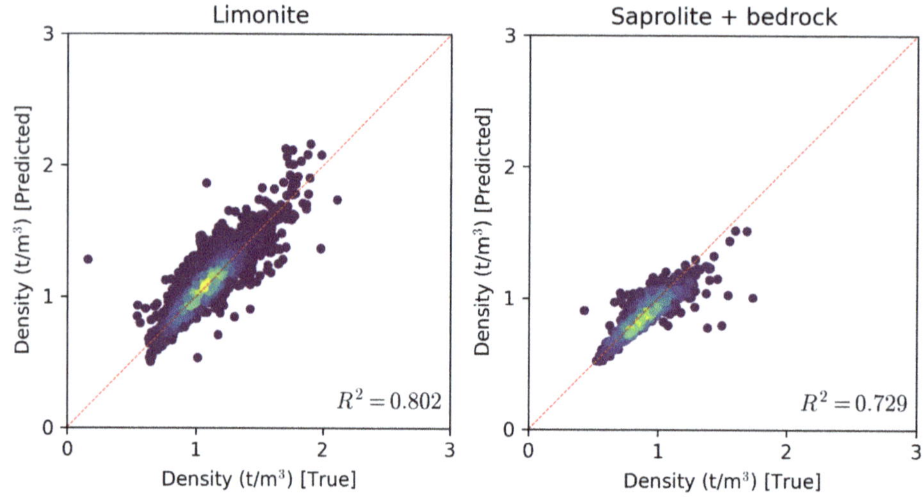

FIG 12 – Scatter plots of measured (true) dry bulk density versus predicted dry bulk density for samples with core recovery greater than 100 per cent.

Quantile-quantile (QQ) plots of the predicted dry bulk density and measured dry bulk density values show the statistical distributions in the limonite are essentially identical (Figure 13).

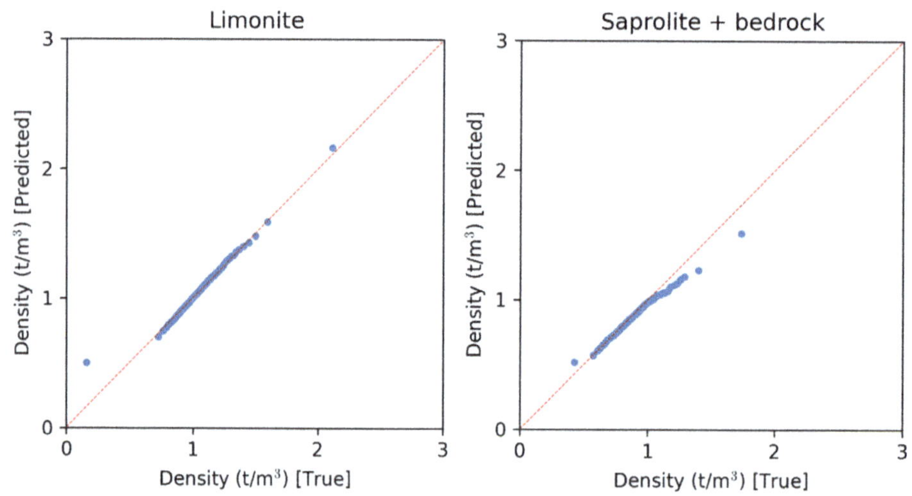

FIG 13 – QQ plots of measured (true) dry bulk density versus predicted dry bulk density for samples with core recovery greater than 100 per cent.

This confirms that there is no bias in the predictions and that the scatter of points evident in Figure 12 is random, unbiased error. In the saprolite, there is a small bias in the samples with measured dry bulk density greater than 1.1 t/m^3: the predicted dry bulk density values are slightly lower than the measured dry bulk densities. The bias is limited to a small portion of the data and is consistent with the outliers evident in Figure 12.

CONCLUSIONS

In common with many young, tropical, nickel laterite deposits, the Hill X deposits are characterised by a wide range of dry bulk density values which follow the theoretically expected general trends, driven by porosity and mineral assemblage.

This study demonstrated that with good quality dry bulk density measurements and assay data, machine learning models can be trained to produce accurate predictions of dry bulk density with reasonable precision. If dry bulk density measurements are missing from the sample database, dry bulk density can be predicted directly from assays with a high degree of accuracy.

A significant proportion of core samples in the limonite and upper saprolite zones at Hill X have core recoveries greater than 100 per cent over multiple consecutive drilled intervals. The authors conclude that the most likely explanation for the extra-long recovery samples is that the pressure differential between the inside of the core barrel and the face of drill bit is sufficient to force material under the drill bit into the core barrel.

When the Cubist models trained with the normal recovery samples were applied to the extra-long recovery samples, the predicted densities of the extra-long recovery samples were strongly correlated with the measured densities and unbiased. They can be treated the same as samples with core recovery less than or equal to 100 per cent.

The authors conclude that the deformed material behaves in a plastic manner and the *in situ* dry bulk density is preserved. The theory that the extra-long recoveries are due to linear expansion of the core in the core barrel or core trays is not supported because expansion would reduce the dry bulk density of the core. Conversely, there was no evidence that the dry bulk density had increased due to the collapse of void spaces during deformation.

The authors postulate that the majority of the deformation of the drill core is due to microscopic mechanisms such as slippage at microcracks, slippage along the close-packed directions of laminar minerals and strain within amorphous mineral phases. These mechanisms would be promoted by the presence of poorly structured types of goethite and smectites and which are an observed characteristic of the extra-long recovery samples.

An important implication for Mineral Resource estimation is that the mass of material recovered in the core barrel can be considered to have been recovered from within the drilled interval and is representative of the chemistry of the drilled interval.

ACKNOWLEDGEMENTS

The authors would like to thank PT. Vale Indonesia and AMC Consultants for permission to present this paper and the peer reviewer for their valuable feedback.

REFERENCES

Ahmad, W, 2008. Nickel Laterites: Fundamentals of chemistry, mineralogy, weathering processes, formation, and exploration, PT Vale Inco internal report.

Kholghifard, M, Ahmad, K, Ali, N, Kassim, A and Kalatehjari, R, 2005. Collapse/swell potential of residual laterite soil due to wetting and drying-wetting cycles, *Natl Acad Sci Lett,* 37:147–153. https://doi.org/10.1007/s40009-013-0221-4

Babechuk, M G, Widdowson M and Kamber, B S, 2014. Quantifying chemical weathering intensity and trace element release from two contrasting basalt profiles, Deccan Traps, India, *Chem Geol,* 363:56–75.

Dominy, S C, Noppe, M A and Annels, A E, 2002. Errors and uncertainty in Mineral Resource and Ore Reserve Estimation: The Importance of Getting it Right, *Explor Mining Geol*, 11(1–4):77–98.

Golightly, J P, 2010. Progress in understanding the evolution of nickel laterites, in *Society of Economic Geologists Special Publication*, 15:451–472 (Society of Economic Geologists).

Lipton, I, 2000. Modelling Bulk Density – the Importance of Getting it Right, in *Proceedings of the Mine Geology 2000 Conference* (The Australasian Institute of Mining and Metallurgy: Melbourne).

Lipton, I and Horton, J, 2014. Measurement of Bulk Density for Resource Estimation – Methods, Guidelines and Quality Control, *Mineral Resource and Ore Reserve Estimation, the AusIMM Guide to Good Practice,* Monograph 30 (Australasian Institute of Mining and Metallurgy, Melbourne).

Manuel, J R and Clout, J M F, 2017. Goethite classification, distribution and properties with reference to Australian iron deposits, in *Proceedings of the Iron Ore 2017 Conference* (The Australasian Institute of Mining and Metallurgy: Melbourne).

Parrish, I S, 1993. Tonnage factor – A matter of some gravity, *Mining Engineering*, 45:1268–1271.

SGS, 2023. Automated mineralogical mapping using TIMA-X FEG GMH System, SGS Technical Report for PT Vale Indonesia (internal report).

In-mine exploration

Extending mine life

The application of mine geology to improve geological model confidence at the Invincible Mine, St Ives Gold Mine

S Marwick[1], K Stinson[2], M Fitzgerald[3], C Muchechetere[4] and M Ngara[5]

1. Senior Geologist: Mining, Gold Fields Ltd, Kambalda WA 6442. Email: sarah.marwick@goldfields.com
2. Project Geologist, Gold Fields Ltd, Kambalda WA 6442. Email: keanu.stinson@goldfields.com
3. Geology Manager, Gold Fields Ltd, Kambalda WA 6442. Email: mike.fitzgerald@goldfields.com
4. Superintendent: Mine Geology, Gold Fields Ltd, Kambalda WA 6442. Email: carlton.muchechetere@goldfields.com
5. Senior Geologist: Mining, Gold Fields Ltd, Kambalda WA 6442. Email: menford.ngara@goldfields.com

INTRODUCTION

Since its discovery in 2012 to the end of 2024, the Invincible orebody produced 2.6 Moz of gold from open pit and underground mining. Mining operations commenced in 2015, and current production rates are around 225 000 ounces per annum. At the end of 2024 the Proven and Probable Mineral Reserve for Invincible was 22 Mt at 3.81 grams per tonne for 2.7 Moz and a combined pre-mined orebody total of 5.3 Moz (production + Mineral Reserve). Gold Fields expects that continued exploration and the improved geological understanding will extend the life-of-mine into the 2040s and beyond (Gold Fields Limited, 2025). The Invincible Mine is part of the St Ives Gold Mine where 16 Moz has been mined with 45 years of continuous operation from over 65 separate deposits (Oxenburgh *et al*, 2017).

Historically, significant economic mineralisation at the Invincible Mine was thought to be only constrained within quartz-carbonate shear vein arrays, hosted within a mudstone sequence. Exploration drilling at Invincible south testing the mudstone sequence intersected significant mineralisation in the footwall, consisting of sub-horizontal extensional veins arrays hosted within coarse sedimentary units controlled by discrete shear zones which was verified by subsequent underground development. The style of this mineralisation and the size potential required a full reassessment of orebody modelling and mine geology practices. This resulted in a shift towards bulk mining methods in these zones, delivering significant productivity gains and reduced unit costs. Over this time the geological story of Invincible has continued to evolve through application of disciplined exploration and mine geology practices.

PROPERTY LOCATION AND GEOLOGY

The Invincible Mine is located approximately 65 km south of Kalgoorlie, 550 km east of Perth, WA (Figure 1). Mineralisation is hosted within a package of mudstone, sandstone and conglomerates that form a stratigraphically continuous NNW-SSE trending, west-dipping package on the eastern margin of the Merougil Basin. Gold mineralisation is hosted within two primary vein styles: steeply west dipping quartz-carbonate breccia/shear veins, and sub-horizontal quartz extensional vein arrays. The interaction of both major and secondary shear zones with lithological horizons of differing competency localises zones of high-grade mineralisation and heavily influence vein style. Mineralised veins have varying levels of wall rock alteration which shows gradation into ore zones. From proximal to distal these are albite ± hematite, phengite, and biotite-carbonate. The highest gold grades are associated with intense albite-carbonate bleaching of the wall rock, characterised by finely disseminated pyrite, in all vein styles, regardless of host lithology.

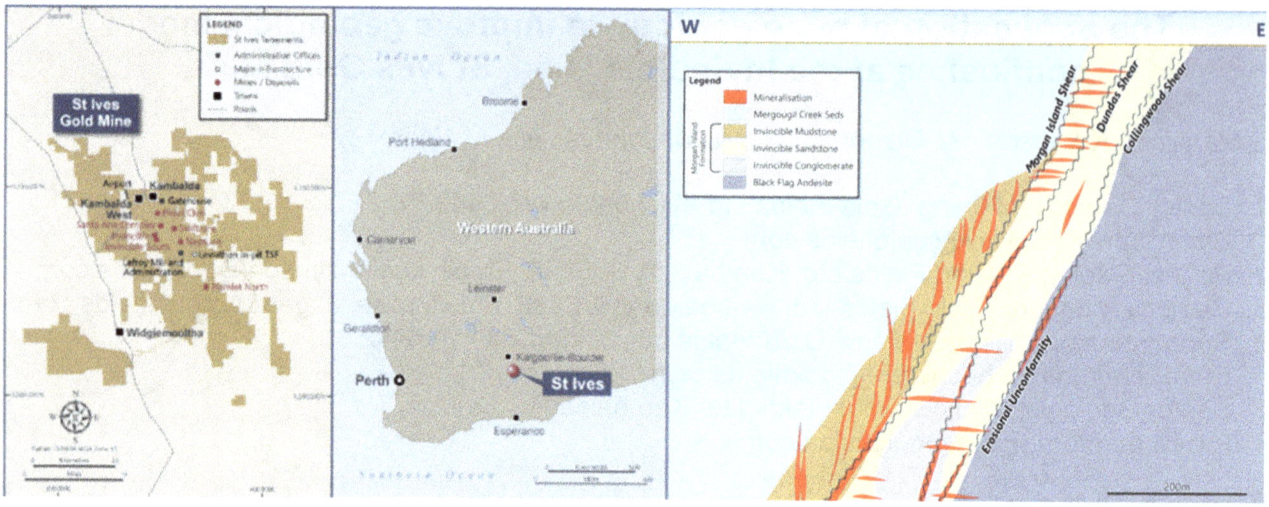

FIG 1 – Site location and geological setting of Invincible (Stafford *et al*, 2025).

EVOLUTION OF THE GEOLOGICAL MODEL AND IMPACTS OF MINE GEOLOGY PROCESSES

Until 2020, Mineral Reserves at the Invincible Mine were defined within quartz-carbonate shear vein arrays hosted entirely within a mudstone sequence. Exploration drilling at the Invincible South part of the mine intersected significant mineralisation in the footwall of the mudstone sequence, consisting of sub-horizontal extensional veins arrays hosted within coarse sedimentary units (Figure 2). Low geological confidence prevented their initial inclusion in the Mineral Reserve. At the same time active underground mining operations confirmed the extensional vein arrays were controlled by primary and secondary shear zones. In early 2021 using the strategic planning process and Business Process Review (BPR), extensional vein mineralisation with bulk stoping methodology was identified as a high value initiative to meet the site 2021 All In Cost (AIC) objectives and deliver (0.5–1 Moz target) upgrade in reserves. Trial mining occurred in Q4 2021 and to date over 300 000 oz has been mined by bulk stoping practices and the reserve locally in this area has increased by greater than 250 per cent. Mine reconciliations to date for this bulk mineralisation have been globally within tolerances of ±5 per cent.

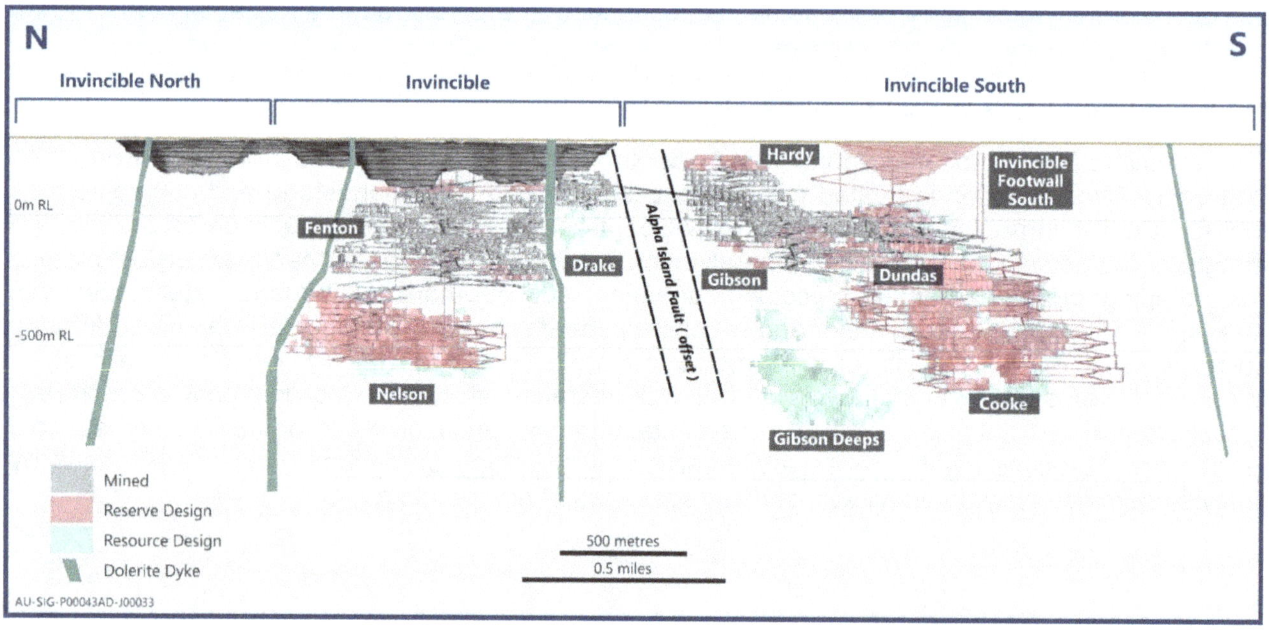

FIG 2 – Long section of the Invincible orebody (Gold Fields, 2025).

CHANGES IN INVINCIBLE MINE GEOLOGY APPLICATION

Mine geology workflows at Invincible have needed to evolve where mineralisation occurs in the footwall extension vein domain. Key changes have been:

- Mapping and Sampling – Move from level backs mapping to wall mapping as the standard workflow in the extensional domain area. Where drive sampling is required, it is done vertically rather than horizontally to ensure a representative sample.
- Drilling – Specified drilling design parameters to ensure optimum intersection of extensional veins, and shear zones within the mineralisation corridor. This has resulted in changes in drill platform strategy.
- Wireframing interpretation methodology.

The improved capture of changes in lithology, vein orientation, veining intensity and structural data by diligent mapping of the underground exposures, along with the expertise of industry structural specialists, enabled the development of a new structural model for the Invincible South area of the mine. This model includes mineralisation controlled by the thrust duplexing of sedimentary units, and the interaction of coarse sedimentary units with both major and secondary shear zones localising high-grade mineralisation. Importantly, the traditional modelling practices for defining these extensional veins have evolved from wireframe and CIK-derived models of individual veins to a bulk modelling approach, with zones of highest vein density wireframed and estimated in a single zone of mineralisation.

The revised modelling approach and increased geological understanding prompted changes to stope design, drilling patterns, and paste fill sequencing, enabling a shift toward bulk stoping extraction in these zones, delivering productivity gains and reduced unit costs. Enhanced understanding of the mineralisation system has led to a significant expansion of the Mineral Reserve in the Invincible South area beyond the traditional mudstone-hosted mineralisation. To date, 50 per cent of the mineral endowment is hosted in areas outside of the mudstone domain.

This case study illustrates how high-quality mine geology can support Mineral Reserve growth and step-change operational outcomes, while providing an exploration framework for similar mineralisation styles within the St Ives camp.

REFERENCES

Gold Fields Limited, 2025. 2024 Mineral resources and Mineral Reserves Supplement to the Integrated Annual Report 2024 [online]. Available from: <https://www.goldfields.com/mineral-resources-reserves-report.php> [Accessed: 24 February 2026].

Oxenburgh, S K, Falconer, M, Doutch, D, Edmonds, P, Foley, A and Jane, M, 2017. Kambalda-St Ives Goldfield, in *Australian Ore Deposits*, Monograph 32, pp 215–222 (The Australasian Institute of Mining and Metallurgy: Melbourne).

Stafford, J, Stinson, K, Fitzgerald M and Zutah, S, 2025. St Ives Gold Mine, Invincible Part 2: An Update on the Exploration and Geology of a World-Class Orebody, in *Proceedings of the Case Histories of Discovery Conference – NewGenGold 2025*, pp 89–106 (Newmont GoldCorp).

Innovation

Rethinking mining geology

Practical Implementation of the KORE Spector System at a high-throughput diamond core gold operation – enhancing efficiency and data quality

S Davies[1], S Reese[2], S Goodfellow[3] and B Cooke[4]

1. Geology Superintendent, Northern Star Resources Ltd, Perth WA 6008. Email: sidavies@nsrltd.com
2. Lead Geologist/Senior Technical Project Manager, KORE Geosystems, Toronto ON M5V 2L1, Canada. Email: sreese@koregeosystems.com
3. Senior Director Product and AI, KORE Geosystems, Toronto ON M5V 2L1, Canada. Email: sgoodfellow@koregeosystems.com
4. Project Geologist, KORE Geosystems, Perth WA 6030. Email: benaike@koregeosystems.com

INTRODUCTION

The Jundee gold deposit, situated within the Yandal Greenstone Belt of Western Australia, comprises a structurally complex narrow-vein lode system. Gold mineralisation occurs in multiple basalt and dolerite units, sedimentary sequences, and intrusive dykes. Northern Star Resources Jundee Operation completes 430 000–460 000 m of diamond drilling annually, generating large volumes of core requiring rapid, accurate and consistent geological and geotechnical logging.

Traditional workflows face challenges in maintaining high-quality Rock Quality Designation (RQD) measurements, standardised lithology logging, and reliable metre marking under operational pressure. These tasks are time-intensive and prone to inconsistency across shifts and personnel. The deposit's scale and geological complexity provide a strong Justification for adopting automation technologies capable of improving efficiency, repeatability, and data quality.

KORE SPECTOR SYSTEM

To address these challenges, Jundee has deployed KORE Geosystems' Spector Geo System, an integrated digital core-logging platform that standardises and automates key geotechnical and geological workflows. KORE provides an integrated Platform-as-a-Service (PaaS) system combining high-resolution optical hardware, Spector Geo software that includes KORE's proprietary Joyce.ai machine-learning engine to deliver a fully integrated, image-based digital logging process.

Automation enabled by artificial intelligence (AI) targets three high-volume tasks: (1) automated metre marking, (2) RQD calculation, and (3) baseline lithological classification. Shifting these tasks to AI inference reduces logging time, improves repeatability, and increases data set completeness. The image-centric workflow enhances cross-shift consistency, reduces subjectivity, and provides geologists with a visual reference linked to each interpreted feature.

Early results show significant improvements in core processing throughput and data standardisation, while reducing operational costs associated with manual data capture.

IMPLEMENTATION

Previous workflow

Core processing at Jundee remains largely manual, sequential, and dependent on technician and geologist expertise. Core is processed in a 12-hour workflow. After core arrives, technicians manually metre mark and orientate core, and geologists undertake high level geological assessment (quick log). RQD measurements are then manually recorded before full lithology, mineralisation, and structural logging is completed and recorded in AcQuire®. These tasks are time-intensive, subjective, and prone to inconsistency.

Core is photographed with a digital camera mount, where lighting and tray alignment introduce further inconsistencies. The workflow concludes with core cutting, sampling, and preparation for laboratory submission.

Core imaging and Metre marking

Digital metre marking, generated by Joyce.ai, provides centimetre-scale accuracy, independent of operator, time of day, or working conditions. Comparative assessments between manual and digital metre marks at Jundee show that metre marks vary by 5–20 cm, with discrepancies arising in fragmented intervals where subjective human judgement varies, while digital calculations apply consistent geometric logic and length rules to every tray, resulting in markedly improved repeatability.

Spector Geo uses driller's run blocks as depth anchors and integrates optical character recognition (OCR) block reading with logic rules, including total core recovery (TCR) validation, to automatically detect inconsistencies and highlight suspect intervals. A key efficiency gain is that digital metre marking is produced concurrently with digital RQD generation, removing the need for a separate manual marking step and embedding depth accuracy directly within the RQD workflow.

The primary limitation observed is poor core preparation prior to imaging. When core is clean, aligned, and properly presented in trays, the digital workflow consistently delivers higher data quality, lower variance, and substantially improved reproducibility compared with any manual method.

RQD process

Manual RQD assessment remains highly subjective, with variability arising from differences in identifying fractures, estimating intact piece lengths, and classifying broken or faulted intervals. This often leads to optimistic rounding and clustering at 100 per cent RQD in high-quality intervals, masking genuine variability and limiting the resolution of geotechnical models.

The KORE team compared manual and digitally derived RQD values across 97 drill holes from Jundee. The study found a mean absolute difference of 7.6 per cent, primarily due to inconsistent fracture interpretation, human input errors, and difficulty estimating lengths within damaged intervals. Manual data sets frequently clustered at the upper end of the scale, while digital measurements produced a continuous, realistic distribution at centimetre resolution. The limitations associated with manual RQD can therefore have a detrimental impact on the data available for mine planning.

In several drill holes, including the 125–180 m interval shown in Figure 1, the AI-derived RQD identified heterogeneities absent from the manual interpretation. Digital RQD generation enhances data fidelity, repeatability, and geotechnical confidence, while being substantially more efficient: up to 83 per cent faster processing than manual methods (Table 1). All segmentation data are preserved, allowing geotechnical teams to interrogate and control the quality of measurements, ensuring accurate and reliable data for decision-making.

As more labelled data are collected, the system's AI capabilities continue to improve, with future enhancements including better recognition of mechanical breaks, fracture orientation, joint sets, and joint roughness proxies.

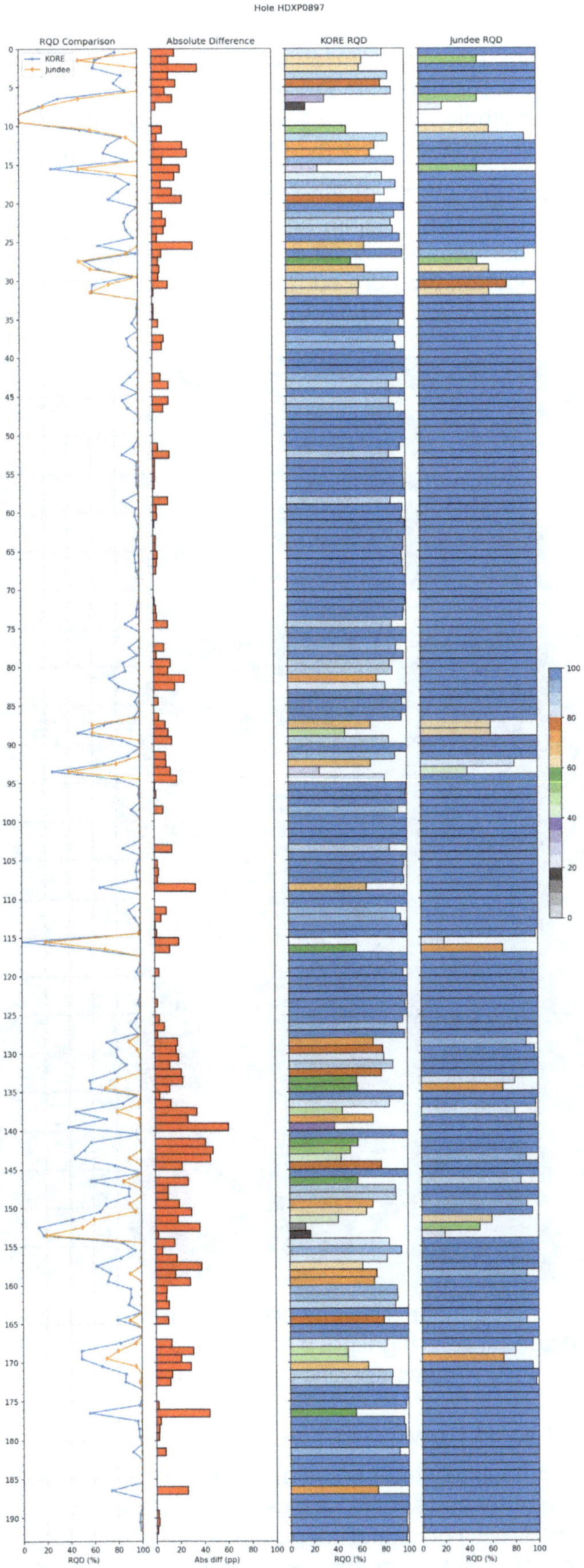

FIG 1 – Calculated RQD per metre for a Jundee drill hole, showing variability between KORE (AI-based segmentation) and Jundee operator (manually derived). From left to right, the figures show RQD downhole calculated for KORE (blue) and Jundee (orange), the absolute difference (KORE RQD – Jundee RQD), KORE RQD values, and Jundee RQD values.

Lithology model

KORE's lithology model is trained directly from labelled intervals created by geologists. As more core is logged digitally, the system accumulates image-and-label pairs, improving the model's confidence and consistency. The Jundee lithology model encompasses over 400 drill holes and 150 000 m of labelled core, with substantially more imagery available to generate training data.

The model now performs strongly on major lithological units (Figure 2), providing consistent first-pass classification, reducing repetitive logging, and improving data quality. Because the system is trained on photographs, its performance is fundamentally tied to what a geologist can observe visually. If a feature is not visible on the core surface, the AI will not infer beyond what the imagery supports.

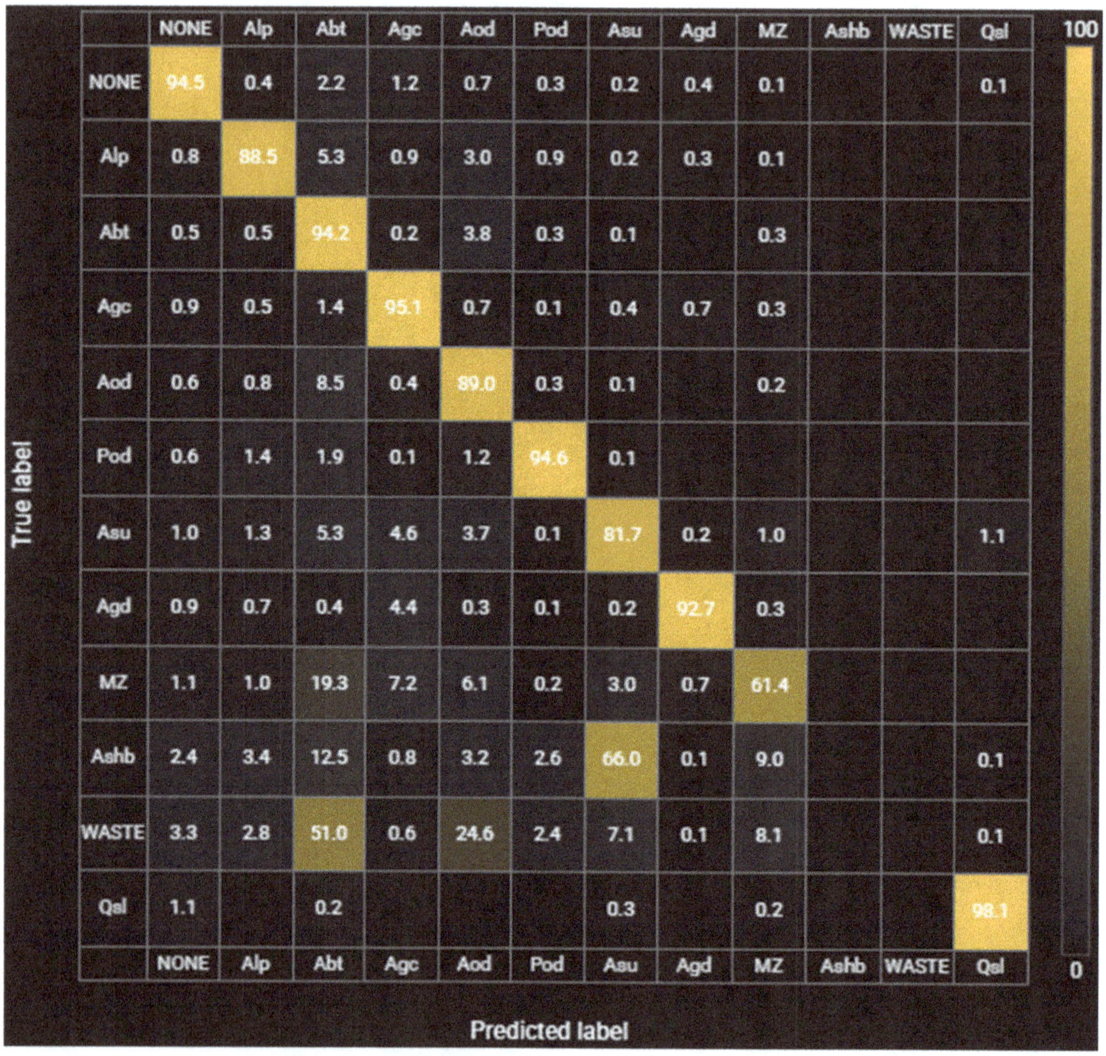

FIG 2 – Confusion matrix for Jundee lithology prediction model. Diagonal cells show correct classifications, while off-diagonal cells highlight class misidentifications. Colour intensity illustrates class-specific performance.

Updated KORE-based workflow

Core will now be scanned immediately after receival at the core shed using KORE optical hardware, producing high-resolution imagery that serves as a central reference. Metre marking, RQD, lithology, and sampling are completed directly on digital images in Spector Geo, where the embedded Joyce.ai copilot automatically generates metre marks and RQD calculations, and provides baseline geological classifications that geologists can review and refine. Sampling intervals are reviewed digitally before being transferred to physical core, ensuring alignment with logged contacts.

All data is exported via the AcQuire® REST API. The resulting workflow is streamlined, consistent, and productive. Planned improvements include streamlined data flows, conveyor-fed scanning to increase throughput, and core processing facility upgrades, advancing toward a fully integrated digital logging workflow (Table 1).

TABLE 1

Comparison of average processing times for 100 m of diamond core, showing time saved following the Spector Geo workflow and percentage difference.

Task	Manual	SPECTOR geo	Time saved	Percentage difference
Metre marking	15 min	5 min	30 min	+86%
RQD	20 min			
Logging	40 min	30 min	10 min	+25%
Sample marking and generating cutsheet	30 min	20 min	10 min	+33%
Photography	15 min	15 min	--	--
Total time spent	120 min	70 min	50 min	+42%

Average processing time for 100 m of diamond core has been reduced from approximately 120 mins to 70 mins, representing a 42 per cent efficiency gain.

An additional benefit of Spector Geo is its web-based platform that synchronises imagery in near real-time, enabling fully remote core logging tied to a consistent visual reference for all interpretations. This flexibility improves workforce utilisation and throughput, allowing a single geologist to log substantially more core per day than is achievable through conventional rack-based workflows.

Under traditional logging practices, ore zones have at times been sampled sub-optimally due to subjectivity, time pressure, and variability in manual workflows, introducing potential grade bias. The adoption of a digital, image-centred logging and review processes reduces this risk by improving repeatability, enforcing standardisation, and enabling review of sampling decisions.

As the volume of drill holes processed through Spector Geo increases, the system benefits from progressively larger and more consistent data sets. This drives improved standardisation of geological logging and strengthens geological model fidelity—particularly across complex contacts—while reducing rework and increasing confidence in resource estimation. These improvements ultimately support more reliable mine design and enhance orebody extraction efficiency.

CONCLUSION

Implementation of KORE's Spector Geo system at the Jundee operation has resulted in measurable improvements in efficiency, data quality and workflow consistency. Automation of metre marking and RQD calculation significantly reduced processing time while improving repeatability and resolution of geotechnical data sets.

The AI lithology model provides a consistent first-pass classification of major rock units, reducing repetitive logging and improving standardisation. While interpretations remain controlled and validated by geologists, the combination of AI-assisted classification and image-based logging reduces subjectivity and improves consistency at lithological contacts.

Overall, the system demonstrates that AI-assisted digital logging can improve data quality, efficiency and scalability in high-throughput underground gold operations.

ACKNOWLEDGEMENTS

We acknowledge the Northern Star Geological Leadership Team for their ongoing support, guidance, and encouragement. Their contributions have been vital.

We extend our gratitude to the KORE Geosystems team for their dedication, expertise, and hard work throughout this project. Their commitment has been instrumental in bringing this work to completion.

Northern Star Resources Ltd is gratefully acknowledged for granting permission to publish this work.

Network effects or shear madness? Using machine learning to model mineralised structures

P E Jankowski[1], M Yurdakul[2], F Bizouerne[3] and A Diallo[4]

1. FAusIMM, Technical Consulting Director, ERM Consulting, Perth WA 6005. Email: phil.jankowski@erm.com
2. MAusIMM, Principal Consultant, ERM Consulting, Spring Hill Qld 4000. Email: metin.yurdakul@erm.com
3. Geology and Mineral Resources Manager, Predictive Discovery Limited, Conakry Guinea. Email: frank.bizouerne@predictivediscovery.com
4. Superintendent Geology, Predictive Discovery Limited, Conakry Guinea. Email: adama.diallo@predictivediscovery.com

ABSTRACT

Predictive Discovery Limited discovered and is developing the Bankan Gold Project, a Palaeoproterozoic shear controlled Birimian deposit in Guinea. A Definitive Feasibility Study based on the current resource model has produced a Probable Reserve of 51.6 Mt @ 1.78 g/t to be mined from three open pits and an underground mine, at an all-in sustaining cost of US$1057 per oz. In the main North-east Bankan (NEB) open pit, the mineralisation is a complex network of mineralised anastomosing shears with intervening low-grade and barren rock; this has been demonstrated on scales from macro to micro.

Structural measurements of oriented core provided a coherent set of shear orientations, which enabled the interpretation of the first order and some of the second order structures as inputs to the resource model.

The current mine plan is for a low-selectivity, conventional drill, blast, truck and shovel approach as interpreting the smaller scale shear network from wide spaced resource data carries significant interpretation risk.

A deterministic interpretation of a trial grade control program was used to derisk the mineral resource estimate and provide realistic ore loss and dilution factors for the resource to reserve conversion; however this single interpretation is based on subjective interpretation choices, similar to the resource model and does not consider interpretation risk. Novel machine learning based approaches have been applied to the grade control data to provide a set of alternative interpretations of the shear network. This ensemble modelling framework enabled calibration and quantification of both geological interpretation uncertainty and grade control estimation risk.

The outcome is a methodology that supports the deployment of a risk-based grade control strategy in routine production, as multiple scenarios can be considered in daily ore waste decisions.

INTRODUCTION

Predictive Discovery Limited's (PDI) Bankan Gold Project is the largest gold discovery in West Africa in a decade (Roberts, Murphy and Nganare, 2021). A JORC classified Mineral Resource of Indicated 83.7 Mt @ 1.54 g/t and Inferred 16.8 Mt @ 2.26 g/t has been defined to date, predominantly at the North-east Bankan deposit. The project is approximately 10 km from the town of Kouroussa (2014 census population: 31 262); Kouroussa is 554 km by road from the national capital of Conakry by the sealed national highways N1 and N2.

Artisanal gold mining in Guinea dates to the third century. There are 200 000 to 300 000 artisanal miners active; production is estimated to be in the 6–8 tpa range. In the Kouroussa region approximately 50 000 miners are active. Mining comprises shallow shafts and diggings in weathered ore, with largely manual crushing and gravity separation processes.

In late 2018, the Kaninko area was highlighted by PDI in a terrane-scale assessment of the Siguiri Basin. Field visits identified widespread artisanal pitting into weathered bedrock with shallow surficial workings in lateritic cover. An exploration program including BLEG stream sediment geochemistry, surficial sampling, mapping, trenching and power auger drilling identified the NEB prospect.

Systematic surface reverse circulation and diamond core drilling commenced at NEB and the nearby Bankan Creek deposit. PDI completed a Definitive Feasibility Study (DFS) for the Bankan Project in June 2025, outlining a 250 koz per annum operation over more than 12 years. The government of Guinea has approved the Project's Environmental and Social Impact Assessment (ESIA) and issued a Certificate of Environmental Compliance, and the Exploitation Permit application is in the final stages.

GEOLOGY

Regional geology

The Siguiri Basin of upper Guinea and south-west Mali is a Palaeoproterozoic volcano-sedimentary basin, and part of the Birimian Supergroup which hosts most of the gold deposits of West Africa.

The Siguiri Basin is interpreted to have been a marine platform, with a lower turbiditic sequence of sandstones with subordinate black siltstone and an upper sequence of limestones and acidic volcanics. The rocks have been metamorphosed to greenschist facies. Late orogenic felsic intrusive plutons and intrusions are found throughout the basin.

The gold deposits are orogenic lode deposits, temporally and spatially related to structures formed during the Eburnean Orogeny between 2200 Ma and 2088 Ma. Prolonged weathering has led to the formation of extensive lateritic duricrusts, and deep saprolite profiles; vertical remobilisation of the gold during lateritic weathering is common and primary gold deposits may be overlain by lateritic or supergene gold deposits.

Primary gold mineralisation in the Siguiri basin is structurally controlled; according to Lahondère *et al* (1999), two mineralised structure orientations are found:

- E-W structures represented by veins attributed to tensional sinistral fractures oriented NW-SE and to the conjugated faults of NE-SW direction.
- NNE-SSW structures which are developed within a dextral normal fault.

These structures may have controlled the distribution of magmatic fluids. Magma was responsible for the formation of the granitic and dioritic intrusions and their related volcanic equivalents and constitutes the source of gold mineralisation (Lahondère *et al,* 1999).

The Bankan project area is largely covered by lateritic duricrusts and is deeply weathered. Outcrops are sparse, and the underlying bedrock geology is known largely from regional scale geophysics and drilling.

The project area is located in an area of greenstones near the south-west margin of the Siguiri Basin, surrounding the intersection of NNW striking and a NW striking structures on the margin of a regional granitic batholith. Numerous anastomosing NNE striking structures have been interpreted from the aeromagnetic data. Smaller granitic intrusions in the greenstones are structurally controlled and provide evidence for significant heat and fluid flow late in the orogenic history, which in other parts of the Siguiri Basin has been demonstrated to be part of the gold mineralisation process. Both the NEB and BC deposits are partially hosted by these granitic intrusions.

NEB deposit geology

NEB developed at the hanging wall contact of a small tonalitic intrusion intruded into a mafic sequence. The tonalite is a coarse grained quartz-plagioclase-K feldspar-hornblende-biotite intrusive. Minor accessory phases include muscovite and sericite; alteration minerals include carbonate, epidote, clinozoisite. The plagioclases may be extensively albitised. The mafic sequence comprises a mixed basalt, andesite and gabbro assemblage, and is dominated by plagioclase-quartz-chlorite, with accessory biotite and hornblende; carbonate alteration and pyrite mineralisation are common.

Mineralisation consists of wide zones of structurally controlled chlorite, silica and sericite alteration with associated pyrite and quartz veining. Sulfide minerals largely comprise pyrite with minor chalcopyrite. In the tonalite the sulfide mineralisation is generally associated with the later stage veining. Higher grade ore is characterised by higher pyrite and covellite, and arsenopyrite and sphalerite; low-grade ore lacks covellite, galena, sphalerite, and bismuth species. Other sulfides that have been noted include tennantite-tetrahedrite, hessite, gersdorffite, bornite and cobaltite.

The mineralisation is structurally controlled by a network of dextral shear zones. The largest and most persistent is the Main Shear Zone (STMZ), a structure dipping approximately 40° to the west at or just above the hanging wall contact of the tonalite. It typically consists of a zone of shearing, strong mylonite fabric and sericite alteration from 4 m up to 36 m thick, often with significant quartz veining; in places it comprises up to four separate mylonite zones; this is likely to represent a zone of anastomosing structures. The STMZ has been intersected over a strike length of at least 800 m and 1150 m down dip; it is open at depth and along strike to the south.

In the footwall of the STMZ, a very well developed second order shear 3 m to 5 m thick (STSZ01) has a very similar structure and alteration to the STMZ, and forms a step over or jog (Micklethwaite, Sheldon and Baker, 2010) from the STMZ to a more weakly developed structure; and hence is a locus for dilation and fluid flow associated with mineralisation.

The STSZ01 nearly outcrops, whereas the STMZ terminates below the surface above its intersection with STSZ01. This fault duplex is interpreted to represent a soft-linked overlapping shear system, where a component of strain is accommodated by rotation or folding between the main bounding shear segments, as well as at the termination of the segments. Below the STSZ01 shear, four other parallel structures have been interpreted, with similar relationships to the STMZ; these are less well constrained by drilling, and have a greater degree of uncertainty in their location and extent.

Higher grades are found in and on the immediate footwall of the shear, with lower grade mineralisation in both the tonalitic footwall and the greenstone hanging wall. An interpreted structural history (Harris and Murphy, 2022) indicates E-W compression with kinematic indicators in the Bankan mylonites consistent with a reverse (oblique dextral) zone. The degree of ductile shearing suggests hundred's metres of west over east displacement. Highest grade mineralisation is commonly ponded below and around the shear zone. Gold was deposited at an advanced stage in the deformation, overlapping with sulfide, chlorite, sericite, K-spar and carbonate alteration phases.

SHEAR ZONE MINERALISATION

Shear zones are tabular zones of strain, that grow primarily by segment linkage as they accumulate strain and displacement, typically resulting in shear zone networks. A shear zone network is characterised by anastomosing patterns and local variations in thickness and finite strain (Fossen and Cavalcante, 2017).

The formation of shear hosted mineralisation systems are related to the processes that operated when those systems were active. There are typically systematic patterns in geometry with implications for permeability enhancement and mineral deposit formation.

Two important geometric elements of shear hosted mineralisation are segmentation, and the development of step-overs between segments (sometimes referred to as jogs). Segments are relatively continuous sections of a structure that localise displacement, then tend to bifurcate into arrays of smaller structures or fracture networks towards their tips. Segmentation can also be observed at multiple scales.

The step-overs represent zones of more distributed strain, they are classified a soft-linked when a component of strain is accommodated by bed rotation or folding, and there is no through going fracture network step-overs (Micklethwaite, Sheldon and Baker, 2010).

DRILLING AND RESOURCE MODELLING

PDI have used several different drilling contractors at the Bankan project, drilling aircore, reverse circulation and diamond core holes. A total of 1052 AC, RC and DD holes for 157 171.49 m have been drilled for the geological modelling, all data has been used; for the resource estimate only the DDH and RC were used. Some of the deeper diamond holes have a RC precollar in expected hanging wall waste and core thereafter. Drill hole spacing is variable with typically 40 m × 40 m in the upper parts of the deposits, and spacings as much as 100 m at the lower fringes. All holes were geologically logged for lithology, alteration, mineralisation minerals, weathering, and veining.

Leapfrog Geo software was used to create models of the lithologies to fill the resource model space. The weathering surfaces were modelled as digital terrain model surfaces; other lithologies were modelled as solid wireframes. Logged shear intervals were coded to the appropriate structure and modelled as lithologies.

Downhole composites were extracted from the database at 3 m lengths using a minimum length of 50 per cent of the composite length. Leapfrog grade shells were extracted from the composite files at nominal 2 g/t Au (High-grade), 0.4 g/t Au (Medium Grade), and 0.2 g/t Au (Low-grade) cut-offs as resource estimation domains. Smoothing parameters were chosen in an iterative process, after reviewing preliminary shells to create an appropriate continuity. The interpreted shear models were used to provide a moderate degree of anisotropy to the grade shells.

The High-grade domain comprised a large zone along the STMZ. The Medium Grade and Low-grade domains are largely in the footwall of the High-grade domain. The ultimate footwall of the Low-grade domain is poorly controlled due to lack of data. Both were also manually post-processed to remove anomalies and shapes based on isolated intersections. As a final post-processing step, the domains were intersected with the base of laterite DTM as an upper constraint.

Gold grades were estimated using Ordinary Kriging. The kriging estimation parameters were chosen from a kriging neighbourhood analysis; a second pass for the Medium Grade domain was implemented to ensure all blocks were estimated. An open pit optimisation and a preliminary underground stope design were used to constrain the reported resource; in general areas with 80 m by 40 m drilling were classified as Indicated. Open pit resource had a 0.5 g/t block cut-off applied.

Compared with previous Inferred Mineral Resources, the infill drilling demonstrated a greater number of internal higher and lower grade structures.

GRADE CONTROL DRILLING AND RESERVE

In 2022, a trial grade control pattern was completed over a section in the centre of the proposed North-east Bankan open pit. The programme comprised 168 reverse circulation drill holes for 14 038 m, drilled on a 10 mY × 7 mX pattern to an average depth of 70 m below the natural surface. Drill holes were sampled and assayed using the same procedures as the resource drilling. The aims of the trial grade control programme were:

- Validate the resource data with closer spaced comparable RC data.
- Validate the overall resource modelling approach.
- Confirming the short-scale mineralisation variability as measured by directional variograms.
- Confirm the grade-tonnage prediction and calibrate the resource to reserve ore loss and dilution factors.
- Provide drill spacing recommendations for upgrading the then current Inferred Resource to Indicated.

To validate the resource and grade control data quality, five pairs of resource and grade control holes within 3 m were identified; from these 99 resource-grade control 1 m downhole paired composites were defined. The statistics of the paired composites show that:

- The resource holes have a greater range than the grade control.
- The resource holes have a higher mean but lower median.
- The mean of the resource is heavily influenced by a single very high value.

To further assess the source of the differences between the two data sets, a comparison was made of the samples where the paired resource composite was >4 g/t and the paired grade control composite was <2 g/t. This comparison showed that most of the differences are in continuous intersections rather than randomly distributed through the drill hole; this suggests that the differences are driven by small high-grade structures.

As a further comparison, the entire 1 m downhole grade control data set (13 657 composites) was compared to the entire 1 m downhole resource data set (932 composites) in the grade control volume; the statistics show that:

- The resource mean is higher than the grade control.
- The GC median is higher than the resource.
- The resource data is much more variable than the grade control.

The combined drill hole database (grade control plus resource data) was used in Leapfrog to produce nested grade shells at 1.8 g/t Au, (high-grade) 0.8 g/t Au (medium-grade) and 0.4 g/t Au (low-grade) cut-offs; the low-grade shell hanging wall and footwall contacts approximate the medium-grade shell used in the resource estimate, however with significant internal waste. Anisotropy was controlled by interpreted grade trends, corresponding to the STMZ and STSZ01 used in the resource model. Grades were estimated by ordinary kriging in a process similar to that used by the resource estimate.

At a 0 g/t Au cut-off, the GC model has 77 per cent of the tonnes, 117 per cent of the grade and 90 per cent of the contained gold of the resource. This reflects changes to the mineralised volume from the greater complexity of internal waste defined by the close spaced drilling. At a 0.4 g/t Au cut-off, the GC model has 82 per cent of the tonnes, 111 per cent of the grade and 91 per cent of the contained gold of the resource. This reflects the somewhat lower tenor of the GC data, but a lower degree of smoothing allowed by the closer spaced drilling and better defined variogram. The 0.4 g/t Au cut-off in the resource model is most comparable to the GC model at a 0 g/t Au cut-off, with the GC having 96 per cent of the tonnes, 98 per cent of the grade and 94 per cent of the contained gold of the resource.

The conclusions of this study were:

- The trial grade control data is generally similar to the resource data, albeit with some higher grade structures not being interpreted in the wider-spaced resource data.
- The short scale grade variability was confirmed by the grade control drilling to be relatively high. The internal grade architecture of the high-grade mineralisation is complex, and likely to be controlled by either the localised structures or variations in lithology.
- Due to the grade and geological variability, any highly selective mining option will carry elevated risk due to local estimation uncertainty and the consequent risk of misclassification.
- The GC grade-tonnage curve overlays the resource model grade-tonnage curve and shows the greater selectivity that has been achieved by the much closer spaced drilling.

STRUCTURAL DATA

The diamond core data set is mostly NQ with minor HQ and HQ triple tube. Holes were downhole surveyed with gyroscopic tools; the Champ Gyro or the Reflex EZ Shot depending on the contractor. Where possible, core is orientated by a downhole orientation tool and structural measurements of features in the core recorded.

In North-east Bankan, a total of 297 foliations and 27 shear zone planes have been recorded (Figure 1). The contours of the poles to the planes have been contoured using the method of Kamb (1959); the contours form an elongated ellipse centred at a dip of 48° to a dip direction of 271°; this aligns with the general interpretation of the shear zones as north-striking.

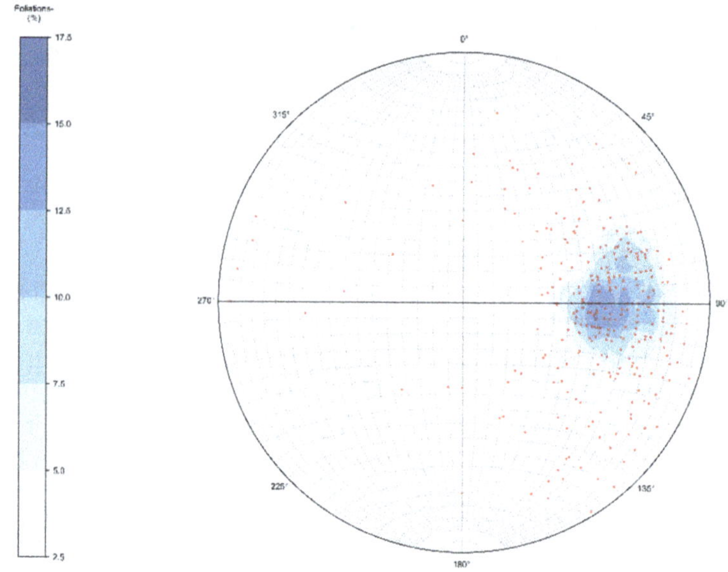

FIG 1 – Measured foliations from North-east Bankan oriented core, n=324.

The spread of the contours is in a range of dips between 30° and 70°, and dip directions between 235° and 305°; this accords with the interpretation of the mineralisation being controlled in an anastomosing network of shears at a range of scales.

METHODOLOGY

Machine learning framework overview

For this study, an ensemble of Artificial Neural Networks (ANN) was used to predict gold grades and quantify estimation uncertainty (Avalos and Ortiz, 2020). The methodology prioritised geostatistical validity through rigorous spatial partitioning and feature engineering designed to capture anisotropic controls on mineralisation. The workflow (Figure 2) proceeds through four distinct stages: (i) data integration and spatial feature engineering, (ii) leakage-free data set partitioning, (iii) deep neural network ensemble training, and (iv) probabilistic prediction and back-transformation.

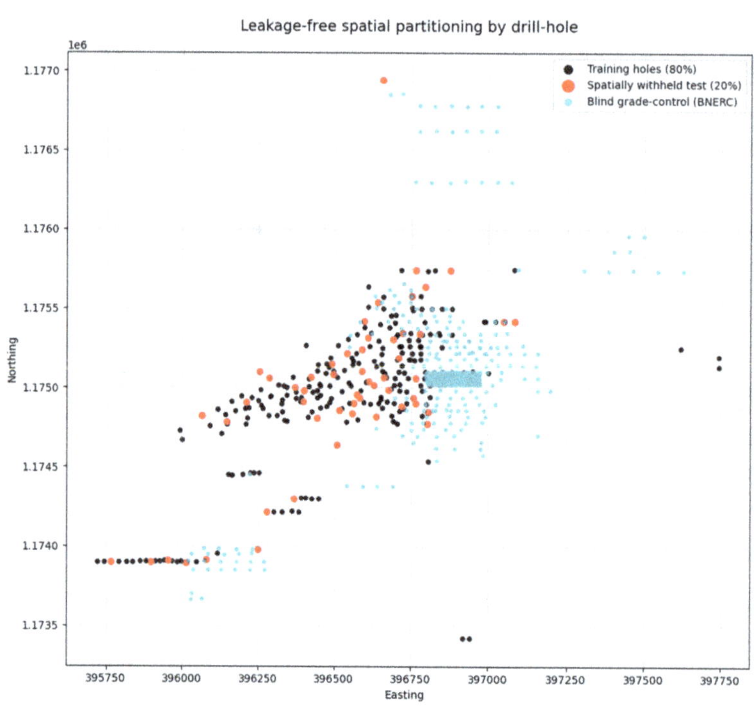

FIG 2 – Plan-view map of drill hole collar locations showing the leakage-free spatial partitioning strategy.

Data integration and feature engineering

Three primary data sets were integrated to construct the feature matrix: RC and diamond drill hole assay intervals (n=90 371), diamond core structural measurements (n=359 foliation and shear planes), and geological logs (lithology and weathering).

Preprocessing and target transformation

To deskew the grade, the target variable was transformed using a logarithmic Log_Au = ln(Au_ppm + 1) Missing values in critical fields (coordinates, hole ID) were removed. Geology intervals were matched to assay centroids via interval lookup, with unmatched sections categorised as 'UNKNOWN' to preserve data density.

Structural feature engineering

Given the structural control on mineralisation, explicit spatial features were engineered:

- **Hybrid Structural Imputation:** To resolve the scarcity of structural data, we implemented a hybrid imputation strategy. Dip and dip-direction values were interpolated using Inverse Distance Weighting (IDW) locally, smoothly transitioning to a regional trend (Dip: 48°, Dir: 271° over a 150 m buffer distance).
- **Trend Domain Clustering:** Structural pole vectors were clustered using K-means (k=4) to generate categorical "Trend Groups", allowing the model to learn distinct mineralisation geometries.
- **Proximity and Density:** Euclidean distance to the nearest major structure and structural density (count within 500 m radius) were calculated and log-transformed to compress dynamic range.

All numerical features were scaled using a RobustScaler to minimise the influence of outliers, while categorical features (Lithology, Weathering, Trend Group) were One-Hot Encoded, resulting in a feature space of >40 dimensions.

Spatial partitioning strategy

To mitigate spatial data leakage—a pervasive issue in geostatistical machine learning (Fouedjio and Klump, 2019)—data set partitioning was strictly conducted at the **drill hole level** rather than the sample level (Figure 3).

- **Exploration Set:** Data from exploration holes were split 80/20 into training and test sets (n_train=40 749) and testing (n=10 261) sets. This ensures that the model is evaluated on unseen locations, not just unseen samples within known locations.
- **Blind Validation Set:** A dedicated grade control data set (prefixed 'BNERC', n=39 361) was entirely withheld during model development. This data set targets the high-grade core and serves as a proxy for operational performance.

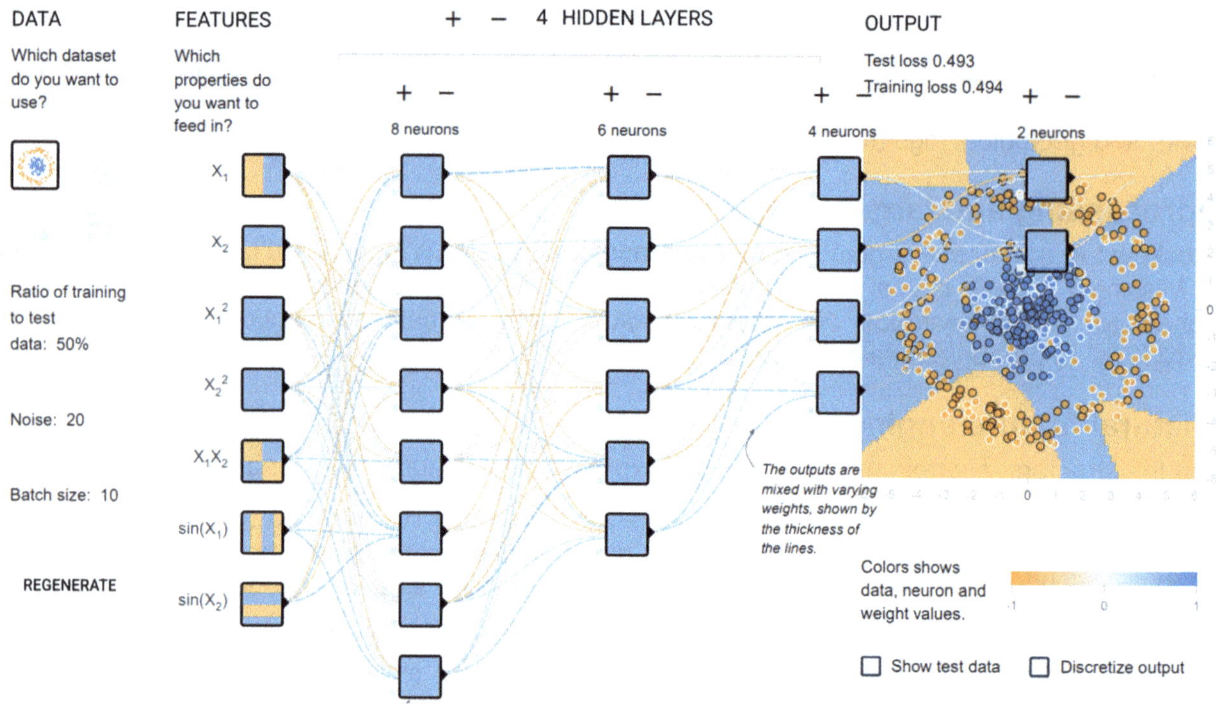

FIG 3 – Illustration of the deep neural network architecture used in the Strong 2025 ensemble model. The production model has four hidden layers with 160-80-40-20 neurons, ReLU activation, and L2 regularisation (α = 0.0001). A proportionally scaled version (8-6-4-2 neurons) is shown using TensorFlow Playground (Smilkov and Carter, 2016).

Deep neural network ensemble

The core predictor is an ensemble of 15 Multi-Layer Perceptrons (MLP) (sklearn.neural_network. MLPRegressor; Pedregosa *et al*, 2011). An ensemble approach was selected to quantify model uncertainty without the computational cost of bootstrap resampling, utilising diverse initialisations to capture the loss landscape (Lakshminarayanan, Pritzel and Blundell, 2017).

Architecture and hyperparameters

Each constituent model features a consistent deep architecture:

- **Structure:** Input Layer, Dense(160), Dense(80), Dense(40), Dense(20), Output(1).
- **Activation:** Rectified Linear Unit (ReLU) for hidden layers.
- **Optimisation:** Adam optimiser with adaptive learning rate.
- **Regularisation:** L2 penalty (alpha=0.0001) and Early Stopping (patience=20 epochs, 10 per cent validation hold-out) to prevent overfitting on high-nugget noise.

Aggregation and uncertainty

Models were initialised with distinct random seeds (100–114). The final prediction is the arithmetic mean of the ensemble outputs in log-space, subsequently back-transformed via Au_pred = exp (Log_Au_mean) – 1.

Associated uncertainty is quantified as the standard deviation of the ensemble predictions in log-space. This metric serves as a proxy for epistemic uncertainty, highlighting areas where the model lacks consensus due to geological complexity or low data density.

Summary plot illustrating the global impact of geological and spatial features on gold grade prediction. Features are ranked by the sum of SHAP value magnitudes (y-axis). The colour scale represents the feature value (red = high, blue = low), and the x-axis indicates the impact on model output (log-space). Note that structural proximity and local structural density) are dominant predictors, confirming the structural control on mineralisation. Figure 4 shows the SHAP feature importance analysis (Lundberg and Lee, 2017).

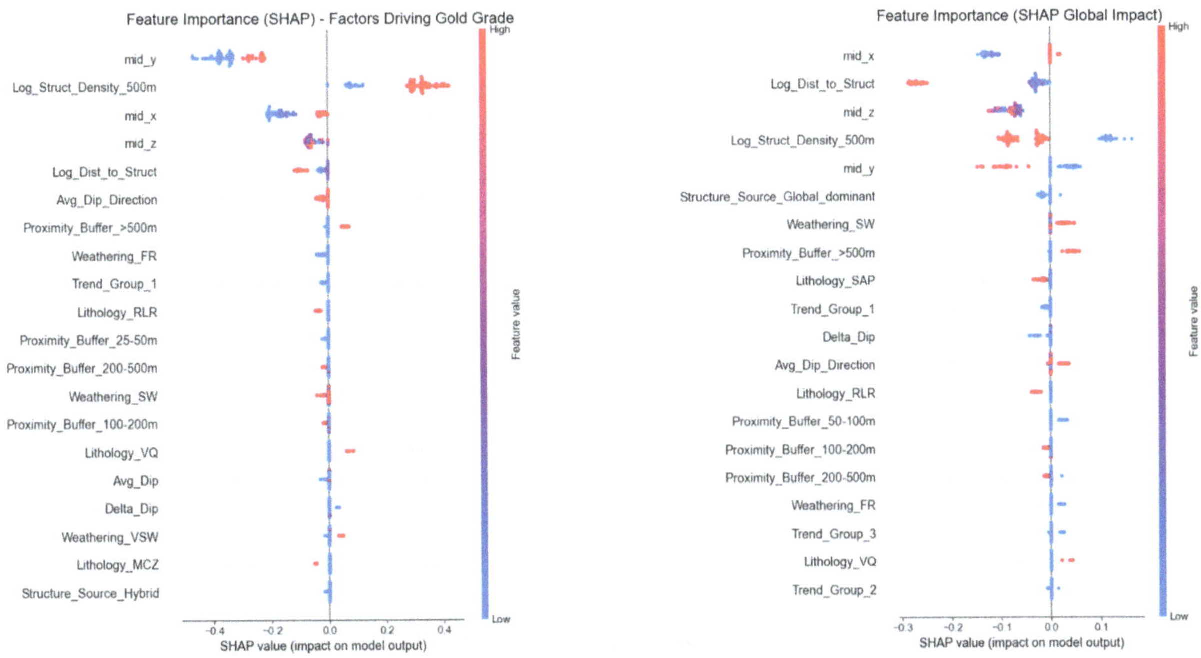

FIG 4 – SHAP – factors driving gold grade and SHAP global impact.

Comparison of predictive accuracy between the log-transformed domain and the back-transformed original scale. (A) Scatter plot of Actual versus Predicted values in log-space (Log_Au), in Figure 5. showing a tight correlation (R^2 approximately 0.15) and reduced variance. (B) Back-transformed predictions (g/t) versus actual assay values, highlighting the model's ability to reproduce high-grade outliers despite the nugget effect.

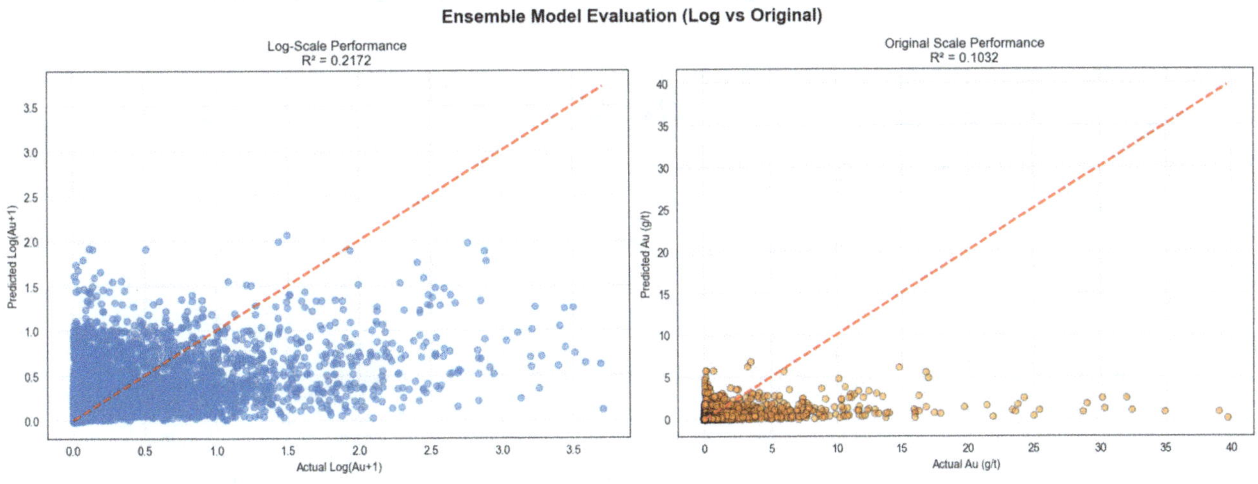

FIG 5 – Ensemble model performance evaluation.

Hexbin density plot comparing actual versus predicted gold grades (N=10 261 test samples) is shown in Figure 6. Darker colours indicate higher data density. The distribution aligns closely with the 1:1 identity line (dashed blue), demonstrating that the model is unbiased across the bulk of the grade distribution. The marginal histograms (top and right) confirm that the predicted distribution reproduces the log-normal shape of the input assay data.

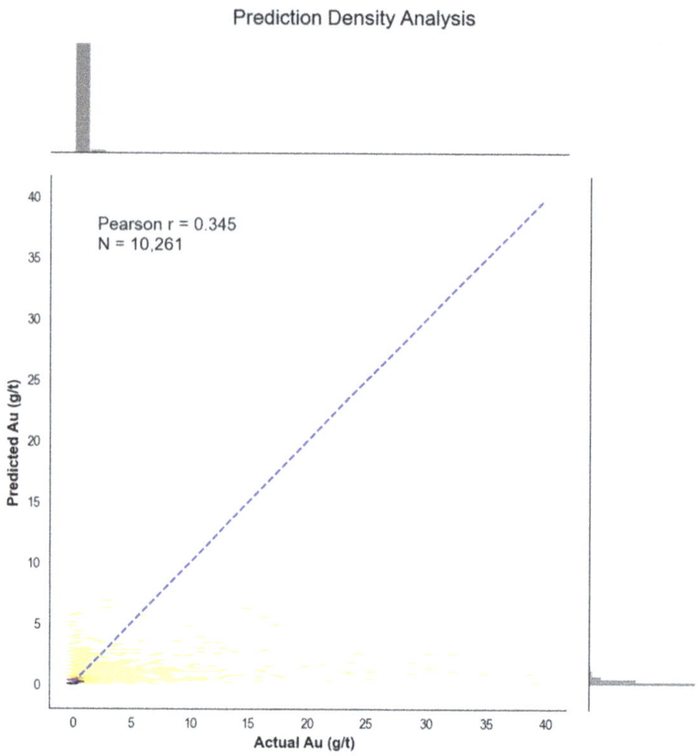

FIG 6 – Bivariate prediction density analysis.

The comprehensive diagnostic dashboard for model validation (Figure 7) comprises: (A) Log-scale accuracy plot. (B) Histogram of residuals (Error = Actual – Predicted), showing a normal distribution centred near zero. (C) Residuals plotted against predicted values to check for heteroscedasticity. (D) Relationship between gold grade and structural proximity, verifying that the model captures the distance-decay effect. (E) Boxplot of residuals grouped by structural trend domain, ensuring no systematic bias exists across different structural orientations. (F) Q-Q plot assessing the normality of prediction errors.

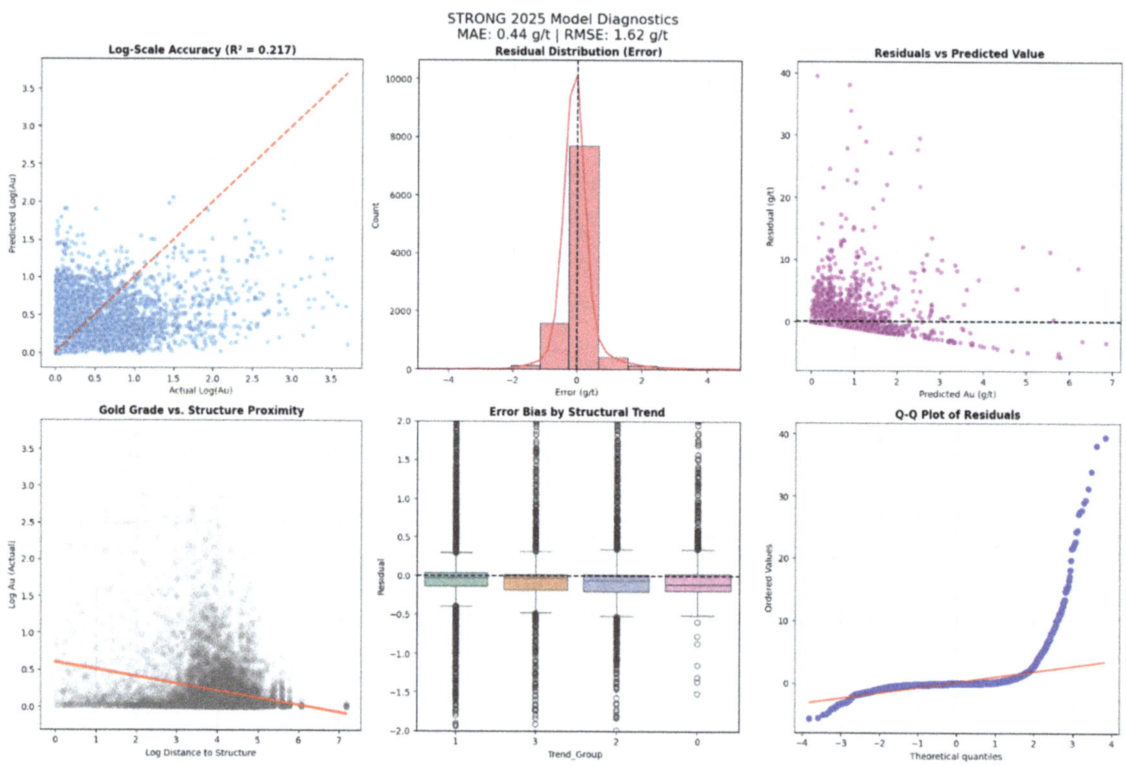

FIG 7 – Comprehensive diagnostic dashboard for model validation.

Comparison of observed (black solid line) versus predicted (red dashed line) gold grades along three representative drill holes from the blind test set is shown in Figure 8. The ensemble model (red) successfully tracks the high-frequency variability and spatial location of mineralised intercepts, demonstrating the model's ability to reproduce spatial continuity downhole. The shaded region represents the prediction uncertainty.

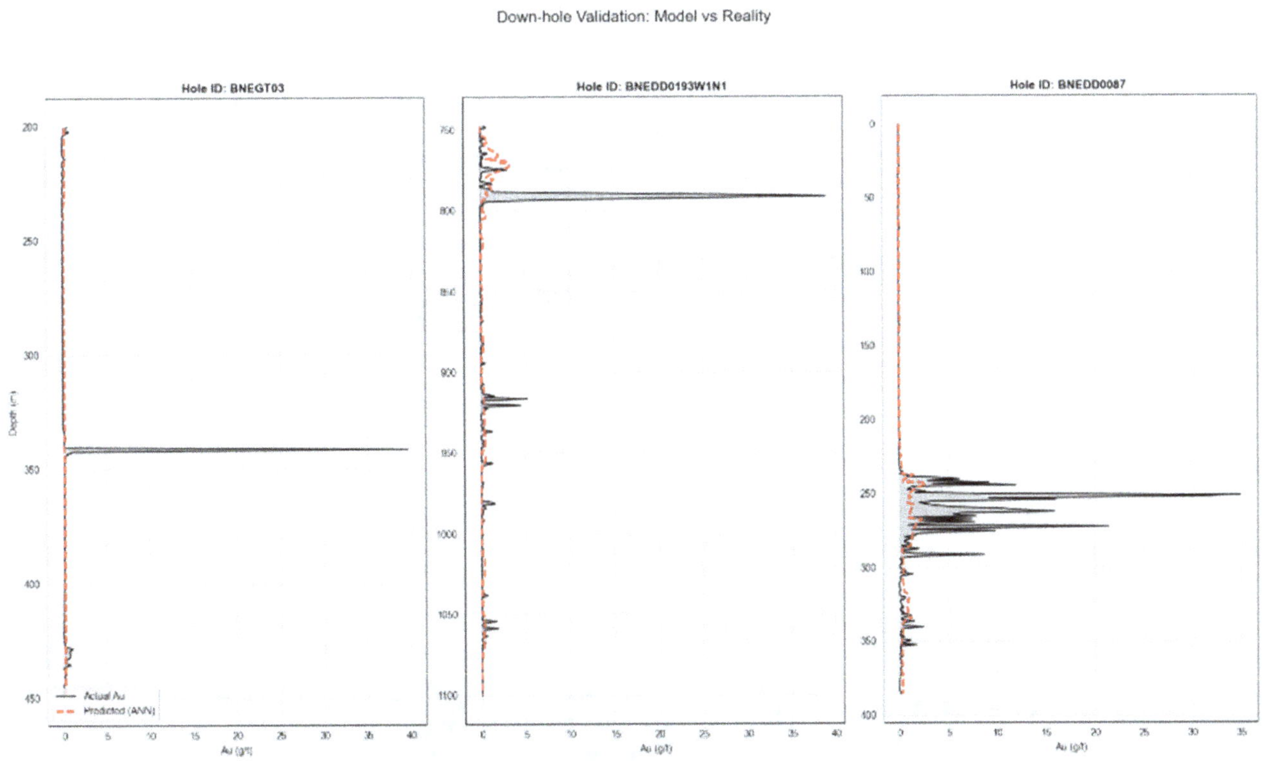

FIG 8 – Downhole validation strip logs.

The plan view (Figure 9) shows the ensemble standard deviation across the test data set. Warm colours (red/orange) indicate areas of high model disagreement, typically associated with complex geological structures or low data density. These zones represent priority targets for future infill drilling to reduce resource risk. (Bottom) Vertical cross-section looking North. Note that uncertainty often increases at depth (lower elevation) where drill hole spacing widens, correctly identifying the limits of the current confidence domain.

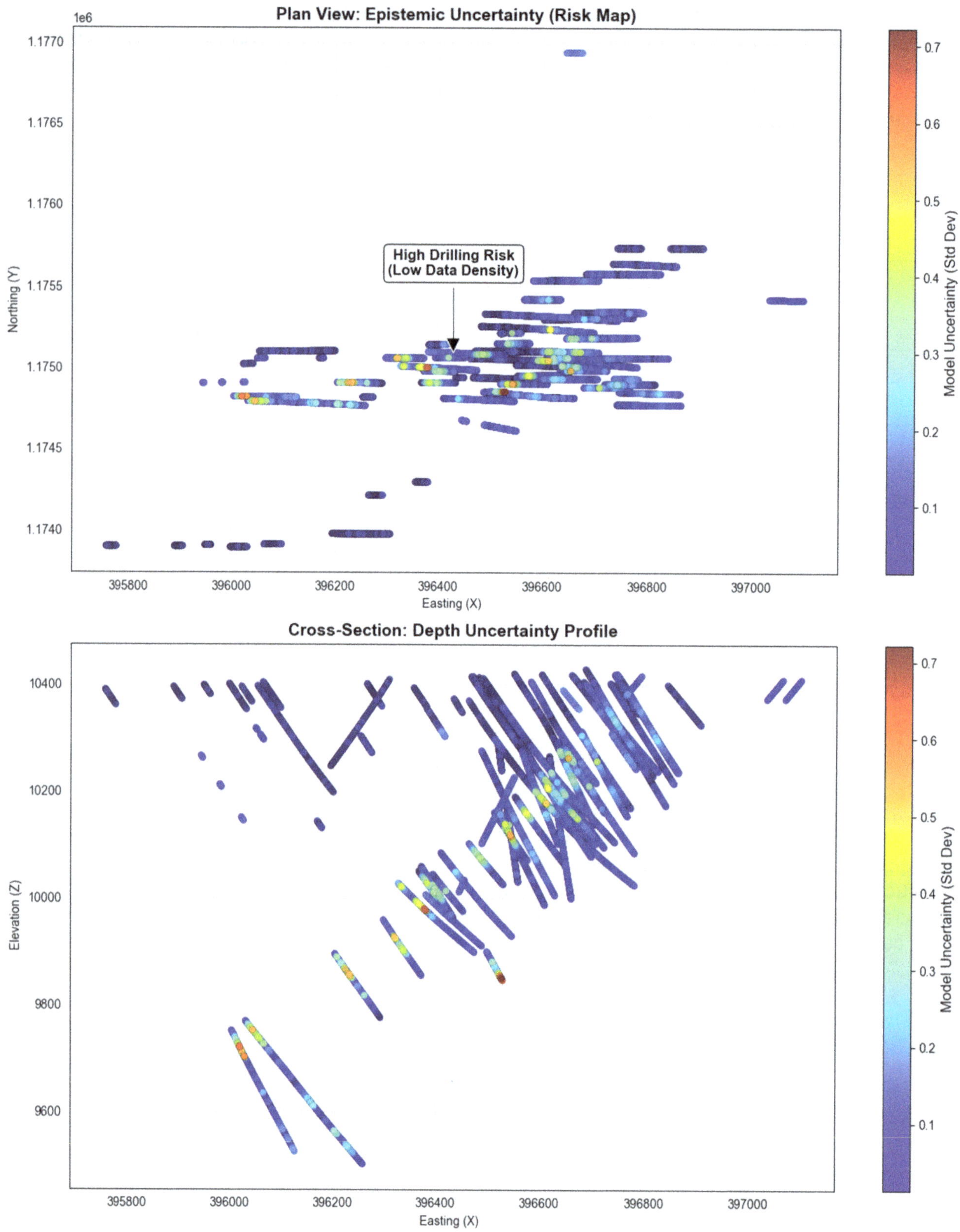

FIG 9 – Spatial distribution of epistemic uncertainty (prediction risk).

Comparative grade control modelling

The synthetic structural measurements were used in Leapfrog to create a finely detailed structural model that provided anisotropy for a set of comparative grade wireframes at the same cut-offs. The comparison between the results of original manually controlled anisotropy and the synthetic anisotropy shows that there is a considerable overlap, however the synthetic control has provide more detailed shapes with greater partitioning of higher grades into the High-grade Domain. As a further check the model was run four further times, however the resulting wireframes from the additional anisotropy runs did not produce significantly different results.

At a zero g/t Au cut-off, the GC with manual anisotropy model has 77 per cent of the tonnes, 117 per cent of the grade and 90 per cent of the contained gold of the resource. This reflects changes to the mineralised volume from the greater complexity of internal waste defined by the close spaced drilling. The High-grade domain defined using GC constrains the influence of the high-grade composites to a greater extent than that of the Resource High-grade domain used in the resource mode.

The re-estimation with the synthetic anisotropy has produced a very similar model, with the greatest differences at the top and the bottom of the model, where relatively minor changes to the assumed anisotropy produced locally significant changes in the wireframe interpretations. The comparison of the contained metal (Figure 10) shows especially close correspondence in the middle of the model, where there is informing data both above and below. Importantly, this is in saprolite where there are no structural measurements at all.

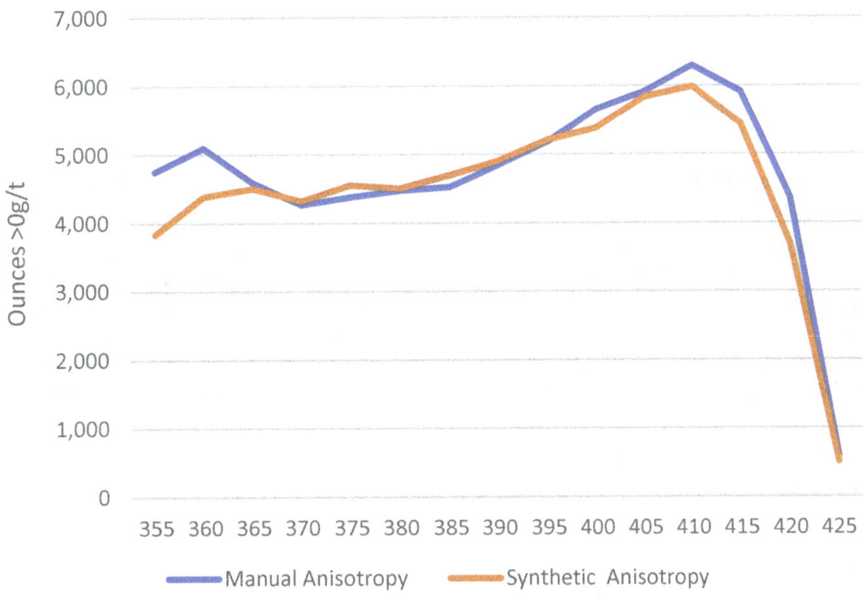

FIG 10 – Contained ounces per 5 m bench, NE Bankan grade control models.

CONCLUSIONS

This study developed a deep neural network ensemble for gold grade prediction and uncertainty quantification in a structurally complex deposit. By integrating multi-source geological data with rigorous feature engineering and a leakage-free, drill hole-level validation strategy.

Performance on the independent grade-control data set (39 361 intervals) closely matched internal validation, confirming operational transferability.

The model's consistent performance on the spatially withheld 'Blind Grade Control' data set confirms its robustness. This indicates that the framework is not merely a theoretical exercise but is sufficiently calibrated for application in short-term mine planning and production definition.

Risk Mapping for Strategic Planning via ensemble standard deviation allows for the mapping of prediction risk. These risk maps distinguish between areas of geological complexity and areas of data scarcity, providing a quantitative guide for optimising future infill drilling campaigns.

The use of the synthetically derived structural measurements assigned to grade data has demonstrated a performance equal to a traditional manual interpretation. Once operational, this methodology will be able to be deployed routinely to assist in the rapid modelling of grade data and optimisation of the grade control model estimate.

ACKNOWLEDGEMENTS

Bankan is the largest gold discovery in West Africa in a decade. The Predictive team applied terrane scale analysis of geology and geophysics to identify the target and apply for the first Exploration Permit in October 2019; this was the lease covering the North-east Bankan deposit. In a short time a team of national and expatriate staff was assembled; and implemented an exploration and resource development program. This paper would not have been possible without their dedication and high quality technical work.

Grateful acknowledgement is also extended to the board and management of Predictive Discovery for permission to publish this paper, and the management of ERM for providing the resources to complete this study.

REFERENCES

Avalos, S and Ortiz, J M, 2020. Recurrent neural networks for drill hole logging and lithology prediction, *International Journal of Mining, Reclamation and Environment,* 34:500–519.

Fossen, H and Cavalcante, G C G, 2017. Shear zones – a review, *Earth-Science Reviews,* 171:434–455.

Fouedjio, F and Klump, J, 2019. Exploring the effect of data leakage on machine learning performance in spatial prediction problems, *Journal of Spatial Information Science,* 19:107–134.

Harris, T and Murphy, B, 2022. Multi-faceted Geophysical program at NE Bankan and Exploration Targeting, unpublished internal predictive discovery report.

Kamb, W B, 1959. Petrofabric observations from Blue Glacier, Washington, in relation to theory and experiment, *Journal of Geophysical Research,* 64:1908–1909.

Lahondère, D, Egal, E, Lacomme, A, Le Berre, P, Costea, C A, Deynoux, M, Guerrot, C, Diabaté, B, Diallo, A, Diallo, A B, Diallo, S, Gaye, F, Iliescu, D, Minthé, D and Feybesse, J-L, 1999. Projet de cartographie géologique du Nord-Est de la République de Guinée et Notice explicative de la carte géologique à 1/200 000, Feuille n° 12 87, Siguiri, 1ère édition, Conakry, Edité sous la responsabilité du Ministère des Mines, de la Géologie et de l'Environnement. (Geological mapping project of the northeast of the Republic of Guinea and explanatory note at 1:200 000 scale, sheet number 12 87 Siguiri , first edition. Conakry, edited under the responsibility of the Minister of Mines, Geology and the Environment).

Lakshminarayanan, B, Pritzel, A and Blundell, C, 2017. Simple and scalable predictive uncertainty estimation using deep ensembles, *Advances in Neural Information Processing Systems,* 30:6402–6413.

Lundberg, S M and Lee, S-I, 2017. A unified approach to interpreting model predictions, *Advances in Neural Information Processing Systems,* 30:4765–4774.

Micklethwaite, S, Sheldon, H A and Baker, T, 2010. Active fault and shear processes and their implications for mineral deposit formation and discovery, *Journal of Structural Geology,* 32:151–165.

Pedregosa, F, Varoquaux, G, Gramfort, A, Michel, V, Thirion, B, Grisel, O, Blondel, M, Prettenhofer, P, Weiss, R, Dubourg, V, Vanderplas, J, Passos, A, Cournapeau, D, Brucher, M, Perrot, M and Duchesnay, E, 2011. Scikit-learn: Machine Learning in Python, *Journal of Machine Learning Research,* 12:2825–2830.

Roberts, P A, Murphy, F C and Nganare, A, 2021. The Bankan gold camp: a greenfields discovery in Guinea, West Africa, in *NewGenGold 2021,* pp 209–221 (Paydirt Media and Keith Yates and Associates: Perth).

Smilkov, D and Carter, S, 2016. TensorFlow Playground – Play with neural networks! (online). Available from: <https://playground.tensorflow.org/>

Using chance-constrained optimisation to assess the value of drilling

R Jeuken[1]

1. Geometallurgist, BHP, Brisbane Qld 4000. Email: rick.jeuken@bhp.com

ABSTRACT

Geological uncertainty presents a significant challenge in mine-planning. While infill drilling can reduce this uncertainty, identifying the most valuable drilling locations remains complex. This paper presents a chance-constrained optimisation approach that integrates drilling decisions into an ore blending model. By incorporating uncertainty directly into the planning process, the model guides drilling toward areas where reducing variance most improves blending outcomes.

Chance-constrained programming enables the blending of uncertain geological properties to maximise plant output while maintaining quality constraints within acceptable risk levels. The model quantifies the value of additional drilling, prioritising locations that enhance blending flexibility. A case study from a coalmine in Queensland, Australia, demonstrates that targeting high-quality blocks for variance reduction yields better results than traditional approaches focused on low-quality areas. This method offers a practical and effective strategy for optimising mine plans under uncertainty and highlights the value of probabilistic modelling in resource planning.

INTRODUCTION

Managing uncertainty is a core challenge in mine planning, particularly geological uncertainty, which affects key variables like grade and tonnage. Ideally you will model ore properties, such as grade parameters or density as random variables, with uncertainty quantified using conditional simulations. In the planning process mine planners use various strategies to manage this uncertainty, often using mathematical programming with fixed risk buffers on properties. Doing this, however, does assume a fixed model of uncertainty.

Reducing uncertainty—often through additional drilling—can improve planning outcomes, though it incurs costs. Selecting optimal drill locations is complex and traditionally relies on evaluating the spacing of predefined patterns and how reducing spacing reduces the uncertainty. This value of information (VOI) framework helps assess whether and where additional data collection is worthwhile. Recent research explores VOI using operations research and machine learning to optimise drilling strategies.

In coal mining, where strip mining and sequential seam extraction are common, the sequence of uncovering and extracting coal can be relatively inflexible over shorter planning horizons. This is due to depth of cover to coal and equipment constraints eg coal uncovered by a dragline is in a fixed sequence. This means that one can use blending models that take in these fixed paths and then incorporate uncertainty using offsets or chance constraints to ensure the properties of produced coal meet quality targets with high probability.

In this paper the author describes a novel approach using chance-constrained programming to directly evaluate the impact of reducing uncertainty on coalmine-plans. The method developed identifies drilling locations that offer the highest return on investment, enhancing decision-making in infill drilling and ore blending.

This paper is an adaptation of the opensource paper Jeuken and Forbes (2024) with details in the references. Many additional references within Jeuken and Forbes (2024) give examples of background and recent work in assessing the value of drilling.

ORE BLENDING OPTIMISATION

Ore blending optimisation involves using mathematical programming techniques to determine the most efficient combination of different ore types to achieve desired quality and processing performance. Mathematical programming, often referred to simply as optimisation, is a powerful mathematical discipline that focuses on finding the best solution from a set of possible choices. This process involves maximising or minimising an objective function, f, which represents a quantifiable measure of performance or value, subject to a set of constraints that define the feasible region of solutions. This section briefly introduces chance-constrained programming and its application to ore blending (see also Jeuken, Forbes and Kearney (2021)).

Chance-constrained programming

A chance constraint introduces probabilistic conditions into the decision-making process in optimisation problems. Unlike traditional constraints, which demand a solution to precisely meet certain conditions or not at all, chance constraints permit the possibility of constraint violation, albeit only with a certain acceptable probability. A chance-constrained optimisation problem is generally of the form:

$$\max f(x)$$

$$s.t. \mathbb{P}\{G(x,r) \leq 0\} \geq 1 - \epsilon$$

where x is a vector of decision variables, r is a vector of random variables, and $\epsilon \in (0,1)$ is the maximum allowable likelihood of failure. The random variables r introduce uncertainty into the system, and the chance constraint ensures that the probability of satisfying $G(x,r) \leq 0$ is at least $(1 - \epsilon)$.

Chance constraints are particularly useful in fields such as finance, engineering, logistics and data science, where decision-making under uncertainty is common. They allow for a more flexible and realistic modelling of problems where strict adherence to constraints is not always possible or practical, and some level of risk is acceptable. By choosing different values of ϵ, a decision-maker can control the risk level associated with constraint violations, balancing between feasibility and conservatism in their solutions.

Some chance-constrained problems are straightforward to solve exactly. For example, if r is multivariate normal, then you can express the problem as a second-order cone program and solve it exactly in a tractable and efficient manner using standard solvers. Alternative techniques are available if the uncertainties are not normally distributed.

Ore blending model

In the study for this paper, the author uses a simplified version of the coal blending model from Jeuken, Forbes and Kearney (2021) to assess how input variance affects outcomes. The focus is on the property of coke strength after reaction (CSR), the detail of which you can find in Diez, Alvarez and Barriocanal (2002). Predicted CSR is a key property of coking coals, impacting the efficiency of coke utilisation in blast furnaces. Therefore, it holds significant importance in any decarbonisation strategy related to steelmaking.

The initial input for the coal blending model is a mining schedule. This schedule quantifies the ore available for extraction and blending at future intervals from various sources. Each of these sources possesses distinct quality parameters, including ash content, volatile matter, and CSR. As the author is illustrating a method in this paper, he will limit the coal model to a single property, CSR, to showcase the effectiveness of chance constraints.

The blending model takes the volume capacity of ore feed into a processing plant as an input. This capacity varies over time based on scheduled maintenance, allowances for breakdowns, and non-specific process delays. Thus, there is a maximum feed capacity, measured in tonnes, for each time period.

The model also contains constraints on CSR to ensure that the CSR of the blended material in each time period meets a minimum threshold. In this example, the author applies the constraints independently for each time period t, and the weighted value of property CSR must exceed a minimum threshold CSR^- at each time. That is:

$$\mathbb{P}\{CSR \geq CSR^-\} \geq 1 - \epsilon, \forall\, t \in \mathcal{T},$$

This simple framework blending model can easily include additional constraints to model mining processes such as capacity or availability limits on equipment, plant, stockpiles etc. The objective function can include costs, values and discounting depending on the problem or time horizon of interest. In this study the author has treated the property CSR as additive. In practice one usually recalculates the CSR of a blend although comparison studies of the project deposit have shown that the error of the assumption of being linearly is minimal.

In the ore blending example of this paper, the distribution of the random parameter CSR is normally distributed, ie the CSR of each block, c, is $CSR_c \sim N(\mu_c, \sigma_c)$. As the author chooses that $\epsilon \leq 0.5$, he can write the chance constraint as a second order conical constraint, which in turn is readily solvable using commercial solvers such as Gurobi (Gurobi Optimization LLC, 2024) or C-PLEX (IBM ILOG CPLEX, 2009).

Value of drilling

The model the author has described consisting of an objective function, inventory constraints, capacity constraints, and a property constraint on CSR, represents a highly versatile blending model. The author has quantified the uncertainty of the property CSR in each block of coal through the standard deviation,σ_c. By employing binary variables in the blending model, you can select blocks in which reducing the standard deviation σ_c will maximise the overall value of the plan. These decisions of variance reduction are directly correlated to the selection of drilling locations.

Thus, in the blending model the author introduces a binary variable, Z_c where $Z_c = 1$ selects block c to have the standard deviation σ_c reduced by a factor $m_c \in (0,1)$. The author has predetermined the values for each m_c. This reduction in variance results from drilling and sampling in or near block c.

The described model is set-up so that you can solve for existing uncertainty in CSR in each block of coal. Alternately you can reduce the variance in each block or set of blocks that aligns with drilling and sampling a hole, or you can set-up the binary variables so that the model itself chooses which hole or set of holes will give the best overall value in the plan. You can also include costs associated with each drilling location in the objective function. And, thus, you can optimise your drilling locations, in the context of how you expect to use your ore in the mine plan.

Blending model formulation

The details of the blending model formulation and the inclusion of the value of drilling is beyond the scope of this paper. You can find the explicit details in the opensource paper Jeuken and Forbes (2024).

CASE STUDY

In this case study, the author evaluates a medium-term coalmine planning scenario with a pre-determined mining path—and thus pre-determined feed available for blending. There is, however, an opportunity to diminish geological uncertainty by drilling additional holes in the short-term. The evaluation that the author presents ostensibly spans a two-year horizon, commencing approximately one year from now, allowing time for planning, drilling, gathering, and analysing additional samples.

The case study is based on a coking coalmine complex in Queensland, Australia. As described in the previous section the author will apply chance-constrained programming to limit the probability that a blended coal property, CSR, falls below a required threshold.

Evaluation data

To maintain confidentiality, the author has applied linear multipliers to the data without altering the problem's structure. The model spans 24 monthly periods over two years and includes 108 coal sources from various pits and seams, each becoming available at specific times, seen in Figure 1.

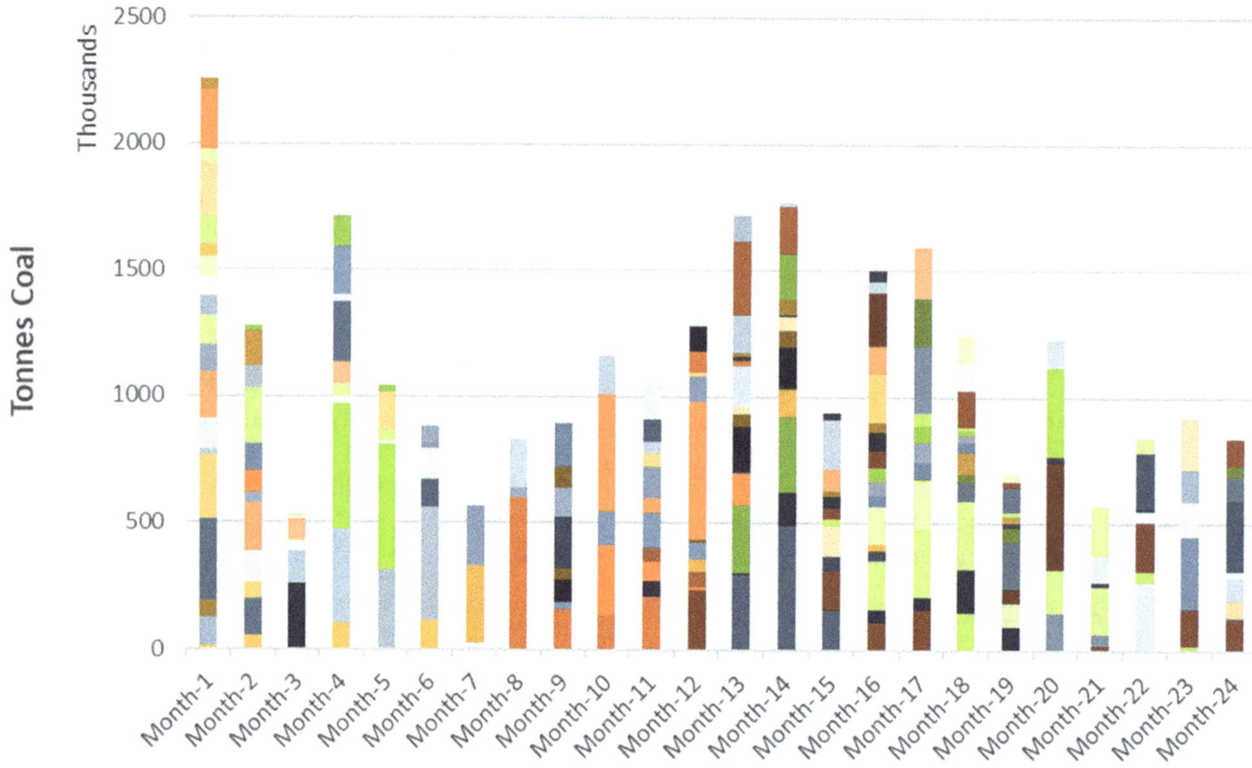

FIG 1 – Total tonnes of coal made available in each period, from 108 separate locations.

Each coal block has an expected CSR value. The minimum threshold for CSR is set at 67 per cent. Analysis shows that approximately 40 per cent of the scheduled coal falls below this threshold, Figure 2a. This situation offers an opportunity to blend up this lower quality material to meet the minimum quality required. The amount of coal that you can blend with an acceptable risk of producing coal below the threshold depends on the variance associated with each block of coal. Figure 2b shows the estimated CSR and standard deviation for each block and illustrates they are not correlated.

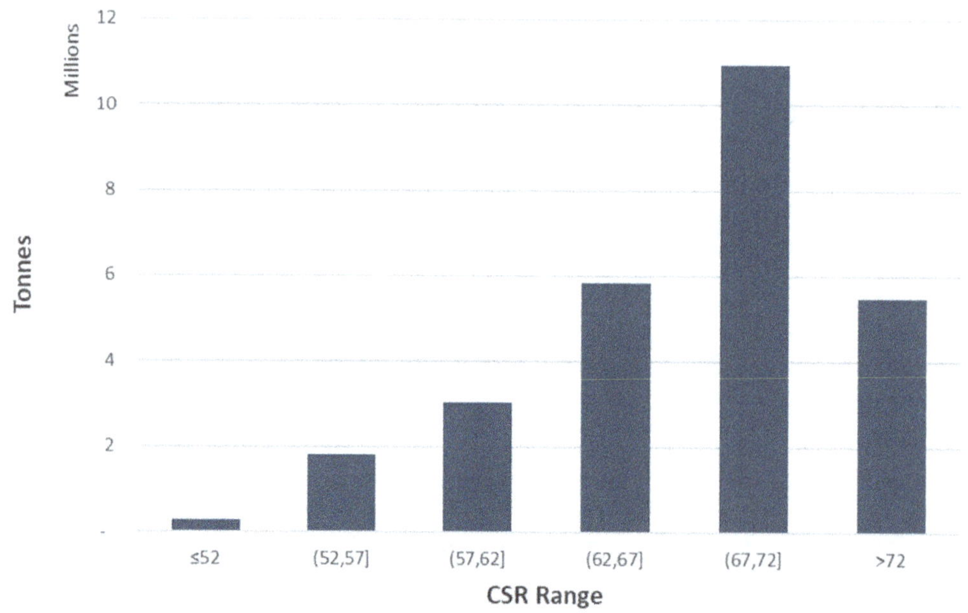

(a)

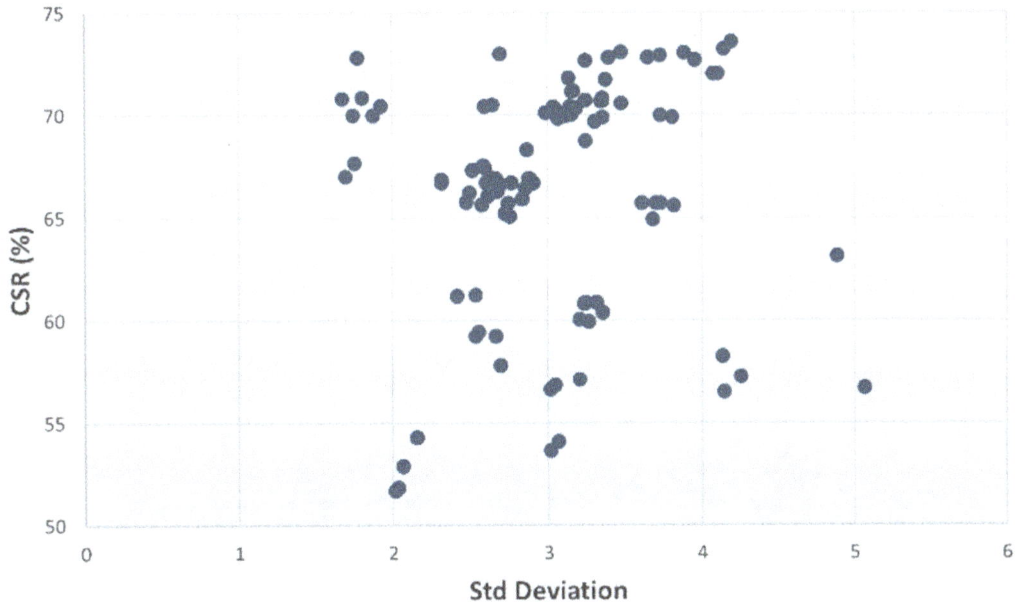

(b)

FIG 2 – (a) Histogram of the tonnes of coal by the expected CSR value of each tonne. (b) Plot of CSR versus Standard Deviation for each block of coal (RHS).

When you drill and sample a hole in a block of coal, the uncertainty of the properties of that block decreases. More broadly, and depending on the variogram range, you would anticipate that for each hole drilled, the variance of all blocks within a neighbourhood diminishes. In this case study, for analytical simplicity, the author assumes that each selection of a block to drill solely reduces the variance of that specific block and the expected CSR value is unchanged.

Evaluation method

The author assesses the value of drilling using two approaches. First, he evaluates each of the 108 source locations individually by reducing its variance and solving the blending model to observe the improvement in objective value. This process identifies which blocks offer the highest return from drilling, producing a ranked list of priorities. Thus, the author iteratively reduced the variance of each block, solved 108 instances of the resulting model, and assessed the outcomes.

Extending the evaluation to test all pair combinations or a larger selection of locations becomes impractical due to the rapidly increasing time required. Instead, the author utilised the model outlined above with binary variables to choose which set of blocks is optimal to drill. This allowed the author to test combinations of k = 1 to 10 blocks and identify the optimal group for variance reduction. The model selects which is the optimal set of blocks where variance reduction adds the most overall value to the plan.

RESULTS

The author first evaluated the impact of reducing the variance of each block individually. For each of the 108 blocks, he measured the increase in revenue from blending after reducing its CSR variance. Figure 3 shows, for each of the 108 blocks, the expected value of CSR alongside the revenue increase when reducing the variance of the CSR value for that block (through drilling). The data has been sorted from the highest expected CSR value to the lowest. It is worth noting that the increase in revenue does not exhibit a strong correlation with the expected CSR of a block, as illustrated in the Figure 3 – likely due to variation in block tonnage.

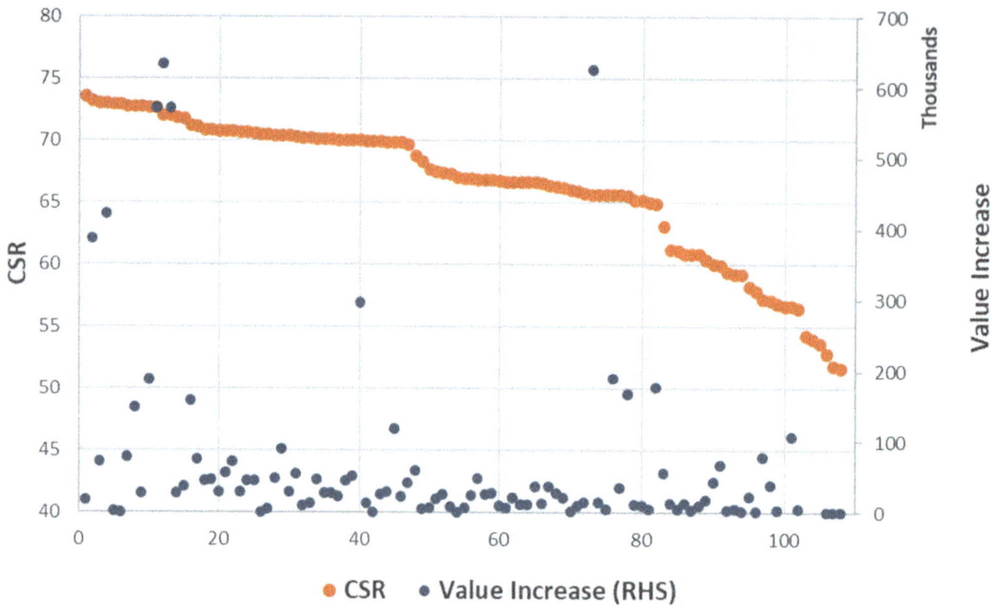

FIG 3 – Plot showing the ordered by *a priori* value of CSR, in orange (the line) and the dollar value increase resultant of reducing the variance of that block, in grey (the dots).

In Figure 4, the standard deviation of the CSR value for each block is presented alongside the revenue increase resulting from reducing the variance of the CSR value for that block (through drilling). The author has again sorted the data from the largest value increase to the lowest. Notably, there is a good correlation between reducing the standard deviation of high-variance blocks and the increase in value.

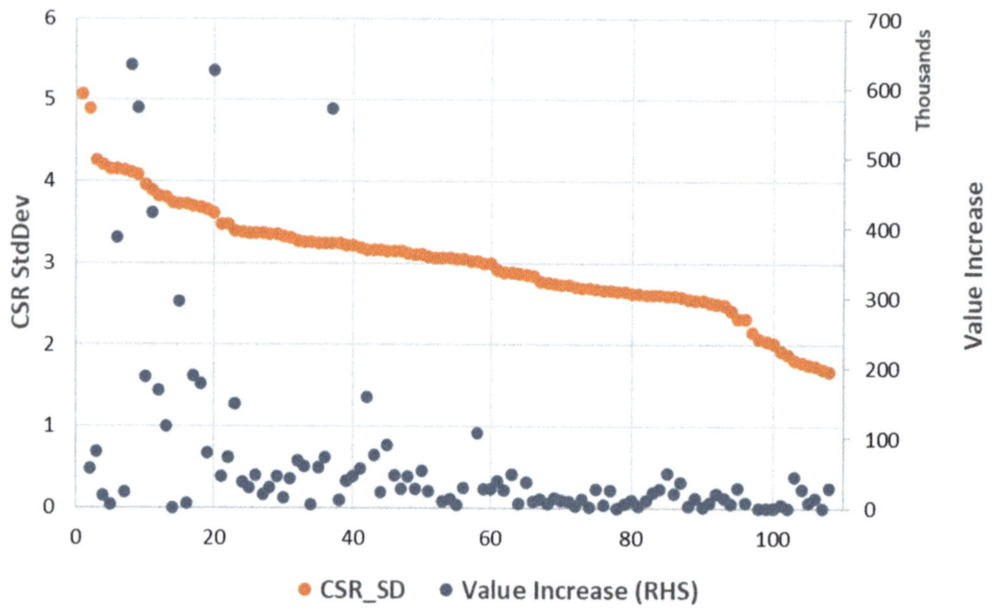

FIG 4 – Plot showing the ordered by *a priori* value of the standard deviation of CSR, in orange (the line), and the dollar value increase resultant of reducing the variance of that block, in grey (the dots).

Figure 5 then shows the top ten blocks ranked by revenue gain from drilling. For each block, there is a couplet containing the expected value of CSR and the standard deviation. Almost all of these blocks have expected CSR values above or near the 67 per cent threshold and high standard deviations (mostly ≥ 3.6).

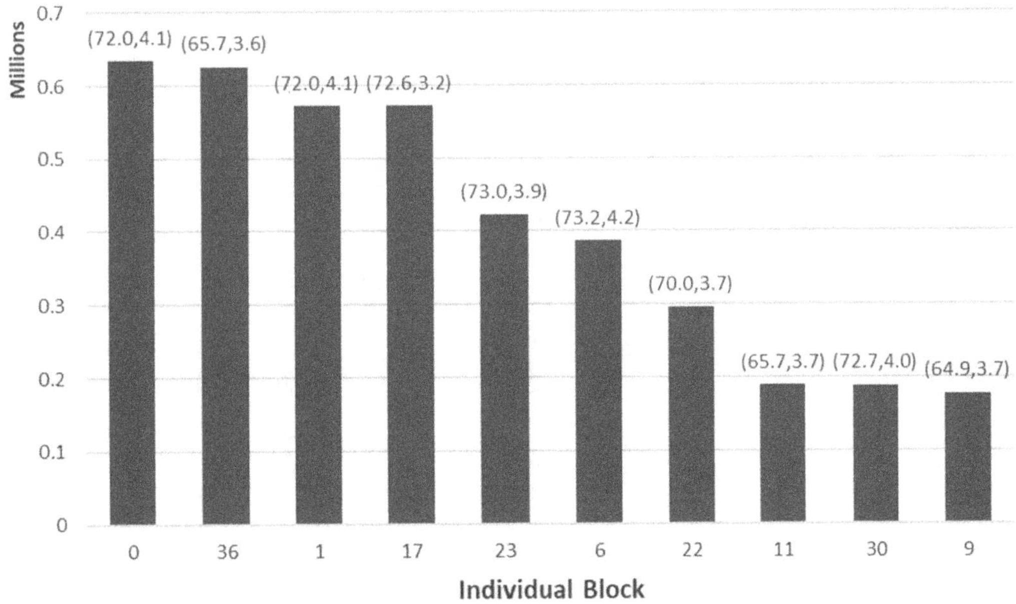

FIG 5 – Plot showing the increase in dollar value (LHS) by individually drilling the 10 blocks that give the largest increase in dollar value. The couplets contain the expected value of CSR (%) and the standard deviation.

The results from the evaluation where the author allows the model to select k = 1 to 10 blocks are as follows. In Table 1 the author shows the high-level use of ore as more blocks have their variance reduced for selected values of k. You can observe directly that with additional drilling, the use of high-grade ore does not change, whereas the use of low-grade ore (CSR <65 per cent) increases.

TABLE 1

Total feed tonnes of ore (in kt) by CSR range of block and as number of holes drilled (k) increases.

CSR range	k=0	k=1	k=2	k=5	k=10
<65%	3642	3752	3850	4135	4363
65–70%	8539	8539	8539	8539	8539
>70%	12 381	12 381	12 381	12 381	12 381

Finally, in Figure 6, you can see the cumulative value of drilling up to the tenth location. It is evident that the incremental value of each additional hole diminishes as drilling approaches 10 locations, with additional holes offering minimal return.

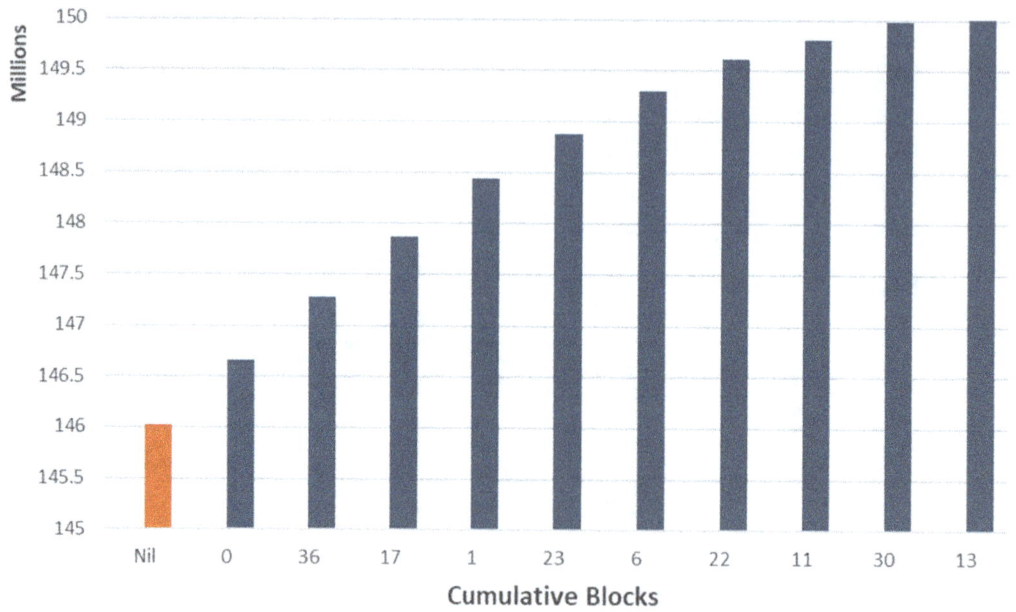

FIG 6 – Plot showing the cumulative increase in dollar value (LHS) by drilling up to k blocks. That is when k = 1, the selected block is block '0'. When k = 2, the selected blocks are '0' and '36'.

Discussion of results

The results of these trials have yielded some interesting points.

- You can see in Figure 5 that most of the value uplift occurs when the variance reduction is on blocks with high initial values of CSR. However, in Table 1 we see that the change in the plan is that the additional tonnes are blocks with low values of CSR.

  This confirms that variance reduction is best targeted at high value ore, allowing you to include more low value ore in a plan, to blend it up. This is in direct contrast to the author's experience in drilling coal deposits where infill drilling often targets low quality coal to pin down its exact value.

- There is diminishing return as you add more locations for drilling. While this is intuitively true, this approach finds what that point is. In this case the author did not include the cost of drilling in the objective function, the evaluation is purely on value uplift. As each location has an incurred cost, allowing a model to select any number of locations will limit itself to a small subset, not when the profit is all consumed, but when the incremental return is below the cost of additional information.

- When using the model to choose the 10 best locations to drill, ie k = 10, then the set of blocks selected is almost identical to the top 10 blocks when evaluating them individually and differing only slightly in order. However, solving all 108 individual cases took 15 mins, while the model-based approach took over an hour per value of k when k>4, highlighting a trade-off between accuracy and computational efficiency.

CONCLUSIONS

This study highlights the critical role of geological uncertainty in mine-planning. The mining industry typically evaluates drilling strategies using regular hole patterns and fixed spacing, assessing outcomes through variance reduction or sometimes using a mine plan and multiple orebody realisations. In contrast, this approach introduces a chance-constrained blending model that selects optimal drilling locations using easy-to-solve formulation. Applied to a coal mining case, this method offers a practical and novel way to manage geological uncertainty.

A key insight from the case study is that reducing the variance of high-quality, high-CSR blocks often yields more value than drilling low-quality blocks—challenging conventional industry practices. High-CSR blocks, when made more certain, can better accommodate lower-quality material, improving overall blending outcomes and profitability. While the case study focuses on CSR this same principle will apply to all properties that are linearly additive (or approximately so) – eg sulfur, volatile matter etc.

This information allows us to straightforwardly determine a threshold for the number of holes to drill, considering external factors such as cost and practical considerations like drilling or labour capacity. By considering these factors alongside the diminishing returns observed in the cumulative value plot, we can make informed decisions regarding the optimal number of holes to drill in practice.

REFERENCES

Diez, M A, Alvarez, R and Barriocanal, C, 2002. Coal for metallurgical coke production: predictions of coke quality and future requirements for cokemaking, *International Journal of Coal Geology*, 50(1–4):389–412.

Gurobi Optimization, LLC, 2024. Gurobi Optimizer Reference Manual, Gurobi Optimization, LLC. Available from: <https://docs.gurobi.com/projects/optimizer/en/current/index.html>

IBM ILOG CPLEX, 2009. V12, 1: User's Manual for CPLEX, *International Business Machines Corporation*, 46(53):157.

Jeuken, R, Forbes, M and Kearney, M, 2021. Optimal blending strategies for coking coal using chance constraints, *Journal of the Operational Research Society,* 72(12):2690–2703. https://doi.org/10.1080/01605682.2020.1811167

Jeuken, R and Forbes, M, 2024. The Value of Drilling—A Chance-Constrained Optimization Approach, *Mining, Metallurgy and Exploration,* 41:2279–2289.

Integrating geological chronology into AI-driven geological modelling – toward geologically coherent domain construction

R Reid[1], S Palmer[2] and J Van Zyl[3]

1. Group Resource Geologist, Harmony Australasia, Brisbane Qld 4064. Email: ronald.reid@harmonyseasia.com
2. Resource Geologist, Eva Copper Mine Pty Ltd, Brisbane Qld 4064. Email: sam.palmer@harmonyseasia.com
3. Resource Geology Specialist, Harmony Australasia, Brisbane Qld 4064. Email: jaco.vanzyl@harmonyseasia.com

ABSTRACT

The application of machine learning (ML) and artificial intelligence (AI) is gaining momentum in the mining industry. These technologies are now widely used to develop digital twins, optimise ore sorting and energy usage, enable predictive maintenance, and analyse large data sets for exploration and modelling. Companies such as Earth.AI, StratumAI, and OreFox are leveraging AI and neural networks to process geochemical, geophysical, and drill hole data, aiming for a first-mover advantage in exploration and targeting. Meanwhile, software providers like Micromine (Grade Copilot) and Maptek (DomainMCF) have introduced AI-powered modules that enable users to model numerical and categorical data directly. These tools are increasingly being used to construct geological and grade models that feed into Mineral Resource estimates.

However, a critical shortcoming remains: the absence of geological chronology. Geological features such as lithology, alteration, and oxidation do not simply co-exist in space. A sequence of geological events governs their distribution. Most AI-assisted tools currently model spatial relationships without accounting for the temporal order of geological processes. This often results in models that appear realistic at a global scale but fail to honour geological constraints when examined in detail.

This paper proposes a methodology that combines the analytical strengths of AI with the interpretive value of geological chronology. By integrating event sequencing into the modelling workflow with several examples, we demonstrate how AI can be used not just to generate plausible models, but to construct geologically valid interpretations that respect the timing and relationships between geological units. This approach improves the robustness of geological models and the reliability of the domains used in resource estimation.

INTRODUCTION

With increasing demand for raw materials, declining average grade, and the move to deeper mineralisation, high-quality resource models with extensive mineralogical information are required more than ever (Jooshaki, Nad and Michaux, 2021). A comprehensive 3D geology model is essential for accurately estimating a deposit's resources and serves as the foundation for the entire resource estimation process, ensuring the accuracy and reliability of the final results (Dumakor-Dupey and Arya, 2021; Erdogan Erten, Yavuz and Deutsch, 2022; Kaplan, Dagasan and Topal, 2021; Reid and Cowan, 2023). A reliable mineral resource estimate is critical to the success of every mining project. However, a poor quality estimate that under- or overestimates mineral grades will have significant consequences and value losses for any mining project (Kaplan, Dagasan and Topal, 2021).

While geological modelling underpins the Resource estimate and involves a comprehensive analysis of often limited sample and geology data sets through spatial information processing, geological interpretation, and geostatistical tools, many practitioners do not seem to fully understand how to effectively generate geologically realistic envelopes with which to constrain Mineral Resources (Cao *et al*, 2024; Li *et al*, 2023; Reid and Cowan, 2023). From the perspective of its usefulness in resource estimation, most geological logging data is just noise (First *et al*, 2024) and can mislead the modelling geologists, adding complexity where none is needed. Therefore, a critical step in the process of domaining for geology, which significantly impacts estimation quality, is to clarify and simplify the geology (Atalay, 2025).

Unfortunately, resource geology teams are strained by ongoing demands from miners to shorten the mine planning horizon and generate updated models in ever-diminishing time frames (Buchanan *et al*, 2023). Sullivan *et al* (2019) note that the end-to-end Resource modelling process, depending on the size and complexity of the model, can take several months, which is a drain on company resources and limits project agility. Production pressures force mine geologists to focus their time on where they have the most impact, and the most time-consuming tasks remain the simplification of the data, the interpretation of the orebody, and its conversion into a block model (Cowan, Spragg and Everitt, 2011). This is made more difficult by the need to produce valid wireframes from sectional and plan interpretations, and is commonly the primary source of deviation between actual and estimated ore tonnages (Battalgazy *et al*, 2023; Cowan, Spragg and Everitt, 2011). In the process, a simplified yet robust geological model is often neglected. Machine learning (ML) methods have been proposed in recent years as an alternative to geostatistical techniques and manual wireframing to alleviate these stressors in Mineral Resource estimation (Mahboob, Celik and Genc, 2023).

In recent years, there has been a boom in research on 3D geological modelling based on machine learning, alongside increasing use of artificial intelligence (AI) in exploration and ore targeting, resource modelling, geometallurgy, and the compilation of Digital Twins (Li *et al*, 2023; Maptek, 2025; Micromine, 2025; OreFox, 2024; PETRA, 2025; Stratum AI, 2025). Artificial neural networks have also gained popularity for resource estimation due to their ability to model complex relationships in sample data (Battalgazy *et al*, 2023). Al-Alawi and Tawo (1998) outline how artificial neural networks are computer models designed to emulate human information processing capabilities, such as knowledge processing, prediction, and control. Deep learning (DL) is a subfield of machine learning that focuses on neural networks with many layers (Al-Alawi and Tawo, 1998; Newby, 2024); although the term was introduced in the mid-1980s, early examples can be traced back to the 1960s, with the first industrial impacts felt in the early 2000s (Sullivan *et al*, 2019).

Machine learning enables computers to learn from historical data and to approximate unknown natural functions that map variables and make predictions without being explicitly programmed (Atalay, 2025; Cao *et al*, 2024; Newby, 2024). This allows the ML algorithm to leverage the statistical properties of known data to generate models that enable the geologist to make judgments and predictions from the input data. The AI algorithms aim to predict the behaviour of complex systems and learn patterns in data without explicit instructions; they can process large amounts of data with greater accuracy and are not influenced by personal biases or preconceived notions, allowing for objective analysis (Buchanan *et al*, 2023; First *et al*, 2024; Jooshaki, Nad and Michaux, 2021). Though claims that ML is 'more accurate' are not necessarily correct, since models are only an expression of the input data, poor-quality input data will yield poor-quality results. In resource estimation, these predictions can reflect a block grade distribution and how it might relate to geological processes (First *et al*, 2023; Li *et al*, 2023), but they do not generate a 'true' model.

The creation of large data sets, the development of more effective algorithms, and the availability of powerful computers have led to a significant increase in the use of machine learning in recent years. These new technologies have the potential to change the processes used by resource modelling teams to generate updated interpretations of their geological data by simultaneously analysing large drill hole data sets (Buchanan *et al*, 2023). However, to date, a significant amount of research has been directed towards other fields in mining rather than resource or geological modelling. There have been a number of studies that have examined new applications of AI and ML in mineral exploration, evaluating their effectiveness, limitations, benefits, and challenges (Hasan, Bin Shafiq and Khatun, 2023; Jooshaki, Nad and Michaux, 2021). Jackson (2023) discusses in the top ten uses for AI in mining, how AI can provide valuable insights that humans may miss, and in his belief, this will enable better decision-making. According to him, AI can help with mining exploration by analysing large amounts of data and identifying on-site targets. However, in his list, exploration ranked only seventh among the top ten uses of AI in mining, while resource and geological modelling were not listed at all.

Artificial neural networks are relatively new to the earth sciences and have only been sparsely demonstrated in this field (Al-Alawi and Tawo, 1998). Neural networks are a form of ML that are a derivative of cognitive computing, inspired by the structure and functioning of the human brain, comprising interconnected neurons organised into input, hidden, and output layers (Newby, 2024). They are versatile and can be used for regression, classification, and pattern recognition, and learn by adjusting the weights of connections between nodes during training with labelled examples (the training data set). These techniques help in identifying patterns and correlations in vast amounts of data, reducing the time and cost involved in geological data analysis (Bergen *et al*, 2019; Hasan, Bin Shafiq and Khatun, 2023).

A key feature of ML algorithms is that they learn from the data without requiring rigid statistical assumptions, yet their performance and accuracy are highly dependent on the quality of the data sets, and one of their inherent weaknesses is a lack of interpretability (Atalay, 2025; Dumakor-Dupey and Arya, 2021; Erdogan Erten, Yavuz and Deutsch, 2022). Further, the functional capabilities of these artificial intelligence methodologies rely upon a significant data set (Avalos and Ortiz, 2020; Dumakor-Dupey and Arya, 2021). The geological ML model development cycle requires understanding the problem, characterising it, and then acquiring the relevant data; one considerable hurdle in using neural networks in geology is that 3D geological units must be built from generally sparse information and with a limited understanding of the geological setting. In addition to the spatial location of drill holes and composite grade, lithological, geochemical, and structural data can be included as dependent variables in the model, thereby improving estimation; such information is often overlooked in standard geostatistical techniques (Dumakor-Dupey and Arya, 2021). That said, machine learning models are sensitive to data complexity, and when the data is too complex, the model may learn every detail, including the noise, negatively affecting the estimation through overfitting (Aydar, 2022). Moreover, most geological data is subject to uncertainty and interpretation, posing considerable challenges in quantifying and addressing any errors within the model (Ghyselincks *et al*, 2025). The application of ML algorithms to resource modelling has returned focus to the fundamental importance of the accuracy of the underlying drill hole database (Pym *et al*, 2022). There is a natural tendency for geologists to record too much detail, focusing on minute details which may not be important to the overall picture instead of looking holistically at the data. This subdividing and splitting complicates the issue; more productive modelling inputs are commonly derived by lumping data together rather than splitting data apart (First *et al*, 2024). This highlights the criticality of developing robust logging processes and training junior staff in the importance of accurate logging (Ghyselincks *et al*, 2025; Pym *et al*, 2022), specifically, logging to ensure the data is supportive for ML processes. With recent advances in ML-based logging, the primary role of the geologist will be to validate the ML-based logs rather than to generate the log itself.

ML's inherent advantage over traditional kriging methodologies for Resource estimation is its ability to leverage non-linear correlation trends, model geological logging data, and identify high-quality data sources by learning from historical data. The deep learning algorithms can be a practical alternative in addressing these issues because they do not require the calculation of experimental variograms and can capture nonlinearity globally by learning the underlying interrelationships present in the data set (Battalgazy *et al*, 2023; First *et al*, 2024). According to First *et al* (2023), rather than relying on a single ML model, the best results are achieved by designing different models and then ensembling (or merging) the most promising ones, which enables better characterisation of ore block clusters for mine planning, targeted drilling, and improved identification of unmined potentially economic clusters. Thus, optimal results are achieved by not relying on any one neural network, but by creating several block models using both machine learning and, in some cases, kriging and then averaging the results (Kaplan, Dagasan and Topal, 2021; Mogilny, First and Yusufali, 2023). Although these ensembled ML models have been shown to be more accurate and perform better than the traditional methods, error assessment between ML and non-ML methods is not always straightforward (Mahboob, Celik and Genc, 2023; Mogilny, First and Yusufali, 2023; Zaki *et al*, 2022).

Despite all the work discussed above, one of the main yet poorly addressed flaws in the machine learning workflow is the lack of a geological chronology in model construction. The application of processes such as DomainMCF (Maptek, 2025) and Grade Copilot (Micromine, 2025) has proved to be a rapid way to produce models based on drill hole and other data. It requires no structural analysis of the modelled categorical parameters, thus contributing to the belief that the ML process is simpler and dispenses with the requirements of a structural or chronological analysis in generating the model (Kapageridis *et al*, 2021). When the ML algorithm assesses the data, it does so holistically, which is claimed to be one of its strengths; however, this can also be viewed as a significant flaw. For example, the emplacement of a porphyry is not determined by the location of a younger volcanic package, which the 'holistic' ML algorithm considers; although the placement and geometry of an older unit may determine the porphyry's shape and location. A single lithologic classification does not account for the complex correlation between lithologic units in geological space and time (Li *et al*, 2023). Thus, for the process to be considered accurate, it must account for the chronological order of the underlying units, whilst ignoring the presence of younger units. In a systematic review of 145 papers on the application of machine learning in exploiting mineralogical data by Jooshaki, Nad and Michaux (2021), there is no mention of modelling chronology, stratigraphy or understanding the timing of the units modelled. The authors undertook a word frequency map of research on ML across a subset of 55 papers in the mining industry, and the keywords stratigraphy, timing, or chronology are not shown at all in these maps (Jooshaki, Nad and Michaux, 2021). The authors believe you cannot generate a realistic model without having a basic understanding of the underlying chronology. The implicit modelling methods recently made popular across the industry make generating geological models so easy that the underlying basis of the models often gets forgotten (Reid and Cowan, 2023), yet these methods require the geologist to consider chronology in using the wireframe output meaningfully; thus the wireframes are generally regarded as essential to the resource modelling process (Cowan, Spragg and Everitt, 2011). However, the ML revolution works by writing the data directly to the block model (Sullivan *et al*, 2019) and thus can be considered a 'direct-to-block' geological modelling method, as first raised by Cowan *et al* (2011) in their significant study on using the power of the RBF interpolant and a geological chronology to directly inform the block model, dispensing with the wireframes altogether. The focus of the Cowan *et al* (2011) study was on creating a geologically attributed block model directly from drill hole geological data, exactly the process that ML seeks to emulate. However, the authors firmly established that a fundamental step in this process is to establish the underlying chronology. Stratigraphy implies that each geological unit embodies a definite geological history. To reconstruct a lithology model with meaningful geological time relationships, a progressive geological modelling strategy using a machine learning algorithm is required.

One study by Li *et al* (2023) proposed to predict the lithology of the study area by using a stratigraphic classifier to create stratigraphic classes that reflect the geological time information of the geological units. This study was a simple one that used the z-axis to assess the data and relied on the younger unit sitting above the older one. Whilst successful, this approach has very restricted use. A further study that attempted to incorporate a geological chronology into the machine learning process was conducted by Ghyselincks *et al* (2025). The authors used a mathematical process to build a Gaussian model of *transformations* (changes that occur in the rocks, such as folding and faulting) and *depositions* (the actual units that comprise sediments, eroded units, or intrusions) in a process they term conditional generation, which they used to build 3D geologically robust random models. Ghyselincks *et al* (2025) used a Gaussian noise map to organise the random noise around the training data set, comprising surface geology and drilling, and a Gaussian diffusion model, commonly used in image generation (Ghyselincks *et al*, 2025). This model was quite successful, but used processes that are not currently available to geological modellers without significant investment in programming skills and time.

In this paper, the authors emulate the establishment of a chronology, as first raised by Cowan, Spragg and Everitt (2011), whilst using the power inherent in the ML process. The process used Micromine's Grade Copilot to build several geological models that encompass all the relevant data, yet are built with a coherent geological chronology.

PROPOSED SOLUTION

The proposed solution to the chronology problem in the ML model is based on the groundbreaking work of Cowan, Spragg and Everitt (2011) where they utilised the power of the original Leapfrog product to build a series of overlapping indicator-based domains by querying the underlying function, which directly informs the block model rather than using the final wireframes. This same workflow can be done using Grade Copilot, ensuring the compositing process is correctly sequenced to exclude younger units at each step and that each model run is executed in the correct sequence.

Using the categorical modelling step on a binary population, or using the numeric modelling on the binary indicator, will generate the same model. Using the indicator gives control over selecting the confidence level for a unit that best represents the unit's actual continuity.

Step 1 is to identify what to model and why. What data is available? How can this data be used to build the model, and what supporting information is available to help the AI improve it? For instance, when building an oxidation model, what additional data could be used to improve it? In addition to the oxidation codes, data such as sulfur assays, detailed density variables, and a topographic surface can be used to determine the distance from surface (rather than using depth downhole, or RL, which may be misleading).

Step 2 is to identify the most significant units to model. The lithology codes need to be appropriately cleaned and simplified into the major components. Attempting to compile an accurate geology model of a deposit using many different rock units, most of which are immaterial to the model's robustness, results in excessive work, slows the process, and increases the likelihood of error. Is it material that the host unit comprises intercalated sandstones and siltstones? It might: a coarsely crystalline quartzite unit will behave very differently from a fine-grained siltstone/shale when deformed, and this may have a significant impact on the underlying geology model and the mineralisation. However, if the stratigraphy comprises intercalated fine- and coarse-grained sandstones with the occasional thin conglomerate unit, perhaps this is better modelled as a single 'homogeneous' package and may remove noise.

Step 3 is to composite the data to ensure consistent sample support. Any assay-based tables and lithology or alteration tables should be composited separately in the first instance. It is essential to provide a sufficient number of data points, as ML methods perform better when using significantly more data points (Atalay, 2025; Erdogan Erten, Yavuz and Deutsch, 2022). To this end, use the smallest composite length that does not result in splitting any numeric data (ie if the average grade interval is 2 m, do not use a 1 m composite length; use 2 m). Moreover, ensure minor lithological intervals are filtered out; a 2 m intersection within a block model comprising 20 × 20 × 20 m blocks is not relevant to the model. The minimum lithology interval length should equate to at least half the block size. Once the various composited files are complete, they are merged into a single composite file using the finest composite length available; longer lithology tables are flagged onto shorter assay composites.

CASE STUDY 1 – HIDDEN VALLEY GEOLOGY

In this study, the requirement was to build a robust geological model to inform the geotechnical study and the Resource model update. At Hidden Valley, modelling the metasomatic contact schist unit found along the pit highwall is vitally important, as this unit is most responsible for pit wall instability.

Geology

The Hidden Valley-Kaveroi deposit is a low-sulfidation carbonate-base metal epithermal gold-silver deposit contained within the Miocene-aged (12–14 Ma) Morobe Granodiorite (Bodorkos *et al*, 2013; Pascoe, 1991; Reid *et al*, 2024). The Morobe Granodiorite intrudes the upper-greenschist Kaindi Metamorphics, which comprise psammites, pelites, metaconglomerates, and marbles (Figure 1). Recent shrimp-based U-Pb dates indicate the Kaindi Metamorphics are Lower Permian in age (Bodorkos *et al*, 2013), a much older sub-unit of the regionally extensive Owen Stanley Metamorphics, which are Mid to Late Cretaceous in age (Bodorkos *et al*, 2013; Corbett and Leach, 1998; Dow, 1977; Pascoe, 1991; Rinne *et al*, 2018). The deposit sits on the south-east edge of the Morobe Goldfield, which is contained within the Bulolo/Wau Graben, a regionally significant extensional graben (Corbett and Leach, 1998). The Morobe Goldfield deposits have been correlated with the Edie Porphyry suite of intrusives and are dated at 3 to 4 million years (Ma) using K-Ar and U-Pb methods (Bodorkos *et al*, 2013; Corbett, 1994; Dow, 1977; Dow, Smit and Page, 1974).

While significant faulting contributes to geotechnical difficulties, the contact between the granodiorite and the metasediment comprises a hornfels-metasomatic schist and poses particular challenges (Figure 1). The schist contains significant amounts of biotite (commonly retrograded to chlorite) and is very strongly deformed. It has a very low competency and is the major cause of wall instability, particularly on the eastern highwall. Understanding the distribution of this schist and modelling it adequately is of great importance to the mine design and construction.

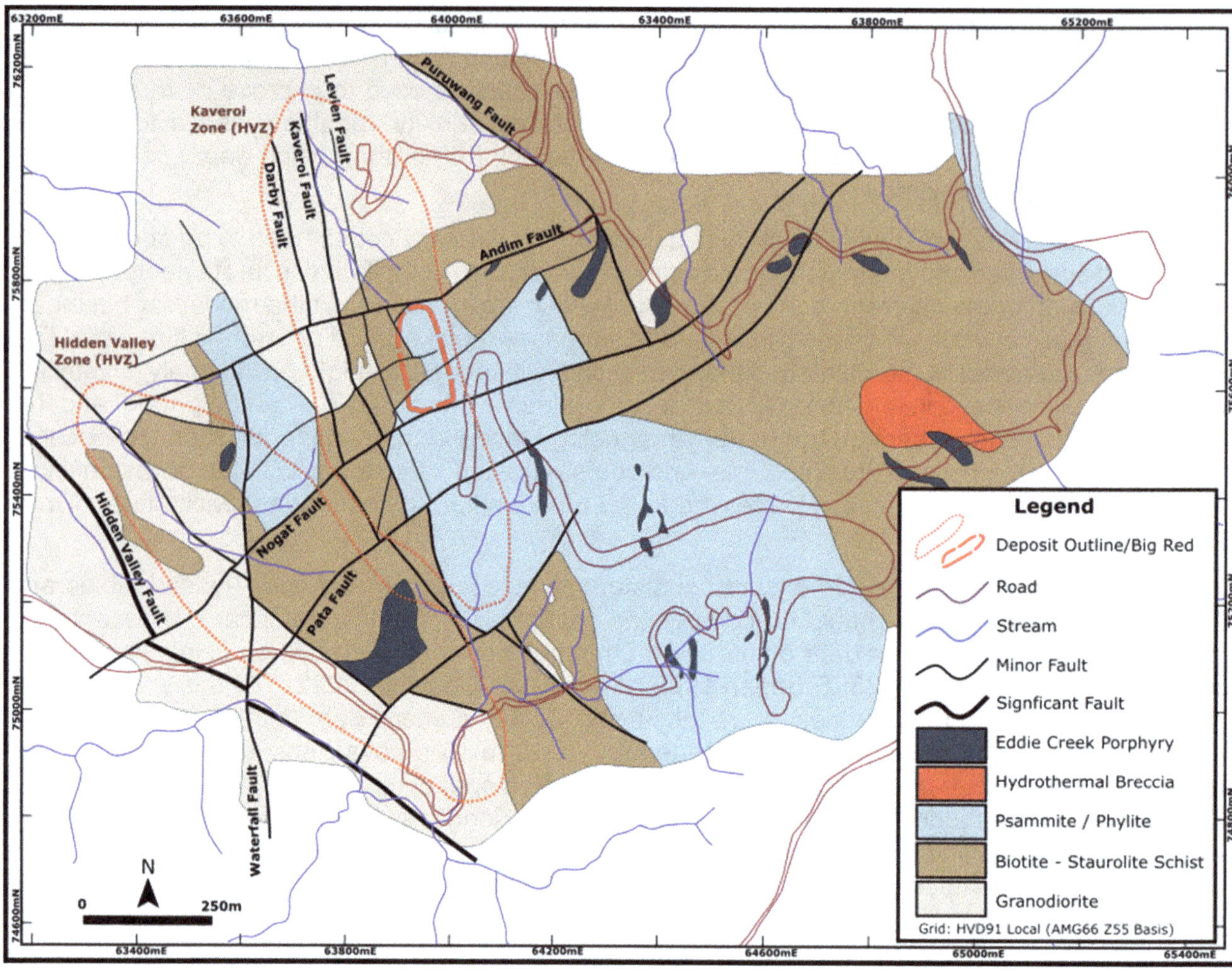

FIG 1 – Local pre-mining geology and structural setting showing the location of the Hidden Valley (HVZ) and Kaveroi (KVZ) orebodies.

Problem statement and methods

The lithology model needs to be constantly updated as new drilling comes in to ensure the schist is adequately mapped. There are two options available with rapid turnaround for mapping the schist unit. We could use the implicit modelling process, which has been the method used previously. While this process is relatively fast, it is limited in the types of information it can use to inform the schist. The second option is to use Micromine's Grade Copilot, a Neural Network methodology that has been rolled out across Micromine's Geology and Resource modelling modules. Using Grade Copilot allows the modeller not only to model the lithology but also to call on other information, such as geochemistry and other wireframes, which may inform the underlying model.

The data was cleaned to ensure the logged material was consistent between holes, and representative of the geology. A control model was built using the standard Copilot categorical modelling methodology, and the model was put aside for later comparison.

The stratigraphic column was used to build a Micromine macro that generated a series of indicator composite files, then ran individual models for each step, and combined the models into the final product. The process followed these nine steps:

1. Composite the geology file to simplify the geology, removing thin units such as narrow pegmatite veins and thin dykes that have little bearing on the overall geology model.
2. Write the geology to the assay file to compile an assay composite file that is informed by the geology.
3. Composite the assay file to a regular support that best represents the deposit; this was the Resource Estimate composite length.

4. Flag the composite file for each indicator following the flow presented in Figure 2, starting with the youngest unit and progressing to the oldest, removing the younger units at each stage. This ensures that older rocks are flagged as 1 (unit of interest) or 0 (older than the unit of interest), while younger rocks will be null.
5. Convert the indicators to text format. This allows the use of the categorical modelling method, although it could be kept as a numerical field to use the numerical modelling method to generate a field of 'probabilities' between 0 (not in the unit) and 1 (inside the unit). Selecting a value of 0.5 on this probability estimate returns the same result as a categorical model.
6. Generate a subset of the composite file by filtering the composite file to exclude younger rock, starting with the oldest unit and progressively extracting a new composite file by filtering unflagged material (filter out the null fields for each unit).
7. Run each Grade Copilot model, starting with the oldest units, and running each unit separately, using the previous model run to inform the next model where necessary; for example, using the Granodiorite-Sediment contact to control the Schist unit, which 'overprints' but follows the underlying contact.
8. Run an expression that writes a unit code for the final model. The expression used in the study is as follows:

    if ([COV_IND]>0.5) then ("COV") *(cover)*
        elseif ([STR_IND]>0.5) then ("STR") *(Structural Breccia)*
        elseif ([FPOR_IND]=1) then ("FPOR") *(Felsic Porphyry)*
        elseif ([IPOR_IND]=1) then ("IPOR") *(Intermediate Porphyry)*
        elseif ([SCH_IND]=1) then ("SCH") *(Schist)*
        elseif ([SED_IND]=1) then ("SED") *(Metasediment)*
        else ("GGD") *(Granodiorite)*
    endif

9. Validate against the original model and the drilling.

Understanding the relationships among the various units is important, not only in terms of age but also from a modelling perspective. At Hidden Valley, the Miocene-aged Morobe Granodiorite intrudes the Cretaceous-aged Owen Stanley Metamorphics, and thus, chronologically, it is younger (Figure 2). However, on the scale of the model, the granodiorite forms the model's floor and is thus the 'oldest' unit, upon which the schist and the metasediments sit. Moreover, the schistose unit is a hornfels unit that sits on the contact between the granodiorite and the metasediments. Stratigraphically, this unit lies 'between' the granodiorite and the metasediments; however, it is patchy at the scale of the model. Using Grade Copilot, this unit can be modelled using the more consistent contact between the granodiorite and the metasediment as a controlling feature. Thus, while the true chronology is metasediment-granodiorite-schist, in model space the chronology is granodiorite-metasediment-schist.

Once the model chronology has been determined, the compositing strategy can be understood. To force the AI model to adhere to the chronology, the units are modelled as binary indicators, that is, either/or, inside/outside. The indicator is flagged for each attribute, starting with the youngest unit (Figure 2). In this instance, the youngest unit is the cover unit (COV), so all the drilling is flagged 1=COV and 0=Not COV. Table 1 describes the flagging process. This process is repeated for every unit until the chronologically oldest unit. A series of composite subfiles is extracted from the master composite file for each indicator and unit, with null values excluded during compilation. The composite files should become smaller as the units age, as all the younger rocks have been excluded (Table 1). This ensures that only the relevant samples are contained within the composite file.

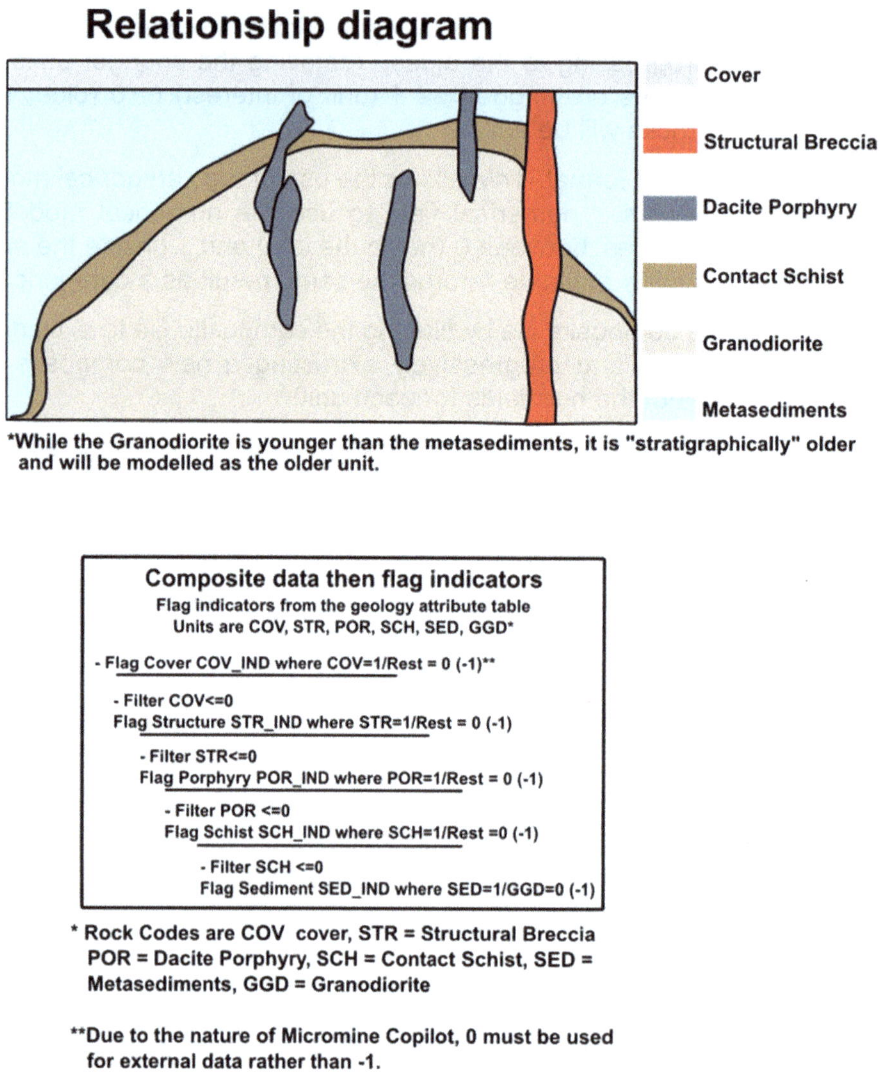

FIG 2 – (top) The project chronology needs to be simplified, and the timing relationships determined to generate an accurate model. (bottom) Once the chronology is created, a series of composite files needs to be developed that progressively eliminates younger rocks. These younger rocks are not converted to a 'waste unit'; they are flagged as null and removed entirely from the data set, so they have no impact on the model.

TABLE 1

The flagging strategy for a hypothetical drill hole shows that indicators require flagging older rocks with a value of 0 and the unit of interest with a value of 1, while ignoring younger rocks. In this case, the GGD is the oldest unit and is the default background unit.

From-To	DH Lithology	COV_IND	STR_IND	POR_IND	SCH_IND	SED_IND	GGD_IND
0–10 m	Cover	1					
10–50 m	Sediment	0	0	0	0	1	
50–65 m	Structure	0	1				
65–125 m	Sediment	0	0	0	0	1	
125–145 m	Porphyry	0	0	1			
145–210 m	Sediment	0	0	0	0	1	
210–220 m	Schist	0	0	0	1		
220–525 m	Granodiorite	0	0	0	0	0	1

The final composite files should decrease in size as the units get older.

Each unit is run separately, from oldest to youngest, to ensure the older units can inform the younger ones when appropriate. Figure 3 is a simplified diagram that shows the impact of running an AI model without a chronology. The intrusive units have been modelled with all units informing the location of all others. The AI basically maps what is the same, what is different, and how these two (or more) units interact. In this approach the younger unit has no impact on the contact between the older units; however, the location of the older unit will affect the younger unit.

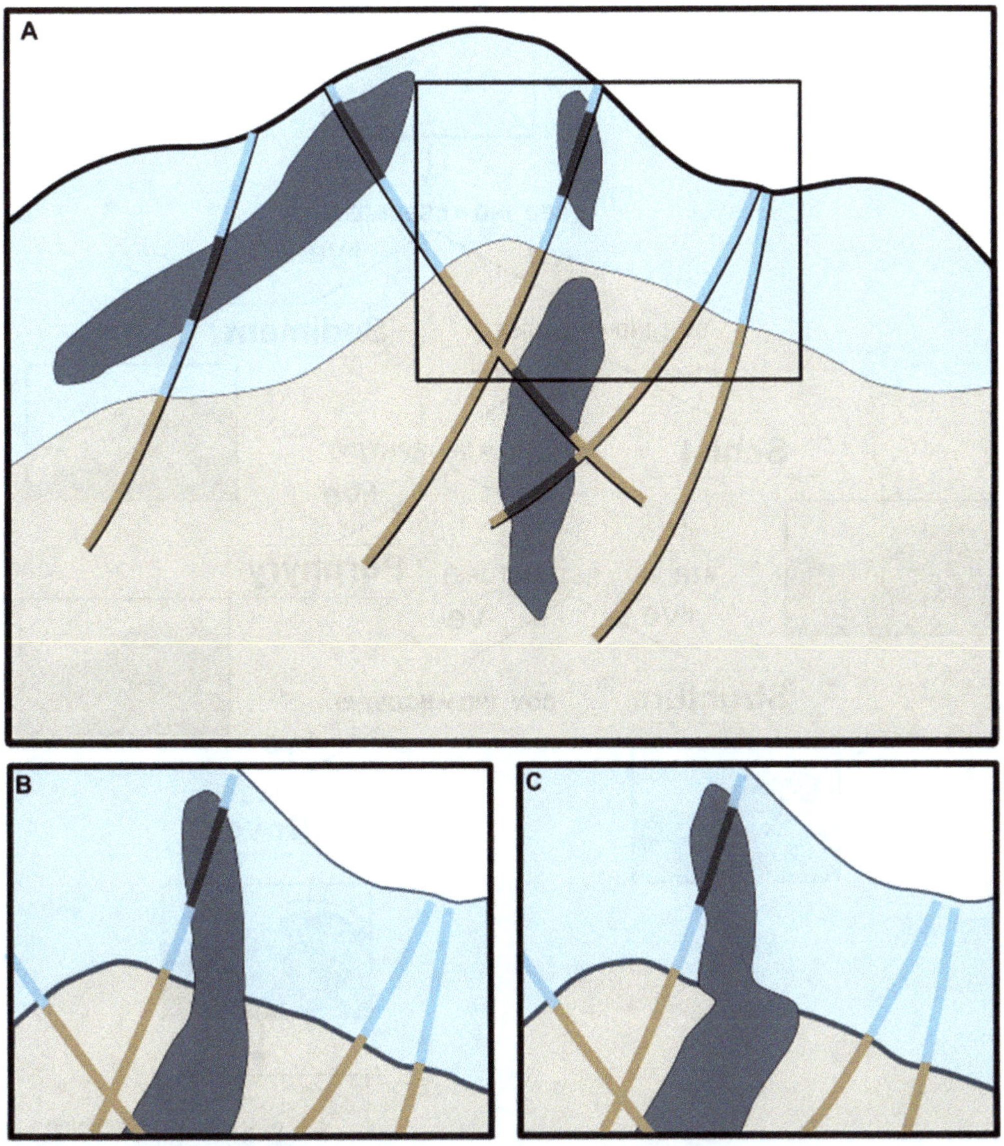

FIG 3 – The location of boundaries in the older rocks (pale red and pale blue) will have a bearing on the younger unit (pink). (a) Shows the model generated using a standard AI algorithm that connects like-for-like material, but 'pushes off' non-like material. Modelling each material separately, however, requires separate decisions about how to model the younger rock and about what impact, if any, the older contacts have on it. (b) Shows that the contact between the pale brown rock and the blue rock has no impact on the pink intrusive, (c) shows the contact between the pale red rock and blue rock does impact the pink intrusive. Understanding the nature of the impact the contact has on the intrusive is vital for a robust model.

Running the older units first allows their results to inform and influence the next step in the process (Figure 4). This can be done by writing the indicator confidence map for the older stage, say the metasediment-granodiorite contact or f(Structure) to the composite file for the next step, in this case, the f(Schist) function, which is controlled by the metasediment-granodiorite contact. This ensures that the older unit model will be used to inform the model of the younger unit. This needs to be done judiciously, as there can be unexpected impacts if this step is misapplied.

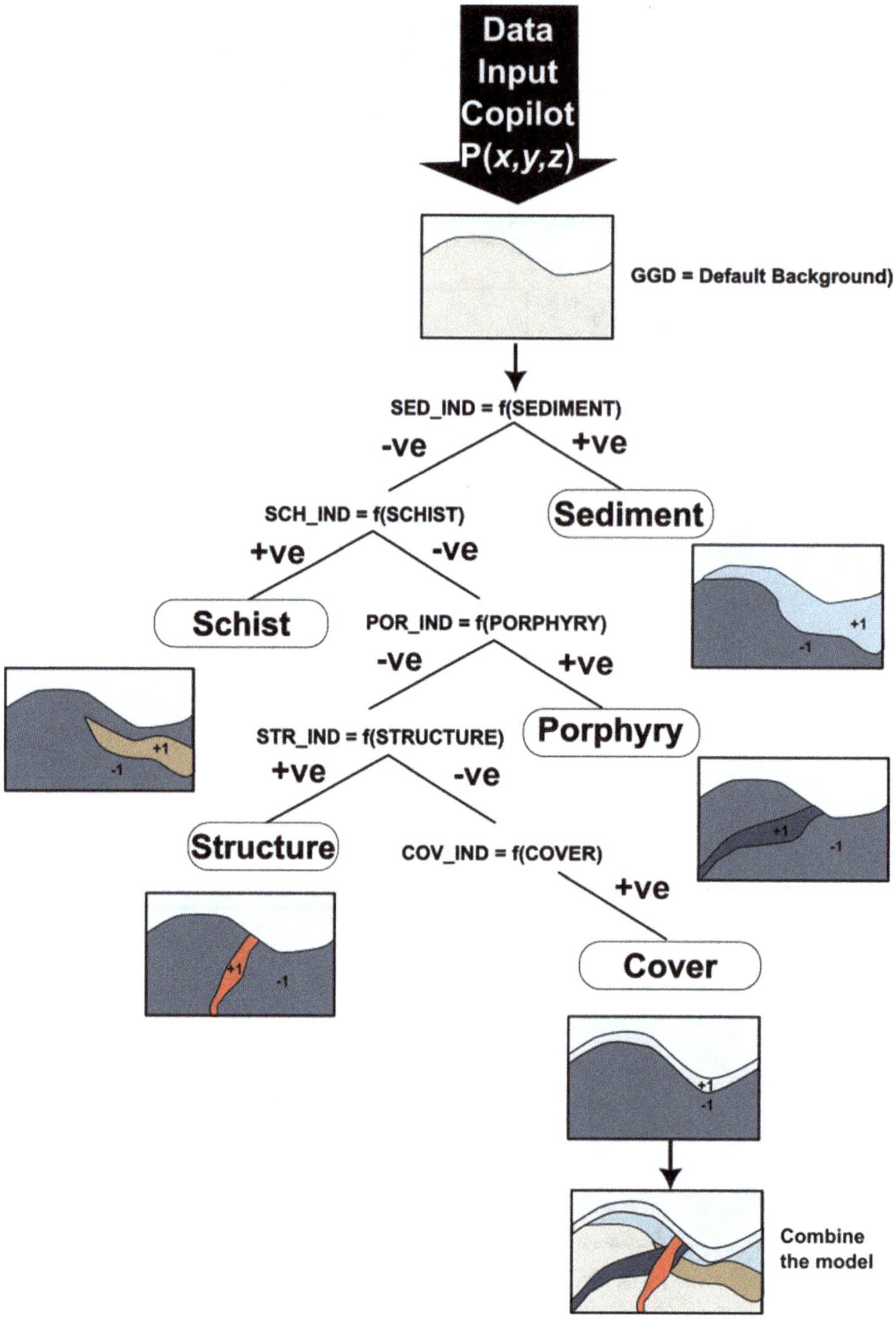

FIG 4 – Building the model requires developing the chronological order of steps and modelling the oldest to the youngest. Oldest to youngest accounts for the fact that while younger rocks have no bearing on the geometry of the older rocks, the location of boundaries in the older rocks will have a bearing on the younger contacts (adapted from Cowan, Spragg and Everitt, 2011).

Additionally, running each unit separately means you can control Grade Copilot using different guides, transformation terms, or tuning to fit each unit independently, whilst retaining the influence of older units and excluding the impact of younger units on older ones.

Discussion

The indicators were run individually into the model, as this allows the model to be validated against the composited indicator file to ensure expectations align with reality. Figure 5 shows the difference between the indicator run as a categorical model using Grade Copilot and the same unit modelled as a numerical model, which generates a probability that the unit is 0 or 1, very similar to an Indicator Kriged model. When run as a categorical model, Grade Copilot can also write the confidence levels to the file, which are similar to probabilities but represent the model's confidence in being either 0 or 1. Having this confidence level allows the user to decide which level of confidence meets their expectation of data continuity. This is shown in Figure 6 where the various levels of confidence indicate the strength of confidence in the schist unit. If the user wishes to be conservative, selecting a lower confidence threshold to flag the schist (eg 0.2) will yield a larger band of schist that can be flagged in the model as an area of concern. In general, the model flags the indicator at a confidence of 0.5.

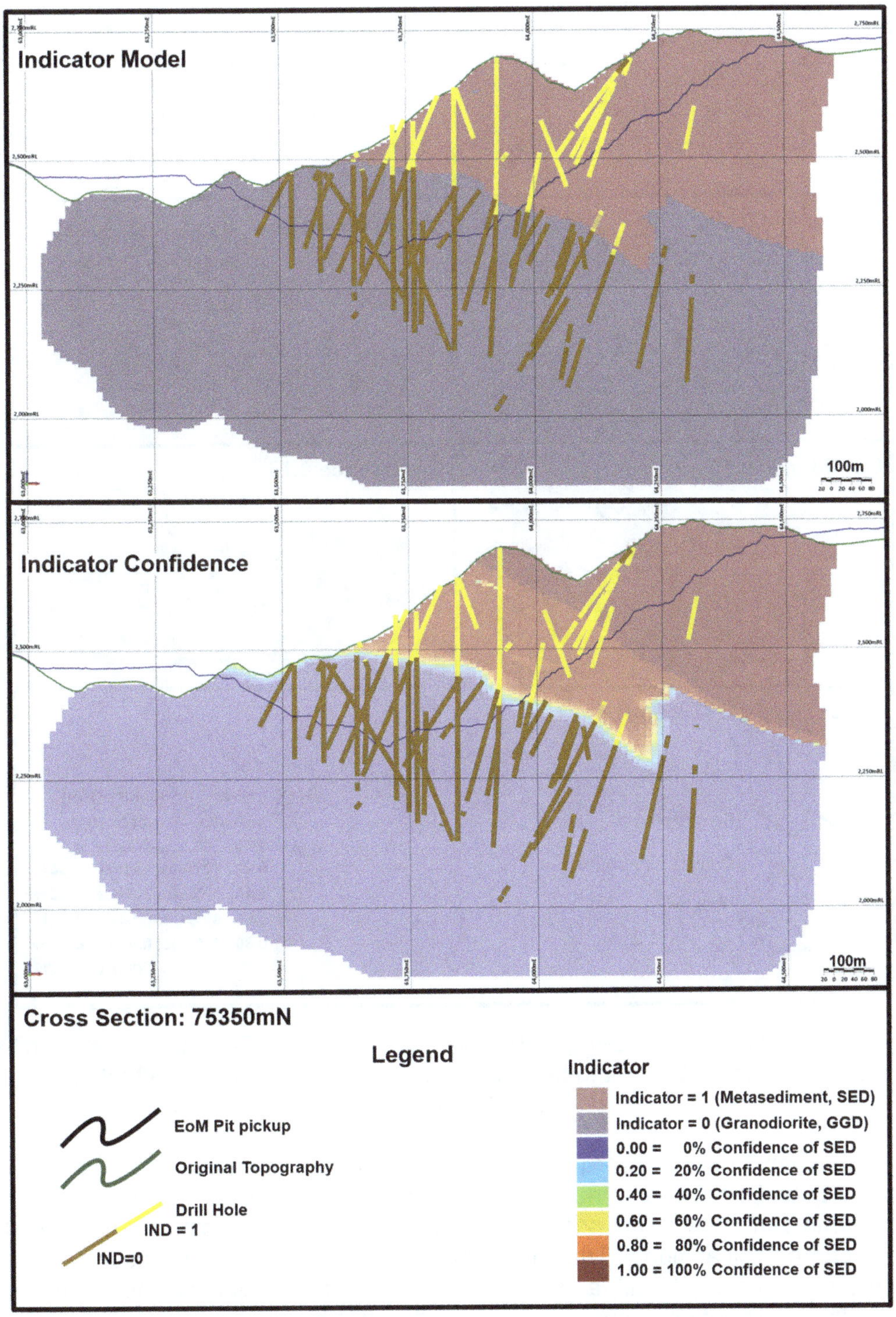

FIG 5 – Section through the model showing the first run, using the composites that only comprise the granodiorite (GGD) and the Metasediments (SED). The model is run as a categorical model, with the confidence of each indicator recorded. The top shows the copilot model's indicator as modelled; the bottom shows the confidence level ('probability') that the block is metasediments.

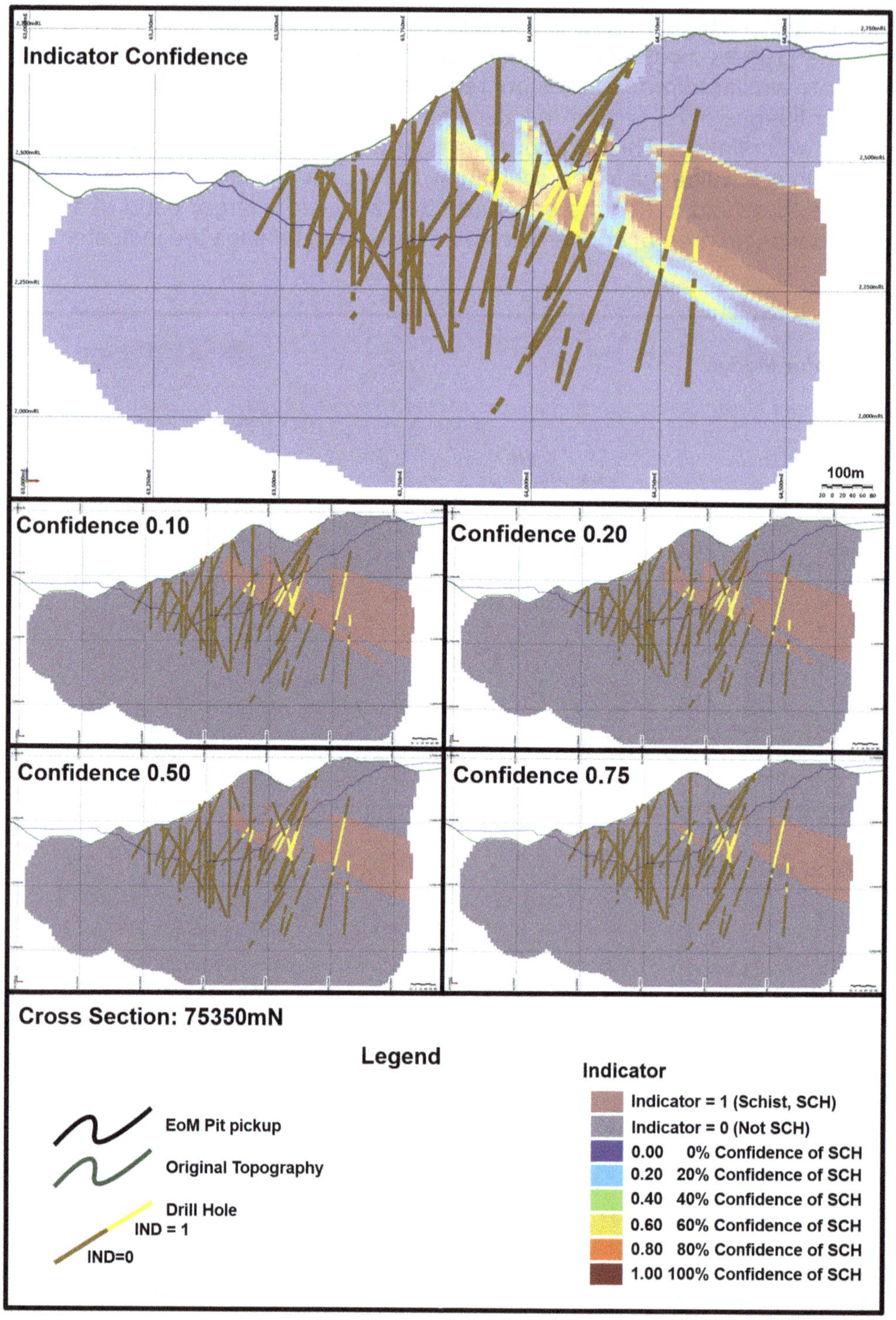

FIG 6 – Indicator maps generated using Copilot for the schist unit. Processing each unit separately allows the modeller to tailor the unit to best represent it on the ground or to build in conservatism into the model.

The final model, when flagged and run, is shown in Figure 7 alongside the standard categorical process and the output from this workflow. It is evident from this comparison that the categorical model shows some variations in the sediment-granodiorite contact and the schist unit that are not reflected in the chronological model. The chronological model is a better representation of the reality observed in the drilling and pit wall, and provides a strong foundation for the future, enabling the operation to outline zones of geotechnical hazard in the pit's highwall.

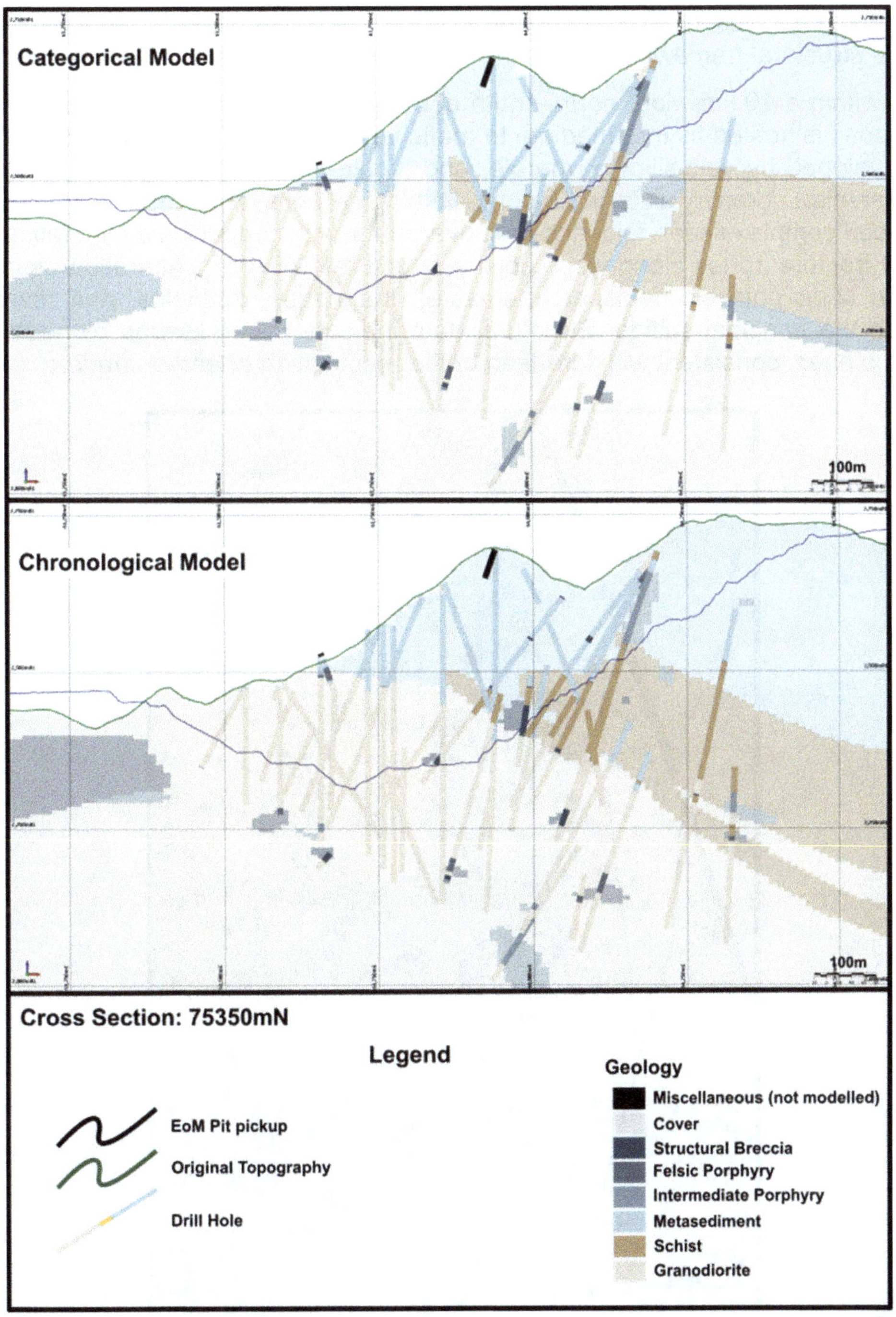

FIG 7 – Comparison of the categorical process with the standard categorical model for section 75350 mN. While on the surface the differences appear minor, there are significant differences, particularly around the contact between the granodiorite and the metasediments, and the impact this has on the schist.

CASE STUDY 2 – LITTLE EVA MULTI ELEMENT

Geology

The Little Eva deposit, the largest copper-gold occurrence within the Eva Copper Project, lies in the northern Mount Isa Eastern Succession, within the Corella Formation of Palaeoproterozoic age (~1.75–1.73 Ga) (Page and Sun, 1998). This formation comprises scapolitic calcareous metasediments, quartzites, and granofels that were intruded by felsic magmas during the Wonga Event (~1.74 Ga) (Foster and Austin, 2008; Neumann, Gibson and Southgate, 2009). Subsequent deposition formed the Knapdale Quartzite, Mount Roseby Schist, Dugald River Shale, and Lady Clayre Dolomite, followed by regional deformation associated with the Isan Orogeny (1.60–1.51 Ga).

The Isan Orogeny produced a complex history of folding, faulting, and metamorphism that established the structural framework controlling IOCG-style mineralisation across the region.

Little Eva sits within a 10 km-wide, north–south corridor bounded by the Rose Bee and Coolullah faults. The deposit is hosted by intermediate to mafic volcanic and subvolcanic rocks of the Corella Formation, enveloped by calc-silicate, marble, and biotite-scapolite schists of the Roseby Schist (Figure 8). Later felsic porphyry units cross-cut all lithological units. Three ductile deformation phases (D_1–D_3) produced complex shear, fold, and fault overprints, with copper-gold mineralisation localised in breccia and fracture zones along west-dipping structures (45–65°). Alteration assemblages are dominated by sodic-potassic feldspar, hematite, magnetite, actinolite and carbonate, with chalcopyrite as the principal sulfide. Structural data indicate mineralisation plunges gently south within foliation planes, consistent with late-Isan brittle reactivation of earlier fabrics.

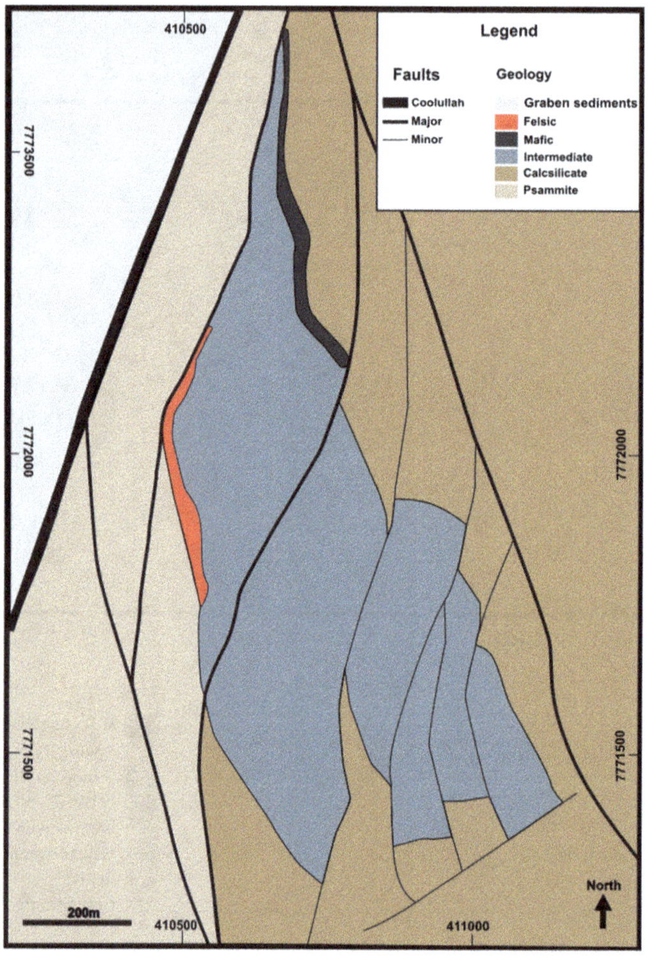

FIG 8 – Geology of the Little Eva Deposit.

In the absence of radiometric age constraints at the Little Eva deposit, the geochronology is inferred from field relationships observed in drill core and mapping. The psammitic sequence forms the background stratigraphy and defines the footwall to the mineralised intermediate volcanic unit, with the contact highly sheared. The boundary between the intermediate volcanic and the calcsilicate unit is more gradational, and the relative timing is uncertain; however, for modelling purposes, the calcsilicate is treated as the older of the two. The mafic volcanic unit appears to cross-cut both the calcsilicate and the intermediate intrusive contact, indicating it is younger than both. The felsic porphyry intrudes all lithologies and is therefore interpreted as the youngest unit in the sequence.

Problem statement and methods

The current geological database for the Little Eva deposit contains 83 unique lithology codes, which have historically complicated domaining and geological modelling. A comprehensive relogging campaign was completed over a two-month period to consolidate descriptions and improve consistency prior to geochemical analysis; however, some inconsistencies remain. To overcome this, the five principal lithological groups were defined using distinct geochemical signatures (Figure 9) then reconciled back to core observations for validation.

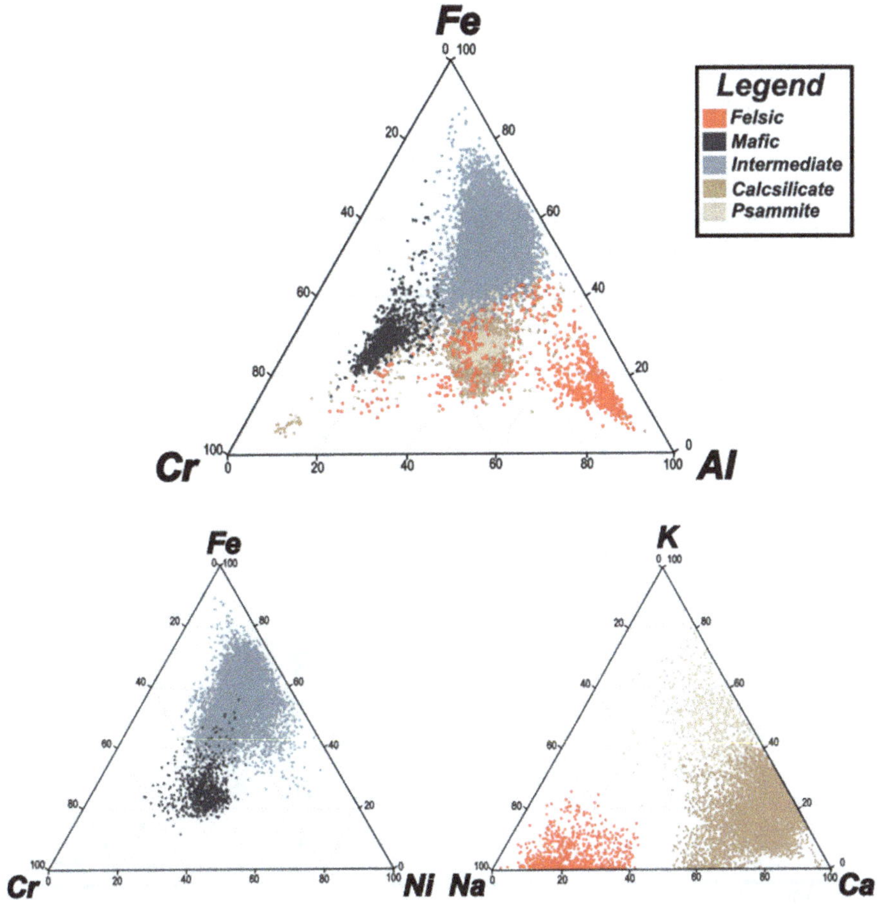

FIG 9 – Ternary diagrams showcasing the geochemical differences of the lithologies at Little Eva. (top) Fe-Cr-Al on all data. (bottom left) Fe-Cr-Ni on the mafic units. (bottom right) K-Na-Ca on the felsic and sedimentary units. *Cr and Ni were multiplied by 1000 for display.

To objectively refine the lithological framework, hierarchical agglomerative clustering was applied in Micromine using major and immobile-element chemistry. Hierarchical clustering groups the samples by progressively merging the most similar data points, producing a structured hierarchy of clusters suitable for resolving geochemically coherent lithological units (Sadeghi, 2025).

The initial clustering phase used Al, Ca, Fe, K, Ti, V, and Cr, which effectively separated mafic from felsic and metasedimentary compositions. Elements such as Fe, Ti, V, and Cr indicated mafic affinity, whereas Al, K, and Ca distinguished felsic and calcareous rocks. A secondary clustering step applied to the mafic subset using Cr and Ni isolated a high-Cr/Ni population interpreted as more strongly mafic hanging wall rocks.

Felsic and metasedimentary units were further differentiated using K, Th, Ca, Al, and Na, allowing clear discrimination of felsic porphyry, calc-silicate, and psammite lithologies. While ternary diagrams (Figure 9) visually demonstrate these trends, the multi-element clustering approach provides a more robust and internally consistent classification.

This workflow produced five coherent and geologically defensible lithological domains, consistent with both field relationships and major-element geochemistry. These refined domains form the basis for subsequent geological modelling within Micromine Grade Copilot.

Discussion

The standard categorical Grade Copilot model generated from the geochemically clustered data set produces a broadly plausible lithological interpretation; however, it does not adequately reproduce the textural and cross-cutting relationships observed in drill core. Incorporating event sequencing into the modelling workflow ensures that these geological relationships, particularly intrusive, cross-cutting, and overprinting events, are appropriately honoured in the final model.

Figure 10 illustrates the contrast between the two approaches. When event sequencing is applied, the contact between the intermediate volcanic unit and the footwall psammite is clearly expressed, consistent with the interpreted geological chronology. The felsic intrusive also forms a more coherent, geologically realistic body, as the model is explicitly allowed to cross-cut the surrounding lithologies.

By embedding a chronology for event sequencing into the modelling process at Little Eva, the interpretation becomes more than just statistically plausible; it becomes geologically defensible. This approach demonstrates how neural network-assisted modelling can honour the relative timing and structural relationships among lithological units, resulting in more robust geological domains and, ultimately, more reliable inputs for resource estimation.

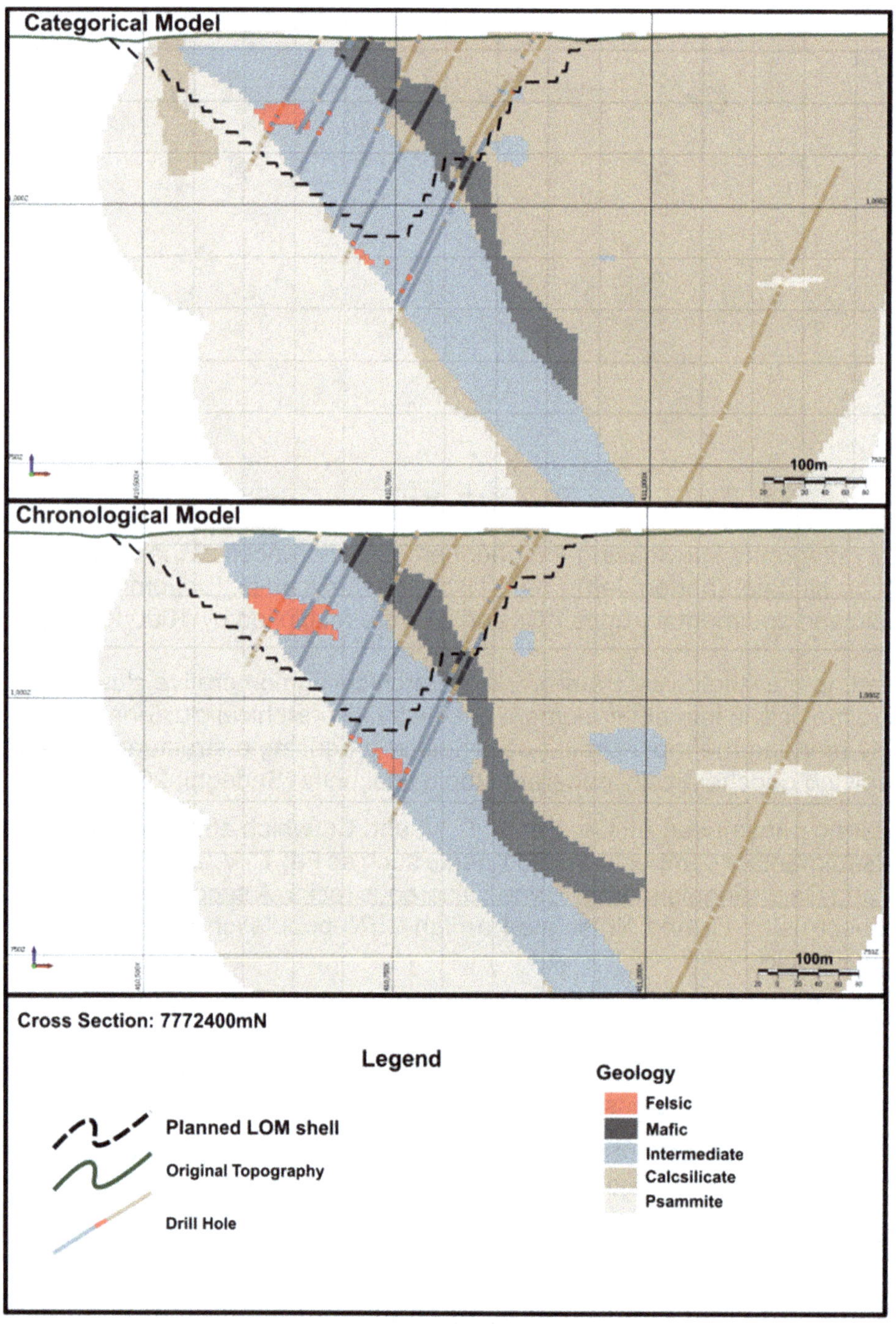

FIG 10 – Comparison to standard run.

CASE STUDY 3 – KERIMENGE OXIDE

Geology

The Kerimenge deposit is located in the southern part of the Bulolo Graben (Figure 11), a north-west-trending down-faulted block formed during the southern extension of the New Guinea mobile belt. The basement rocks of the Bulolo Graben are the Owen Stanley Metamorphic rocks and Morobe Granodiorite. The Owen Stanley Metamorphics are clastic metasedimentary rocks with minor limestone and metamorphosed basic volcaniclastic rocks (Denwer, 1997, 1993), locally known as the Kaindi Metamorphics. These basement rocks have been intruded by the Edie Porphyry suite and by eruptions and redeposition of andesitic to dacitic volcanic rocks from the Bulolo Volcanics (Denwer, 1993). The Otibanda Formation unconformably overlies the older units.

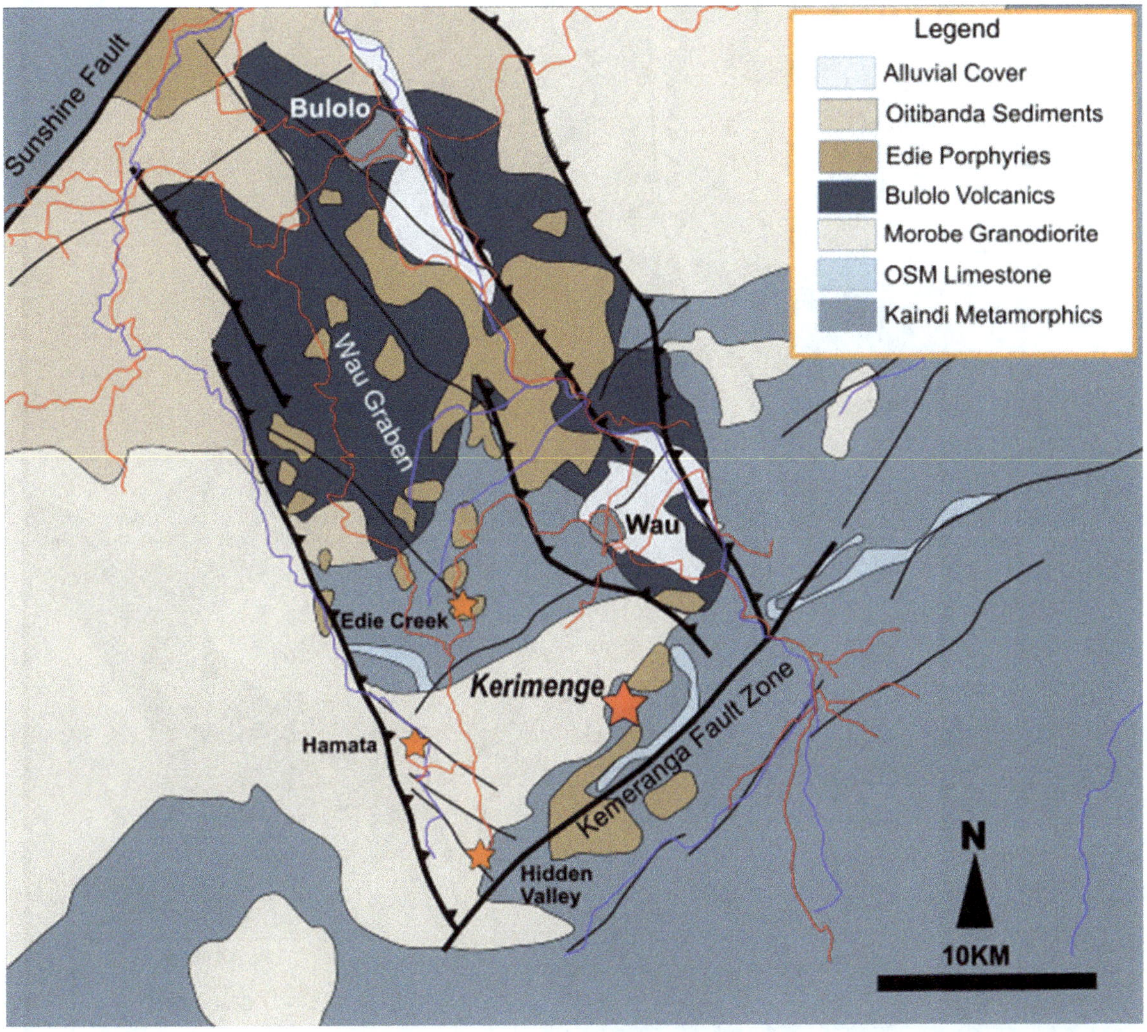

FIG 11 – Regional geological map showing the location of Kerimenge in relation to the Hidden Valley and Hamata deposits.

The basement rocks at Kerimenge consist of the Kaindi Metamorphics, comprises interbedded calcareous to pelitic schist, phyllite, and marble (Denwer, 1993). The Kaindi formation is subsequently intruded by the Kerimenge porphyry sill, which is the feldspar-amphibole-biotite phyric quartz andesitic/dacitic end member of the Edie porphyry suite (Denwer, 1993). The sill is the primary host to the mineralisation.

The mineralisation at Kerimenge occurs within the porphyry, hosted in a series of crackle breccias and silicified fractures (Figure 12). The crackle breccias are most prominent in the southern half of the deposit, especially at fault intersections, and often contain higher-grade gold. Manganocarbonate stockworks become increasingly dominant further to the north. Denwer (1997, 1993) and referenced workers identify four stages of gold mineralisation, progressing from quartz-pyrite to quartz-pyrite-gold, then quartz-arsenopyrite-gold-manganocarbonate, and finally manganocarbonate with or without quartz.

Weathering at Kerimenge plays a vital role in metallurgical recovery, with fresh material showing poor recovery. The current oxidation model comprises four weathering horizons: complete oxidation (OX), partial oxidation (POX), fracture-oxidised (FOX), and fresh/unweathered (FR).

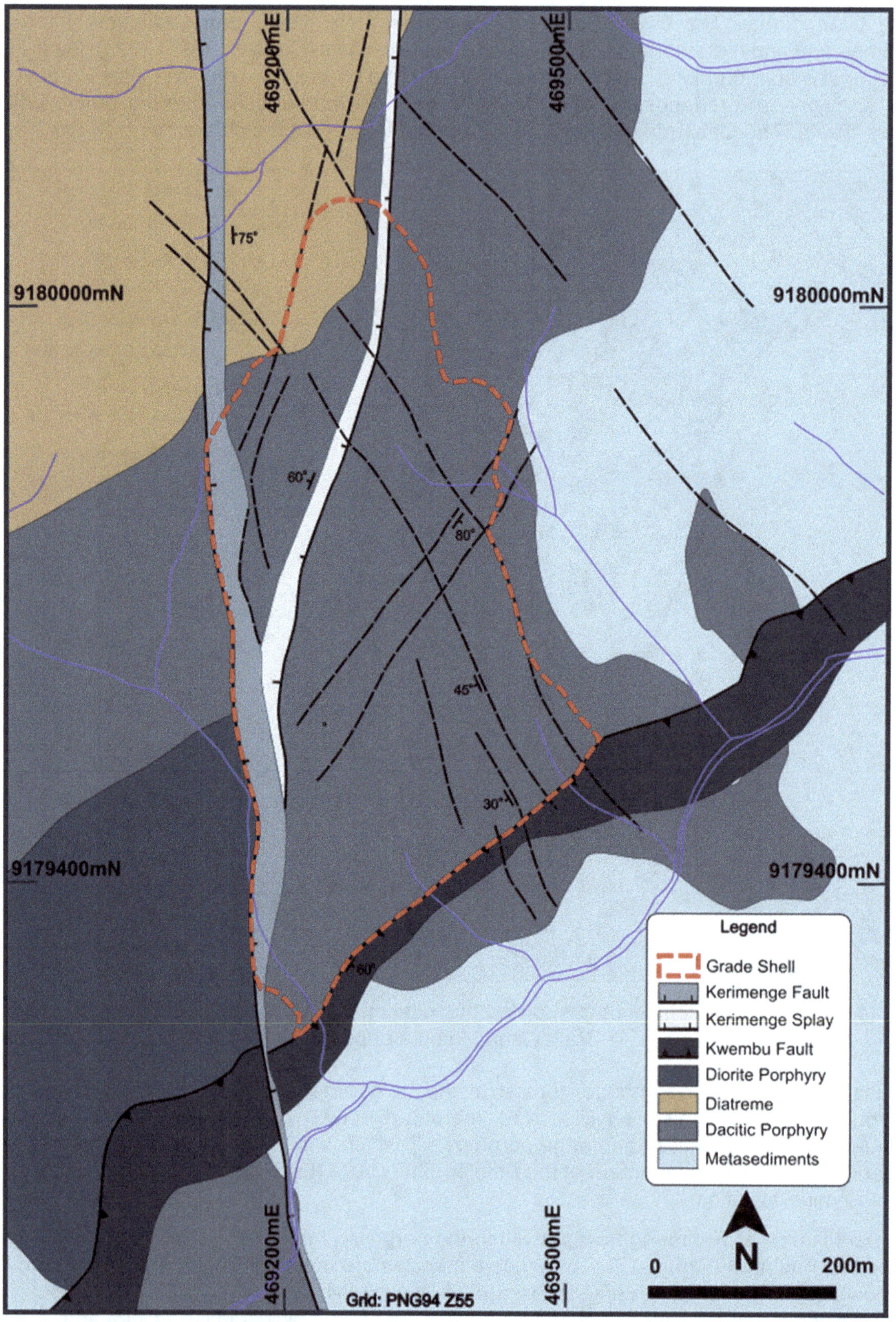

FIG 12 – Geology map of the Kerimenge deposit.

Problem statement and methods

Accurate characterisation of weathering profiles (oxidation domains) is essential for mine planning, metallurgical prediction, and grade control. At Kerimenge, metallurgical recoveries are linked to the degree of weathering/oxidation of the host rocks. However, developing a reliable oxidation model remains challenging due to several factors:

- The geological and geochemical data may be limited
- Structural and rheological contrasts influence the weathering/oxidation profile
- Deep oxidation 'keels' linked with major structures may cause localised variability that is hard to model with conventional techniques.

Conventional deterministic and implicit methods may not effectively identify the complex, multidimensional, nonlinear relationships between geochemistry, structural features, and subsurface depth. An effective oxidation model thoughtfully considers all these factors. As mining advances or more boreholes are drilled, the observed profiles during logging or mining align with the predicted weathering patterns.

To overcome the limitations and develop a predictive weathering and oxidation model, the authors explored the use of a neural network-based workflow using Micromine's Grade Copilot. The objective was to develop a workflow using Grade Copilot, to predict weathering at Kerimenge, accounting for geochemistry, structure, and the relative timing of weathering.

Exploratory data analysis (EDA) was performed to assess the relationship between geochemistry and oxidation state, using visual and statistical analyses. Scatter plots were generated comparing the depth below topography against carbon (C), sulfur (S) and iron (Fe) (Figure 13a–13c). Additionally, carbon, sulfur and iron were plotted against each other and categorised based on observed weathering conditions (Figure 13d–13F). The terrain at Kerimenge is exceptionally rugged and using downhole depths to determine depth below surface would not yield reliable comparisons, so the depth below the topography was calculated.

In the Distance-to-topography versus C and S plots (Figure 13a and 13b), a relationship between complete oxidation, depth, C, and S content is observed; however, as depth increases and weathering diminishes, the distinction becomes less clear. In the distance-to-topography plot versus Fe, the depth relationship is evident; however, it is less distinct because these plots rely only on total Fe and do not differentiate between $Fe2+$ and $Fe3+$. Comparing S against Fe in Figure 13d reveals two trends: a positive correlation between S and Fe concentration, and a dense cluster of points between 3 and 5 per cent Fe as weathering intensity decreases, while Fe concentration remains relatively steady, and sulfur concentration increases. Although correlations are observed in the S versus C (Figure 13e) and C versus Fe (Figure 13f) scatter plots, clear relationships between geochemistry and oxidation are not obvious.

In addition to the scatter plots, Fe, C, and S were plotted on a ternary diagram to investigate whether plotting the three elements together could reveal any additional relationships (Figure 14). In Figure 14a, the samples with low Fe:S and Fe:C ratios cluster near the 100 per cent Fe apex and show a strong correlation with the logging data (Oxidised). The fresh material also displays clustering; however, this cluster exhibits a trend, which can also be seen in Figure 14b, with data trending from a Fe:S ratio of 55:45 towards a Fe:C ratio of 80:20. In Figure 14b, a third cluster is identified, corresponding to a Fe:S ratio of 55:45 and a spread towards the Fe apex, which correlates with fracture oxidation. Samples with Fe:C and S:C ratios above 80:20 display a significant spread with no correlation between logged weathering and geochemistry.

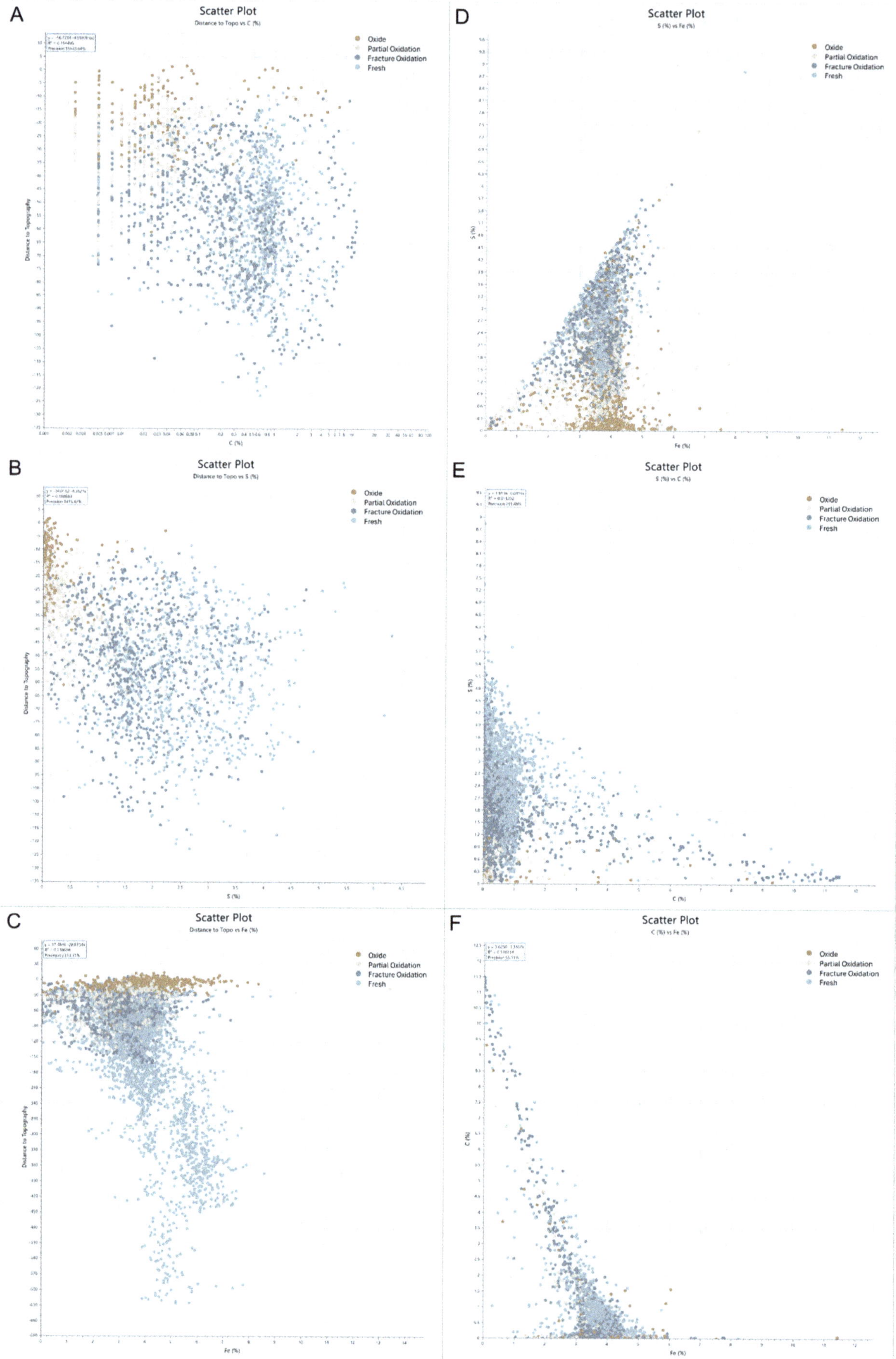

FIG 13 – Scatter plots of distance-to-topography versus C, S and Fe (a to c), and S versus Fe, S versus C and C versus Fe (d to f).

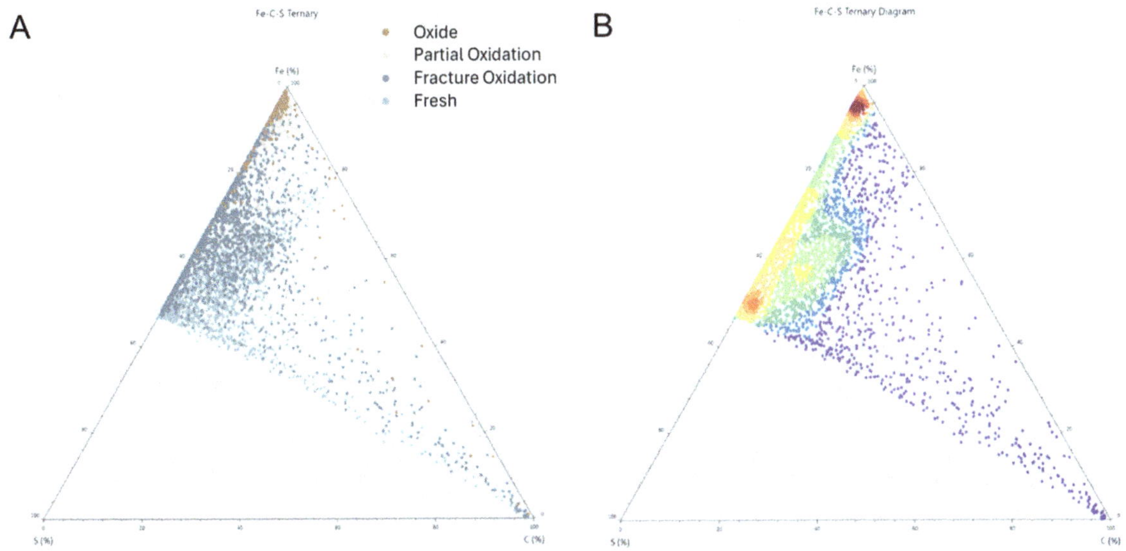

FIG 14 – Ternary diagrams of Fe-C-S. (a) coloured by logged weathering and (b) coloured using point density percentiles.

The exploratory data analysis highlights the complexity of the relationship between geochemistry and weathering. Figure 13 demonstrates a complex relationship influenced by multiple processes. The data form diffuse, overlapping domains in the geochemical space, with clustering reflecting multi-variable control on weathering intensity. The patterns in the data indicate that it is well-suited for modelling weathering using Grade Copilot and neural networks.

Workflow

The workflow for modelling weathering at Kerimenge is outlined below and aligns with the Micromine process flow. The workflow is based on the premise that thorough data validation has been completed before starting the modelling. The data was merged into a single file and composited to the resource composite length. The indicator variables for each weathering code were flagged into the composite file using the following workflow:

a. OX_IND = 1 when Oxide = OX,

b. POX_IND = 1 when Oxide = POX,

c. FOX_IND = 1 when Oxide = FOX,

d. FR_IND = 1 when Oxide = FR

A Structural Trend based on topography and faults was created, since weathering is affected by both distance from the topographic surface and faulting. Grade Copilot was used to interpolate the geochemical proxies for weathering: Fe, C, and S into the model using the Structural Trend.

The weathering indicators were interpolated sequentially using Grade Copilot in the determined chronological order: FR→FOX→POX→OX. The chronological order ensured each successive oxidation state effectively overprinted the first and had increasing weathering intensity. The indicator confidence values were assessed to determine the cut-offs for assigning the degree of Oxidation. Each weathering classification was determined by selecting the category with the highest confidence.

Discussion

Two models were created in Micromine using Grade Copilot: a basic categorical model based on default settings and a chronological indicator model. In the categorical models, all Oxidation categories were modelled simultaneously, with the neural network establishing relationships among each category. Micromine's Grade Copilot allows users to specify guide surfaces, structural trends, and/or other guide fields to provide as much data as possible for the neural network to learn in building the model. This approach was not used in the basic categorical model but was employed to guide the chronological model.

Figure 15 contains sections of the outputs of the categorical and chronological models. Both models honour the input data where drilling occurs. In the basic categorical model, the data do not exhibit any trends, as observed during the EDA. In contrast, the chronological model used more of the available data. Furthermore, the basic model extends the weathering into areas with no supporting data, resulting in large unsupported zones of oxide, partially oxidised material, and fracture-oxidised zones.

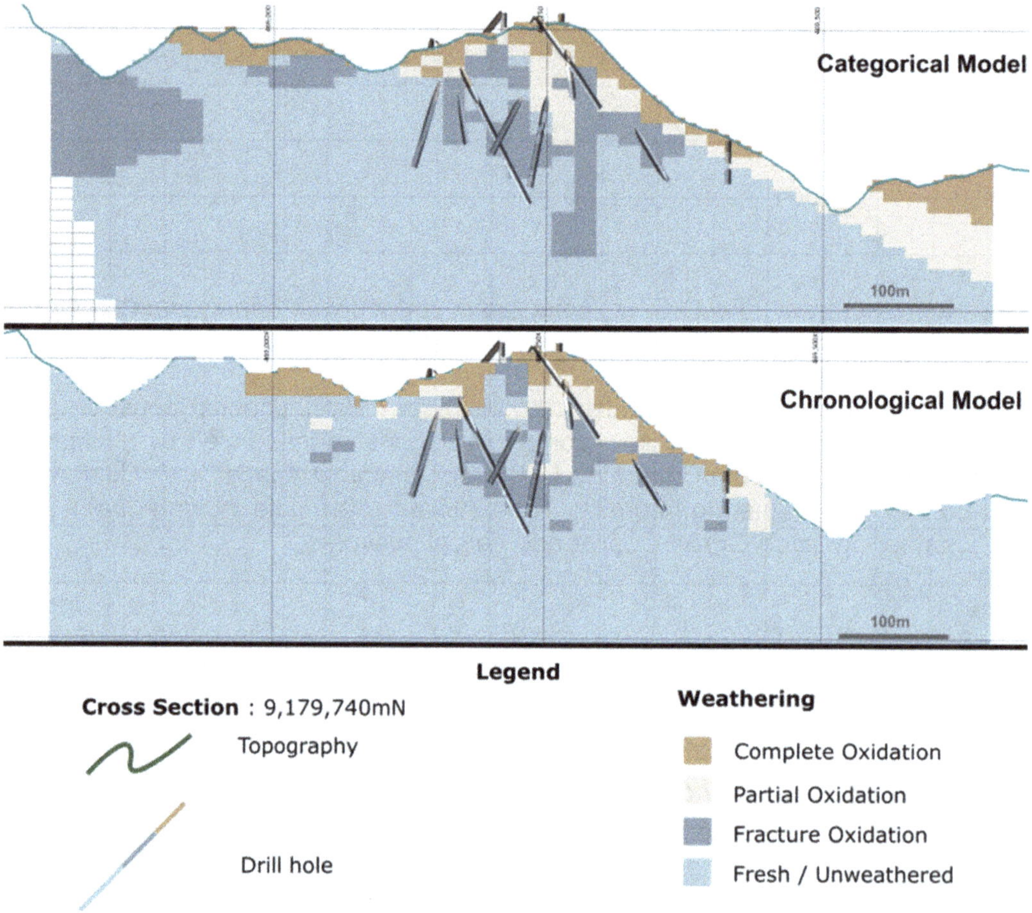

FIG 15 – Comparison of the grade copilot categorical model with the chronological weathering model for Kerimenge.

The basic categorical model produces a more continuous model that seems to capture the variable nature of weathering better. It is well recognised that weathering is influenced by underlying geology and structures, and going forward, this should be used to guide the neural network in generating a weathering model.

CONCLUSION

The resource model may underpin the mine plan, but the geological model underpins everything. The case studies presented here demonstrate that without a clear understanding of a deposit's geological chronology, the modelling process falters and the resulting geological model is fundamentally compromised. Integrating chronology into the machine learning workflow gives the algorithm the structural context it needs to generate models that are both geologically defensible and operationally reliable.

Looking ahead, the geological community will benefit greatly from software developers incorporating the chronology directly into machine learning enabled modelling platforms. In particular, tools should allow practitioners to:

- define a geological chronology that the software uses sequentially when constructing models
- assign 'draping,' 'cutting,' or 'cross-cutting' behaviours to units so the software can enforce appropriate interaction rules:
    - 'draping' units lie conformably on those beneath (eg sedimentary or volcanic sequences)
    - 'cutting' units sit unconformably and may erode the underlying geology
    - 'cross-cutting' units transect earlier formations (eg intrusions or mineralised zones).

Embedding these capabilities will markedly enhance the reliability, geological fidelity, and long-term usability of machine learning models, ensuring that future resource estimations are built upon models that honour the true history of the rocks themselves.

ACKNOWLEDGEMENTS

The authors extend their thanks to Harmony for allowing us to publish this paper and supporting us in presenting at the 2026 International Mining Geology Conference.

REFERENCES

Al-Alawi, S M and Tawo, E E, 1998. Application of Artificial Neural Networks in Mineral Resource Evaluation, *Journal of King Saud University – Engineering Sciences,* 10:127–138. <https://doi.org/10.1016/S1018-3639(18)30692-5>

Atalay, F, 2025. Effect of Domaining in Mineral Resource Estimation with Machine Learning, *Minerals,* 15(4):22. <https://doi.org/10.3390/min15040330>

Avalos, S and Ortiz, J M, 2020. Recursive convolutional neural networks in a multiple-point statistics framework, *Computers and Geosciences*, 141:25. <https://doi.org/10.1016/j.cageo.2020.104522>

Aydar, A, 2022. Resource Estimation of Roll Front Uranium Deposits by Using Traditional and Machine Learning Methods for Nichols Ranch Uranium Deposit in Wyoming, Masters Thesis (unpublished), Colorado School of Mines, Colorado, United States of America.

Battalgazy, N, Valenta, R, Gow, P, Spier, C and Forbes, G, 2023. Addressing Geological Challenges in Mineral Resource Estimation: A Comparative Study of Deep Learning and Traditional Techniques, *Minerals,* 13(7):29. <https://doi.org/10.3390/min13070982>

Bergen, K J, Johnson, P A, De Hoop, M V and Beroza, G C, 2019. Machine learning for data-driven discovery in solid Earth geoscience, *Science,* 363(6433):1299. <https://doi.org/10.1126/science.aau0323>

Bodorkos, S, Sheppard, S, Saroa, D, Tsiperau, C U and Sircombe, K N, 2013. New SHRIMP U-Pb zircon ages from the Wau-Bulolo region, Papua New Guinea, Geoscience Australia and Mineral Resources Authority, Papua New Guinea. <https://doi.org/10.11636/Record.2013.025>

Buchanan, G, Burden, C, Absolom, M and Gole, A, 2023. A machine learning approach to geological modelling of an Irish-type Zn-Pb deposit, in *Irish-Type Zn-Pb Deposits around the World* (eds: C J Andrew, M W Hitzman and G Stanley), presented at the Irish-type Zn-Pb deposits around the world, pp 385–392 (Irish Association for Economic Geology: Dublin). <https://doi.org/10.61153/uymp7702>

Cao, X, Liu, Z, Hu, C, Song, X, Quaye, J A and Lu, N, 2024. Three-Dimensional Geological Modelling in Earth Science Research: An In-Depth Review and Perspective Analysis, *Minerals,* 14(7):40. <https://doi.org/10.3390/min14070686>

Corbett, G J, 1994. Regional structural control of selected Cu/Au occurrences in Papua New Guinea, in *Proceedings of the Geology, Exploration and Mining Conference* (ed: R Rogerson), pp 57–70 (The Australasian Institute of Mining and Metallurgy: Melbourne).

Corbett, G J and Leach, T M, 1998. *Southwest Pacific Rim Gold-Copper Systems: Structure, Alteration and Mineralization* (Society of Economic Geologists: Phoenix, Arizona). <https://doi.org/10.5382/SP.06>

Cowan, E J, Spragg, K J and Everitt, M R, 2011. Wireframe-Free Geological Modelling – An Oxymoron or a Value Proposition?, in *Proceedings Eighth International Mining Geology Conference*, pp 247–260 (The Australasian Institute of Mining and Metallurgy: Melbourne).

Denwer, K P, 1997. The Kerimenge-Lemenge Gold Prospect, Morobe Goldfield, Papua New Guinea, in *Geology, Exploration and Mining Conference Proceedings,* pp 71–74 (The Australasian Institute of Mining and Metallurgy: Melbourne).

Denwer, K P, 1993. Geology and mineralisation of the Kerimenge-Lemenge Prospect, Wau District, Papua New Guinea, Masters Thesis (unpublished), University of Tasmania.

Dow, D B, 1977. A Geological Synthesis of Papua New Guinea (No. 201), Bureau of Mineral Resources, Geology and Geophysics, Commonwealth of Australia.

Dow, D B, Smit, A J and Page, R W, 1974. 1:250,000 Geological Series – Explanatory notes, Wau, Papua New Guinea, Bureau of Mineral Resources, Geology and Geophysics, Dept. of Minerals and Energy, Commonwealth of Australia.

Dumakor-Dupey, N K and Arya, S, 2021. Machine Learning—A Review of Applications in Mineral Resource Estimation, *Energies* ,14(14):29. <https://doi.org/10.3390/en14144079>

Erdogan Erten, G, Yavuz, M and Deutsch, C V, 2022. Combination of Machine Learning and Kriging for Spatial Estimation of Geological Attributes, *Natural Resources Research,* 31(1):191–213. <https://doi.org/10.1007/s11053-021-10003-w>

First, D M, Mogilny, D, Rajcoomar, Y H and Vedrik, S, 2024. Leveraging geological logging as pathfinders and proxies in machine learning resource and geometallurgical modelling, in *Proceedings International Mining Geology Conference 2024,* pp 47–65 (The Australasian Institute of Mining and Metallurgy: Melbourne).

First, D M, Rajcoomar, Y H, Mogilny, D, Sucholutsky, I and Yusufali, F, 2023. Interpreting Machine Learning ML Neural Network Patterns, *Resourcing the Green Transition* (Society of Economic Geologists, London, Great Britain).

Foster, D and Austin, J, 2008. The 1800–1610Ma stratigraphic and magmatic history of the Eastern Succession, Mount Isa Inlier and correlations with adjacent Paleoproterozoic terranes, *Precambrian Research,* 163(1):7–30. <https://doi.org/10.1016/j.precamres.2007.08.010>

Ghyselincks, S, Okhmak, V, Zampini, S, Turkiyyah, G, Keyes, D and Haber, E, 2025. Pre-print, Synthetic Geology -- Structural Geology Meets Deep Learning. <https://doi.org/10.48550/arXiv.2506.11164>

Hasan, S, Bin Shafiq, S and Khatun, L, 2023. Exploring the potential of artificial intelligence and machine learning in mineral exploration: a review article, *International Research Journal of Modernization in Engineering Technology and Science,* 5(10):3265–3280. <https://doi.org/10.56726/IRJMETS45281>

Jackson, A, 2023. Top 10: Uses of Artificial Intelligence in Mining, *Mining Digital.* <https://miningdigital.com/articles/top-10-uses-of-artificial-intelligence-in-mining> [Accessed: 16 July 2025].

Jooshaki, M, Nad, A and Michaux, S, 2021. A Systematic Review on the Application of Machine Learning in Exploiting Mineralogical Data in Mining and Mineral Industry, *Minerals,* 11(8):816. <https://doi.org/10.3390/min11080816>

Kapageridis, I, Albanopoulos, C, Sullivan, S, Buchanan, G and Gialamas, E, 2021. Application of Machine Learning to Resource Modelling of a Marble Quarry with Domain, *Materials Proceedings,* 5(1):12. <https://doi.org/10.3390/materproc2021005012>

Kaplan, U E, Dagasan, Y and Topal, E, 2021. Mineral grade estimation using gradient boosting regression trees, *International Journal of Mining, Reclamation and Environment,* 35:728–742. <https://doi.org/10.1080/17480930.2021.1949863>

Li, H, Wan, B, Chu, D, Wang, R, Ma, G, Fu, J and Xiao, Z, 2023. Progressive Geological Modeling and Uncertainty Analysis Using Machine Learning, *International Journal of Geo-Information,* 12(1):97. <https://doi.org/10.3390/ijgi12030097>

Mahboob, M A, Celik, T and Genc, B, 2023. Review of machine learning-based Mineral Resource estimation, *Journal of the Southern African Institute of Mining and Metallurgy*, 122(11):1–10. <https://doi.org/10.17159/2411-9717/1250/2022>

Maptek, 2025. Maptek – Machine learning assisted domain modelling. <https://www.maptek.com/products/domainmcf/> [Accessed: 19 November 2025].

Micromine, 2025. Introducing Micromine Origin Grade Copilot – Neural Network Modelling for Resource Geologists, *Micromine.* <https://experience.micromine.com/grade-copilot> [Accessed: 19 November 2025]

Mogilny, D, First, D M and Yusufali, F, 2023. Precise Domaining of Carbonate Quartz Vein Structures Using Deep Learning and Novel Pattern-Matching Filters, in *Proceedings of the 26th World Mining Congress Conference*, pp 1497–1505. (World Mining Congress: Brisbane).

Neumann, N L, Gibson, G M and Southgate, P N, 2009. New SHRIMP age constraints on the timing and duration of magmatism and sedimentation in the Mary Kathleen Fold Belt, Mt Isa Inlier, Australia, *Australian Journal of Earth Sciences,* 56(7):965–983. <https://doi.org/10.1080/08120090903005410>

Newby, E, 2024. A Geologist's Guide to AI Buzzwords, *Micromine – Technology for Mining.* <https://www.micromine.com/geologists-guide-to-ai-buzzwords/> [Accessed: 21 August 2025].

OreFox, 2024. OreFox AI – Empowering Mining Decisions with AI and Data Science. <https://orefox.com/> [Accessed: 19 November 2025].

Page, R W and Sun, S, 1998. Aspects of geochronology and crustal evolution in the Eastern Fold Belt, Mt Isa Inlier, *Australian Journal of Earth Sciences,* 45(3):343–361. <https://doi.org/10.1080/08120099808728396>

Pascoe, G J, 1991. Hidden Valley Gold project development summary, 1987–1991, in *Proceedings of the PNG Geology, Exploration and Mining Conference* (ed: R Rogerson), pp 69–76 (The Australasian Institute of Mining and Metallurgy: Melbourne).

PETRA, 2025. Orebody learning software for mining, *PETRA.* <https://www.petradatascience.com/> [Accessed: 19 November 2025].

Pym, F A, Crook, K E, Hetherington, P M and Murphy, M P, 2022. Machine learning in resource geology – why data quality is critical, in *Proceedings International Mining Geology Conference 2022,* pp 149–170 (The Australasian Institute of Mining and Metallurgy: Melbourne).

Reid, R, Magautu, D, Lucien, M, Gagau, L, Mondo, M and Mudinzwa, E, 2024. Hidden between the data, small but significant high-grade structural lodes, in *Proceedings International Mining Geology Conference 2024*, pp 221–230 (The Australasian Institute of Mining and Metallurgy: Melbourne).

Reid, R J and Cowan, E J, 2023. Towards quantifying uncertainties in geological models for mineral resource estimation through outside-in deposit-scale structural geological analysis, *Australian Journal of Earth Sciences,* 70(7):990–1009. <https://doi.org/10.1080/08120099.2023.2217882>

Rinne, M L, Cooke, D R, Harris, A C, Finn, D J, Allen, C M, Heizler, M T and Creaser, R A, 2018. Geology and geochronology of the Golpu porphyry and Wafi epithermal deposit, Morobe Province, Papua New Guinea, *Economic Geology,* 113(1):271–294. <https://doi.org/10.5382/econgeo.2018.4551>

Sadeghi, B, 2025. Clustering in geo-data science: Navigating uncertainty to select the most reliable method, *Ore Geology Reviews,* 181:106591. <https://doi.org/10.1016/j.oregeorev.2025.106591>

Stratum AI, 2025. Stratum AI. <https://www.stratum.ai> [Accessed: 19 November 2025].

Sullivan, S, Green, C, Carter, D, Sanderson, H and Batchelor, J, 2019. Deep learning – a new paradigm for orebody modelling, in *Proceedings International Mining Geology Conference 2019*, pp 181–189 (The Australasian Institute of Mining and Metallurgy: Melbourne).

Zaki, M M, Chen, S, Zhang, J, Feng, F, Khoreshok, A A, Mahdy, M A and Salim, K M, 2022. A Novel Approach for Resource Estimation of Highly Skewed Gold Using Machine Learning Algorithms, *Minerals,* 12(7):900. <https://doi.org/10.3390/min12070900>

Preserving geological collections for teaching and skills development – a digital petrographic atlas of Australian coals

S Rodrigues[1], T Congo[2] and J S Esterle[3]

1. Honorary Fellow, School of the Environment, The University of Queensland, St Lucia Qld 4067; Director, CarbonMat, Brisbane Qld 4000. Email: sandra.rodrigues@carbonmat.com.au
2. Honorary Fellow, School of the Environment, The University of Queensland, St Lucia Qld 4067; Researcher, KTH Royal Institute of Technology, Department of Materials Science and Engineering, Brinellvägen 23, 100 44 Stockholm, Sweden. Email: congo@kth.se
3. Emeritus Professor, School of the Environment, The University of Queensland, St Lucia Qld 4067. Email: j.esterle@uq.edu.au

ABSTRACT

Despite a global need for geoscience to underpin the exploration and stewardship of minerals, energy and natural resources, and the manufacture of current and future materials, university teaching programs are declining in many countries. In addition to expertise and teaching skill, there is also an irreversible loss of invaluable geological collections. Geological teaching collections vary by subdiscipline and range from core, hand samples, polished and thin sections, and extracted concentrates of rocks, minerals and fossils. This paper presents one strategy for geological knowledge and sample conservation in a digital age. We focused on our own discipline where in recent years, the teaching of coal geology and organic petrology has sharply declined across Australian universities. Organic petrology not only supports a strong metallurgical coal industry, it also records plant evolution and climate signatures through time, informs source and burial history models for hydrocarbons and sedimentary hosted minerals, and underpins development of future carbon materials. The Australian Coal Association Research Program (ACARP) supported a prototype development of a **Digital Petrographic Atlas of Australian Coals** designed to support self-directed learning and teaching of organic petrography. We used the Hilgers Fossil™ system to collect, compile, view and analyse high-resolution, reflectance-calibrated images of representative coal samples. We utilised the university and government geological survey collections to ensure open file access and focused on key drill core samples from the Early and Late Permian coal measures of the Bowen Basin with selected Mesozoic coal from Queensland. A total of 85 polished coal blocks were retrieved, freshly polished and scanned to capture petrographic variability in composition and rank within and between coals in different geological settings. The scans were used to develop an **Illustrated Atlas** as well as a **Digital Library of Scanned Samples** that can be uploaded to a software viewer that emulates the optical microscope for each user to **learn and practice** petrographic analysis. More broadly, the product serves as a pilot for the digital preservation of teaching collections in both universities and government agencies, that can be utilised for continued teaching and skills development, both within and outside the university environment.

INTRODUCTION

Learning petrology requires specialised microscopes in the classroom and a number and variety of samples, resulting in substantial physical collections. Even before COVID, universities started to rationalise geoscience programs and people sought ways to digitalise their rock and thin section collections to reduce storage space but also move to online teaching. Geological surveys and museums recognising the value of their physical collections, their fragility and storage hunger, have started to conserve thin section collections through scanning and archiving the samples in relation to their geological context (eg <https://www.wa.gov.au/government/announcements/seeing-rocks-new-light-thin-section-scanners>).

In our discipline of organic petrology, polished blocks of resin mounted coal are examined in reflected light using optical microscopes equipped with oil immersion objectives. The main aim is usually to identify and quantify macerals, which are the organic equivalent to minerals in rocks (Stopes, 1935). In a teaching environment, this approach requires access to multiple microscopes to accommodate an entire class, as well as substantial effort in sample preparation. Polished blocks must be prepared in advance and routinely re-polished to maintain surface quality.

When geoscience classes became oversubscribed relative to available microscopes and prepared petrographic samples, we trialled the development of a digital teaching collection using the Diskus Fossil Viewer™ software. A small suite of eight teaching samples from different Queensland basins was selected and scanned to produce a database of reflectance calibrated images that could be uploaded and viewed within the software. Installed on desktop computers, Diskus Fossil Viewer™ emulates the camera output and software interface of a coal petrographic microscope, allowing analysis of vitrinite reflectance and maceral composition. Preliminary trials showed that rather than struggling with technique, students were able to learn the material and compare and discuss their results as professionals would. The digital and distributive nature of the exercise came to its fore during COVID when teaching went remote.

In Australia, thousands of coal and dispersed organic matter samples exist that were prepared and analysed in support of the exploration for coal and hydrocarbons (documented in Harrington *et al*, 1989; McKillop, 2016; Edwards and Buckler, 2024). Results of those analyses are often published in journal papers and books, searchable in well completion reports, or compilations thereof (eg Mutton, 2003; Juniper, 2010); however, sample curation is a potential issue. The success of digital teaching prompted a broader recognition that this approach could serve not only as an effective teaching tool but also as a means of conserving the vast sample collections resident in universities, government geoscience and research agencies. Supported by the Australian Coal Association Research Program (ACARP) we developed a prototype digital atlas for Australian coals with a focus on Queensland's Bowen Basin. The atlas comprises three principal components described below.

THE DIGITAL PETROGRAPHIC ATLAS

The illustrated atlas

Coal is a heterogeneous sedimentary rock composed primarily of organic constituents, known as macerals, whose physical and chemical properties evolve with increasing thermal maturation (Taylor *et al,* 1998). These systematic changes form the basis of coal's classification, behaviour, and suitability for different applications. An understanding of coal properties has been fundamental to the development of metallurgical coal industries in Australia and internationally, has informed models of sedimentary basin evolution used in hydrocarbon and mineral exploration, and continues to guide the emerging use of coal as a precursor for advanced carbon materials.

In this context, several excellent illustrated textbooks for learning coal and organic petrology (eg Stach *et al,* 1982; Taylor *et al,* 1998; Wagner, Malumbazo and Falcon, 2018) and web-based tools (eg <https://energy.usgs.gov/photoatlas/>) are available. However, with the exception of the textbook by Diessel (1992), none focus specifically on Australian coals. We therefore developed an atlas centred on Australian coal examples, while providing sufficient background to enable users to understand variations among coals worldwide, with different geological ages and depositional histories.

We prepared a printable, heavily illustrated atlas that presents concepts of coal formation and the resulting mappable variations in coal composition and rank. These variations occur at multiple scales, within and between seams at mine sites and across a basin, and reflect changes in depositional environment and burial history. The systematics, definitions and criteria for the recognition of different of coal macerals and minerals are presented and used to organise over 40 annotated plates containing over 300 images selected from reflectance calibrated scans of 85 samples within the digital library. A partial example of images is shown in Figure 1. In addition to the photomicrograph plates, the atlas explains how to prepare samples and conduct petrographic analyses to determine maceral composition and how to take vitrinite reflectance measurements for estimations of rank or thermal maturity, and includes some examples of exercises for users.

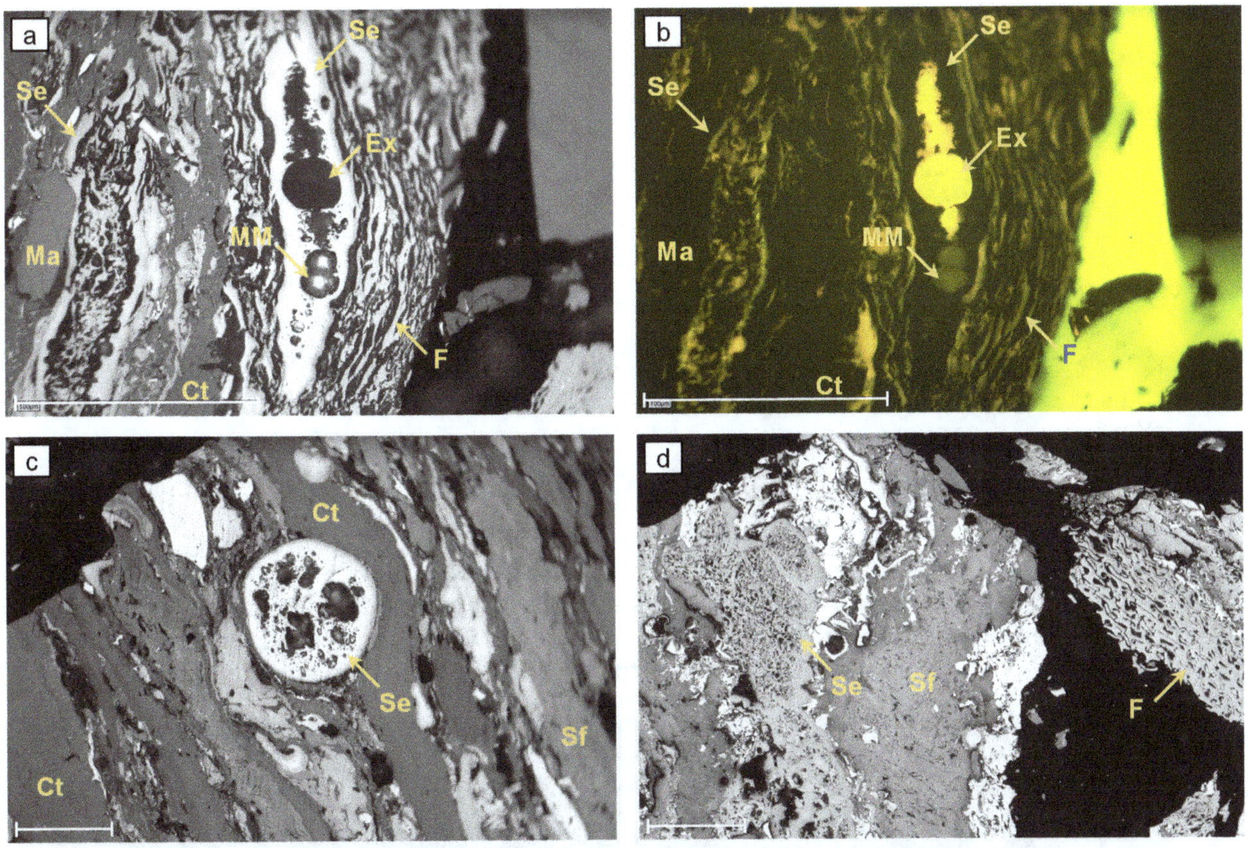

FIG 1 – Example of photomicrograph plates within the atlas. (a) An elongated vesicular secretinite (Se) with vesicles filled with exsudatinite (Ex) and mineral matter (MM): Reids Dome Beds, Bowen Basin, PS 2781, Rr = 0.72 per cent. (b) Blue light image of the same region as plate (a) showing the highly fluorescing nature of the exsudatinite (Ex) allowing it to be distinguished from the mineral infilling in the secretinite (Se): Reids Dome Beds, Bowen Basin, PS 2781, Rr = 0.72 per cent. (c) A round secretinite (Se) showing accommodation of the collotelinite (Ct) around it: Reids Dome Beds, Bowen Basin, PS 2786, Rr = 0.72 per cent. (d) Cluster of vesicular secretinites (Se) with relatively low reflectance intensity compared to previous examples: Aries-Castor Seam, Rangal Coal Measures, Bowen Basin, PS 1892, Rr = 0.73 per cent.

Digital library of scanned coal samples

The digital library was designed as a curated, expandable resource that captures the range of coal types and ranks relevant to teaching and comparative petrographic analysis. Sample selection was guided by the input of ACARP industry monitors, with an emphasis on metallurgical coals, but prioritised well-documented archival material rather than commercially traded products. To this end, the library draws primarily on the extensive and systematically curated coal petrographic block archive of the Queensland Geological Survey (QGS), supplemented by The University of Queensland's teaching collection.

This archive comprises thousands of polished blocks derived from exploration drilling, each accompanied by open file stratigraphic information and petrographic data, including coal rank (vitrinite reflectance) and maceral group composition. Using the archived material allows variation in coal quality to be examined in their geological context, enabling learning exercises that demonstrate how coal targeted for metallurgical use or unconventional gas production can vary in composition, rank, and therefore utility. From the QGS archive, we selected 19 key drill cores (Figure 2) that illustrated variation in both rank and type in different geological settings, including Early and Late Permian coal measures of the Bowen Basin and selected Mesozoic coals. A total of 85 polished coal blocks were retrieved, freshly re-polished and scanned using a Leica DM6000 microscope equipped with the Hilgers Fossil ™ system. The motorised stage, controlled by the Hilgers software, allows scanning of the entire polished block surface, producing over a thousand reflectance-calibrated images per sample which can be exported as a zip file (Figure 2a and Figure 3).

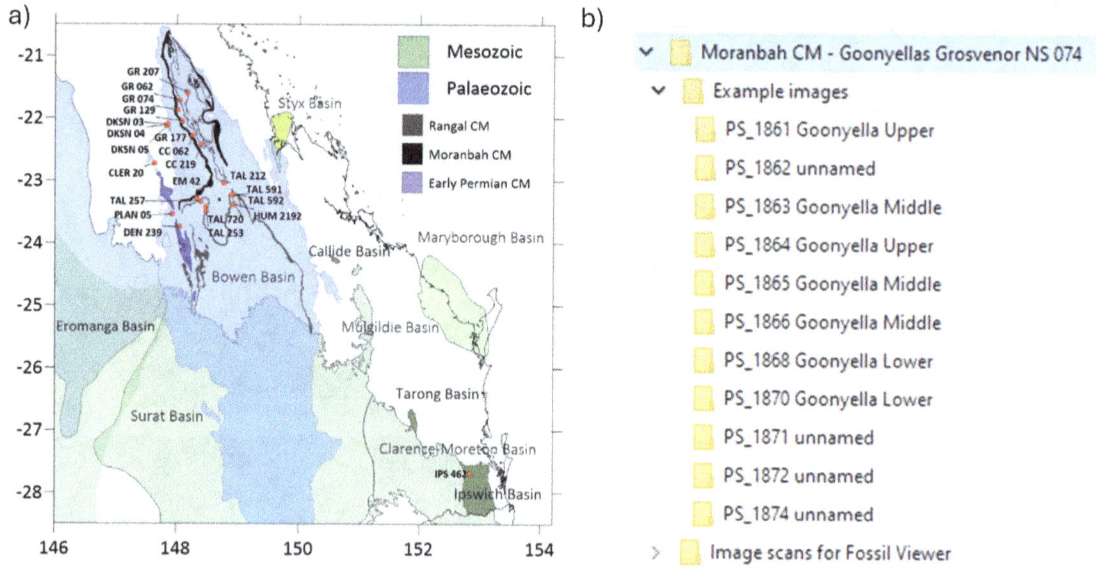

FIG 2 – (a) Location of selected GSQ drill holes sampled for the digital atlas in the different coal measures of the Permian Bowen Basin and the Triassic Ipswich Basin. (b) Example of the digital library structure for the Grosvenor NS 074 borehole.

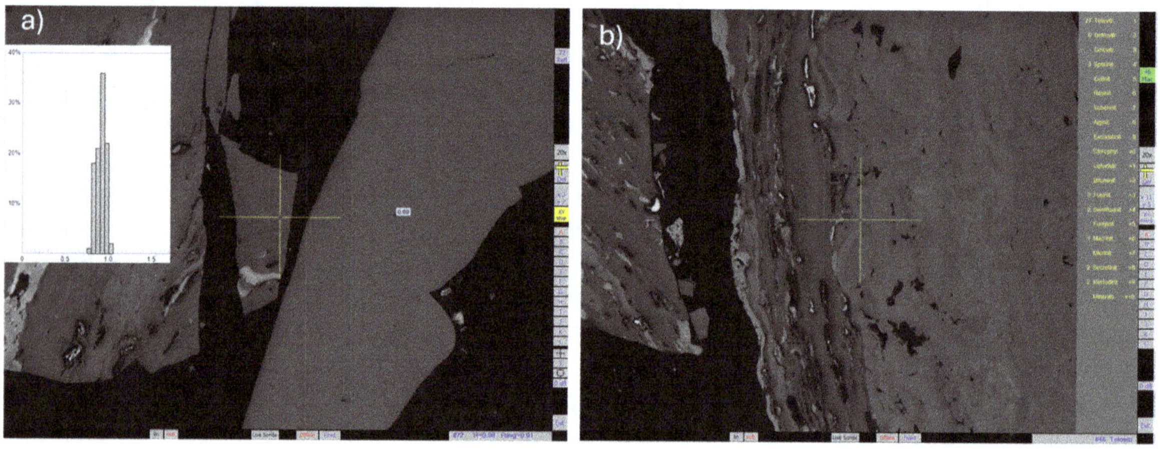

FIG 3 – Screenshot of the Hilgers Fossil Viewer™ software showing (a) vitrinite reflectance measurements and histogram (inset); (b) maceral analysis with a classification list on the right.

These compressed files are compatible with a digital viewer that emulates the optical microscope. This enables users to navigate the scanned surface, identify and point count macerals, and measure and record reflectance values in a manner analogous to conventional petrographic analysis. The scanned specimen database is distributed together with the Hilgers Fossil Viewer™, providing access to the digital library without the need for specialised laboratory infrastructure.

Learning exercises

In addition to supporting self-guided learning on the recognition of different macerals, the quantification of their abundance within samples, and the measurement of coal rank, the atlas places each specimen within its geological context. Each sample is geo-located within a specific borehole and referenced to the relevant coal measure and coal seam; for example, the Early Permian Reids Dome coals or the Late Permian Moranbah or Rangal coal measures. These units display strikingly different characteristics that are used to determine coal quality and end-use suitability.

Spreadsheet-based exercises are provided to allow users to compare their vitrinite reflectance measurements with those obtained the authors under the same measurement conditions, and with values recorded in the QGS borehole information. Similarly, users can compare their maceral classifications with those of the authors and with published record. Through these exercises, users may encounter operator subjectivity, as well as the inherent challenges in applying strict classification criteria to organic matter that has evolved through geological time, burial and upheaval.

While digital teaching and learning exercises can be highly effective and economically favourable, it is important to emphasise that petrographic competence should also include the ability to physically examine a specimen, learn how to prepare the sample, how to use different microscope systems and how to recognise when petrographic results don't seem quite right. This takes time and mentoring, and a community of practice not afraid to 'round robin'. With the decline of coal and organic petrologists in Australian academia, the atlas project provides a tool that might be used to not only conserve samples and teach but reverse the decline.

SUMMARY

The decline of geoscience teaching programs globally presents a dual challenge: the loss of disciplinary expertise and the irreversible loss of physical geological collections that underpin both education and research. Petrological teaching collections, including cores, hand specimens, polished blocks, and thin sections, represent decades of investment and irreplaceable geological context. Once dispersed or discarded, their scientific value cannot be recovered, even where analytical results remain archived. In this project, we present a prototype Digital Petrographic Atlas of Australian Coals that addresses both challenges by combining high-resolution, reflectance-calibrated digital petrography with geological context and structured learning exercises. Using archived material from university and government collections, the atlas preserves access to representative coal samples while supporting self-directed learning and formal teaching in organic petrology. The digital library enables users to practise petrographic analyses in a manner analogous to conventional microscopy, while removing the logistical constraints associated with specialised equipment and physical sample handling. Beyond its immediate application to coal geology and organic petrology, this work demonstrates a scalable approach to the digital preservation of geological teaching collections. While digital archives cannot replace physical samples, they extend their educational and scientific lifespan and ensure continued access to geological knowledge embedded within them. As geoscience programs continue to contract and physical collections face increasing pressure, such approaches offer a practical pathway to safeguarding geological heritage and maintaining essential skills for future generations of geoscientists.

ACKNOWLEDGEMENTS

This project could not have been possible without the support of the Australian Coal Association Research Program ACARP C33065 and the excellent collection from the Geological Survey of Queensland (GSQ) conserved at the Queensland Exploration Data Centre in Zillmere. We also thank the Vale UQ Geoscience Program for infrastructure and in-kind support of this project.

REFERENCES

Diessel, C F K, 1992. *Coal-Bearing Depositional Systems,* 721 p (Springer: Berlin, Heidelberg).

Edwards, D and Buckler, T, 2024. Organic Geochemistry (ORGCHEM) Schema, Australian Source Rock and Fluid Atlas, Geoscience Australia, Canberra.

Harrington, H J, Brakel, A T, Hunt, J W, Wells, A T, Middleton, M F, O'Brien, P E, Hamilton, P E, Beckett, J, Weber, C R, Radke, S, Totterdell, J M, Swaine, D J and Schmidt, P W, 1989. Permian coals of Eastern Australia, *Bulletin 231 Parts 1 and 2*, Bureau of Mineral Resources, Geology and Geophysics.

Juniper, L, 2010. Quality of Australian Black Coals, ACARP Report C17053.

McKillop, M, 2016. Coal maturity trends within the Bowen Basin, An update of the vitrinite reflectance study by Beeston 1986, Queensland Geological Record 2016/04.

Mutton, A J (Compiler), 2003. *Queensland Coals 14th Edition,* Queensland Department of Natural Resources and Mines.

Stach, E, Mackowsky, M-Th, Teichmüller, M, Taylor, G H, Chandra, D and Teichmüller, R, 1982. *Stach's Textbook of Coal Petrology*, 3rd ed, Borntraeger, 535 p.

Stopes, M C, 1935. On the petrology of banded bituminous coal, *Fuel*, 14:4–13.

Taylor, G, Teichmüller, M, Davis, A, Diessel, C, Littke, R and Robert, P, 1998. Organic Petrology, Gebrüder Bornatraeger, Berlin-Stuttgart, 704 p.

Wagner, J, Malumbazo, N and Falcon, R M S, 2018. Southern African Coals and Carbons: Definitions and Applications of Organic Petrology, *Struik Nature*, 248 p.

PhotonAssay™ – techniques and workflows to establish fitness for purpose

R Sterk[1], *J Nessler*[2], *M van de Ven*[3], *M Knorsch*[4] *and D Woestenburg*[5]

1. Principal Consultant, RSC, Dunedin 9600, New Zealand. Email: r.sterk@rscmme.com
2. Consulting Exploration Geologist, RSC, EMEA. Email: j.nessler@rscmme.com
3. Senior Resource Consultant, RSC, Dunedin 9600, New Zealand. Email: m.vandeven@rscmme.com
4. Consulting Geoscientist, RSC, Perth WA 6005. Email: m.knorsch@rscmme.com
5. Consulting Statistician, The Hague, Netherlands. Email: dionwoestenburg@hotmail.com

ABSTRACT

Analytical techniques based on gamma activation analysis (GAA) are rapidly gaining popularity in the mining industry for analysis of Au, Cu, and Ag. Since being introduced commercially by Chrysos Corporation in 2018 under the brand name PhotonAssay™ (CPA), 130 companies have reported Au exploration results to major stock markets using data collected with this technique, and 12 companies have used such results to support a publicly reported mineral resource estimate (MRE).

While it has been established that, under best-practice operating conditions, results from CPA are accurate and precise, test results may not prove accuracy for every style of mineralisation and any individual CPA instrument. Therefore, many operators carry out their own test work, comparing CPA assays with results obtained by conventional techniques (eg fire assay). The authors' experience with comparative studies around the world is that they are often not designed for rigorous scientific testing, and hence do not meet the objective of drawing robust conclusions on the quality of CPA data. Additionally, the issue of materiality, key to every mining and exploration project, is rarely discussed, and studies do not define a data quality objective (ie when is something fit for purpose).

In this paper, we add detail to previously published recommended testing programmes, setting out the sample selection process, and detail all critical steps during sample preparation, including the collection of true duplicates for precision benchmarking.

Based on several real-world examples, we then provide a step-by-step process to assess the validity of the testing data, accuracy and precision, and whether any observed differences are both statistically significant and material to the project.

INTRODUCTION

Gamma activation analysis (GAA) is a non-destructive method that uses X-rays for the rapid analysis of Au, Ag, and Cu in geological and metallurgical process samples (Tickner *et al,* 2017; Tremblay *et al,* 2019). Since its commercial introduction by Chrysos Corporation (Chrysos) in 2018, GAA has emerged as a fast and cost-effective alternative to conventional analytical techniques, such as lead fire assay (FA) or aqua regia digest followed by atomic absorption spectrometry (AAS) or inductively coupled plasma mass spectrometry (ICP-MS). Gamma activation analysis is now widely known under its Chrysos brand-name: PhotonAssay™.

Frequently cited benefits of Chrysos PhotonAssay™ (CPA) include: the ability to analyse larger samples (400–600 g) of coarse crushed (2–3 mm) material; a highly automated process increases sample throughput (~70 samples per hour, per unit) that minimises sample-handling mistakes; the lack of ecotoxic chemicals such as lead and cyanide; and an ~80 per cent lower carbon and energy footprint compared to a standard 30–50 g FA (Chrysos, 2022, 2025). CPA has wide applicability, spanning the entire mining value chain from exploration, resource development, grade control, and processing plant control (Tremblay *et al,* 2019; Dominy *et al,* 2024).

As with any emergent technology, there is a requirement to prove that data produced are of comparable or superior quality to that produced by conventional approaches. There are several published case studies that focus on the comparison of results from CPA and FA in different settings (Tickner, Lannan and Preston, 2021; Dominy *et al,* 2024; Hitchman *et al,* 2024). Comparative studies can provide justification for a switch to CPA and may be undertaken by companies independently or in collaboration with commercial laboratories or Chrysos. However, the authors have found that company-led comparative studies commonly lack a sound statistical foundation. Most studies do not address the important relationship between the size of the study and the confidence required for decision-making. As a result, many such studies do not succeed in reaching an objective conclusion; instead, subjective statements are made that do not aid decision-making.

Junior explorers and major mining companies are adopting CPA, and market announcements on major exchanges are starting to rely on results obtained through CPA. The authors used data from opaxe (www.opaxe.com) to analyse 130 stock market releases that announced results from CPA, including results from 12 companies that used CPA Au data to support a publicly reported MRE up until 2026. The review indicates that most companies relied on subjective conclusions to support the switch to CPA from FA.

This paper adds to the growing literature on best-practice for vetting CPA by providing a standard-practice, step-by-step guide to reliably test the performance of CPA compared to other analytical techniques for Au, and we present a simple flow chart to statistically test and determine whether CPA is fit for purpose.

CPA can be used to analyse a range of elements; however, it is mostly used for Au analysis. Therefore, this paper centres on the comparison between Au analyses obtained by CPA and FA. Our comparative workflow and statistical methods described are applicable to other elements such as Ag and Cu. The authors note that Chrysos is not the only provider of GAA analysis for geological samples (eg Baltic Scientific Instruments). However, the authors have only reviewed Chrysos' facilities, and therefore this paper is focused on PhotonAssay™. The principles underpinning the proposed test programme and described in this paper are transferrable.

HOW CPA WORKS

Gamma Activation Analysis relies on high-energy (>6 MeV) X-rays to excite transitions in nuclides of target elements, and detect the characteristic gamma-rays that are emitted from the nuclides when they decay from their excited state (Tickner *et al,* 2017). The gamma-ray count is proportional to the concentration of the element in the sample, and can be used inversely to determine the element concentration. Matrix or particle-size effects are largely insignificant, as the highly penetrative X-rays allow for the measurement of the entire sample volume contained within a 320 mL plastic jar, and the gamma-ray signal is emitted by the excited Au atoms even when Au is contained in solid particles.

Calibration steps are required to ensure that the CPA instrument delivers consistent results. System-inherent variations (eg changing activation energy or detector efficiency) are corrected by measurement of a reference disc with bromine-salt, placed underneath each jar. Bromine activates in a very similar way to Au, but produces a distinct lower-energy signal that can be used to correct for signal variance.

While the gamma-ray signal is directly proportional to Au concentration, the normalisation slope, referred to by Chrysos as the k-value, is determined by analysis of so-called *k-cal standards*. These standards are custom-made Au-bearing glass frits, with a superior chemical and physical stability to pulverised rock certified reference materials (CRM), and are analysed on a daily basis. Additionally, a monthly check of an instrument's performance is carried out by analysing a suite of high-quality CPA-specific CRM (*super k-cal*). After every major change of components, maintenance and/or calibration events, a change test suite is analysed for confirmation of the required operating performance that includes blanks, glass frit jars with various fill levels, and CRM of a wide range of grades, masses, densities, and matrices.

Elevated concentrations of U, Th, and Ba interfere with Au analysis, as they increase the background of the gamma-ray signal. For example, above combined U+Th concentrations of 15 ppm, or Ba concentrations of 3500 ppm, the lower detection limit of the CPA increases from 0.01 ppm to 0.03 ppm, and measurement precision decreases from 11.5 per cent to 18 per cent at 0.1 ppm but there is minimal impact on precision for Au concentrations above 0.3 ppm (Chrysos, 2023).

DATA QUALITY PRINCIPLES

Quality and materiality

A logical outcome of comparative CPA-FA studies is an objective statement on the *quality* of the CPA data, with quality conventionally expressed in terms of the accuracy and precision of the data. However, the scope of such studies is rarely specific about the data quality objective (DQO). In other words, it is not clear *how* accurate and precise the data need to be to support the transition to CPA. For instance, Hitchman *et al* (2024) in their study on the feasibility of transferring to CPA at the Fosterville deposit, Australia, state that 'success would be achieved if photon assaying [....] is analytically accurate and precise' (p 341), without stating what would trigger such a statement to be validated; ie when is something accurate enough and precise enough? In other words, when is something good *enough*?

Defining a DQO is critical in any assessment of data quality (Abzalov, 2008; Pitard, 2013; Dominy *et al,* 2024) and part of formulating a DQO is translating the quality requirements into a testable hypothesis. In most studies the authors reviewed, an explicitly stated DQO is absent, but it is implied that the CPA data need to be *at least as good as the quality of FA data*. However, such a statement also comes with caveats. There is an intuitive margin of error where the results would still satisfy the DQO. For example, if the precision of FA duplicate pairs is 12.1 per cent (as defined by the root-mean-square coefficient of variance (CV) approach (Stanley and Lawie, 2007; Abzalov, 2008)), then if the precision of CPA duplicate pairs is 12.5 per cent (ie CPA has slightly worse precision), this would not necessarily lead to a rejection of the hypothesis that the CPA data are at least as good as the FA data; it is within an intuitive (subjective) margin of error.

Well-known statistical tools can provide guidance; Student t-tests and equivalent tests remain the gold standard in many disciplines and industries to indisputably state differences between populations using well-known conventions. Key CPA papers such as Tickner, Lannan and Preston (2021), in their first comparison of results from standard FA versus the results of CPA, use such tests to conclude that CPA is accurate. Arguably, if no statistically significant difference is found, one can state that one method is as accurate as the other. However, a small, yet statistically significant difference between the data for two different techniques does not always mean that the data are not suited to the purpose or objective; such an outcome may occur, especially with a large data set. This concept is not just true in the mining industry, but also in other industries, and the concept of statistical significance versus practical significance is a well-studied subject area (Gelman and Stern, 2006; Peeters, 2016; Miller, 2023).

In the mining industry, there is arguably a disconnect between statistical test results and the practical ramifications of these results from a materiality perspective. An example of this is the outcome of a classic 'laboratory umpire' study, where a company submits a few hundred samples to another laboratory each quarter to compare the assay results. In the authors' experience many of these studies indeed find a small, yet statistically significant difference using conventional statistical testing methods. However, rarely has a 'small' difference between two laboratories stopped a mine in its tracks or led to re-assaying of all samples. This leads to a dead end and is arguably one of the reasons why many practitioners skip any statistical testing and go straight to subjective 'expert interpretation'.

The issue of materiality is key to every mining and exploration project, in any commodity, and at any part of the grade curve; yet, linking quantitative accuracy and precision thresholds to associated project risk, or even worse, project failure, is difficult. Full model-to-mill reconciliation is the only quantitative comparison available to provide a feedback loop to any decision made in the complex process of data collection, estimation, mining, and processing. It is clear that a small change to a part of a sampling or analytical process will have an almost undetectable impact on the monthly mine reconciliation data. This is not helped by the poor state of most mine reconciliation systems, which are often based on Excel sheets, and which, apart from noise in analytical data, also embed noise from the modelling, mining, handling, and processing steps, and do not have any defined performance thresholds themselves. The consequence is that strictly following the scientific method, ie quantitative thresholding of accuracy and precision using common statistical tests, may not lead to practically relevant conclusions about data quality. Qualitative comments and interpretations that focus on materiality may need be added to the conclusions.

Accuracy

Testing the DQO that the CPA data quality is *at least as good as the quality of FA data* would be possible if the true grade of the material being assayed was known as a basis of comparison. For example, if the true grade of a sample is 5.00 g/t Au, and if the FA result is 4.69 g/t Au, and the CPA result is 4.81 g/t Au, then, objectively, the CPA result is at least as good (ie as accurate) as the FA result. In this example, if the FA result has been set as an acceptable benchmark to allow resource classification to meet production risk objectives, then by inference, the CPA results should be acceptable also. Of course, the true grade of a geological material is never known, and hence making any claim on the accuracy of the CPA data requires asking: Compared to *what*?

Since the true grade of a geological sample is unknown, it is impossible to determine an exact bias of the CPA method, and CPA performance must be compared to the best-known alternative (ie FA to extinction). However, the destructive nature of FA methods complicates this, as the sample can only be analysed once, and comparing CPA results against a benchmark FA analysis from a single laboratory would, in the best scenario, only provide a relative bias, with no guarantee that the benchmark is not biased itself.

In all comparative studies reviewed by the authors, samples analysed by CPA were sent to only one laboratory for routine FA analysis, and these studies can therefore only assess the relative bias (ie the bias between two individual laboratories). This is not necessarily a fatal flaw, as the routine FA laboratories typically have a track record of supporting robust resource estimates, and results from that laboratory have been accepted across a significant period of time as providing fit-for-purpose results. However, this experimental design does put constraints on the conclusions that can be drawn from such a testing programme. As discussed in Tickner, Lannan and Preston (2021), 'if [...] the systematic or random errors [of the benchmark FA method] are not negligible compared to the errors of the new assay technique being evaluated [CPA], then any comparison of grade results will include the errors of both methods' (p 2). To illustrate this, if there was a consistent and statistically significant bias introduced by scooping the 50 g pulp from the 300 g lot for FA, this would impact the comparison between FA and CPA. This problem could be addressed by using FA to extinction in the comparison study. However, if the bias was consistent, and these FA values are from the mine's routine laboratory, then this bias would also be present in the data on which the MRE is based, and which were accepted as fit for purpose. Therefore, this bias would have no bearing on the evaluation of quality of the CPA data.

For most comparative studies, there are several additional sources of information available that provide important context to appraise materiality in the comparison of the accuracy of CPA against FA.

- First, if there are frequent umpire tests conducted on the routine FA laboratory itself, then these data are important to set the framework of reference for accuracy. If the tenor of bias observed between the routine FA laboratory and the umpire FA laboratory is similar to the bias observed between the routine FA laboratory and the CPA laboratory, then this provides an argument for accepting the CPA results. In other words, if the routine FA umpire results demonstrate a small and statistically significant bias, then a bias of a similar nature in the CPA versus FA comparison would validate the CPA results by proxy.
- Second, FA laboratories routinely test a large number of FA-certified CRMs. The bias of these CRMs sets a semi-quantitative framework of reference for acceptance of any assay bias, in which the results of a comparative study should be viewed. For instance, if 12 different CRMs routinely demonstrate biases between -3 per cent and +3 per cent, and the assays were accepted in the database, and were used to inform and classify a high-confidence MRE, then, arguably, a bias within that range is also acceptable for the CPA data.

Precision

Precision is a concept, rather than a well-defined quantity (Hyslop and White, 2009). There is no universally accepted metric, and this makes the communication of acceptance levels for precision equally challenging. The reader is referred to the work of Stanley and Lawie (2007), who demonstrate that there are five different variations of precision definitions that are used in geological applications to represent precision: Coefficient of Variation (CV), Relative Precision (RP), Relative Variance (RV), Absolute Relative Difference (ARD), and Half Absolute Relative Difference (HARD).

Ultimately, which one is 'correct' comes down to convention, or one's interpretation of which is the correct correction factor to use. The authors favour expression of the precision of paired grade data by the CV, obtained using the 'root-mean-squared' formula for duplicate data, as presented by Stanley and Lawie (2007), and shown below in Equation 1.

$$RMS\ CV\ (\%) = 100 \times \sqrt{\frac{2}{N}\sum_{i=1}^{N}\left(\frac{(a_i - b_i)^2}{(a_i + b_i)^2}\right)} \tag{1}$$

This aligns with reviews of precision also presented in Abzalov (2008) and Smee *et al* (2024), and should, in the authors' opinion, be the one and only universally applied measure of average precision of paired grade data.

The authors note that, until 2025, Chrysos itself preferred to present precision values calculated across different parts of the grade range. This is because the data are inherently heteroskedastic; precision decreases abruptly with increasing grade close to the detection limits (Smee *et al,* 2024), which makes the calculation of a single average CV value challenging. In the authors' experience, clipping data below ~3× the level of quantification (LOQ) generates a data set for which the CV is a robust and repeatable precision metric that can be confidently used in comparative studies. The LOQ is the lowest analyte concentration that can be quantitatively detected with a stated accuracy and precision; LOQ = limit of detection (LOD) + 10 x σ_{blank}.

Like the framework of accuracy introduced before, it is again important to establish what *exactly* is compared when precision is calculated in this study. For standard FA, precision is calculated from paired analysis on pulverised samples. These are often referred to as 'pulp duplicates'. For CPA, the final aliquot usually contains material at a larger grain size and of a larger sample mass compared to FA. It is acceptable to compare duplicates created at this final CPA aliquot stage with those created at the final aliquot stage of FA, as both represent the outcome of 'routine' processes.

A DQO for the precision of the CPA data needs to be specified by the geologists managing the technical aspects of the programme. What is the maximum tolerance value of precision, as measured by the CV of duplicate pairs, beyond which the data are not considered fit for purpose? As with accuracy, it can be safely stated that the precision of the CPA data must be at least as good as that of the FA data. Additional information may be contributed by looking at historical duplicate data, particularly when these data can be linked to resource reconciliation.

Another powerful precision DQO can be set using the variogram, which can be generated without the need for extensive duplication of both FA and CPA samples, as long as a sufficient amount of samples have been analysed both by FA and CPA (ie several thousand). This method was applied by Hitchman *et al* (2024) in their review of the performance of FA versus CPA at the very nuggety Fosterville deposit; they demonstrated a lower nugget on normal-score variograms for CPA (~0.4) compared to FA (~0.5), which is objectively better and confirmed CPA's superior performance, in terms of precision, to FA.

Last, precision is dependent on the natural inherent variability of the sample and sampling/analytical errors. Specific to CPA, elevated concentrations of U, Th, and Ba, as well as the 'positional heterogeneity' of Au particles in jars lead to a decrease in precision in CPA analysis (Dominy *et al,* 2024). Therefore, it is important to benchmark the precision of CPA analysis in the full context of all these variables.

COMPARATIVE CPA-FA TEST PROGRAMME

Sample selection

The success of a comparative study between CPA and FA is contingent on the correctness of the sample selection process. For instance, in many of the comparison studies reviewed by the authors, minimum cut-off grades were applied to samples drawn from the database in order to select samples ('original') for testing, which introduces an artificial bias close to the selected cut-off grade. This happens, because the 'original' sample is *always* above that cut-off grade (as that is what it was selected on), but the 'duplicate' *may* return a grade that is lower (Long, 2015). On the contrary, randomised sample selections commonly lead to a very large amount of unmineralised samples, due to the heavily skewed grade distribution in almost all Au data sets, often leading to extreme waste of funds and time working on samples that are not mineralised. The correct approach is to randomly select samples from intervals that were logged as mineralised (Long, 2015) or are contained in modelled ore zones.

Any statistical analysis should only be undertaken on data that are from the same population, as the analysis of precision or accuracy will be flawed if distributions demonstrate any multi-modality. In this context, separate sample populations or domains could be represented by samples from different mineralisation stages, or different parts of a deposit. Even though CPA is relatively agnostic with respect to host rock type, FA is not, and any differences need to therefore be calibrated to relevant domains. For this reason, domains of a project/mine site need to be considered during sample selection, and results must be reviewed separately for each domain (see also approach, comments, and context in Dominy *et al* (2024)). An equal number of samples should be selected from each domain. The authors recommend 200 samples per domain as a reasonable starting point for a robust statistical orientation study.

Sample preparation and analysis workflow

To correctly review any differences between FA and CPA results, sample preparation and analytical processes should mirror those of standard and routine processes. This includes the equipment used, comminution sizes, sample weights, and protocols for splitting, digestion, and analysis of samples. The authors have reviewed CPA-FA studies where the laboratory equipment was optimised to achieve the 'best possible results'. Though tempting, it is important to note that the comparison is not that of best-behaviour CPA against best-behaviour FA. The point is to establish that routine CPA quality matches that of routine FA.

The authors propose a sample preparation workflow that provides the data and checks and balances needed to confidently compare results from FA and CPA (Figure 1). The following considerations require particular attention.

1. Crush, pulverise, and split samples as per usual sample preparation practices for FA samples, with the aim of comparing any accuracy or precision with routine FA processes (steps 2, 3, 7, 8, 13, and 15 in Figure 1).
2. Carry out two routine CPA analyses per crushed sample (step 5 in Figure 1) and three routine CPA analyses per pulverised sample (step 10 in Figure 1). The duplicate pairs will provide information on the routine CPA precision by comparing the two (note that the calculation of variance needs to be corrected for the use of three pulp pairs versus two crushed pairs). Prevent overfilling the CPA jars or scraping the excess material (coned) off the top (step 4 in Figure 1).
3. The recombined CPA pulp sample (step 11 in Figure 1) was previously split using an LSD splitter from the bowl, which does not represent FA routine, therefore the pulp must be milled again for ~5–10 seconds to 'get back on the flow chart' of the routine original FA process (step 12 in Figure 1). Otherwise, the calculated variance of the FA pulps will not be comparable with the routine process. This is important as the study aim is benchmarking and acceptance-testing based on routine accuracy and precision, hence, the routine process must be honoured.

4. If samples travel between CPA and FA laboratories and potentially segregate, insert another step of 5–10 seconds milling to re-homogenise the sample (between steps 14 and 15 in Figure 1).
5. Carry out two routine FA analyses. The duplicate pair will provide information on the routine FA precision by comparing the two.
6. At the umpire laboratory, ensure that samples are first re-homogenised and subsequently split as per routine umpire laboratory processes. The FA charge weight should be the same as for the primary FA (steps 19–21 in Figure 1).

Quality control

Quality control (QC) protocols are essential for determining that the sample preparation and analytical processes deliver consistent results that are fit for the purpose of informing the comparison between the CPA and FA. Therefore, QC steps should include screen size tests, analysis of certified reference materials (CRMs), and analysis of coarse blank samples, for both CPA and FA. If the geologists managing the technical aspects of the testing programme insist that only client-inserted CRMs can be trusted (instead of relying on laboratory-inserted CRMs, which are arguably fit for purpose; this is a discussion that goes beyond the scope of this paper), then typical client CRM insertion rates of 5 per cent are insufficient, and should be increased to one in five for a comparative study, to generate enough control points for robust statistical analysis (5 per cent of 200 study samples would yield only 10 CRM assays). For smaller sample sets with <400 samples, CRM insertion should be adjusted to obtain at least 25 CRM results per individual CRM for CPA and FA. CRMs should be the same for both CPA and FA. Such insertion rates require a considerable number of CRM jars to not halt the continuous operation of the CPA instrument. The authors recommend inserting three unique medium- to high-grade CRMs. Several CRM manufacturers offer CRMs certified for Au by both CPA and FA, which should be preferred over CRMs certified for just FA.

Heterogeneity – an added bonus

In addition to conclusions on accuracy and precision, the testing programme in Figure 1 allows important information on the heterogeneity and Au deportment to be collected. Having precision averages for two groups of samples of similar size, but of different grain size, allows determination of the sampling constants K and alpha, which in turn can then be used to determine the optimum size of the split and the grain size, while remaining below a defined target threshold of precision. For a worked-out example of how to calculate these factors and use them to make decisions, see Minnitt (2016) and Dominy and Minnitt (2012).

EVALUATION OF RESULTS

This section outlines the series of recommended steps for the review and analysis of paired CPA-FA results, and other routine QC data, to confidently determine the performance of CPA compared to FA (schematised in Figure 2). The authors used assay data from several real-world examples to illustrate different evaluation steps. It should be noted that none of the real-world data sets used here to present workflows are indicative of expected accuracy or precision outcomes; they are used to demonstrate workflows only, and they are not 'case studies'.

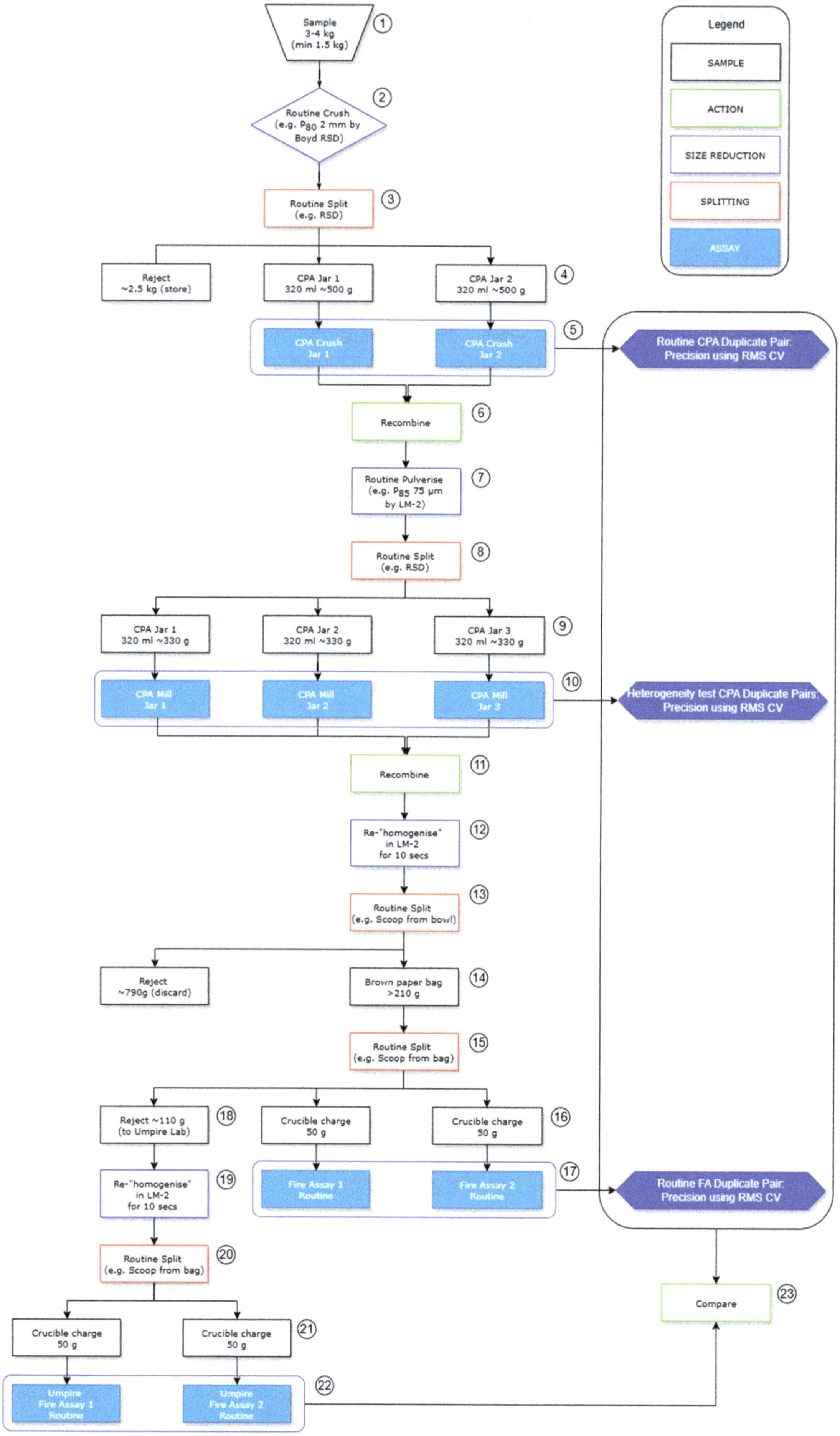

FIG 1 – Sample preparation and analysis flow sheet for comparative CPA versus FA analysis.

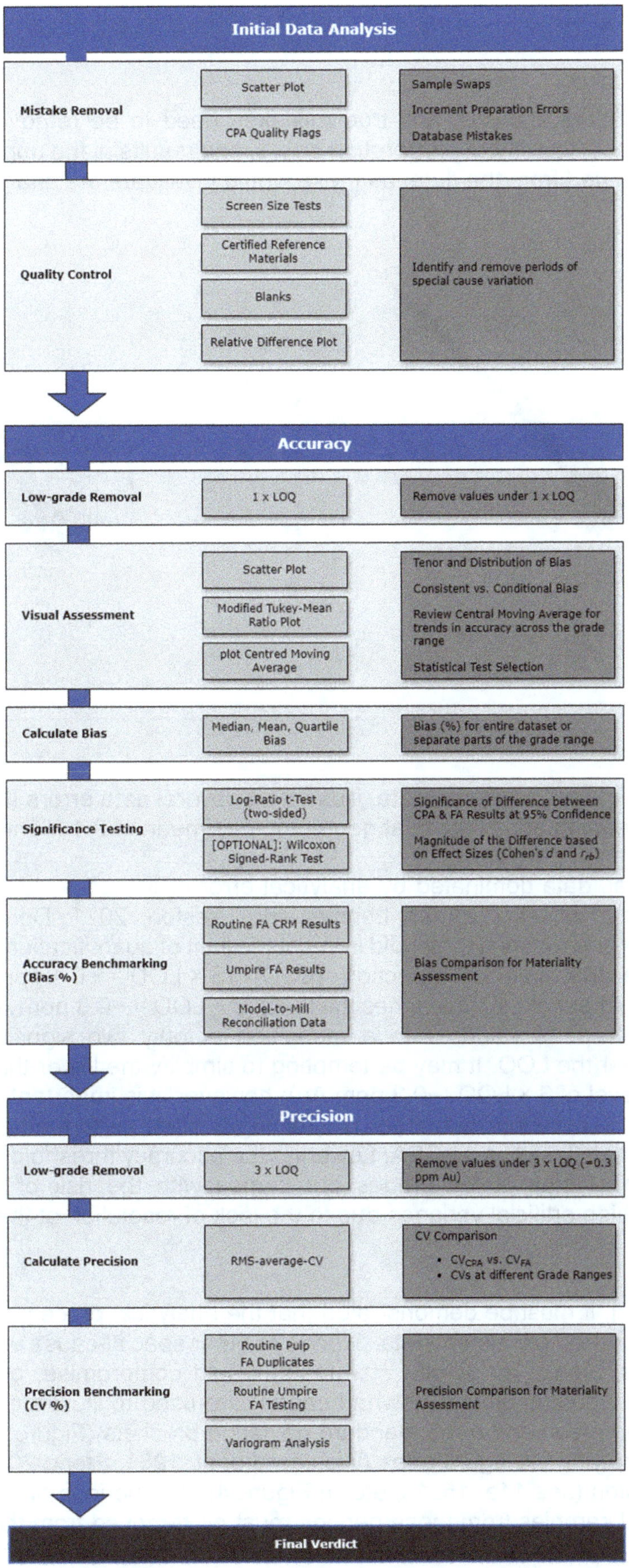

FIG 2 – Schematic data review and analytical process for comparative CPA-FA studies. Each light grey box represents a process step (left) that contains items to use, plot, calculate, or analyse (centre) for the corresponding data assessment(s) (right).

Initial data analysis

Basic validation

As with any statistical data analysis, any true mistakes need to be removed from the data first (Figure 3). Sample swaps, increment preparation errors, and results at the upper detection limit must be identified and eliminated from the data, as these would invalidate the analysis.

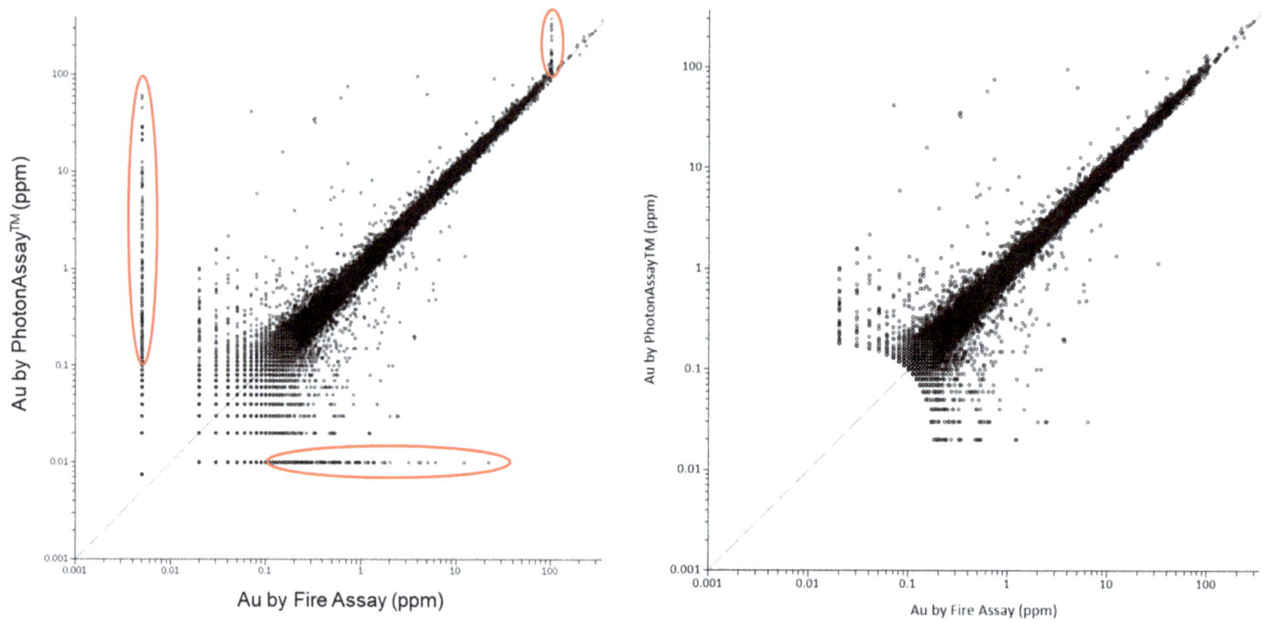

FIG 3 – Scatterplot of raw, unfiltered data, illustrating several data errors (left); and data after cleaning and clipping at geometric pair mean of 0.1 (right).

Following error removal, data dominated by analytical error or limited by low resolution should be excluded from further analyses (Tickner, Lannan and Preston, 2021; Dominy *et al,* 2024). For accuracy assessments, the removal threshold is set at the limit of quantification (LOQ), approximated here as three times the limit of detection (LOD) (3 × LOD = ~0.1 ppm Au). For precision assessments, a higher threshold of three times the LOQ (3 × LOQ = ~0.3 ppm Au) is applied because the analytical system reports values with a resolution of only two significant digits, artificially increasing variance near the LOQ. It may be tempting to simplify the lower-threshold approach and set both at the same level of 3 × LOQ (~0.3 ppm Au); however, it is important to get an indication of bias at low-grades, since resource estimation domaining is often performed at this level, and plant tailing samples may be analysed using CPA. Lowering the accuracy threshold to 3 × LOD (~0.1 ppm Au) allows evaluation of bias at low-grades but comes with the risk of Type-II errors ('false negatives'), given the high artificial variance due to the lack of resolution at these levels.

Quality control

Following data cleaning, it must be demonstrated that the analytical processes (both CPA and FA) were in control and delivered consistent data. Trends or other special-cause variation (as evidenced by results from CRMs, blanks, or screen size tests) would compromise, or even invalidate, the comparison. For CRMs, control plots (Shewhart charts) are used to illustrate the results per CRM, relative to their certified mean and three standard deviation brackets (Figure 4). Additional insights can be gained from applying Westgard rules (Westgard *et al,* 1981; Sterk, 2015) to identify periods of special-cause variation (see 14a, 15u1s, etc, in Figure 4). Any periods of special-cause variation must be identified, and samples from these periods must be removed from the data set, especially if special-cause variation is evident across different CRMs.

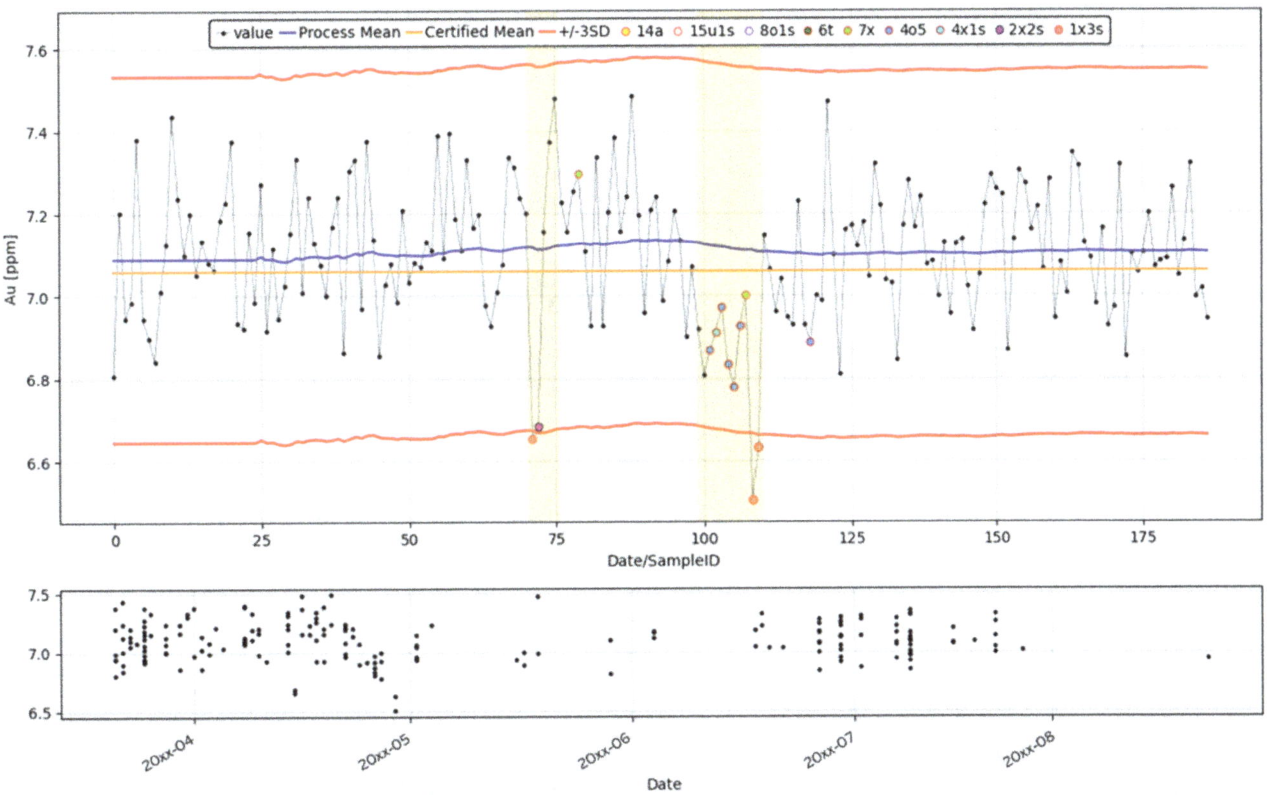

FIG 4 – Example of Shewhart control plot for one Au CRM, with periods of special-cause variation highlighted in yellow.

Accuracy

Basic visual assessment

Before accuracy can be calculated and assessed for significance, it is important to assess whether any bias depends (ie is conditional) on the mean concentration. If conditional, an average bias result may not hold across the full range of paired means.

With the cleaned data from periods where processes were in control, a good place to start this assessment is a visual evaluation of basic scatter of CPA and FA measurements (Figure 5, left). It is important to use the log scale for both measurements, allowing relative differences to be interpreted consistently from low to high grades. The authors have found that QQ plots are widely used to describe the accuracy of paired data across the grade range by comparing the quantile distribution of the two measurements. As simple and practical as this approach is, it treats the two measurements as independent, which obscures important information about the distribution of each pair, and therefore its utility is limited.

Due to scale dependence, the absolute numerical difference between CPA and FA is larger when the assay values are higher, and the ratio of the results is a more consistent way to quantify any bias (it is scale-invariant). However, analysing ratios is problematic because ratios are intrinsically asymmetric; all negative differences between original and duplicate are expressed as ratios between zero and one, while positive differences between original and duplicate are expressed as ratios higher than 1.0. A logarithmic transformation solves this issue (eg the logarithm of 2.0 is 0.301; the logarithm of 0.5 is -0.301). A Tukey mean-ratio plot (Figure 5, right) as used in Tickner, Lannan and Preston (2021) plots the geometric mean grade of the pair versus the ratio between CPA and FA (log-transformed to correct for asymmetry), and helps identify systematic deviations. A simple centred moving average function can be plotted to highlight any trends.

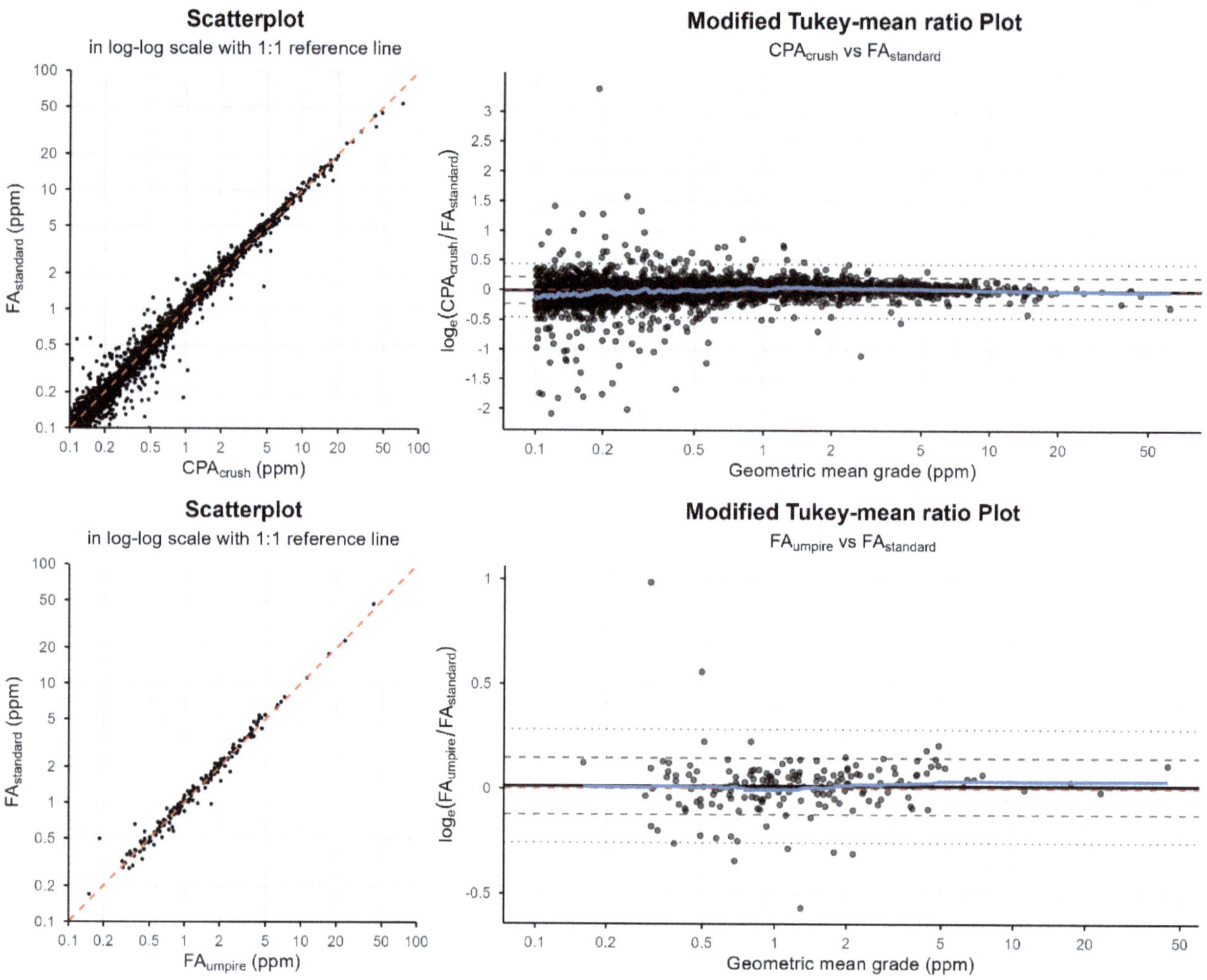

FIG 5 – Scatterplot (left) and modified Tukey plot (right) for inspecting differences between FA and CPA_{crush} versus $FA_{standard}$ (top) and FA_{umpire} versus $FA_{standard}$ (bottom). The black solid horizontal line represents the mean value of the ratio (ie the average bias across the entire grade range, see Bias Calculation section); the horizontal black dashed and dotted lines represent one and two standard deviations from the mean respectively. The blue line is the centred moving average of log ratios by grade level with a window of 101 pairs (edge-truncated). The plots present results from one study and are not indicative of typical CPA outcomes.

When the log-ratio is not conditional to the grade level, the moving average trendline follows a straight horizontal line. If the moving average deviates from a straight horizontal line in the Tukey plot, the bias is conditional and any bias calculation or significance testing should be evaluated separately for different grade ranges. In this case, the calculation of a *single* summary statistic of the average bias across the entire grade range could be misleading, especially if the bias is negative at one end and positive at the other end (in which case the *average* bias could be close to zero, suggesting a high degree of accuracy that is not representative of the entire data set).

Bias calculation

The mean or median bias can be calculated, either across the whole grade range (if the bias is not conditional), or for each part of the grade range where bias is consistent. The bias can be expressed as a percentage using the transformation:

Equation 2 turns the average of the log-ratios back into 'normal space', and by subtracting one leaves the residual, which is the bias, expressed as a percentage.

$$Bias\ (\%) = (e^{\,log(ratio)} - 1) \times 100 \quad (2)$$

Significance testing

Statistical tests should then be used to determine whether any calculated biases between CPA and FA are statistically significant (Table 1). The log-ratio t-test (Tickner, Lannan and Preston, 2021) and Wilcoxon Signed-Rank test (Wilcoxon, 1945; Napier-Munn, 2014) can both be used for the accuracy-significance testing of CPA and FA data, and are both simple to implement using Excel. The log-ratio t-test can be used if the bias is multiplicative in nature and provides more interpretable effect sizes (see below), whereas the Wilcoxon Signed-Rank test is useful when the data have extreme outliers and where the bias is additive in nature, but ignores the magnitude of the differences. The authors recommend using the log-ratio t-test in the first instance and use the Wilcoxon Signed-Rank test to double check results.

TABLE 1

Overview of statistical tests for accuracy suitable for comparison of CPA versus FA measurements.

Test name	Null hypothesis	Assumed data distribution	Unit effect type	Test formula	Effect size
log-ratio t-test (two-sided)	Mean of logarithm of CPA/FA ratios is zero	ratio: lognormal (asymmetrical) log(ratio): normal (symmetrical)	(Log-) ratio*multiplicative*	$Y_{LR} = \log_e\left(\frac{Y_{CPA,i}}{Y_{FA,i}}\right)$ $t = \frac{\bar{Y}_{\mathbf{LR}}}{\hat{\sigma}_{Y_{LR}}/\sqrt{n}}$	$Cohen's\ d = \frac{\bar{Y}_{LR}}{\hat{\sigma}_{Y_{LR}}}$
Wilcoxon Signed-Rank test	Sum of ranks of positive differences equals sum of ranks of negative differences	Non-Normal (asymmetrical)	Rank *additive*	$R_i = rank(\lvert D_i \rvert),$ $W^+ = \sum_{i:D_i>0} R_i,$ $W^- = \sum_{i:D_i<0} R_i$ $W = \sum_{i=1}^{n} sign(D_i) \times R_i$	$r_{rb} = \frac{W^+ - W^-}{W^+ + W^-}$

Notes. D = difference; $\hat{\sigma}$ = sample standard deviation (as estimation of the population standard deviation σ); R_i = rank score of the absolute paired difference; W^+ = sum of the positive ranks; W^- = sum of the negative ranks; $sign(D_i)$ = indicator if the difference is positive (= 1), zero (= 0) or negative (= -1); r_{rb} = rank-biserial correlation (effect size of Wilcoxon); $\log_e$ = natural logarithm; Y_{LR} = pair log-ratio; n = number of paired observations. Overline above value ($\bar{Y}_{LR}$) indicates the sample mean.

The log-ratio t-test evaluates whether the CPA/FA ratio is equivalent to 1, since $\log_e(1) = 0$. The Wilcoxon Signed-Rank test evaluates whether the distribution of the difference between paired CPA/FA data is symmetric around a central value, and tests whether this centre value differs significantly from zero. The results of the selected statistical tests are assessed by evaluating the resulting *p*-value against the selected threshold ($\alpha = 0.05$).

In addition to determining whether a bias is statistically significant, the effect size can be used as a quantitative reflection of the magnitude of a phenomenon (Kelley and Preacher, 2012). There are many different effect sizes, the most well-known one being the correlation coefficient (Pearson's r), and the coefficient of determination (R^2). For the log-ratio t-test, the effect size is commonly reported as Cohen's *d* (Cohen, 1988) and represents a measure of the standardised difference between two means, calculated as the logarithm of the mean ratio (here between CPA and FA measurements) divided by the standard deviation of those differences (eg Lakens, 2013; Hassan *et al,* 2020). Cohen's *d* measures how many standard deviations the mean of the log ratios deviates from a hypothesised difference. Under the null hypothesis, the mean of the log ratios is assumed to be zero, ie no difference between CPA and FA ($\mu_D = 0$).

The effect size metric is important as it can further help with the discussion around the materiality of any differences observed. It can be appraised by comparison against common qualifications used in the literature (Table 2); however, these conventions should be considered as general guidance and do not dictate precise or required interpretations (Thompson, 2007; Lakens, 2013).

TABLE 2

Thresholds for interpreting (absolute) Cohen's *d* effect size (Cohen, 1992).

Effect size	| Cohen's *d* |
small	>0.20
medium	>0.50
large	>0.80

Beyond an average bias and effect size calculation, a graphical examination of effect sizes across grade ranges provides valuable insight into the relevance of the conditional nature of the differences between analytical methods. The authors recommend applying one-sample t-tests on the centred moving averages of the log-ratios using a moving window of 101 paired observations. For each window, Cohen's d can be calculated based on the log-ratio test. The number 101 is a reasonable choice, because for 101 observations per window (100 degrees of freedom), a two-sided t-test becomes significant at α = 0.05 when effect size is small (Cohen's d > 0.20 or < -0.20, see Equation 3):

$$\left|t_{d=.20;n=101}\right| = |d| \times \sqrt{n} = .20 \times \sqrt{101} = 2.01 > t_{crit,.975,df=100} = 1.985 \qquad (3)$$

The question then is: how large does the total data set of duplicates (N) above 3 × LOQ need to be to get a meaningful window length of n pairs to result in a statistically reliable assessment of bias across the total grade range? To determine this, simulations were performed following the approach of Tickner, Lannan and Preston (2021). Paired assay data ($N_1 = N_2 = 50$) were generated for a log-normal grade distribution with σ = 2, yielding 14.1 per cent of geometric pair means exceeding 10 units after filtering cases under 0.01 ppm. The authors recommend that more than half the window width (n = 101) lies above this threshold, meaning at least 51 observations. With higher grades being less frequent, this requires a minimum data set of 362 duplicate pairs (≥LOQ) to adequately evaluate grade-dependent bias. For initial scoping studies where lower confidence is accepted, we recommend N ≥ 121 and window length n = 33.

The combination of the centred moving average trend line (Figure 6, left), the plotting of the quartiles of the bias calculations (Figure 6, centre) and Cohen's d across the grade range (Figure 6, right) allows a more-informed interpretation of the bias across the grade range. It allows users to quickly determine in which range the bias is statistically significant and relevant, as per their effect sizes (ie small, medium, large).

It should be noted that due to the symmetrical nature of the centred moving average, information is lost at the minimum and maximum grades. Therefore, edge truncation is applied, and the test is performed on the available observations within each window (keeping at least half of the window's observations).

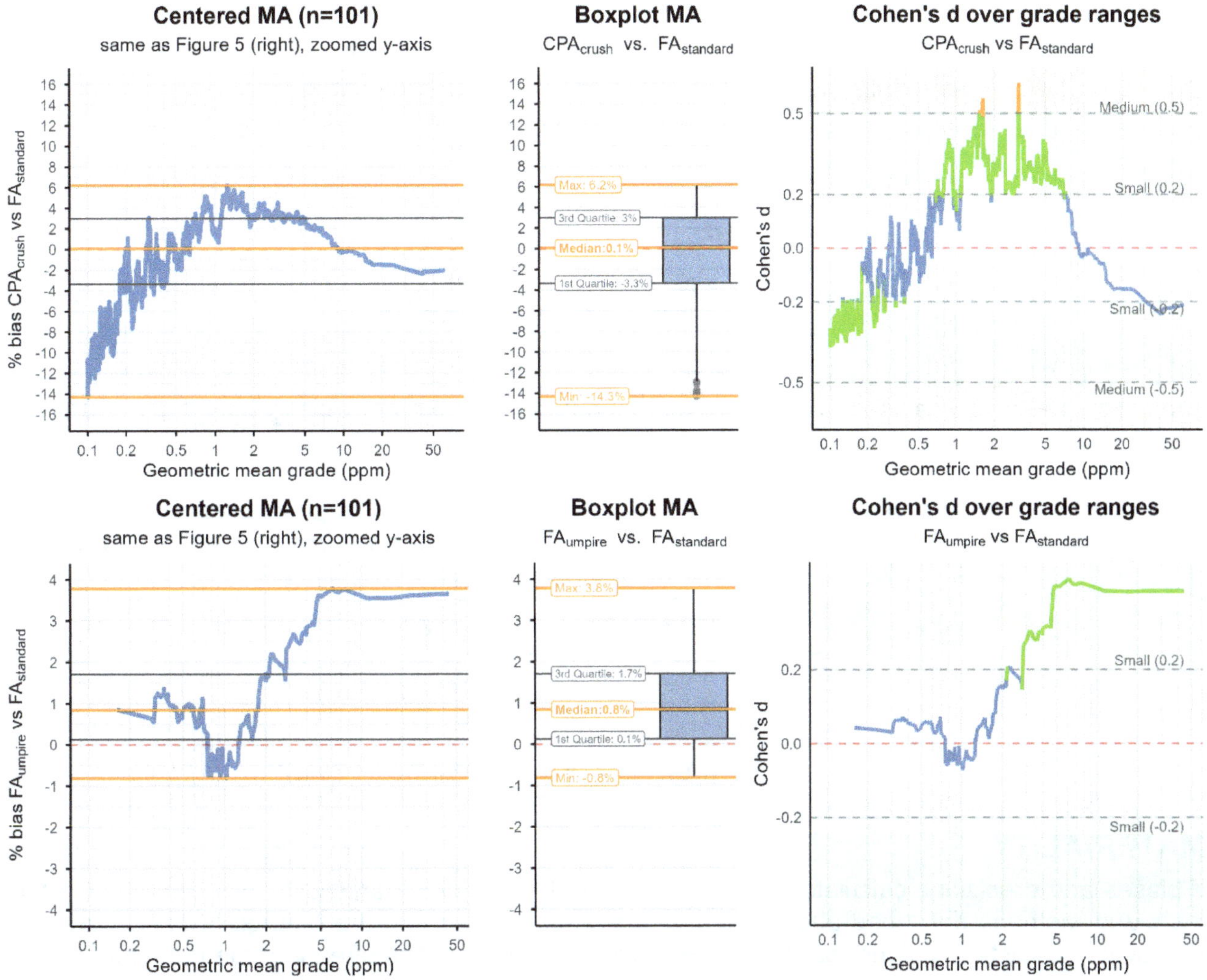

FIG 6 – Visual inspection of mean bias and effect size (Cohen's d) across grade levels for CPA_{crush} versus $FA_{standard}$ (top) and FA_{umpire} versus $FA_{standard}$ (bottom). Left: centred moving averages (MA) with a moving window of 101 observations (edge-truncated). Centre: range and quartile distribution of values by mean grade. Right: effect size and significance across mean grade levels (green (small effect) and orange (medium effect) segments of the line indicate $p < 0.05$, two-sided log-ratio test).

Precision

The precision, expressed as the CV, is calculated from the results of pairwise analysis of CPA duplicate pairs and FA duplicate pairs, using the RMS-CV approach (equation 1), and for pairs with an average grade above 3 × LOQ (Table 3). Precision should be calculated separately for each geological domain. A decision on whether the CPA precision is fit for purpose can then be made based on a simple table-based comparison of precision between CPA crushed, pulverised and FA analyses (Table 3). If the CV for the CPA is better (lower) than that of FA, then it can be stated that the precision of CPA is fit for purpose.

TABLE 3

Precision for three duplicate data groups. In this case, the CPA precision is fit for purpose. NB: These results are from one study and not indicative of broader findings.

Duplicate group	Precision
CPA_{coarse}	12.1%
CPA_{fine}	8.7%
FA	12.5%

If the CV for CPA is similar but slightly higher than that of FA (Table 4), then a Feltz-Miller test (Feltz and Miller, 1996) can be used to test whether that difference is statistically significant at $\alpha = 0.05$.

TABLE 4

Precision for three duplicate data groups. In this case, whether the CPA precision is fit for purpose depends on the outcome of statistical testing. NB: These results are from one study and not indicative of broader findings.

Duplicate group	Precision
CPA_{coarse}	11.8%
CPA_{fine}	9.7%
FA	11.2%

MATERIALITY

If biases and precisions calculated between CPA and FA are not statistically significant using the tests and methods described above, then it can be concluded that the method is fit for purpose. However, if statistically significant biases *are* found (which happens often in large data sets and is the rule rather than the exception), or when CPA precision values are statistically significantly higher than FA, this does not necessarily mean that CPA is not fit for purpose. This presents a practical problem that requires further assessment (see earlier comments and literature references in the section on 'Quality and Materiality').

An appraisal of materiality of any observed differences can be made by comparison of the results against a specific data quality objective/threshold (DQO) rather than testing for parity using the methods described above.

One of the key challenges with studies of this kind is that often no objective is set before the testing is completed (ie a DQO/tolerance threshold does not exist). Industry benchmarks are equally difficult to find, because reporting codes such as the JORC Code (2012) leave such judgments for the Competent Persons to decide, and papers, such as the one by Abzalov (2008) on benchmarks for precision, are rare. If it is not clear what the thresholds are before the test is started, then it is difficult to expect practical conclusions from any significance test.

However, for both accuracy and precision, there are some options that practitioners can draw from to define a DQO, and the authors recommend the following options:

- Accuracy and precision thresholds/tolerances, as set by the geologists (Qualified/Competent Persons) managing the technical aspects of the programme.
- Industry benchmarks (for instance Table 4 in Abzalov (2011) for precision).
- Inferences of DQOs from other routine FA tests (CRM results and routine umpire tests).

Accuracy

In the absence of a DQO set by the geologist, or DQOs from industry benchmarks for accuracy (the authors are not aware of any), an 'accuracy framework' can be established on the basis of the overall performance of FA CRMs, tested over a long period, as part of the routine FA quality monitoring process. For example, if a bias ranging from -1.0 per cent to +3.0 per cent was determined for eight different CRMs that cover the grade range, and the project or mine has been operating successfully and consistently with these long-accepted analytical biases, this range of bias provides a solid framework of reference for the assessment of the materiality of other analytical biases. In simple words: if one would accept a bias of +3.0 per cent for routine CRMs, it would be reasonable to also accept a bias of 3.0 per cent for CPA versus FA. The performance of CRMs can be included as single points in the Tukey plot (Figure 7, left), or presented by a box and whisker plot, showing the median, quartiles and min-max values of all available CRMs over a long period of time (Figure 7, right).

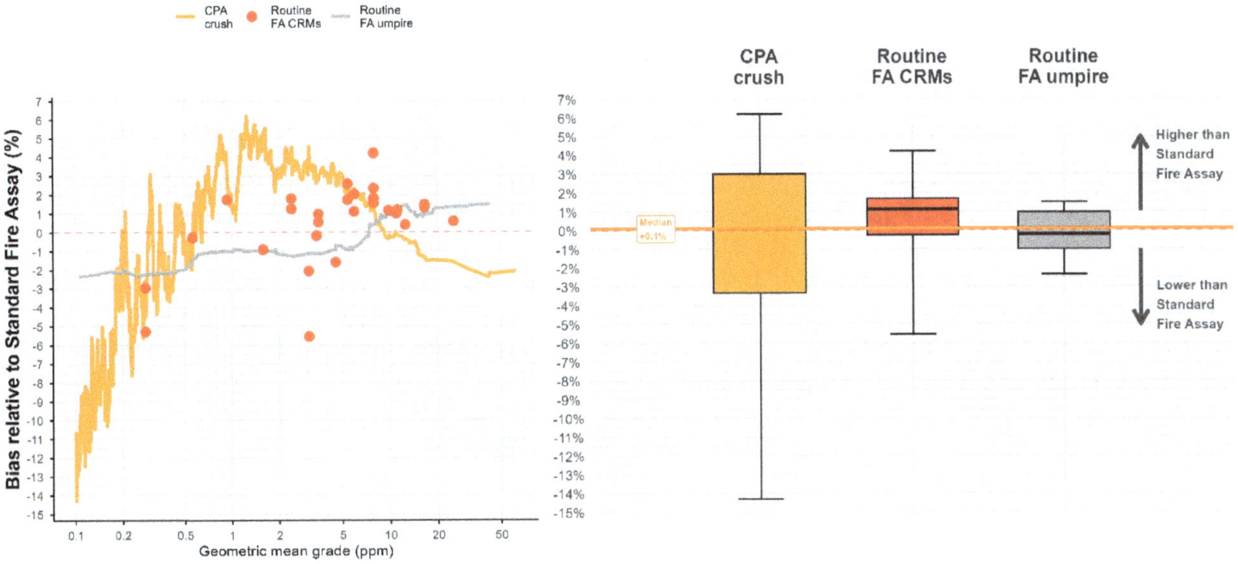

FIG 7 – 'Accuracy Framework', showing CPA versus FA, CRMs for standard FA, and FA umpire against standard FA. The boxplots for CPA versus FA and FA_{umpire} versus $FA_{standard}$ represent the distribution of the (geometric) moving average (N = 101, edge-truncated), expressed in percentage difference. Horizontal dashed line is the standard fire assay '0 per cent baseline' that is compared to. Thick orange line (right) is the median for the CPA analysis.

Similarly, the bias determined for the routine FA laboratory process, relative to the overall performance of regular umpire testing, can be referenced in this accuracy framework. If the project or mine has been operating successfully and consistently within, say a -2.5 to +1.0 per cent umpire laboratory bias, then by inference, such a bias becomes acceptable. The testing programme recommended in Figure 1 includes a set of data from a second FA laboratory, in case routine umpire testing is not available for the project. The results of the umpire testing (ie FA_{umpire} versus $FA_{standard}$) can be plotted in a similar fashion as the CPA versus FA and included as a second line in the Tukey plot (Figure 7, left), or by a box and whisker plot, showing the median, quartiles and min-max values (Figure 7, right).

A pragmatic way to review the biases for CPA versus FA, CRMs, and FA_{umpire} versus $FA_{standard}$ together is by plotting them all together as box and whiskers (Figure 7, right). This, of course, obfuscates the trend of the bias across the grade range, but at least allows the practitioner to set reasonable DQOs from this framework to test against.

An example of the accuracy framework is presented in Figure 7 for a fictitious data set. The routine FA performance based on FA CRMs and umpire FA testing demonstrates a median bias of +1.0 per cent and -0.1 per cent, respectively. The range of biases observed by these two references (as defined by the quartiles) is -1.0 per cent to +1.5 per cent, and -1.0 per cent and +1.0 per cent, respectively. The geologist can use these defined metrics to specify sensible DQOs; for instance, in this case it could be set between -1.0 per cent and +1.5 per cent, based on the interquartile ranges for CRMs and routine umpire testing (Figure 7, right).

When DQOs are then set (eg 'the bias must not be outside -3.0 per cent or +3.0 per cent'), based on either industry norms, Competent/Qualified Person requirements, or an evaluation of routine CRM results and umpire test results (this section), two complementary statistical approaches can then be used to make statistically objective claims: (a) one can examine whether the 95 per cent confidence interval of the log ratio (derived from the log t-test) contains the threshold values (eg log(1.03) or log(1/1.03) for a 3.0 per cent difference); or (b) one could use a formal TOST (two one-sided tests) procedure to test for equivalence within a specified range (Napier-Munn, 2014).

Precision

Similarly, other than identifying statistically significant differences between CPA and FA pulp duplicate (precision) performance, further information on the materiality of any difference in the precision of CPA and FA results can be drawn through evaluation against industry benchmarks (Table 3 on page 143 in Abzalov (2013)), or against a DQO set by the geologist (ie 'the precision must be better than 15 per cent'). In other words, the CPA precision may be statistically significantly different than the routine FA pulps, but still good enough as it is well within those thresholds.

To further test the impact and materiality of the CPA precision, experimental variograms based on FA data can be compared against variograms based on CPA data — this is only possible if the test data set is large enough. If the experimental variograms (downhole, major, semi-major, and minor) demonstrate comparable values for each lag value (Figure 8), they would result in near-identical modelled variograms. Therefore, the weights applied during grade interpolation (eg kriging) would result in near-identical estimates. If CPA is used to analyse plant samples, then a similar exercise on a time-based variogram can be performed.

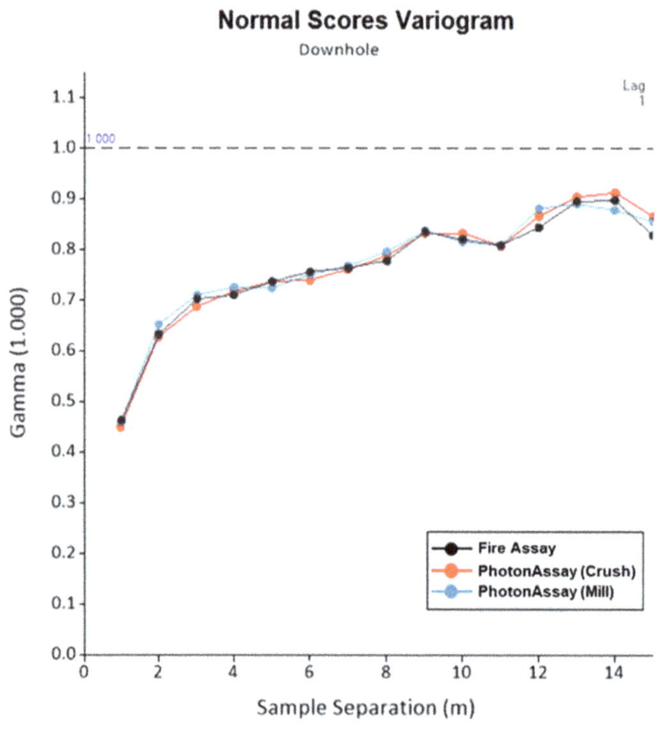

FIG 8 – Example of very similar experimental variograms based on crushed CPA, pulverised CPA, and FA data, demonstrating that results from all three analyses (ie CPA_{crush}, $CPA_{pulp\ (mill)}$ and FA) would result in near-identical modelled variograms, and thus in near-identical mineral resource estimates (from a *precision* perspective).

This approach can also be used to test the impact of using crushed versus pulverised samples for CPA (Figure 8). For example, the authors note that even though an RMS CV difference of 12 per cent for FA versus 8.7 per cent for pulverised CPA appears significant, and appears to justify the cost of further pulverising the CPA samples, it is easy to point out that it has a negligible impact on the resource estimate, and therefore likely not worth the additional expense.

FINAL REMARKS

Whether CPA data are fit for purpose for a project should be evaluated in the context of the deposit geology, its objectives, and associated DQOs. The workflow presented in this paper provides practical steps and highlights critical items for consideration when deciding between CPA and conventional methods.

Conclusions on accuracy and precision, and suitability of CPA to replace FA, can only be drawn when clear DQOs are agreed to before the study starts. If third parties are requested to undertake a comparative study, then they can only determine outcomes if they have been provided with such DQOs by the client.

Last, the authors note that fire assaying itself is not a perfect method, and comparing test results from a single laboratory should be treated with caution. Any observed differences need to always be regarded in that light, and DQOs set accordingly. Dominy *et al* (2024), in their conclusions on CPA review, boldly state that 'coarse-gold assaying with FA is flawed', and make reference to work by Royle (1989), Pitard and Lyman (2013), Dominy (2014, 2017), Lyman, Robertson and Day (2016), and Pitard (2017) and Dominy *et al* (2024) to support that statement. A bias at high-grade was also observed by Hitchman *et al* (2024) in their review of the performance of FA versus CPA at the very nuggety Fosterville deposit in Australia, with CPA results being higher than FA at high grades. Screen-fire test work confirmed that this bias was introduced by the sub-sampling of the pulp using routine low charge weights for the fire assay; a well-known issue.

REFERENCES

Abzalov, M, 2008. Quality control of assay data: a review of procedures for measuring and monitoring precision and accuracy, *Exploration and Mining Geology*, 17:131–144.

Abzalov, M, 2011. Sampling Errors and Control of Assay Data Quality in Exploration and Mining Geology, in *Applications and Experiences of Quality Control* (ed: O Ivanov), chapter 31 (IntechOpen: London). https://doi.org/10.5772/14965

Abzalov, M, 2013. Measuring and Modelling of Dry Bulk Rock Density for Mineral Resource Estimation, *Applied Earth Science*, 122(1). https://doi.org/10.1179/1743275813Y.0000000027

Chrysos, 2022. Chrysos PhotonAssay™ Measurement Performance for PAAU02, Performance Note PN-001, Chrysos Corp. internal report, p 2.

Chrysos, 2023. PhotonAssay Measurement Performance for Gold – Impact of, U, Th and Ba, Technical Note TN-106:6, Revision 2, p 6.

Chrysos, 2025. Emissions, Energy and Waste Assessment – PhotonAssay and Fire Assay, GHD internal report, p 51.

Cohen, J, 1988. The effect size, *Statistical Power Analysis for the Behavioral Sciences*, pp 77–83.

Cohen, J, 1992. A Power Primer, *Psychological Bulletin*, 112:155–159.

Dominy, S C and Minnitt, R C A, 2012. Application of Heterogeneity Testing to Coarse Gold Ores, in *Proceedings of the Sampling Conference 2012*, pp 143–160 (The Australasian Institute of Mining and Metallurgy: Melbourne).

Dominy, S C, 2014. Predicting the unpredictable – evaluating high-nugget effect gold deposits, *Mineral Resource and Ore Reserve Estimation Guide to Good Practice*, pp 659–678 (Australasian Institute of Mining and Metallurgy: Melbourne).

Dominy, S C, 2017. Sampling coarse gold-bearing mineralisation – developing effective protocols and a case study from the Ballarat mine, Australia, in *Proceedings of the World Conference on Sampling and Blending*, pp 71–84 (Australasian Institute of Mining and Metallurgy: Melbourne).

Dominy, S C, Graham, J C, Esbensen, K H and Purevgerel, S, 2024. Application of PhotonAssay™ to Coarse-Gold Mineralisation – The Importance of Rig to Assay Optimisation, *Sampling Science and Technology*, January 2024(1):2–30. https://doi.org/10.62178/sst.001.002

Feltz, C J and Miller, G E, 1996. An asymptotic test for the equality of coefficients of variation from k populations, *Statistics in Medicine*, 15:11.

Gelman, A and Stern, H, 2006. The difference between 'significant' and 'not significant' is not itself statistically significant, *The American Statistician*, 60:328–331.

Hassan, A, Huang, D, Mustafa, E K, Mahama, Y, Damos, M A, Jiang, Z and Zhang, L, 2020. Statistical inference and residual analysis for the evaluation of datum transformation models developed on 3D coordinate data, *Journal of Applied Geodesy*, 14:65–75.

Hitchman, S P, Simbolon, S, Symons, D J, Hoare, J P and Carpenter, J B, 2024. Implementation of Chrysos Photon Assay Method at Fosterville Gold Mine, Victoria, in *Proceedings of the International Mining Geology Conference*, pp 341–345 (Australasian Institute of Mining and Metallurgy: Melbourne).

Hyslop, N P and White, W H, 2009. Estimating precision using duplicate measurements, *Journal of the Air and Waste Management Association*, 59:1032–1039.

JORC, 2012. Australasian Code for Reporting of Exploration Results, Mineral Resources and Ore Reserves (The JORC Code) [online]. Available from: <http://www.jorc.org> (The Joint Ore Reserves Committee of The Australasian Institute of Mining and Metallurgy, Australian Institute of Geoscientists and Minerals Council of Australia).

Kelley, K and Preacher, K, 2012. On Effect Size, *Psychological Methods*, 17:137–152. https://doi.org/10.1037/a0028086.

Lakens, D, 2013. Calculating and reporting effect sizes to facilitate cumulative science: a practical primer for t-tests and ANOVAs, *Frontiers in Psychology*, 4:62627.

Long, S, 2015. What reporting Portable XRF or any other Assay Results to 2012 JORC Code Standards should mean, *AIG News*, pp 11–16.

Lyman, G J, Robertson, I G and Day, T, 2016. The simple facts about sampling gold ores, in *Proceedings of Mill Operators Conference 2016*, pp 159–172 (Australasian Institute of Mining and Metallurgy: Melbourne).

Miller, J E, 2023. Beyond Statistical Significance: A Holistic View of What Makes a Research Finding "Important", *Numeracy*, 16:6.

Minnitt, R C A, 2016. Calibrating K and Alpha in Gy's Formula: A New Approach, *Mathematical Geosciences*, 48:211–232.

Napier-Munn, T, 2014. Statistical methods for mineral engineers – How to design experiments and analyse data, Queensland Australia, report, Julius Kruttschnitt Mineral Research Centre.

Peeters, M J, 2016. Practical significance: Moving beyond statistical significance, *Currents in Pharmacy Teaching and Learning*, 8:83–89.

Pitard, F F and Lyman, G J, 2013. Single and multi-stage Poisson processes: a case study for gold exploration, in *Proceedings of the World Conference on Sampling and Blending*, pp 371–385 (Gecamin: Santiago).

Pitard, F F, 2013. Guidelines for acceptable allotted sampling uncertainty, in *Proceedings of the World Conference on Sampling and Blending*, pp 89–98 (Gecamin: Santiago).

Pitard, F F, 2017. From errors to uncertainty – a clarification for proper use by the Theory of Sampling, in *Proceedings of the World Conference on Sampling and Blending 2017*, pp 355–358 (Australasian Institute of Mining and Metallurgy: Melbourne).

Royle, A G, 1989. Splitting gold assay pulps containing coarse gold, *J Leeds Univ Min Assoc*, 89:63–68.

Smee, B, Bloom, L, Arne, D and Heberlein, D, 2024. Practical applications of quality assurance and quality control in mineral exploration, resource estimation and mining programmes: a review of recommended international practices, *Geochemistry: Exploration, Environment, Analysis*, (2024)24(2):geochem2023-046. https://doi.org/10.1144/geochem2023-046

Stanley, C R and Lawie, D, 2007. Average relative error in geochemical determinations: Clarification, calculation and a plea for consistency, *Exploration and Mining Geology*, 16:267–275.

Sterk, R, 2015. Quality control on assays: addressing some issues, *AusIMM New Zealand Branch Annual Conference, Dunedin*.

Thompson, B, 2007. Effect sizes, confidence intervals and confidence intervals for effect sizes, *Psychology in the Schools*, 44:423–432.

Tickner, J, Ganly, B, Lovric, B and O'Dwyer, J, 2017. Improving the sensitivity and accuracy of gamma activation analysis for the rapid determination of gold in mineral ores, *Applied Radiation and Isotopes*, 122:28–36.

Tickner, J, Lannan, O and Preston, R, 2021. Comparing different assay methods for the determination of mineral ore grade, *Journal of Geochemical Exploration*, 226:106780.

Tremblay, C D, Tickner, J, Treasure, D, Oteri, A and Wheeler, G, 2019. PhotonAssay – efficient and bulk gold analysis in the modern world, in *Proceedings of the International Mining Geology Conference 2019*, pp 88–98 (Australasian Institute of Mining and Metallurgy: Melbourne).

Westgard, J O, Barry, P L, Hunt, M R and Groth, T, 1981. A multi-rule Shewhart chart for quality control in clinical chemistry, *Clinical Chemistry*, 27:493–501.

Wilcoxon, F, 1945. Individual comparisons by ranking methods, *Biometrics Bulletin*, 1:80–83.

Seeing through the shotcrete – enhancing geological modelling with FaceCapture Mapping at Renison

J van Balen[1], M Recklies[2], B Ridgers[3], C Carter[4] and R Detert[5]

1. Production Geologist, Bluestone Mine Tasmania Joint Venture Pty Ltd, Renison Bell Tas 7469. Email: johanna.vanbalen@bluestonetin.com.au
2. Resource Geologist, Bluestone Mine Tasmania Joint Venture Pty Ltd, Renison Bell Tas 7469. Email: martin.recklies@bluestonetin.com.au
3. Senior Production Geologist, Bluestone Mine Tasmania Joint Venture Pty Ltd, Renison Bell Tas 7469. Email: ben.ridgers@bluestonetin.com.au
4. Manager of Resource Development and Planning. Bluestone Mine Tasmania Joint Venture Pty Ltd, Renison Bell Tas 7469. Email: colin.carter@bluestonetin.com.au
5. Manager of Partner Technical Services, Mine Vision Systems, Inc., Pittsburgh PA 15206, USA. Email: ryan.detert@minevisionsystems.com

ABSTRACT

The Renison Mine, situated on Tasmania's West Coast, is owned and operated by Bluestone Mines Tasmania Joint Venture Pty Ltd. Operating as Australia's largest primary tin producer the deposit was initially discovered in 1890, with modern underground production beginning in the late 1960s and continuing to present day. The mine reaches a depth greater than 1200 m below surface and currently has a projected mine life of 10 years. The Renison Mine is geologically complex and, at depth, ground conditions have become more dynamic. This has necessitated the installation of dynamic ground support that includes frequent shotcreting of headings to mitigate seismic risks, resulting in more variable and narrower windows for collecting geological data such as backs mapping, structural information and heading photos. Mine Vision Systems (MVS) FaceCapture Mapping (FC Mapping) was recently introduced into the underground mine geology workflow as a means of collecting geological information through a high-resolution 3D georeferenced laser scan in this increasingly time constrained environment. FC Mapping combines LiDAR scanning with high-resolution photography.

This is the first use of MVS technology in Australia, and to date FC Mapping is proving valuable with scans being efficiently integrated into existing geological modelling processes. The MVS system is a portable unit with all scan processing performed within the unit in real time. Resultant file sizes are small, adding to the flexibility of integration to Leapfrog and Surpac.

The introduction of the FC Mapping has not come without challenges. Strategic considerations include, underground conditions, integration into the underground operational workflow, reference point error, and understanding the quality of scans required for geological modelling versus geological backs mapping. This paper provides a summary of Renison Mine's integration of the FC Mapping, how this process has contributed to our geological decision-making, the challenges encountered, improvement opportunities, and future works envisioned.

INTRODUCTION

The Renison Mine (Renison), located on Tasmania's West Coast, is Australia's largest primary tin producer, with underground operations extending to depths exceeding 1200 m and a projected mine life of approximately 10 years (Metals X Limited, 2025a, 2025b). Renison is owned and operated by Bluestone Mines Tasmania Joint Venture Pty Ltd (BMT), a 50/50 partnership between Metals X Ltd and YTPAH Pty Ltd (Bluestone Mines Tasmania Pty Ltd, 2023). Discovered back in the 1890s (Kitto, 1994; Pink and Crawford, 1996), the economic potential of the area was only partially realised with some surface extraction occurring prior to underground mining commencing approximately 70 years later in the 1960s (Pink and Crawford, 1996). Fast forward to today and Renison is a globally significant tin producer estimated to supply approximately 3 per cent of the world's tin (Bluestone Mines Tasmania Pty Ltd, 2023; Metals X Limited, 2025b) with the 2024 production of 11 006 t of tin-in-concentrate setting a new Renison production record (Metals X Limited, 2025c).

The geological environment at Renison is notably complex, characterised by multiple mineralisation styles, structurally controlled ore zones, and highly variable lithologies encountered during ongoing operations (Kitto, 1994; Pink and Crawford, 1996). This complexity presents significant challenges for the collection of geological data. As mining progresses to greater depths, the risk of seismic events increases due to a decline in rock mass quality and an increase in induced stress. Consequently, the importance of accurate geological mapping and data collection becomes critical for ensuring resource confidence, supporting mine planning, and informing operational decision-making processes. In response to increasingly dynamic ground conditions at depth, shotcrete is routinely applied as part of Renison's Ground Control Management Plan. While effective for ground control, this practice imposes time constraints and restricts opportunities for geological data collection within narrower operational time frames as shotcreting occurs in-cycle with development.

To overcome these challenges, FC Mapping has been implemented at Renison to facilitate the acquisition of high-resolution, georeferenced 3D scans of mine headings. This technology enables the collection of detailed structural and lithological data, which can be efficiently incorporated into the mine's existing geological workflows.

This paper details the process of integrating FC Mapping into Renison's underground mining operations, examining its influence on the quality of geological data and modelling. It also discusses the challenges encountered during implementation and the subsequent improvements made to enhance operational efficiency and strengthen resource confidence.

GEOLOGICAL SETTING

Regional geology

The Renison deposit lies within the north–south trending Palaeozoic Dundas Trough, bordered by the Proterozoic metasediments of the Tyennan Block to the east and the Rocky Cape Block to the west (Denholm, 2018; Kitto, 1994; Seymour, Green and Calver, 2006; Solomon, 1981). Tin mineralisation is associated with the late Neoproterozoic to Palaeozoic sedimentary and volcanic sequences of the Success Creek Group and Crimson Creek Formation, which were brittlely deformed during the Devonian intrusion of the Pine Hill Granite. This intrusive event produced the Federal–Bassett Fault (The Fed), the major structural control on mineralisation at Renison (Denholm, 2018; Kitto, 1994).

Mine sequence geology

The mine sequence at Renison is situated within the upper approximately 80 m of the Success Creek Group and the lower approximately 60 m of the overlying Crimson Creek Formation (Bluestone Mines Tasmania Pty Ltd, 2024b; Kitto, 1994; Solomon, 1981). The Success Creek Group, which forms the oldest part of the sequence, originated in subaerial to shallow marine environments. This group is composed of mudstone, sandstones, siltstones, shales, conglomerates, greywacke, and carbonates (Denholm, 2018).

Overlying the Success Creek Group is the Crimson Creek Formation which includes siltstone, sandstone, fragmental tuffaceous material, conglomerates, cherty-iron formations, carbonates, and volcanic detritus (Denholm, 2018; Kitto, 1994). The thickness of the mine sequence is variable, ranging from approximately 50 m to 200 m, but typically averages around 140 m (Bluestone Mines Tasmania Pty Ltd, 2024b).

Located near the top of the mining sequence, The Fed serves as the primary structural control for mineralising fluids, which tend to preferentially alter dolomitic units (Denholm, 2018; Kitto, 1994). It is important to note that while The Fed does not always host mineralisation, it is frequently associated with mineralised zones (Denholm, 2018; Kitto, 1994). Figure 1 demonstrates an overview of the mine sequence geology.

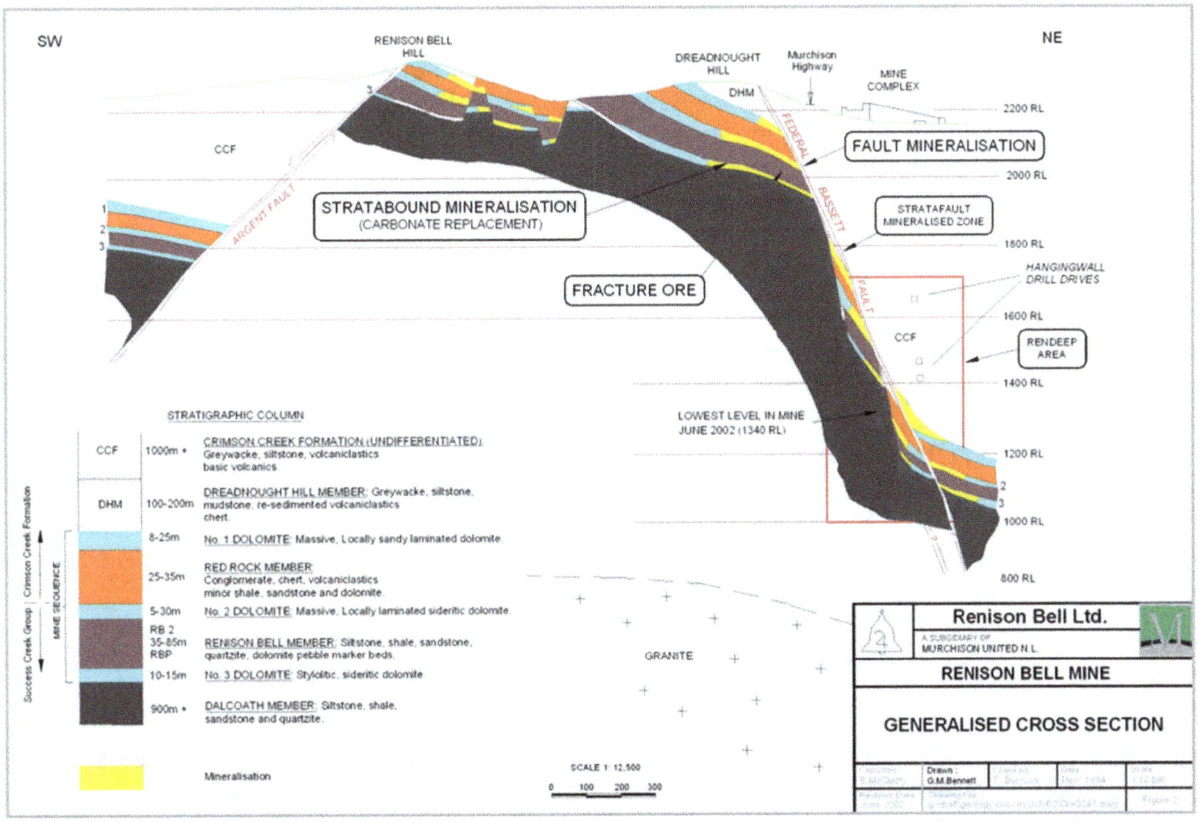

FIG 1 – Renison (formerly, Renison Bell Ltd) Generalised Cross-section showing stratigraphic sequence.

Renison has traditionally been classified as a distal skarn tin deposit. Within the mining sequence, four distinct styles of mineralisation are recognised (Kitto, 1994):

- **Skarn Mineralisation** – historically the primary mineralisation style, characterised by replacement of dolomite horizons by massive to semi-massive cassiterite and pyrrhotite ± pyrite, arsenopyrite, talc, siderite, calcite, and quartz; cassiterite is the primary tin-bearing mineral. Generally, with increasing depth skarn mineralisation becomes less prominent as iron and sulfur decreases, inversely talc content and magnesium increases.
- **Fault Mineralisation** – whilst The Fed is the primary mineralised fault at Renison, second-order faults can host mineralisation, including structures splaying off The Fed. This style contains lower pyrrhotite and cassiterite content than skarn mineralisation but proportionally more quartz ± copper, arsenopyrite, bismuth, fluorite and tourmaline. Tin is present by association to the fault boundary.
- **Fracture Mineralisation** – generally occurs within the unmineralised clastic units as a series of fractures, or breccias, comprised of quartz-pyrrhotite veins and/or disseminated pyrrhotite veins. Typically, arsenopyrite, bismuth and tourmaline levels are elevated too. Tin is present similarly to fault mineralisation.
- **Stratafault Mineralisation** – a hybrid style occurring in zones of complex faulting where two or more subparallel structures converge, resulting in mineralisation of units that are not typically mineralised, from dolomitic to non-dolomitic lithologies. This style is often irregular and is more common at depth, correlating with increased structural complexity.

Given the complexity of the geology and the variability in mineralisation styles at Renison, accurate geological mapping is fundamental to understanding ore distribution. As mining progresses to greater depths, geological conditions become increasingly complex, which further underscores the need for precise mapping techniques.

MAPPING CHALLENGES AT DEPTH

Geological mapping plays a pivotal role in the collection of data at any mine site, with the timely and precise gathering of geological information essential for establishing resource confidence, informing mine design, and supporting ongoing operational planning. Historically at Renison, geological mapping of walls and backs was performed by direct inspection, where geologists would record lithology, structures and geotechnical observations at an active development heading. This approach was effective during shallower mining phases where ground conditions were more competent and headings could be visited repeatedly and with less time constraint, allowing for thorough data collection.

However, as mining has progressed to greater depths, the operational constraints have changed significantly. Deep mining at Renison is now characterised by significantly restricted access windows, a result of increased operational demands and more challenging ground conditions. While conventional mapping techniques are still utilised in certain situations, deeper mining with the escalation in ground stress, the decline in rock mass quality, higher levels of seismic activity, increased talc content, and increased structurally-controlled ground behaviour have necessitated the adoption of dynamic, energy-absorbing ground support systems that include in-cycle application of shotcrete (Bluestone Mines Tasmania Pty Ltd, 2024a).

Shotcreting generally occurs after a development heading has been bogged and prior to the installation of bolts and mesh. Whilst this significantly improves stability and safety, it decreases the time available for geologists to map the freshly exposed rock.

Prior to the implementation of FC Mapping technology, geological technicians at Renison would photograph development headings before the application of shotcrete. This process enabled geologists to preserve and reference some geological information, even after the rock surface was no longer directly visible and accessible.

Despite providing a visual record, relying solely on two-dimensional photographs presented significant limitations. The lack of depth in these images often obscured structural and lithological details, making it challenging to accurately interpret geological features. As the use of shotcrete became increasingly prevalent photo-based mapping gradually supplanted traditional mapping practices.

To overcome the shortcomings associated with two-dimensional imaging, FC Mapping was introduced. This advanced technology enables the capture of high-resolution 3D meshes and imagery, providing geologists with a much more detailed and accurate representation of the exposed rock. Other technologies were trialled at Renison in 2024 however the proposed changes to existing workflows, new data collection methods and file sizes were incompatible with the geology team's needs.

MVS FACECAPTURE TECHNOLOGY

Originating in the US, the MVS FC Mapping system is a handheld mobile LiDAR scanner designed to enhance geological data collection by integrating multiple sensors and producing highly detailed 3D data. This system captures data in .obj and .las file formats. This enables comprehensive digital mapping and modelling of mine headings and can be directly used in modelling software such as Leapfrog or Surpac. As a result, FC Mapping is now being adopted in regions like Canada, Australia, South Africa, New Zealand and South America.

System components and operation

The core of the FC Mapping system is its sensor head, which houses a LiDAR unit, a high-resolution camera, LED lighting, and onboard computing powered by a GPU processor. Operation is managed via a wirelessly connected rugged tablet, which allows users to visualise 3D data in real time during collection. Figure 2 depicts the complete set-up, individual components, and the system in practical use. Users can review collected data on-site and integrate the outputs with other 3D mapping software, enhancing workflow efficiency and data accessibility.

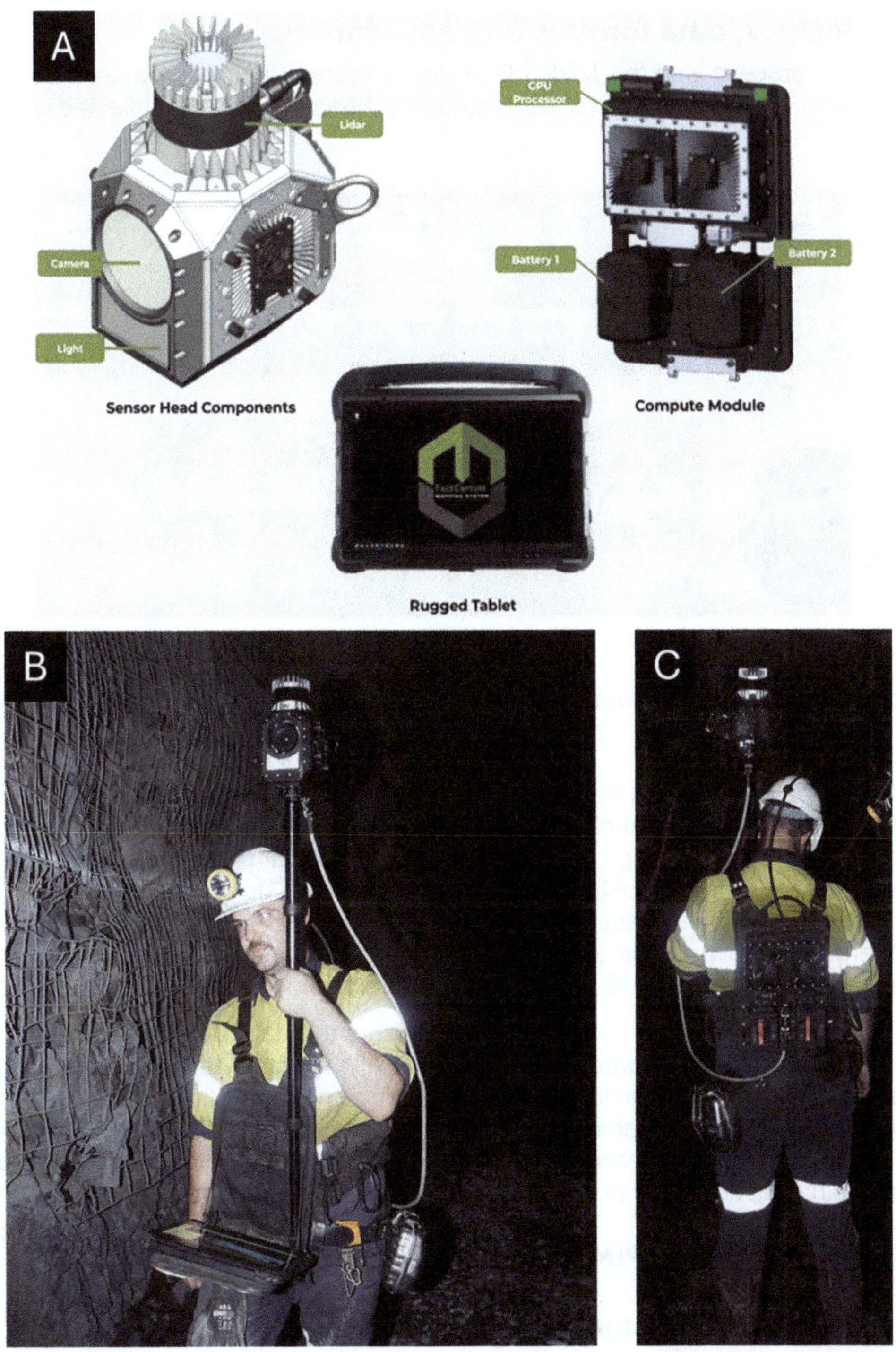

FIG 2 – Top (A) The MVS FC Mapping System components; Bottom (B to C) shows the system in use underground.

Data collection and georeferencing

To ensure spatial accuracy, the system utilises pre-existing survey control points for real-time georeferencing of collected data. These control points are referenced accurately in the underground space using known points and then are provided in CSV format which is then directly uploaded on the FC Mapping tablet. All georeferenced data is stored within the onboard computer for subsequent use in digital mapping and modelling tasks. Additionally, the system features onboard cloud-to-cloud alignment tools, referred to as 'alignment scanning', which allow scans to be collected over longer distances without requiring survey markers for every scan. Georeferenced base scans, which require survey markers, are still required within every 30 m to limit error propagation and maintain overall scan quality. This approach ensures reliable georeferencing while minimising disruptions to ongoing production activities.

Workflow efficiency, data formats and applications

The data collection process with the MVS FC Mapping system typically requires less than 10 mins per use. Frequently, a comprehensive 3D mesh of a heading can be collected in under 3 mins (Figure 3).

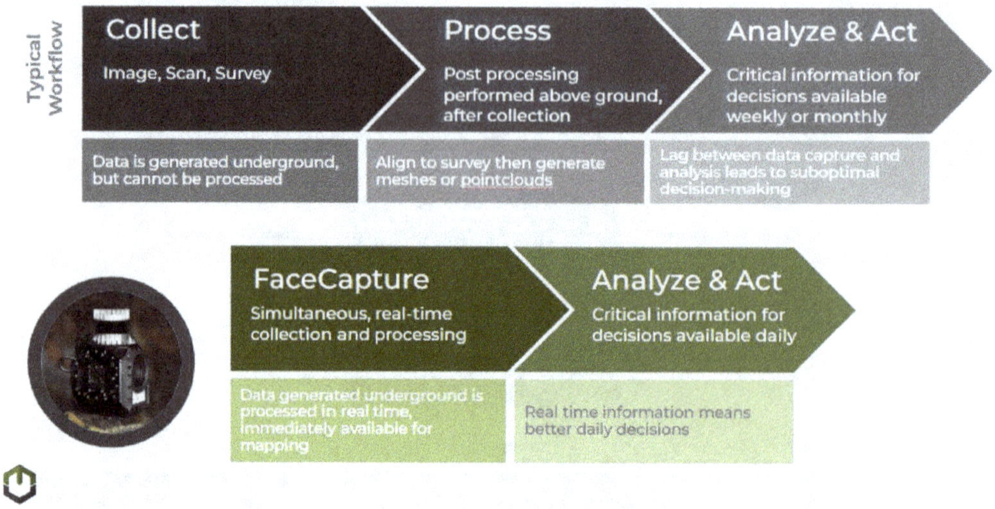

FIG 3 – A comparison of historic data collection workflow and updated workflow using the MVS FC Mapping System.

Data generated by the FC Mapping system exists in two principal formats. A 3D photo-overlaid mesh is produced from the captured point cloud of the development drive and exported as an.obj file. This mesh is used for the visual mapping of geological features and for taking orientation measurements. While traversing the heading, the LiDAR scanner also records a point cloud, which can be exported in .las, .laz, or .ply formats. These point clouds are used for volumetric calculations and for conducting overbreak and underbreak analyses. A typical scan yields a point cloud of less than 20 MB and a mesh of less than 5 MB, which can then be downloaded for further analysis.

Once underground data collection is complete at Renison, the data sets are uploaded via a shared network folder on the tablet. This process ensures that all collected data is readily available for use in digital mapping and geological interpretation, supporting operational planning and resource evaluation. While the Renison geology team is the primary user of the FC Mapping outputs, engineering and short- and long-term planning workgroups also benefit indirectly through activities such as mapping interruptions and updates to geological and mine models.

MVS FACECAPTURE MAPPING INTEGRATION

Implementation and scheduling

The MVS FC Mapping system was incorporated into the Renison underground geology workflow in April 2025. The initial focus was on integrating the technology into existing procedures with minimal disruption to operations. To optimise data quality and maintain efficiency, scans were scheduled once fired dirt was removed, after scaling and bogging of scats but before shotcreting, aligning with established underground development cycles. This ensured headings were safe, dust-free, and accessible for scanning. Where headings are scaled, washing down is already undertaken; however, where scaling is not required, washing down of headings must be completed as an additional step prior to FC Mapping. Geological technicians were already following established workflows for collecting photographs at this stage of the development mining cycle, so it was logical for them to carry out FC Mapping, integrating it naturally into this part of the process.

Scanning guidelines

Whilst numerous development headings are active each shift (approximately 10), only a subset of these require in-cycle shotcrete. Accordingly, specific guidelines were established:

- All Area 5 (A5) and Leatherwood (LD) ore development drives are to be FC Mapped after each advancement cut. Exceptions may occur for short rounds (approximately 2 m or less) or stripping activities where data overlap or duplication is likely. Development accesses are to be scanned where practical due to their potential to host mineralised pockets or significant structures, however these remain lower priority relative to other geological technician tasks. Infrastructure areas (eg fuel bays, workshops, electrical cuddies) are not scanned unless specifically requested by the on-shift geologist.
- Central Federal Bassett (CFB) ore development is scanned only at the request of the on-shift geologist.
- Historical development, rehabilitation areas, declines, and infrastructure do not require routine FC Mapping unless requested.

These guidelines reflected both the geological complexity and ground-support requirements of Renison. For A5 and LD, both of which are ore zones in deeper parts of the mine, dynamic support regimes are required, and shotcrete application is routine. By contrast, CFB is a more geotechnically stable ore zone higher in the mine with less frequent shotcrete application required. Where FC Mapping was not required, geological technicians continued to collect photographs as per pre-MVS workflows. Based on operational capacity and geological workload, initial expectations were to scan approximately 75 per cent of all applicable headings, with this target subject to ongoing review. Post-shift, scan data were uploaded for review as part of the geologist-led QA/QC process. Figure 4 demonstrates examples of data collected during the integration phase.

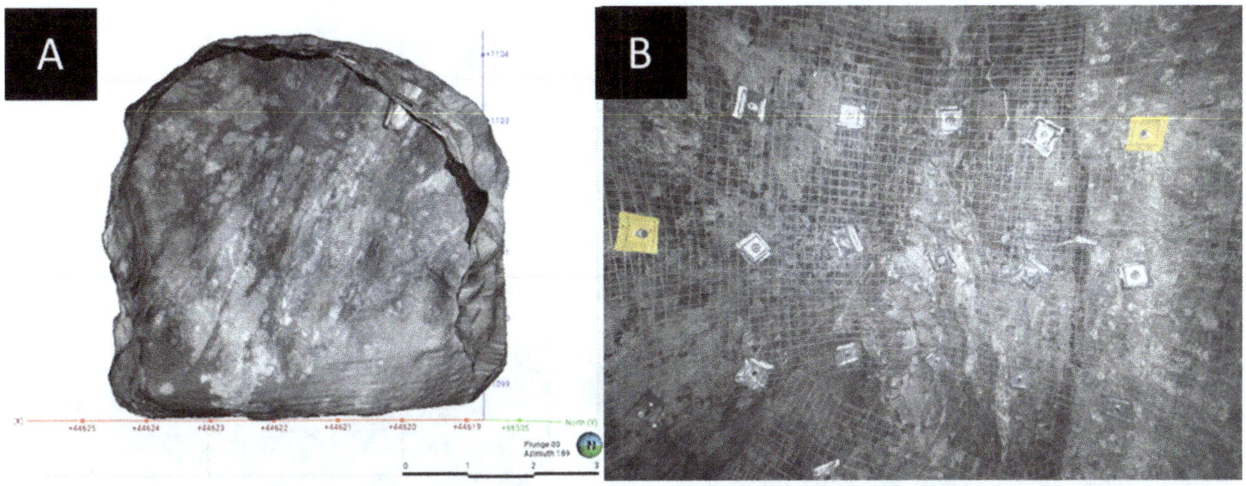

FIG 4 – Left (A) shows an example of the whole heading MVS FC scan (ie walls, backs and face as a single data file); Right (B) is a backs only MVS scan. Location: Area 5.

Quality assurance and process improvement

Establishing a structured QA/QC process was crucial during implementation of FC Mapping, given the newness of the technology and the fact that geological technicians were not the primary end-users of the scan outputs. Each shift, scans were reviewed by the on-shift geologist, with feedback provided to technicians to support ongoing improvement of scanning quality and procedures. This workflow was supported by the development of an FC Scan Tracker to record key metadata for each heading. QA/QC and geological validation were primarily conducted in Surpac, while photo mapping was completed in both Surpac and Leapfrog, and final geological modelling was mainly performed by the resource geology team in Leapfrog.

Challenges encountered

Several challenges emerged during implementation, notably scan accuracy and reliability. Survey marker quality and placement were critical; markers needed to be placed within 15 to 30 m of the scanning area in a triangular geometry. Ideally, survey points should be spaced around 30 to 40 m, with 30 m being desirable for supporting both FC Mapping and monitoring heading development against plan. However, in practice, the Renison survey team typically establishes points at 60 to

70 m intervals due to demanding survey schedules, meaning additional attention was required to achieve the desired spacing for mapping purposes. Incorrectly placed, damaged, or non-visible markers resulted in base scans that could not be georeferenced or increased the likelihood of drift errors. Whilst the system's cloud-to-cloud alignment tool can extend tolerable distances between control points, errors in the initial reference scan or scans spaced too far apart can still propagate drift and prevent proper georeferencing. Renison uses two types of survey markers – survey stations (for geological updates) and laser points (for development advance) – which are not named using the same convention, complicating the selection of appropriate survey points on the MVS tablet. Environmental factors, such as dust (reducing image clarity and point-cloud density) and poor lighting, also affected scan quality. Scans are conducted before the shotcrete rig arrives, as capturing equipment in the heading can introduce distortion or alignment errors, often rendering the data unusable. Scanning with the rig present is discouraged as it can slow down the mining cycle. Additionally, certain drive geometries, resultant from heading firings, sometimes caused stretching or distortion in the point cloud when surfaces were captured from suboptimal angles. Figure 5 demonstrates examples of dust effects within scans captured. When these factors were not well controlled, scan accuracy and overall quality declined.

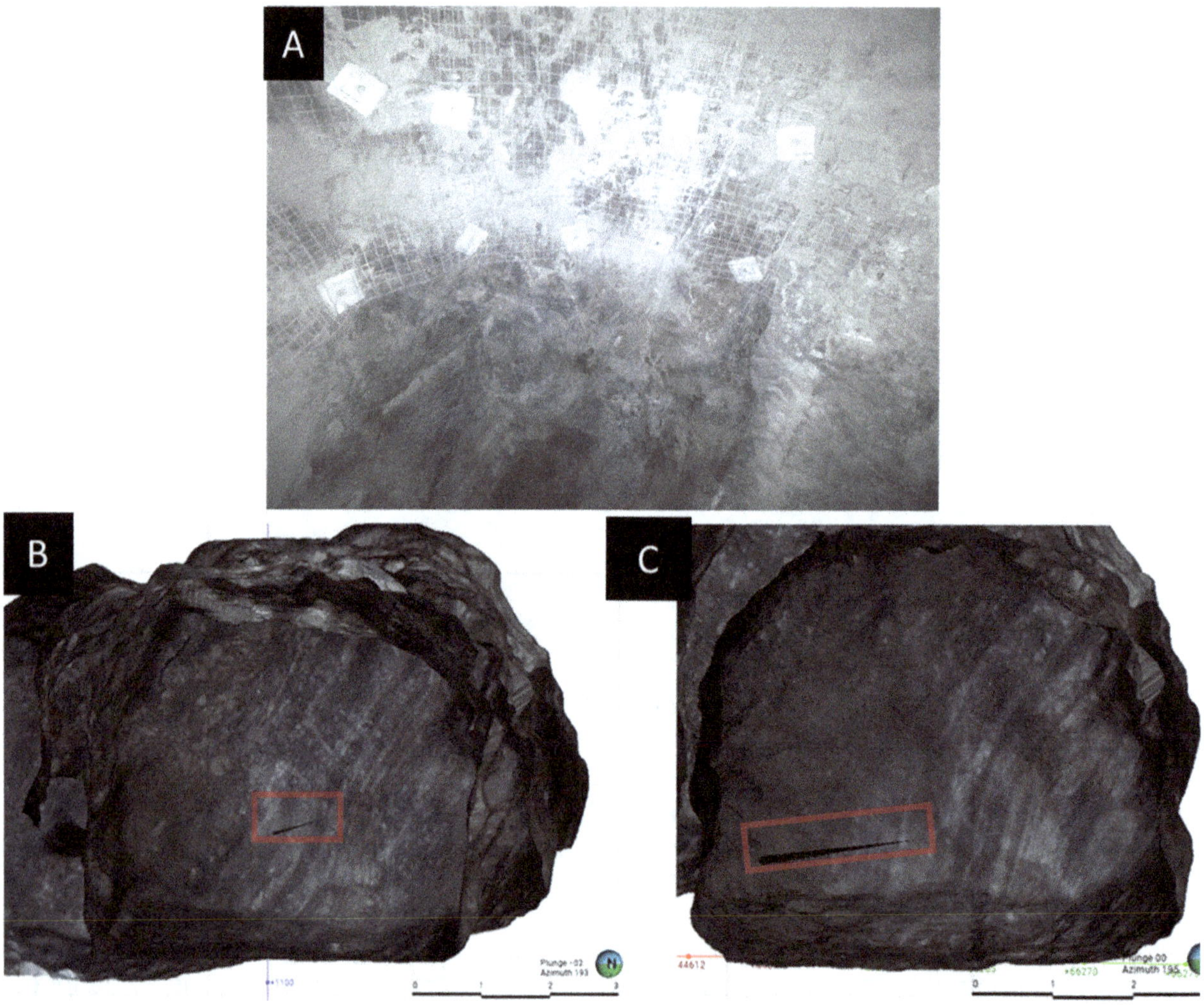

FIG 5 – Top example (A) shows backs with dust *in situ* at time of data collection and bottom examples (B to C) show 'spots' or artificial points due to airborne dust (obscurities shown by red outline). Location: Area 5.

Process enhancements

To address these issues, several improvements were introduced:

- Technicians received physical location maps to assist with identifying survey markers underground.

- Data housekeeping was reinforced and emphasised, as the MVS software does not permit renaming construction files directly; renaming to capture the heading cut number had to be done during scan export.
- During QA/QC, Geologists manually checked the location and cut number of scans to ensure consistent data management.
- Thorough washing of headings prior to scanning was reinforced to reduce dust, which can create artificial points in the data.
- Scans are scheduled prior to any heavy equipment arriving in the heading to avoid introduction of errors and disruption to the mining cycle.
- Additional training covered lighting management, recommending lower lighting settings for higher quality face captures, with further enhancements applied during QA/QC if required.
- To reduce distortion and improve structural definition, geological technicians were directed to collect two complementary scans wherever possible: a wall and face scan, and a backs-only scan. These data sets improved the clarity and interpretability of resulting 3D meshes.
- Standard 2D photographs continued to be captured to maintain minimum data standards.
- Georeferencing limitations were managed within the QA/QC process, ensuring technicians had access to sufficient survey markers, as incorrectly georeferenced base scans could lock alignment scans to the original data set, propagating drift and misalignment issues and potentially requiring rescan.

All improvements were implemented in consultation with both the production geology and resource geology teams, ensuring scan outputs met the requirements for mapping, modelling, and broader geological interpretation workflows.

MVS FACECAPTURE – GEOLOGICAL MODELLING IMPROVEMENTS

FC Mapping technology was introduced at Renison for several reasons including to enhance geological understanding and improve resource model accuracy. As Renison's first use of underground 3D face-capture technology, FC Mapping allows for more detailed data collection in conjunction with traditional mapping methods, especially in areas with frequent shotcrete activities, working effectively alongside conventional mapping approaches. Figure 6 shows conventional mapping used complementary to FC Mapping scans to compare against the modelled geology of an ore drive. In Leapfrog, datapoints can link directly to the FC Mapping scans to dynamically influence the geological model.

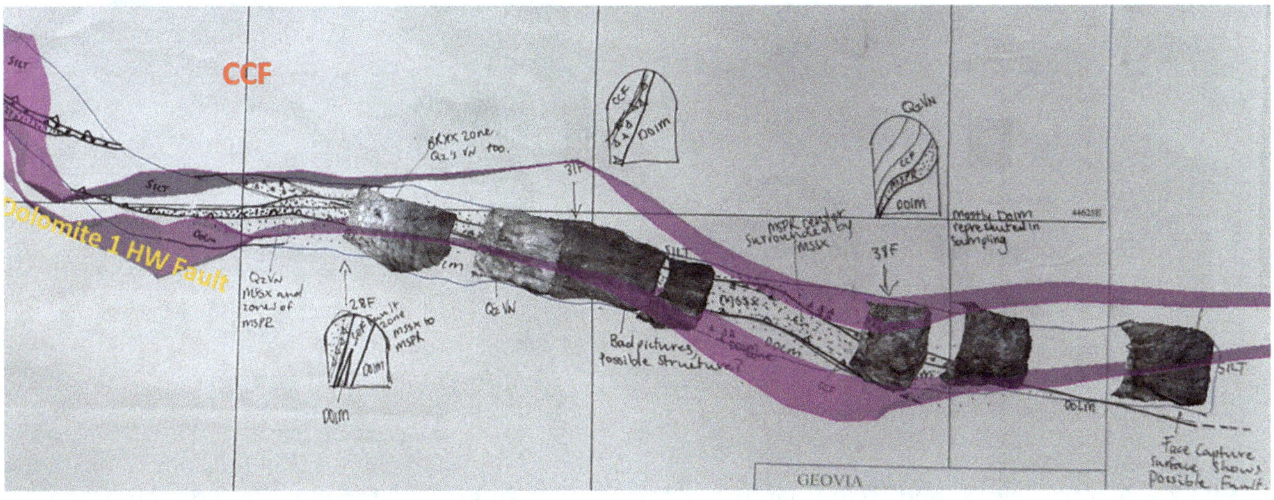

FIG 6 – Conventional backs mapping combined with MVS data. The main structure of the drive has been included for reference along with geology codes/comments.

The first resource models incorporating MVS data were released in June 2025 and were based on scans collected since the initial implementation in April 2025. The incorporation of the data immediately highlighted the complicated nature of transfer zones at a drive-width scale, which previously available data only inferred. These transfer zones are areas between major faults that act as dilutional zones containing stratafault mineralisation. This new information helped to explain local variations in orebody thickness enabling more informed and timely decisions on the placement of development ore drives.

Once the data has undergone its initial QA/QC in Surpac and deemed of acceptable quality, it is imported into Leapfrog Geo. Next, surveyed development drive strings provided by the mine survey department are used to make any further necessary corrections ensuring the data can be integrated into existing data sets (Figure 7 provides an overview). This integration encompasses other geological data sets, structural modelling, and face-sampling results, allowing for more informed domain boundary definitions. Even in instances where explicit geological contacts cannot be delineated, the 3D face data provides important context for establishing the geological setting and constraining model assumptions. In the example shown (Figure 7) for development face A5_1095_4650_DOS face 33, the MVS imagery successfully assisted in confirming the footwall contact bedding angle, adding confidence to the structural interpretation.

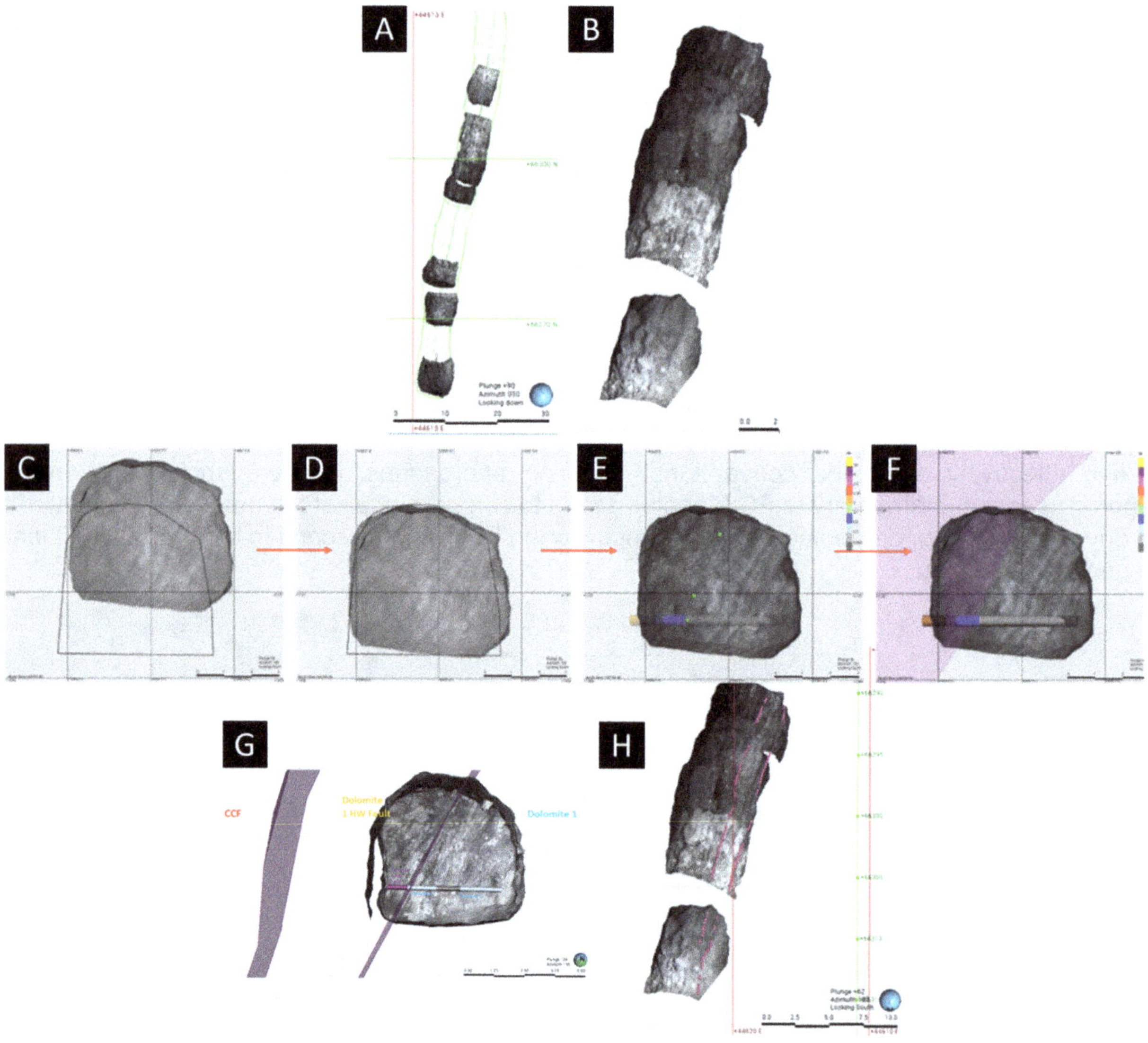

FIG 7 – A plan view (A and B) in Leapfrog of the MVS scan against the surveyed development drives. From there, (images C to F) the process of outstanding QA/QC corrections to ensure compatibility with existing data. Resultant (images G and H) interpretations can therefore be completed and integrated.

Overall, MVS was found to integrate easily into the existing resource-modelling workflow and has since been used across multiple model releases. Limitations were also identified, including that image alignment must currently be repeated for each Leapfrog project, introducing potential discrepancies. Any changes made to the.obj files in Leapfrog Geo is project specific and therefore cannot be saved, exported and/or transferred to other projects. Additionally, photo resolution can affect mineral boundary clarity. Which as discussed in previous sections, is mainly due to dynamic ground conditions, dust and lighting impacts.

At this stage, a direct cost–benefit or specific monetary value has not been explicitly defined, as the primary benefit of FC Mapping is improved geological confidence. By increasing data coverage while maintaining efficient underground mining cycles, the use of this system reduces data loss in shotcreted headings and improves the understanding of structures and heading-scale geological complexity, which in turn has a direct impact on geological interpretation and grade estimation.

MVS FACECAPTURE MAPPING IMPROVEMENTS

While the current implementation of MVS FC Mapping has already proved valuable for geological interpretation and modelling workflows at Renison, the process remains dynamic and continually evolving as both capability and user confidence with the technology develop. The integration of FC Mapping has necessitated ongoing enhancements, and certain aspects of the workflow are anticipated to undergo further refinement in the future. Through workflow optimisation, software adjustments, improved data management practices, and collaboration with MVS for technical support and recommendations, several future improvements have been identified:

- **Capture Quality** – established early on as an important part of the process, ongoing work is required to ensure image quality remains at a high standard. Ongoing work includes ensuring headings are washed down and airborne dust is at a minimum. Part of this includes improving time management where applicable so that washing of headings can be achieved and dust can settle.
- **Data handling improvements** – several manual steps currently exist between capture, storage, QA/QC, and modelling. Whilst file transfer is straightforward, renaming constructions and individual alignment scans both syn- and post-capture would reduce opportunities for human error and scan duplication. Although this function does not exist with the current MVS software, the inclusion of this capability is considered a desirable enhancement for future software updates. In the meantime, geological technicians must continue to rename files during the scan export phase and ensure the FC Scan Tracker is updated.
- **Management of survey markers** – while internal survey data procedures are outside this rollout's scope, establishing a standardised method for geology to collect and manage survey data is essential. Creating unique naming conventions will enhance data consistency and usability. Existing mapping templates could be improved, for instance, by using macros with raw CSV survey files. All generated survey marker names should be formally recorded, with potential integration into the FC Scan Tracker.
- **Re-georeferencing scans** – currently, if a base scan is incorrectly georeferenced, alignment scans referencing it remain locked to the original reference points, propagating errors such as drift and misalignment. Introducing the ability to re-georeference scans after initial capture would reduce rework, prevent unnecessary base scans, and improve workflow efficiency. Currently this improvement is dependent on future MVS software functionality.
- **Point cloud rectification during capture** – the current MVS software does not allow for artificial points generated in scans to be detected or removed during collection. Ideally, visualising anomalies as they arise would enable either: (1) immediate rescanning, or (2) removal of erroneous points. Since these options are not available with present capabilities, adding such functionality would lessen post-processing demands and elevate the quality of the collected data.
- **Process workflow improvements** – QA/QC completed by the production geology team is currently performed in Surpac with scans subsequently imported into Leapfrog for modelling.

Migrating QA/QC to a dedicated Leapfrog project would reduce inefficiencies caused by multiple format conversions and repeated imports, particularly if cloud-based Leapfrog compatible software (eg Seequent Central) was adopted at Renison.

- ***In situ* digitisation and direct capture improvements** – with the ability to use the MVS tablet and software to digitise geological features underground immediately post-capture geologists and geological technicians can annotate features directly onto the raw data. In future, a dedicated notes section could allow both roles to record relevant observations concurrently, eliminating the need for later interpretation and improving integration into existing data frameworks.
- **Enhanced Leapfrog integration and modelling use** – expanding the number of FC scan data that are used alongside drill hole data, traditional mapping, wireframes, and assays to inform domain boundaries and structural interpretation. To ensure data confidence, it is necessary to establish a minimum required number of scans. Furthermore, a corresponding minimum number of these scans should be identified for inclusion in each model to maintain consistency and reliability in geological interpretation. Once survey markers and georeferencing are verified, scans can be used to directly snap wireframes and geological surfaces, enhancing spatial accuracy and model reliability.
- **Operational expansion and consistency targets** – expanding on established scanning guidelines to increase underground coverage. The initial goal was to capture approximately 75 per cent of suitable headings, but this target is flexible and should align with operational needs and geological capacity. As such, the revised aim is to achieve approximately 95 per cent coverage, or as high as is reasonably practical.
- **Gen2 MVS unit implementation** – a second-generation MVS unit (Gen2) is planned to improve hardware reliability, sensor stability, and capture efficiency. Advantages include a lighter unit, smaller tablet or mobile software, and enhanced ergonomics for users. While this upgrade may resolve minor early deployment issues (eg intermittent sensor head performance), a period of implementation and adjustment is expected. Lessons learned from the first-generation unit will inform the successful rollout of Gen2.
- **Share experiences and insight** – opportunities exist to leverage insights and experiences from a neighbouring operation, which have since implemented FC Mapping, to help support further enhancements.

SUMMARY

The Renison Mine in Tasmania, Australia's largest primary tin producer, faces increasing geological complexity and operational challenges as mining extends beyond 1200 m depth. The routine use of shotcrete for ground support has constrained traditional geological data collection, prompting the adoption of Mine Vision Systems (MVS) FaceCapture Mapping (FC Mapping). This handheld LiDAR and high-resolution imaging technology enables rapid, high-quality 3D scans of mine headings, even within tight operational windows. Since its integration in April 2025, FC Mapping has improved the accuracy and availability of geological data for modelling and decision-making, with outputs efficiently incorporated into software like Leapfrog and Surpac.

Implementation required alignment with existing workflows, development of scanning guidelines prioritising ore zones with dynamic ground support needs, and establishment of rigorous QA/QC processes. Key challenges included ensuring scan accuracy, particularly georeferencing with survey markers, managing environmental variables such as dust and lighting, and training geological technicians for optimal data capture and management. Process enhancements addressed these issues through improved survey marker tracking, data housekeeping, additional training, and workflow refinements.

Ongoing improvements focus on streamlining data handling, standardising survey marker management, enabling post-capture re-georeferencing, enhancing point cloud quality, and integrating QA/QC within Leapfrog. The mine aims to expand scan coverage to over 95 per cent of relevant headings and plans to upgrade to a Gen2 MVS unit to further boost reliability and efficiency. The financial value of the system continues to be demonstrated and refined; while benefits in terms of specific dollar values are difficult to quantify, the financial value is realised through improved geological confidence and data coverage without impacting the mining cycle. Overall, the integration of FC Mapping at Renison has enhanced geological interpretation and modelling, supporting safer and more informed mining operations.

ACKNOWLEDGEMENTS

The authors would like to acknowledge the support of Bluestone Mines Tasmania Pty Ltd and Mine Vision Systems Inc. in providing access and insights for this paper. We are grateful to all colleagues who contributed to the development of the paper, and in particular to the geological technicians who provided practical insights into the implementation and application of the system, as well identifying opportunities for its enhancement. Finally, we thank all others whose assistance contributed to the successful execution of this piece of work, and the ongoing enhancements of Face Capturing Mapping at Renison.

REFERENCES

Bluestone Mines Tasmania Pty Ltd, 2023. *Our Company – Bluestone* [online]. Bluestone Mines Tasmania Pty Ltd. Available from: <https://bluestonetin.com.au/company-overview/> [Accessed: 22 Nov 2025].

Bluestone Mines Tasmania Pty Ltd, 2024a. Ground Control Management Plan Revision 9.

Bluestone Mines Tasmania Pty Ltd, 2024b. *Our Work – Bluestone* [online]. Bluestone Mines Tasmania Pty Ltd. Available from: <https://bluestonetin.com.au/about-us/our-work/> [Accessed: 22 Nov 2025].

Denholm, J, 2018. The trace element chemistry and U-Pb geochronology of Tasmanian cassiterite, Honours Thesis (unpublished), University of Tasmania, Hobart.

Kitto, P, 1994. Structural and Geochemical Controls on Mineralisation at Renison, Tasmania, PhD Thesis (unpublished), University of Tasmania, Hobart.

Metals X Limited, 2025a. 2024 Renison Ore Reserve Update [online]. Available from: <https://minedocs.com/28/MLX-Renison-Ore-Reserve-Update-01312025.pdf> [Accessed: 22 Nov 2025].

Metals X Limited, 2025b. 2025 Consolidated Financial Report [online]. Available from: <https://www.metalsx.com.au/wp-content/uploads/2025/08/22_MLX_ASX_Appendix-4D-and-Half-Yearly-Report-and-Accounts_25-August-2025.pdf> [Accessed: 22 Nov 2025].

Metals X Limited, 2025c. Australia's Premier Tin Producer – Presentation September 2025 [online]. Available from: <https://www.metalsx.com.au/wp-content/uploads/2025/09/23_MLX_ASX_Investor-Presentation_19-September-2025.pdf> [Accessed: 22 Nov 2025].

Pink, K and Crawford, P, 1996. *Renison: The Slumbering Giant*, 1st ed (Tasmania: Advocate Printers).

Seymour, D B, Green, G R and Calver, C R, 2006. The Geology and Mineral Deposits of Tasmania, in *Tasmania Geological Survey Bulletin*, 72nd ed [online] (Mineral Resources Tasmania). Available from: <https://www.mrt.tas.gov.au/__data/assets/pdf_file/0007/229633/GSB_72.pdf> [Accessed: 22 Nov 2025].

Solomon, M, 1981. An introduction to the geology and metallic ore deposits of Tasmania, *Economic Geology*, 76(2):194–208.

The unified rock value framework – integrating measured and sensed data with machine-learning for value-centred, behaviour-informed ore control

W Ware[1], B Crawford[2] and S Coward[3]

1. Global Practice Lead Geometallurgy, IMDEX Digital Earth Knowledge, Perth WA 6000.
   Email: wendy.ware@imdexlimited.com
2. Global Principal Geoscientist, IMDEX Digital Earth Knowledge, Melbourne Vic 3000.
   Email: brenton.crawford@imdexlimited.com
3. Director, Interlaced Systems, Perth WA 6000.
   Email: stephen.coward@interlacedconsulting.com

ABSTRACT

Ore control decisions sit at the narrowest and most value-critical point in the mining value chain, yet grade remains the dominant routing criterion because it is available at operational resolution and cadence. While necessary, grade alone is an incomplete proxy for value. It does not capture rock behaviour, which governs mining response, recovery, throughput, cost and risk, and its continued use contributes to plant instability, reconciliation drift and systematic value erosion between strategic planning intent and operational execution.

This paper argues that value-focused ore control requires rock behaviour to be represented explicitly at the point of decision. Building on the Primary–Response Framework and materials science principles, it introduces the Unified Rock Value Framework (URVF), a decision-centred framework that links intrinsic rock characteristics to behavioural response and value across the mine-to-market chain. URVF integrates direct measurement, rock sensing and data-driven modelling to enable behavioural prediction at ore control resolution and to carry those predictions forward into operational, metallurgical and value models that inform routing decisions.

The framework addresses a persistent gap in current practice, where sensing, modelling and machine-learning applications often operate as isolated tools that improve local prediction accuracy but remain disconnected from planning workflows, execution and value assessment. URVF provides a consistent behavioural and value structure that aligns spatial modelling, planning, execution and reconciliation across planning horizons. Recent advances in sensing technologies and hybrid geostatistical and machine-learning methods provide the technical enablers, while disciplined implementation focused on decision clarity, data governance, cross-functional capability and trust supports sustained adoption.

By reframing ore control from a grade-based classification task to a value-optimisation problem, URVF provides a practical pathway to stabilise operations, improve recovery, narrow reconciliation variance and realise value that is currently left unrealised.

INTRODUCTION

Ore control sits at the junction between resource development and the rest of the mining value chain. It is where geological interpretation, planning intent and operational execution converge, and where routing decisions made on short time frames (hours-to-days) determine whether value is realised or irreversibly lost. For the purposes of this paper, ore control refers to the integrated short-term material management system that links *in situ* rock definition to operational routing decisions and performance reconciliation. It encompasses: (i) delineation and classification of material through grade control drilling, geological interpretation and short-term modelling; (ii) translation of this information into executable mine plans and schedules; (iii) physical execution through drill-and-blast, load-and-haul and destination control; and (iv) reconciliation of predicted versus realised performance to maintain and improve system reliability.

While specific workflows vary by commodity and mining method, the governing objective remains consistent: to allocate material in a manner that maximises value under prevailing processing, market and operational constraints.

Despite its systemic role, ore control in most operations remains dominated by grade-based rules. Grade persists as the primary routing criterion because it is measurable, auditable and available at operational resolution and cadence. It provides a simple and defensible basis for decision-making under time pressure (Dowd and David, 1976). However, grade was never intended to represent the full value potential of the rock, and its continued use as a proxy reflects historical data constraints rather than suitability.

Figure 1 situates ore control within the broader mining value chain. At this junction, material is directed either toward value creation through processing and sale, or toward waste streams and long-term liabilities. These decisions influence not only revenue, but throughput stability, recovery, operating cost, product quality and exposure to environmental and social risk. Errors introduced at this stage are difficult to correct later, making value erosion at ore control structural rather than incidental.

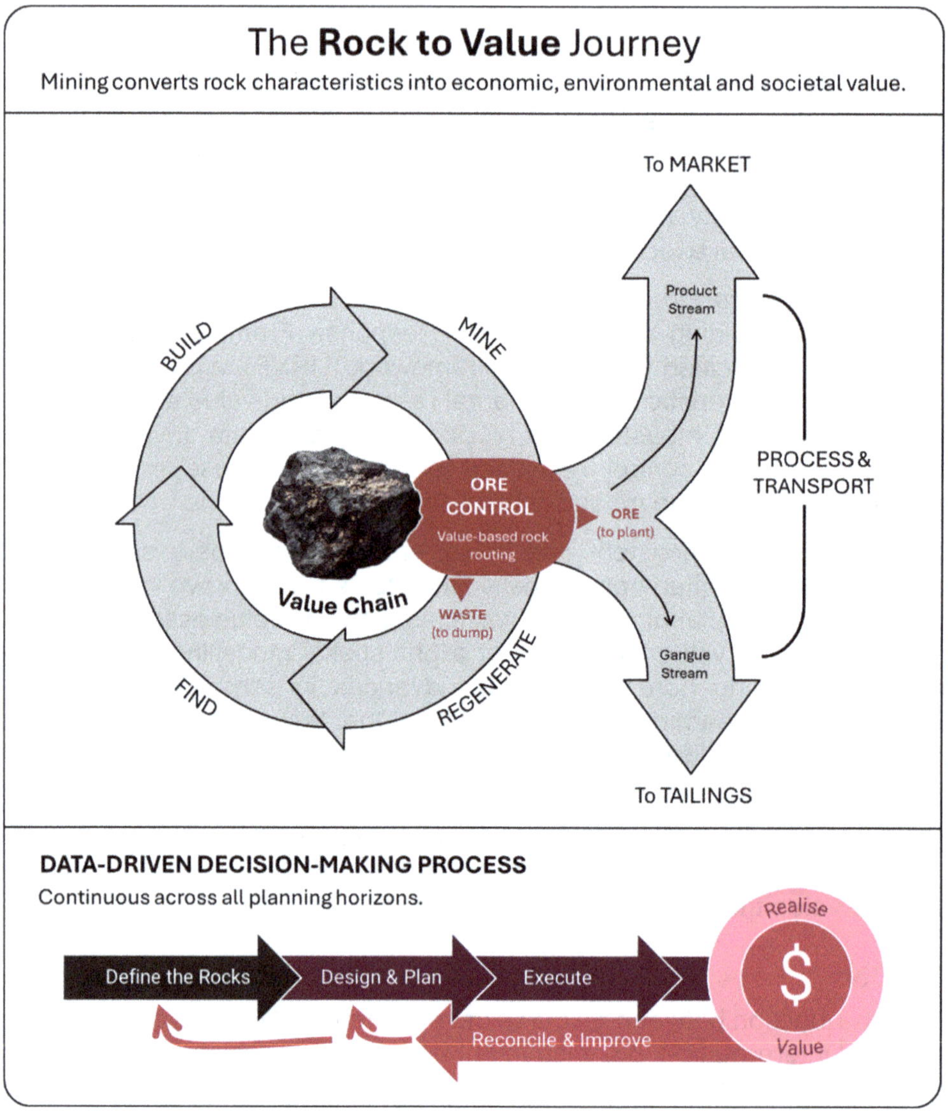

FIG 1 – Ore control is the rock-routing node within the rock-to-value system.

Over the past two decades, geometallurgy has materially improved understanding of orebody behaviour at deposit scale and reduced uncertainty in long-term planning. Strategic programmes have informed flow sheet design, scheduling and life-of-mine optimisation. However, this behavioural insight has struggled to translate into ore control practice. Data generated through strategic programmes are typically sparse, slow to update and too coarse to support decisions at the spatial resolution and tempo required for operational routing.

At the same time, the operational data landscape has shifted. High-density sensing, routine in-field measurements and automated data capture now generate substantially more geological and process-relevant data at ore control cadence. In parallel, advances in machine-learning and digital workflows have accelerated the conversion of this data into predictive models of hardness, lithology and processing response.

In response, many operations have deployed sensors, predictive analytics and decision-support tools to improve local behavioural estimates. Yet these capabilities are often implemented as standalone technical solutions. They improve prediction accuracy in isolation but remain weakly coupled to short-term design, planning and value-based decision frameworks. The result is a proliferation of high-resolution insights that are disconnected, do not persist across planning horizons and do not consistently influence routing, blending or execution.

The remaining challenge is therefore not data availability or analytical capability, but integration. Specifically, the industry lacks a framework that links rock characteristics, behavioural response and value impact within a single structure spanning characterisation, modelling, planning, execution and reconciliation.

This gap motivates the Unified Rock Value Framework (URVF). URVF reframes ore control as a behaviour- and value-driven decision system rather than a grade-based classification task. Its contribution rests on three integrated elements. The framework:

1. Establishes a consistent behavioural and value structure that persists from strategic resource modelling to operational routing, addressing the disconnect between long-term geometallurgical understanding and short-term execution.
2. Translates materials science principles and the Primary–Response Framework into decision-ready behavioural and value metrics at ore control scale, shifting from descriptive characterisation to operational prediction.
3. Embeds organisational alignment, data governance and reconciliation as core components, recognising that technical capability alone is insufficient without trust, continuity and cross-functional integration.

The sections that follow examine how grade-based ore control erodes value, then establish the geometallurgical foundations required to represent behaviour explicitly. The URVF and its integrated workflows are subsequently introduced. The paper concludes by outlining practical implementation requirements, with emphasis on decision clarity, organisational alignment, governed data flows and learning through reconciliation. Together, these elements provide a disciplined structure for aligning rock knowledge, behaviour and value at the point where decisions are made.

LIMITATIONS OF GRADE-BASED ORE CONTROL

Morley (2024) shows that grade-based ore control can fail to deliver value even under stable mining and plant conditions. The underlying issue is not operational instability, but the simplifying assumption that grade, a fixed commodity price, nameplate throughput and an implicit recovery are sufficient to estimate value. In reality, throughput and recovery vary with rock characteristics such as hardness, mineralogy and texture, meaning that material of similar grade can contribute very differently to value (Morley, 2024).

When grade is used as the primary routing criterion, ore control decisions are effectively reduced to a revenue proxy. This collapses routing to a binary accept-reject decision and masks the multivariate factors that determine whether a tonne will ultimately create or destroy value. Recovery and processing cost both vary with rock behaviour, and it is these behavioural drivers that dominate operational performance at the routing point.

A more appropriate operational expression of value therefore considers margin rather than revenue:

$$Operating\ margin\ per\ tonne = \left(\frac{G}{1\ tonne} \times R \times P\right) - C_{process} \quad (1)$$

Where:

G = grade

R = recovery

P = commodity price

$C_{process}$ = processing cost per tonne

This formulation makes the limitation of grade-only ore control explicit. Grade describes concentration. Behaviour governs recovery, cost, risk and timing. When grade is used in isolation, it provides an incomplete and often misleading proxy for value.

The impact of this limitation is amplified by behavioural heterogeneity (Baizhiyen *et al,* 2024). Geological and geotechnical attributes that govern processing response, such as rock strength, whole rock geochemistry, mineralogy and textural variability, often vary at scales far finer than those resolved in resource models (Deutsch *et al,* 2016; Tiu *et al,* 2023). Within a single ore control interval, behaviour-critical variability can change materially and is not necessarily correlated with grade. Grade-only domaining collapses this variability into broad classes, obscuring the true drivers of throughput, recovery and product quality. The result is operational blind spots precisely where variability has the greatest impact on performance.

Throughput provides a practical illustration of how behaviour-driven variability translates directly into value, particularly when assessed on a net mill hour basis. When planning assumptions are aligned with actual rock behaviour, most plants operate within a relatively narrow throughput reconciliation tolerance of approximately ±5 per cent, reflecting anticipated variability in hardness, circuit conditions and operating practice (Wills and Finch, 2015). Sustained deviations outside this range are rarely incidental. They commonly indicate that short-term geological models and ore control rules have not adequately captured the behavioural variability governing how ore breaks, grinds and moves through the plant.

At operational scale, these deviations are rarely the result of a single event. Rather, they arise from repeated misallocation of material whose throughput response differs from plan. A shovel cut classified on grade alone may appear compliant, yet if its breakage characteristics reduce milling rate by even 1–3 per cent, the impact accumulates across shifts, benches and campaigns. In a large base-metal operation processing several thousand tonnes per hour, a sustained 1 per cent divergence between planned and realised throughput equates to tens of millions of dollars per annum in gross metal revenue. At 3 per cent or 5 per cent, the variance increases to hundreds of millions, as illustrated in Table 1.

TABLE 1

Illustrative annual gross revenue sensitivity to sustained throughput variance

Throughput variance (%)	Gross revenue impact (US$M/yr)	Likely net value implication
+5%	+215	Often negative after recovery and cost penalties
+3%	+129	Frequently neutral to negative
+1%	+43	Marginal
-1%	-43	Negative
-3%	-129	Negative
-5%	-215	Negative

Assumptions: 4500 t/h; 8760 h/a; 1.3 per cent Cu; 85 per cent recovery; Cu price US$9881/t.

Gross revenue impacts do not account for recovery loss, dilution, operating cost escalation or timing effects, which typically reduce net value when throughput exceeds design capacity.

Apparent gains from exceeding planned throughput are equally deceptive. Pushing tonnage beyond design capacity often incurs recovery losses, grade dilution and higher operating costs. Over longer horizons these penalties typically offset or reverse the gross revenue uplift. Conversely, sustained underperformance directly erodes revenue and undermines budget guidance. What begins as minor behavioural misclassification at block scale therefore compounds into material annual variance.

Grade-only ore control does not dampen this volatility because it does not represent the behavioural drivers controlling throughput stability.

Grade remains necessary, but it is not sufficient. Value is governed by how rock behaves through blasting, handling, processing and delivery to market, and these behavioural drivers are not represented in grade-based ore control. Addressing this limitation requires an approach that explicitly links intrinsic rock characteristics to behavioural response and translates that response into decision-relevant measures of value. This is the role of geometallurgy. By connecting what the rock is, how it behaves, and how that behaviour influences recovery, throughput, cost and risk, geometallurgy provides the technical foundation for behaviour-informed, value-driven ore control at operational scale.

GEOMETALLURGY AND ROCK BEHAVIOUR

As Beniscelli (2011) observed, geometallurgy is an interdisciplinary activity that studies the cause-effect relationship between relevant geological and metallurgical variables throughout the mining business value chain. Its purpose is to understand how geological characteristics and variability translate into processing behaviour and, ultimately, value.

In strategic programmes, this is achieved through direct measurement including mineralogical and geochemical characterisation and bench-scale metallurgical testing. Once sufficient spatial coverage is achieved, these variables can be estimated across the deposit and incorporated into long-term planning to reduce uncertainty inflow sheet design, resource evaluation and resource-reserve conversion (Beniscelli *et al,* 2000; Boisvert *et al,* 2013; Lamberg, 2011; Ehrig *et al,* 2014). As McKay *et al* (2016) emphasise, these data sets remain the most reliable basis for understanding orebody behaviour at deposit scale and form the backbone of life-of-mine value optimisation. Figure 2 summarises the conventional strategic geometallurgy workflow.

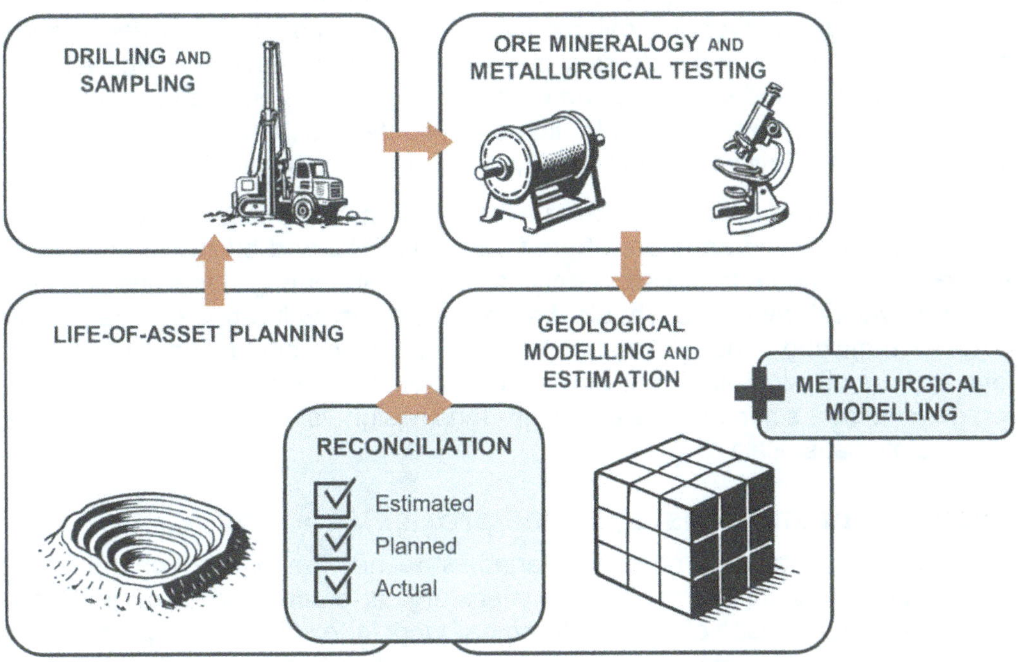

FIG 2 – Strategic geometallurgy workflow.

Strategic geometallurgical workflows are well suited to long-horizon decisions but operate at a different spatial support and decision cadence to ore control (McKay *et al,* 2016). Strategic programmes typically resolve variability at domain to deposit scale and update over months to years. Ore control decisions, in contrast, are made at the scale of benches, blast patterns and shovel cuts, often at metre or sub-metre support and on hours-to-days time frames. Applying strategic models directly at these operational scales can obscure behaviourally important heterogeneity and reduce confidence at the face (Dominy *et al,* 2018; Jang and Topal, 2020).

This mismatch arises from three factors. First, sampling differs. Strategic geometallurgy relies on sparse, high-fidelity diamond core, whereas ore control depends on abundant but low-fidelity reverse circulation (RC) and blasthole drill-chip samples that are unsuitable for direct measurement of many response variables. Second, data-to-decision timing differs. Strategic models update too slowly to capture bench-scale changes that materially affect production, while ore control requires insight generated at operational cadence. Third, decision context differs. Behavioural knowledge from strategic studies rarely translates cleanly into short-term planning, dispatch, routing and reconciliation, and weak feedback loops allow model drift, unstable feed and avoidable value loss.

McKay *et al* (2016) defined the need to address this gap as tactical geometallurgy, arguing that behavioural understanding must reach the point of decision at ore control cadence if value is to be realised rather than merely planned. Bridging this gap requires methods that translate high-fidelity strategic understanding into behaviourally valid predictions at operational resolution under practical sampling and data-integration constraints.

The primary-response framework

Coward *et al* (2009) proposed the Primary-Response Framework (PRF) to address the systematic misuse of geometallurgical data in spatial models. The framework distinguishes between primary variables, which are intrinsic rock attributes such as bulk mineralogy, multi-element chemistry and multi-scale texture that can be directly measured and are generally additive, and response variables, which are behavioural outcomes expressed when energy, fluids, time or stress are applied, including throughput, fragmentation, grindability, liberation, recovery, rheology and weathering.

The central contribution of the PRF was to demonstrate that response variables behave fundamentally differently from primary variables. Response variables are non-additive, scale-dependent and dependent on test conditions. Treating them as additive or directly estimable can lead to biased models and unreliable predictions. The PRF therefore provides the foundational logic for sampling, test design and geological modelling and estimation in strategic geometallurgy programmes (Coward *et al,* 2009). Its physical basis lies in materials science, where rock behaviour is governed by composition and structure, and variability in primary variables controls the properties that determine how material breaks, grinds, liberates, floats, leaches or degrades (Wyllie, 1992; Kakani and Kakani, 2004). This will be expanded upon in the next section.

While essential, the PRF does not by itself resolve the requirements of ore control. It does not specify how behavioural insight can be generated from fragmented RC and blasthole material, nor how it can be delivered at the hours-to-days cadence required for routing decisions. It also does not address how behavioural response should be expressed in value-relevant terms suitable for operational decision-making. These constraints motivate a combined approach in which direct measurement anchors physical and behavioural characteristics, sensing provides data with dense spatial coverage, and data science translates high-density signals into spatially explicit behavioural predictions at operational scale.

Direct measurement and sensing-based proxies

Direct measurement of primary and response variables defines the inherent characteristics of rock and their associated behaviours in mining and metallurgical processing. X-ray diffraction (XRD), automated mineralogy, ICP-based chemistry, Bond ball work index (BBWi), drop weight index (DWi), SAG mill comminution (SMC), and bench-scale flotation and leach tests provide reliable methods for quantifying these variables.

These methods are rarely suitable for ore control. Many require intact samples, controlled preparation and multi-day laboratory cycles. Metallurgical response tests typically require kilograms of representative material and turnaround times measured in weeks to months. RC drill and blasthole samples are used extensively in ore control. These modes of sampling frequently lack the integrity and mass required for reliable measurement, and the numbers and volumes of samples generated often exceed practical on-site laboratory throughput at operational cadence. These constraints are summarised in Table 2.

TABLE 2

Typical requirements for common direct measurement methods used in strategic geometallurgy.

Test type	Typical sample mass	Sample type required	Turnaround time	Approximate cost	Key behavioural insight
BBWi or DWi	5–20 kg	Intact core; competent, representative interval	1–3 weeks	Moderate to high	Throughput, energy response
Batch flotation	2–10 kg	Intact core preferred; well-mixed composite acceptable	1–4 weeks	High	Recovery, concentrate quality
AutoSEM	30–100 g	Intact core or minimally disturbed split	1–3 weeks	High	Mineral associations, grain size, liberation
QXRD	10–50 g	Intact or carefully homogenised aliquot	1–7 days	Moderate	Mineralogy, alteration
Archimedes or pycnometry	50–300 g	Intact core or undisturbed chips	1–3 days	Low to moderate	Density, mass-volume relationships, porosity
ABA or NAG	50–300 g	Intact core or unbiased pulverised aliquot	1–2 weeks	Moderate	Acid generation, environmental risk

Sample integrity is often the most restrictive constraint for direct measurement; many tests require intact or minimally disturbed material to avoid bias from fragmentation, segregation or moisture loss. Values are indicative and depend on laboratory workflows, equipment availability and site logistics.

Where a primary or response variable can be measured directly at the required sample support and cadence, it should be. Where direct measurement cannot meet ore control requirements due to sample integrity, mass constraints, turnaround or cost, sensing provides a practical alternative. Sensors can be used to generate high-spatial density signals that reflect composition, structure and texture, and can be collected rapidly and consistently as part of drilling, logging and sampling workflows. When these are calibrated or trained against direct-measurement data sets from strategic programmes, these signals act as proxies for primary variables and predictors of response. This requires comparable sensing data to be collected across both strategic and tactical campaigns, and strategic test work to measure the response variables relevant to the deposit and value chain. Figure 3 summarises the major sensing data streams available across resource drilling and ore control drilling and shows how they underpin an integrated spatial characterisation workflow.

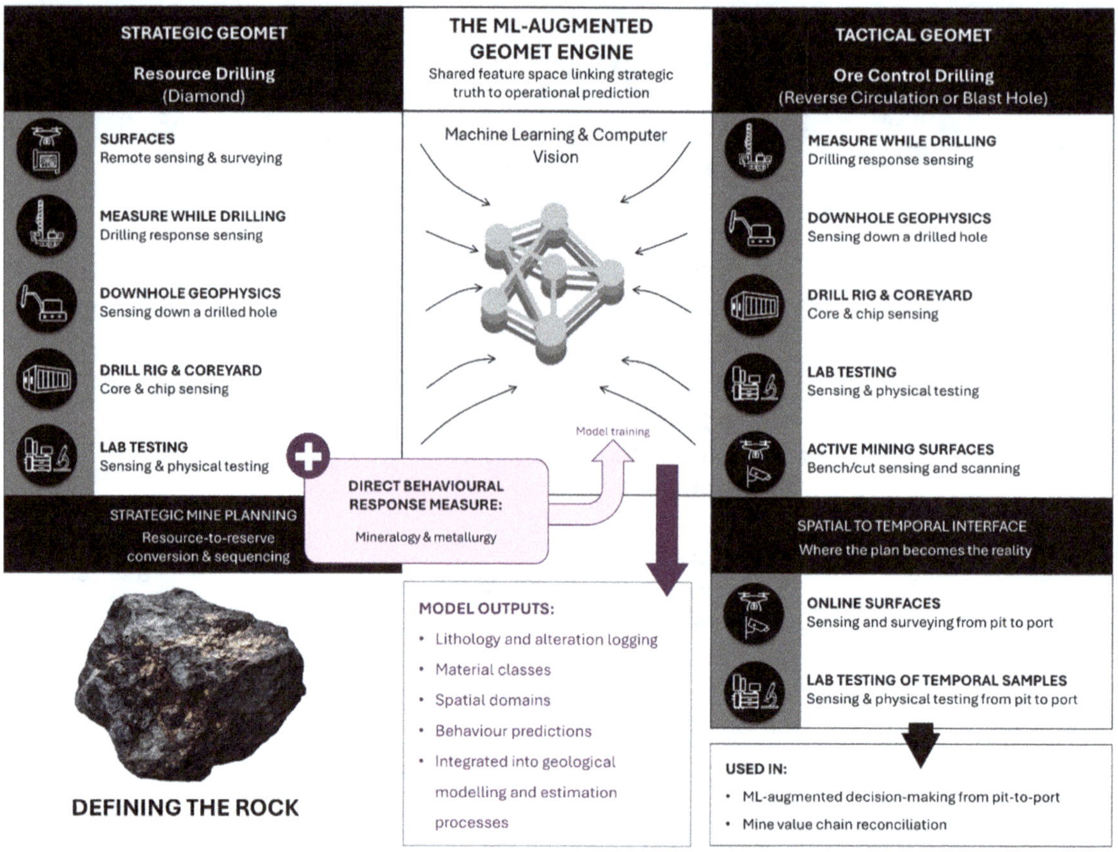

FIG 3 – The 'Define the Rocks' stage of the Unified Rock Value Framework. Strategic direct measurements establish behavioural truth, while shared sensing across campaigns enables calibrated machine-learning extrapolation of behaviour to operational scale and integration into geological modelling.

Measure-while-drilling (MWD) provides an early mechanical response signal during drilling. Parameters such as penetration rate, torque, vibration, pressure and acoustic response vary with changes in rock strength, fracture intensity, fabric and lithological boundaries. The data are continuous along hole depth and, once instrumentation is installed, are acquired at minimal marginal cost. Calibrated MWD can delineate lithological contacts (Silversides, Battalgazy and Dutch, 2025), classify geotechnical rock mass classification and identify domains relevant to comminution behaviour, including crushability and grindability (Germiquet and Minnitt, 2016; Vezhapparambu, 2016). It therefore supports early domaining of mechanical response, informing downstream predictions of diggability, fragmentation and milling throughput. The same signals are also used to support blast design and optimisation.

Downhole geophysics, including density, resistivity, magnetic susceptibility and spectral gamma, provides continuous petrophysical signals that respond to bulk composition, porosity, alteration intensity and sulfide content (Fallon, Fullagar and Sheard, 1997; Fullagar and Fallon, 1997; Dowd and Pardo-Iguzquiza, 2006; Hatherly, 2013; Kitzig, Kepic and Grant, 2018; Jang and Topal, 2020). These measurements extend rock characterisation beyond visual logging by quantifying physical properties that influence mechanical response and mineral processing behaviour. When calibrated against geological and metallurgical data sets, petrophysical parameters can support classification of lithological and alteration domains and provide proxies for density, hardness and comminution performance. Vatandoost (2010) documented early systematic use of petrophysical data for comminution prediction, demonstrating their value as behavioural indicators when interpreted within geological context. At ore control scale, such signals strengthen domain resolution and improve confidence in routing and blending decisions where direct metallurgical data are unavailable.

High-resolution photography of RC or blasthole chips can be transformed into quantitative descriptors of rock fragments through computer vision (Merrill Cifuentes, 2024). Metrics such as colour, particle-size distribution, angularity and eccentricity retain information that is otherwise lost during drilling and support lithological classification, domaining, alteration mapping and behavioural prediction. Commercial platforms such as Datarock Chip demonstrate high-throughput deployment at low marginal cost (Datarock, 2025).

Spectral measurement of production drill chips provides rapid chemical, mineralogical and alteration signatures (Linton *et al,* 2013; Harraden *et al,* 2019; Johnson, Browning and Pendock, 2019; Egaña *et al,* 2020). VNIR-SWIR spectroscopy has demonstrated strong capability for predicting modal mineralogy, clay and carbonate chemistry, alteration intensity and several metallurgically-relevant features linked to hardness, flotation response and acid consumption. Hyperspectral imaging further enriches the feature set by capturing spatial texture and structural information at high resolution.

Routine assay workflows are also being extended to capture behaviour-relevant data without disrupting sample processing times and result turnaround. Approaches include incorporating major and minor oxide geochemistry as a stable compositional fingerprint (Rollinson, 1993; Schouwstra *et al,* 2013), capturing comminution indicators during preparation milling or embedding comminution measures into preparation flows using Geopyörä-style devices (Matus, 2020; Bueno *et al,* 2024; Govender, 2025), and adding FTIR analysis on pulps for predictive mineralogy and metallurgical response (Dehaine *et al,* 2021; Butcher *et al,* 2023; Govender, 2025; Bennett and Munro, 2024).

Operational sensors at the mining face, on conveyors and in plant feed streams add a spatio-temporal dimension. Face imaging, belt spectra, online XRF, particle-size scanners and conveyor-mounted hyperspectral systems capture fragmentation, moisture, dilution, blending performance and evolving mineralogical trends. These measurements complement drilling-scale sensing by tracking how material changes after blasting and handling.

Individually, each sensing technology samples a different facet of the rock system. Together they provide a multi-dimensional representation at operational resolution. Early implementations at Garpenberg (Tiu, 2017; Tiu *et al,* 2024) and Mogalakwena (Germiquet and Minnitt, 2016; Govender, 2025) show that when these data sets are calibrated and embedded into decision workflows, they can support domaining, routing decisions and feed stabilisation at production scale. Sensor selection remains context-specific, constrained by orebody characteristics, decision requirements, operational flexibility and turnaround limits. The consistent principle is that sensing provides spatial resolution and commonality across both strategic and tactical geometallurgical workflows, and direct measurement anchors physical response.

Historically, the limiting factor was data processing. High-density, multi-type sensing streams exceeded the capacity of traditional estimation workflows, leaving much of their behavioural value unused. That constraint has materially reduced as data-science methods have matured. Machine-learning approaches now allow diverse sensing signals to be fused, calibrated against ground-truth data sets and translated into spatially explicit behavioural predictions suitable for operational decision-making.

Machine-learning as the integrator of behaviour

Machine-learning has been widely applied across the minerals value chain, with demonstrated success in predicting response variables from petrophysical data (Vatandoost, 2010; Kitzig, Kepic and Grant, 2018; David, Govender and Mainza, 2018), hyperspectral data sets (Tusa *et al,* 2025; Acosta *et al,* 2020), texture-based comminution behaviour (Tiu, 2017; Tiu *et al,* 2021; Merrill Cifuentes, 2024), flotation response (Jaimez-Salgado, Ortiz and Rodríguez, 2021), and spatial ML-augmented geostatistical estimation (Avalos and Ortiz, 2019; Erten *et al,* 2025; Govender, 2025). Comprehensive reviews by McCoy and Auret (2019) and Shirmard *et al* (2022) document the breadth and maturity of these applications.

Machine-learning is particularly well suited to ore control because, when trained and calibrated on direct-measurement data sets, it can infer both primary and response variables from high-density sensing data (Tusa *et al,* 2025; Johnson, Browning and Pendock, 2019; David, Govender and Mainza, 2018). At operational resolution, these predictions can be propagated through simulation (Morales *et al,* 2019; Stone *et al,* 2018), enabling routing alternatives to be evaluated explicitly in terms of value and risk rather than grade alone (Goodfellow and Dimitrakopoulos, 2016).

Despite these advances, most implementations remain deployed as standalone solutions. As noted by Lishchuk *et al* (2020) and Ortiz *et al* (2020), geometallurgical implementations incorporating machine-learning approaches improve local prediction accuracy but remain weakly connected to planning workflows, operational process models and value assessment. The result is a landscape of isolated technical capabilities, or 'islands of insight', that rarely translate into sustained operational improvement or consistent value delivery.

The limiting factor is therefore not model performance, but integration. Behavioural predictions only create value when they can be carried forward into downstream operational models, applied in decision-making and reconciled against realised outcomes. Achieving this requires machine-learning to operate within a disciplined structure that links characterisation, prediction and decision-making through consistent behavioural and value logic.

Within the Unified Rock Value Framework, machine-learning fulfils this integrative role. It provides the connective layer that fuses direct measurement, sensing and materials science into a coherent behavioural system supporting classification, domaining, prediction and uncertainty propagation within a single decision context. In doing so, machine-learning shifts from a collection of local optimisation tools to an enabling capability that makes behaviour visible, comparable and actionable at the point of ore control decision-making.

THE UNIFIED ROCK VALUE FRAMEWORK

The Unified Rock Value Framework operates across multiple modelling layers spanning spatial representation, behavioural prediction and value translation. Because these layers serve different purposes and operate at different planning horizons, clarity in terminology is essential. Although all models rely on data, they differ in representational focus, temporal scope and decision role.

Table 3 summarises the functional modelling layers within the Unified Rock Value Framework and their respective roles. Spatial estimation models represent the distribution of rock properties in three dimensions. When behavioural attributes are incorporated alongside geology and grade, these block models become geometallurgical in function, but they remain spatial geological models at their core. Operational and metallurgical response models translate those spatial variables into processing behaviour under defined mining and plant conditions. Planning and value models convert technical outcomes into time-based production scenarios, financial metrics and risk exposure. Reconciliation is not a separate model type, but a structured comparison between predicted and realised performance that informs refinement across all layers. Data science models, including machine-learning and hybrid geostatistical–machine-learning approaches, enhance these layers by learning multivariate relationships and improving classification, domaining and prediction.

Distinguishing these modelling layers is essential because value erosion often arises not from errors within a single model, but from misalignment between them.

The conceptual foundation for URVF draws from the materials science tetrahedron, a cornerstone of materials chemistry and engineering in which processing, structure, properties and performance occupy four interconnected vertices (Ellis, 1993; Howell, 2005; Callister Jr and Rethwisch, 2020). These vertices describe a causal chain: processing creates structure, structure controls properties, and properties determine performance. In a geometallurgical context, the material entering this system is rock. As rock moves from mine to market it is progressively transformed by heat, pressure, time, moisture, oxidation, comminution and mixing (Olson, 1997; Gupta and Yan, 2016). Its intrinsic characteristics, defined by primary variables such as mineralogy, chemistry and texture, govern how these transformations unfold and how the material ultimately performs along the value chain.

Geologists define what the rock is. Geometallurgists work across mining, processing and marketing functions to determine how that rock responds when energy, fluids, time and stress are applied. These responses control throughput, recovery, cost, product quality and risk, and therefore determine value. Geometallurgists are the rock to value insight translators, ensuring that the right rock data are available at the right format and time for decision-makers.

TABLE 3

Functional modelling layers within the Unified Rock Value Framework.

Model layer	Representation focus	Typical content	Functional role within URVF
Spatial estimation models (geomet-enabled block models)	Distribution of rock properties in three dimensions	Lithology, structure, grade, mineralogy, density, hardness and other primary or response variables	Provide the spatial backbone for behavioural prediction at resource and ore control scale
Operational and metallurgical response models	Behaviour of rock under defined mining and processing conditions	Fragmentation response, throughput models, recovery models, process parameters	Translate spatial variables into performance outcomes
Planning and value models	Time-based and economic translation of technical outcomes	Schedules, blending plans, revenue, cost, NPV, ESG metrics	Convert behavioural predictions into production and value decisions
Data science and hybrid models	Multivariate relationship learning and uncertainty propagation	Machine-learning, feature extraction, hybrid geostatistical–ML approaches	Enhance classification, domaining, prediction and integration across model layers
Reconciliation frameworks	Performance comparison and feedback	Plan versus actual, production accounting, metallurgical balancing	Evaluate model performance and refine assumptions across all layers

Figure 4 adapts the materials science tetrahedron for rock characterisation and value-based decision-making. The central core represents the primary variables that define the rock, summarised as elements, minerals and textures. The outer nodes represent response domains associated with structure, properties, processing and performance along the Mine to Mill chain. The connecting links represent predictive geometallurgical relationships that translate intrinsic rock characteristics into expected behavioural responses. In this framing, rock characterisation is not an isolated analytical task. It is the causal bridge between what the rock is, how it behaves and how that behaviour influences operational and commercial outcomes.

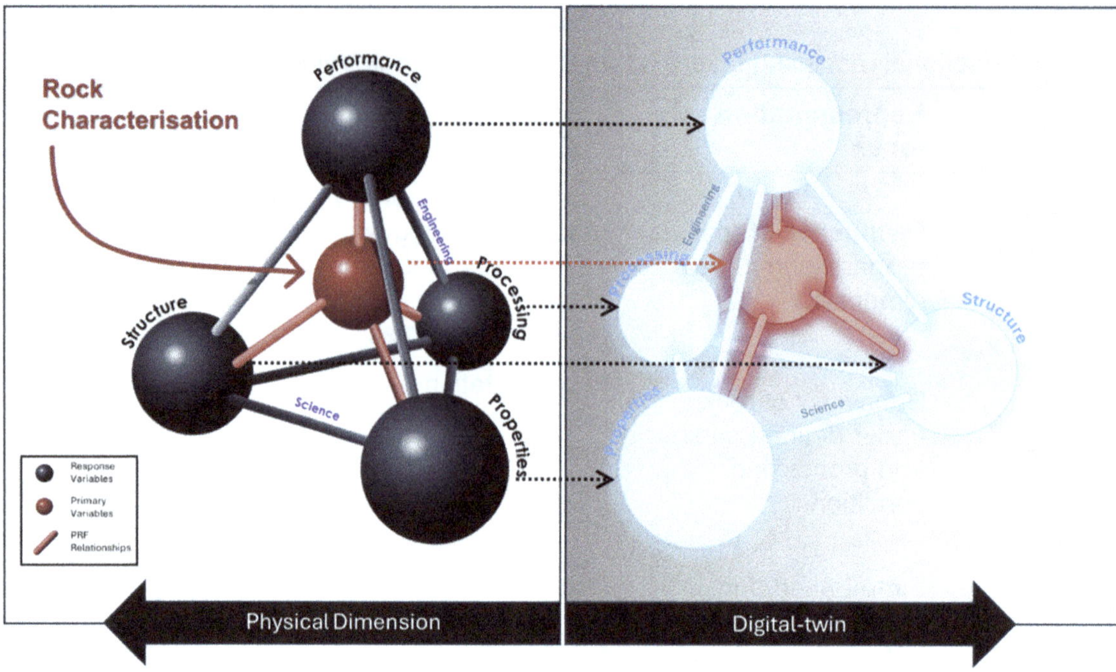

FIG 4 – The unified rock definition framework, integrating the Primary-Response concepts (Coward *et al,* 2009), materials science principles (Ellis*,* 1993), the sustainability and criticality extension (Donahue, 2019) and the digital twin formulation (Deagen *et al,* 2022).

Recent extensions to the classical materials science tetrahedron are directly relevant to value creation in mining systems. Donahue (2019) introduced an explicit sustainability and criticality dimension, recognising that material performance cannot be assessed in isolation from environmental and societal constraints. This perspective aligns closely with the geology-to-ESG risk work of Parbhakar-Fox and Baumgartner (2023), which frames geological characteristics as upstream drivers of downstream environmental and social outcomes.

The Unified Rock Value Framework (URVF) builds on these foundations by explicitly incorporating value and risk. In doing so, it extends the tetrahedron beyond material behaviour to encompass economic outcomes and ESG exposure. The resulting framework provides a holistic view of rock value, including the mechanisms through which value may be created, preserved, or destroyed across the mining value chain.

Deagen *et al* (2022) further advanced the concept through a digital-twin interpretation, pairing the physical tetrahedron with an information tetrahedron linked by continuous data flows. URVF adopts this dual-tetrahedron concept and extends it into the value domain. Within URVF, the material tetrahedron is enclosed by a value envelope that captures both economic performance and ESG risk. Machine-learning and data-science methods underpin the digital twin of the rock system, providing the connective tissue between primary variables, material response, and value metrics at the spatial resolution and temporal cadence required for operational decision-making and ore control.

URVF as a unified system-of-systems

To translate this conceptual structure into operational capability, URVF functions as a unified operating system in which four components work together. Direct measurement anchors behavioural truth through mineralogical, geochemical, metallurgical and petrophysical characterisation from intact core. Materials science provides the mechanistic framework for phenomenological models that represent process flexibility, value chain constraints to generate valid performance predictions. Sensing rapidly captures high-density operational signals that express aspects of structure, composition and texture at ore control scale. Machine-learning fuses direct and sensed data, performs feature extraction, classification, domaining and prediction, and propagates uncertainty to support value- and risk-informed routing.

Through this integration, primary variables can be estimated at operational resolution, response variables can be inferred within a physics-informed digital environment, and value metrics can be calculated for every block, dig line or shovel cut.

Routing decisions can then be based on expected behaviour and value contribution rather than grade alone, improving classification accuracy, stabilising plant feed and strengthening reconciliation across the mine-to-market value chain.

From prediction to decision readiness

Spatial modelling is a critical component of URVF, but it is not sufficient on its own. Erten *et al* (2025) demonstrate that hybrid geostatistical–machine-learning approaches, which combine co-kriging structures, machine-learning algorithms and human steering, outperform standalone methods. Their results show clear accuracy and interpretability gains when primary and secondary information are fused and when human–AI interaction is explicitly incorporated. These advances strengthen the spatial backbone of URVF and support more reliable behavioural prediction. Early operational examples reported by Tiu *et al* (2024) and Govender (2025) show how such hybrid approaches can be embedded into prediction for short-term mine planning and reconciliation workflows.

Decision readiness, however, extends beyond spatial estimation. It requires translating predicted rock characteristics into the physical and operational transformations that ultimately control value. Lechuti-Tlhalerwa, Coward and Field (2019) provide one of the few documented examples of an integrated geometallurgical value-chain model, spanning mining and blasting, hauling, stockpiling and blending, comminution, dense media separation, recovery circuits and materials handling. Their work highlights the importance of linking *in situ* rock properties to downstream responses such as fragmentation, diggability, stockpile residence effects, moisture dynamics, additivity behaviour, comminution response, liberation, recovery performance and circuit configuration, and ultimately to economic and ESG outcomes.

URVF binds these layers within a single, consistent behavioural structure. By doing so, variables estimated in spatial models can be carried forward coherently into operational, metallurgical and value models without conceptual drift. This progression from prediction to decision readiness is what enables behavioural insight to influence routing, blending and processing decisions in a timely and value-relevant manner.

The URVF integrated workflows

The URVF workflow is summarised in Figure 5 outlining the integrated rock-to-value ecosystem and formalises how characterisation, modelling, planning and reconciliation interact within a consistent behavioural and value framework within three unified workflows ecosystems.

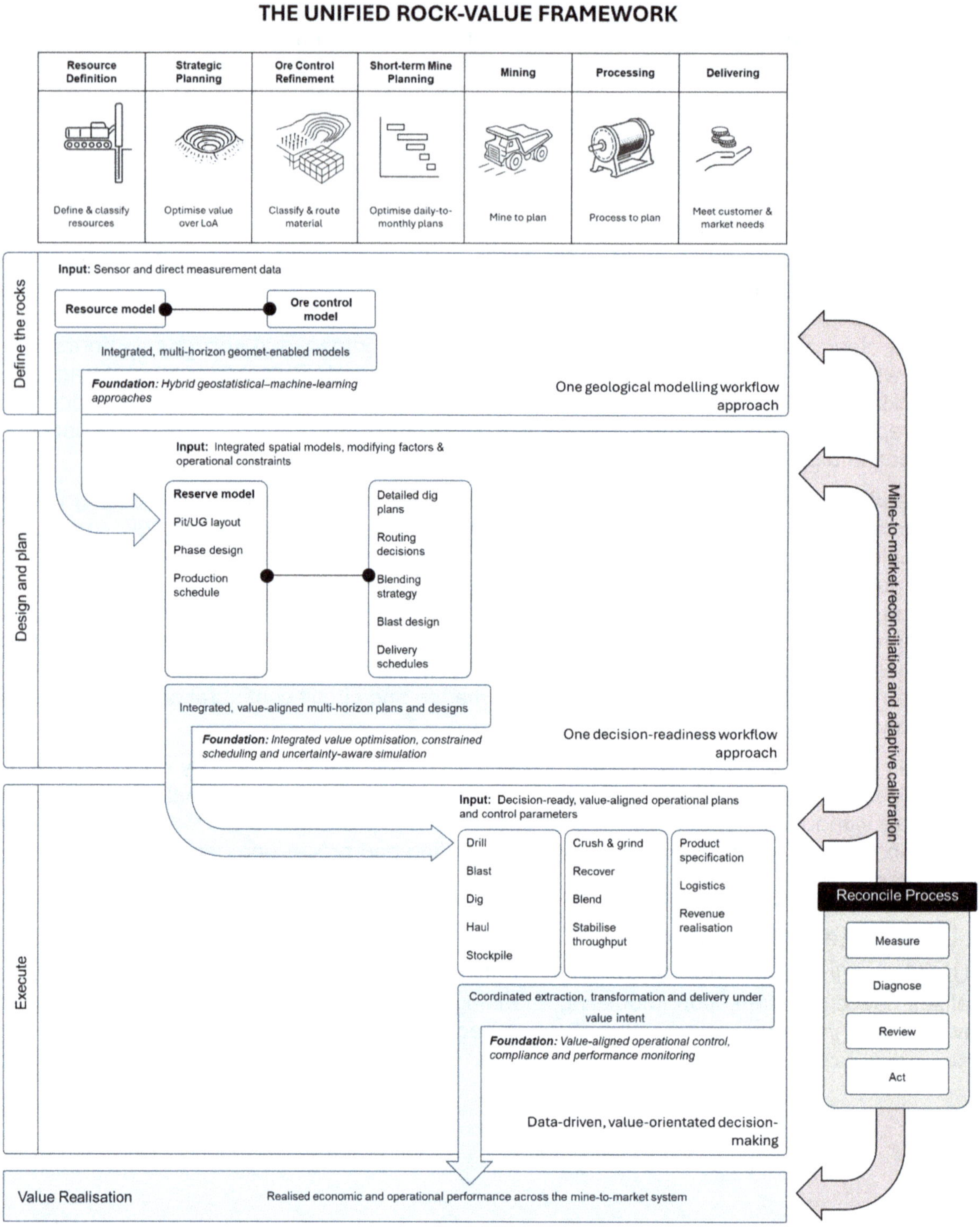

FIG 5 – Generalised schematic of a URVF-ready, integrated rock-to-value ecosystem.

Integrated rock characterisation: Commonality is created across resource definition and ore control through shared sensing, such as measure-while-drilling and downhole geophysics, and shared analytical measures such as XRF and FTIR. Direct measurement from strategic geometallurgical programmes provides the behavioural truth against which predictions are trained, allowing behaviour to be extrapolated where direct measurement is not feasible.

Integrated modelling: Spatial geological and geomet-enabled block models provide the primary and behavioural variables required for downstream translation. These variables are carried coherently into operational and metallurgical response models, where they are expressed as throughput, recovery, quality and cost outcomes under defined processing conditions. Planning and value models then propagate these behavioural outcomes into production schedules, revenue, cost and ESG metrics. Machine-learning supports classification, domaining and prediction across these layers, while hybrid geostatistical–machine-learning approaches provide a mechanism for integrating diverse data sources and expert knowledge (Erten *et al,* 2025).

Integrated planning, execution and reconciliation: Consistent behavioural and value definitions must persist across all planning horizons. When long-term and short-term planning are disconnected, value can be eroded at execution, even when upstream technical work is strong. URVF addresses this by ensuring that the same behavioural assumptions and value logic guide design, planning and execution across blasting, digging, hauling, stockpiling and processing. Spatio-temporal data from mining, haulage and plant sensors feed reconciliation loops that compare realised performance with expectation. Reconciliation frameworks such as those described by Parker (2012) and Morley and Arvidson (2017) provide a structured analyse, review and act cycle that progressively sharpens predictions and reduces operational uncertainty.

Through these integrated workflows, the URVF provides a single end-to-end behavioural system anchored in materials science, strengthened by direct measurement, extended by tactical sensing and made operational through data-driven modelling. It supplies the missing link between strategic understanding and tactical execution, enabling value-aligned ore control that is consistent across benches, shifts and planning horizons.

IMPLEMENTATION REQUIREMENTS

The URVF delivers value only when implemented as an operating system rather than as a technical study. Successful adoption depends less on any individual analytical method and more on decision clarity, organisational alignment, disciplined data systems and a managed pathway to trust. These elements enable behavioural insight to be translated into consistent, value-aligned operational decisions.

Decision-led workflow design

Implementation should begin with explicit clarity on which decisions matter, what value they influence and what uncertainty must be reduced for those decisions to be made confidently. Daily routing, blending, cut-line interpretation, stockpile management and feed sequencing each operate on different time horizons and affect different value levers. They therefore require different information, at different resolution and cadence.

Designing workflows backwards from these decisions ensures that data acquisition, sensing and modelling effort is focused on information that materially changes outcomes rather than on what is simply available. This aligns sampling density and sensing cadence with the tempo of ore control decisions, preserves consistency in value metrics across planning horizons and avoids unnecessary operational burden. Decision-led design is the primary safeguard against analytical effort becoming disconnected from value delivery.

Capability, alignment and trust

Decision-led design only works when geology, planning, mining, processing and marketing are aligned. Deep disciplinary expertise remains essential, but URVF implementation relies on a shared understanding of how rock characteristics drive behaviour and how behaviour translates into value.

URVF is a systems engineering problem. It is enabled by multidisciplinary teams comprising people with T-shaped capability profiles, meaning deep expertise in one domain combined with enough breadth to integrate, collaborate and make trade-offs across domains (Rogers and Freuler, 2015; Delicado, Salado and Mompó, 2018; Trogstad, Kokkula and Van Der Aker, 2021). In a geometallurgy context, this requires geologists and geometallurgists to link mineralogy, chemistry and texture to processing response and economics; planners to apply behavioural variables consistently across long-, medium- and short-term horizons; mining teams to understand how routing and execution affect downstream stability and blending; and processing and marketing to use upstream rock information to manage recovery, quality and risk.

Trust is built through transparency and routine learning. Teams need visibility of inputs, assumptions and uncertainty, supported by clear visualisation of actual versus planned outcomes. Confidence measures should enable adaptive decisions under uncertainty rather than forcing false precision. Trust accelerates when prediction is routinely compared with outcome: structured reconciliation across drilling, blasting, mining delivery, stockpiles and plant response shifts scepticism into adoption and reinforces learning over blame.

Clear decision rights, defined interfaces and shared terminology are non-negotiable. Where planning horizons remain siloed, value erosion persists at execution, making cross-horizon alignment a requirement rather than an aspiration.

Data governance and system connectivity

People make decisions, but only when data are accessible, trusted and connected. URVF does not require more data. It requires data with clear lineage, spatial and temporal integrity and consistent quality control.

Practical enablers include robust sampling quality assurance, consistent quality assurance and quality control across spatial and operational data sets, and unique identifiers that preserve sample history as material moves through the value chain. Data structures must also represent material transformation after extraction. Blasting alters fragmentation, handling affects moisture and fines, stockpiles change oxidation and reactivity, and blending introduces non-linear behaviour. If these transformations are not captured, the link between predicted behaviour and realised performance weakens rapidly, particularly across planning horizons.

A staged implementation roadmap

URVF implementation is most effective when approached as a staged learning process rather than a single deployment. The initial objective is to identify the ore control decisions that have the greatest influence on value, understand the behavioural drivers that affect those decisions, and demonstrate impact quickly. Focusing early effort on a small number of high-leverage decisions helps establish confidence and supports organisational adoption.

Once early impact is established, attention shifts to improving data readiness for those decisions, followed by progressive increases in confidence, spatial coverage and integration across workflows. This approach allows capability to mature without disrupting operations, while ensuring that each stage remains anchored to decision relevance and value creation.

Figure 6 presents an illustrative implementation roadmap that translates these principles into a practical sequence of activities. While durations will vary by operation, the progression is broadly consistent.

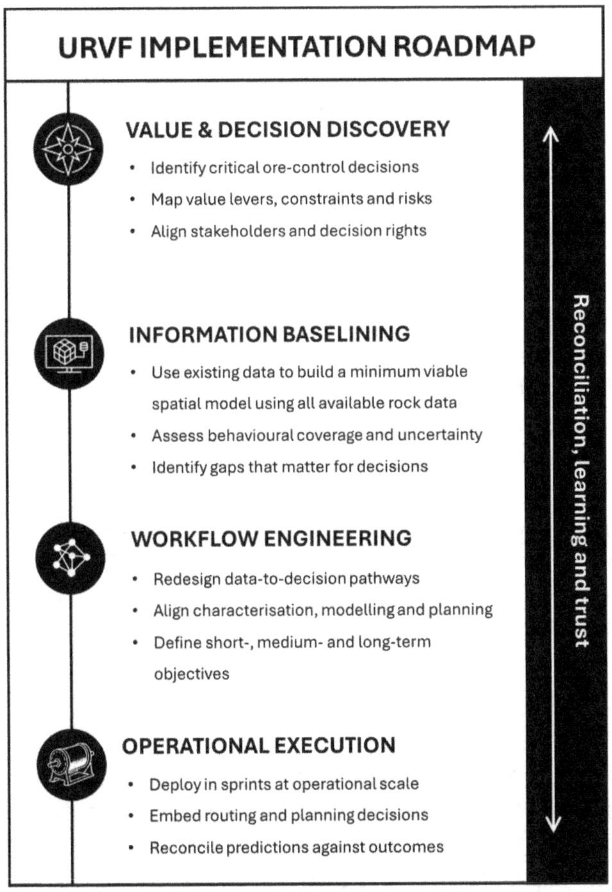

FIG 6 – Illustrative staged implementation roadmap for URVF-based ore control.

Implementation typically begins with value and decision discovery, focused on clarifying stakeholder objectives, decision constraints, value drivers and risk exposure. This establishes a shared understanding of where behavioural uncertainty is eroding value and where intervention will matter most.

The next stage involves information baselining, where existing data are collated and used to construct a minimum viable spatial and behavioural model. The intent is diagnostic rather than optimised performance, enabling the operation to assess what can already be achieved and where targeted enhancement will most effectively reduce uncertainty.

Workflow redesign follows, aligning rock characterisation, modelling, planning and execution so that behavioural and value information flows coherently across the system. Strategic objectives are explicitly connected to near-term delivery horizons to support incremental and measurable value realisation.

The final stage focuses on operational deployment supported by feedback and reconciliation loops. As predictions are compared with realised performance, the system is progressively refined, strengthening confidence and improving alignment across planning horizons over time.

DISCUSSION

The persistent gap between strategic planning intent and operational execution is not primarily a data problem, nor a modelling problem, but a systems problem. Grade-based ore control persists not because its limitations are unknown, but because it has historically been the only information available at operational density and decision tempo. While recent advances have improved access to behavioural signals at ore control scale, most operations still lack a coherent way to connect these signals to planning, execution and value realisation. The URVF addresses this gap by providing a disciplined structure that links rock characteristics, behaviour and value consistently across planning horizons.

Positioning relative to existing practice

Strategic geometallurgy has demonstrably reduced long-term uncertainty through direct measurement and spatial modelling of behaviour-critical variables. Tactical geometallurgy, as articulated by McKay *et al* (2016), identified the business requirement to bring behavioural understanding closer to the point of operational decision-making. More recently, sensing technologies and machine-learning models have enabled increasingly detailed behavioural prediction at short-range scales.

However, in most operations sensing technologies are deployed as inputs to discrete technical solutions rather than as components of an integrated system. Different sites apply different tools: hyperspectral data may delineate lithological variation, measure-while-drilling signals may support geotechnical or comminution domaining, downhole geophysics may provide density or alteration proxies, and FTIR may inform mineralogical or recovery-related insights. Although each application improves local characterisation or prediction, the resulting behavioural signals are rarely carried coherently through modelling and planning workflows. Where spatial coverage is sufficient, response variables such as recovery may be estimated statistically; where data are limited, modifying factors are introduced or broad assumptions adopted, for example assigning a flat recovery to a lithological unit. In this way, behavioural understanding becomes piecemeal and inconsistently translated into operational and economic terms. The result is a set of isolated capabilities that enhance local prediction accuracy but do not deliver sustained operational improvement because they are not systematically integrated across characterisation, modelling, planning, execution and value assessment.

URVF does not replace existing geometallurgical practice. Rather, it provides the missing connective framework between strategic understanding and operational execution. It preserves the physical discipline of the Primary–Response Framework while extending it through tactical sensing, data-driven modelling and explicit value-centred decision logic. In doing so, ore control is reframed from a classification exercise into a decision system that explicitly manages behavioural response and value under operational constraints.

Why integration matters more than prediction accuracy

A central implication of this work is that improving prediction accuracy alone is insufficient to enhance inclusive value delivery. Many operations already possess technically robust geological models and sophisticated metallurgical simulations, yet still experience feed instability, reconciliation drift and value erosion. URVF highlights that value loss most often arises at the interfaces between systems, planning horizons and disciplines rather than within any single model.

By enforcing consistent behavioural and value definitions from characterisation through to execution, URVF reduces interpretation drift and cross-functional misalignment. The framework prioritises decision-readiness over model completeness, recognising that a timely behavioural prediction capable of informing routing can generate greater value than a highly accurate estimate delivered too late to influence operational decisions (Godoy and Dimitrakopoulos, 2004).

Conceptual status and limits of applicability

URVF is presented as a unifying conceptual and operational framework rather than a fully deployed, single-instance system. To the authors' knowledge, no operation has yet implemented URVF in its entirety. Instead, most operations exhibit partial alignment with its components, often unknowingly, through isolated sensing initiatives, geometallurgical models or reconciliation practices.

The contribution of URVF lies in making these connections explicit and providing a structure through which they can be deliberately aligned and scaled. Its effectiveness depends on the availability of at least some strategic direct-measurement data to anchor behavioural relationships. In orebodies with extremely limited test work or highly stochastic behaviour, early application may focus on stabilisation and learning rather than optimisation.

The framework is intentionally modular. Not all operations require the same sensing density, modelling complexity or organisational change. Attempting to deploy all components simultaneously, without decision clarity or readiness, risks recreating the very complexity URVF is designed to manage.

Implications for planning and reconciliation

URVF has important implications for reconciliation practice. Traditional reconciliation often focuses on variance attribution after the fact. URVF embeds reconciliation as a continuous learning mechanism, where behavioural assumptions are routinely tested against realised performance and fed back into models and decisions.

The framework also challenges the conventional separation of long-term and short-term planning. When these functions operate on different behavioural definitions, value erosion can occur even in technically strong operations. URVF provides a common behavioural and value language that allows planning horizons to differ in resolution without diverging in logic.

Future directions

While this paper focuses on ore control, the principles underlying URVF extend naturally to broader value-chain optimisation, including waste management, ESG risk control and closure planning. Future work should focus on operational case studies, formal quantification of value-at-risk associated with behavioural uncertainty, and tighter integration of market and price dynamics into routing logic.

As hybrid geostatistical and machine-learning approaches continue to mature, the technical barriers to implementation are diminishing. The remaining challenge lies in disciplined system design, organisational alignment and trust. Addressing these elements is likely to deliver greater value uplift than further incremental improvements in model accuracy alone.

CONCLUSION

Grade-based ore control remains necessary in modern mining operations, but it is insufficient to manage value under conditions of increasing geological complexity, operational intensity and performance variability. Value is governed by how rock behaves as it moves through blasting, handling and processing, yet behavioural variability is rarely represented at the scale and cadence at which ore control decisions are made. This misalignment contributes directly to throughput instability, reconciliation drift and avoidable value erosion.

The Unified Rock Value Framework (URVF) addresses this gap by providing a behaviour-centred structure that links intrinsic rock characteristics to processing response and translates that response into decision-relevant measures of value. By integrating materials science principles, the geometallurgical Primary–Response Framework, direct measurement, sensing and data-driven modelling, URVF enables behavioural prediction at operational resolution without sacrificing physical meaning or value relevance.

URVF reframes ore control from a grade-based classification task into a decision-optimisation problem. Routing decisions are evaluated on expected value contribution, accounting for recovery, throughput, cost and risk, rather than grade alone. In doing so, the framework shifts emphasis from isolated prediction accuracy to decision readiness, improving classification consistency, stabilising plant feed and strengthening reconciliation across the mine-to-market value chain.

Importantly, URVF is presented as a unifying framework rather than a single deployed solution. Many operations already implement elements of this approach in isolation, through sensing technologies, geometallurgy models or reconciliation practices. URVF makes these connections explicit and provides a disciplined structure through which they can be aligned, scaled and sustained across planning horizons.

The timing for this shift is significant. Advances in sensing, data infrastructure and machine-learning now allow behavioural insight to be delivered at ore control cadence using data that are already routinely collected. What was previously constrained by sampling density, turnaround time and integration complexity is now technically achievable within operational workflows.

Adopting URVF requires changes in practice rather than wholesale system replacement. Implementation is decision-led, depends on organisational alignment across planning horizons and relies on disciplined data governance and structured feedback loops. When implemented as an operating system rather than a technical study, URVF provides a practical pathway for embedding geometallurgical insight into daily decision-making.

By unifying rock characterisation, behavioural prediction and value modelling within a single framework, URVF supplies the missing link between strategic understanding and tactical execution. It provides a foundation for more predictable, resilient and value-aligned ore control, supporting improved performance across benches, shifts and planning horizons.

ACKNOWLEDGEMENTS

The authors acknowledge the support and contributions of colleagues at IMDEX Digital and Interlaced Systems, whose geoscience, processing, geometallurgical, data-science and strategic advisory teams continue to advance the practical application of integrated rock characterisation and decision-focused workflows across the mining value chain. The depth of technical capability, independent thinking and collaborative culture across these organisations has been central to the development and framing of the concepts presented in this paper.

Sincere thanks are extended to industry partners and collaborators who have engaged openly with the ideas explored here and provided valuable operational insight. Finally, the authors recognise the many professionals encountered throughout their careers across geoscience, mining, processing, planning and marketing, whose experience and judgement continue to shape how rock behaviour, decisions and value are understood in practice.

REFERENCES

Acosta, I C C, Khodadadzadeh, M, Tusa, L, Ghamisi, P and Gloaguen, R, 2020. A machine learning framework for drill core mineral mapping using hyperspectral and high-resolution mineralogical data fusion, *IEEE Journal of Selected Topics in Applied Earth Observations and Remote Sensing*, 13:4011–4024.

Avalos, S and Ortiz, J M, 2019. Recursive convolutional neural networks in a multiple-point statistics framework, *Mathematical Geosciences*, 51:917–949.

Baizhiyen, N, Dowd, P, Xu, C and Lewis, D, 2024. Numerical Investigation of the Applicability of Preferential Grade Deportment by Size, *Minerals*, 14:670.

Beniscelli, J, 2011. Geometallurgy – Fifteen Years of Developments in Codelco: Pedro Carrasco Contributions, in *Proceedings of the First AusIMM International Geometallurgy Conference,* pp 3–7 (The Australasian Institute of Mining and Metallurgy: Melbourne).

Beniscelli, J, Carrasco, P, Dowd, P, Ferguson, G and Talcanaza, E, 2000. Estimation of resources and conversion to reserves, Protocols for the assessment, reduction and management of risk, in *Proceedings of MassMin 2000*, pp 9–16 (The Australasian Institute of Mining and Metallurgy: Melbourne)..

Bennett, D W and Munro, P D, 2024. Practical geometallurgy-and let there be light, in *Proceedings of the Mill Operators Conference*, pp 1–14 (The Australasian Institute of Mining and Metallurgy: Melbourne).

Boisvert, J B, Rossi, M E, Ehrig, K and Deutsch, C V, 2013. Geometallurgical modeling at Olympic Dam mine, South Australia, *Math Geosci*, 45:901–925.

Bueno, M, Almeida, T, Lara, L, Powell, M and Delboni, H, 2024. Accuracy and Precision of the Geopyörä Breakage Test, Preprints.org. Available from: <https://www.preprints.org/manuscript/202407.0934> [Accessed: 30 March 2026].

Butcher, A R, Dehaine, Q, Menzies, A H and Michaux, S P, 2023. Characterisation of Ore Properties for Geometallurgy, *Elements,* 19:352–358.

Callister Jr, W D and Rethwisch, D G, 2020. *Materials Science and Engineering: an Introduction* (John Wiley and Sons).

Coward, S, Vann, J, Dunham, S and Stewart, M, 2009. The primary-response framework for geometallurgical variables, in *Proceedings of the Seventh International Mining Geology Conference*, pp 109–113 (The Australasian Institute of Mining and Metallurgy: Melbourne).

Datarock Chip, 2025. Machine-learning workflows to extract value from chip imagery, 2025. Datarock. Available from: <https://datarock.com.au/datarock-chip-software/> [Accessed: 12 Oct 2025].

David, D, Govender, I and Mainza, A N, 2018. The integration of wireline petrophysical measurements and geometallurgical studies to predict comminution characteristics, *Minerals Engineering*, 122:120–132.

Deagen, M E, Brinson, L C, Vaia, R A and Schadler, L S, 2022. The materials tetrahedron has a 'digital twin', *MRS Bulletin*, 47:379–388.

Dehaine, Q, Tijsseling, L T, Rollinson, G K, Buxton, M W N and Glass, H J, 2021. Geometallurgical characterisation with portable FTIR: application to sediment-hosted Cu-Co ores, *Minerals*, 12:15.

Delicado, B A, Salado, A and Mompó, R, 2018. Conceptualization of a T-Shaped engineering competency model in collaborative organizational settings: Problem and status in the Spanish aircraft industry, *Systems Engineering*, 21:534–554.

Deutsch, J L, Palmer, K, Deutsch, C V, Szymanski, J and Etsell, T H, 2016. Spatial Modeling of Geometallurgical Properties: Techniques and a Case Study, *Nat Resour Res*, 25:161–181.

Dominy, S C, O'Connor, L, Parbhakar-Fox, A, Glass, H J and Purevgerel, S, 2018. Geometallurgy – a route to more resilient mine operations, *Minerals*, 8:560.

Donahue, C J, 2019. Reimagining the materials tetrahedron, *J Chem Educ,* 96:2682–2688.

Dowd, P A and David, M, 1976. Planning from estimates: sensitivity of mine production schedules to estimation methods (eds: M Guarascio, M David and C Huijbregts), *Advanced Geostatistics in the Mining Industry*, pp 163–183 (Springer: Netherlands).

Dowd, P A and Pardo-Iguzquiza, E, 2006. Core-log integration: optimal geostatistical signal reconstruction from secondary information, *Applied Earth Science*, 115:59–70.

Egaña, Á F, Santibáñez-Leal, F A, Vidal, C, Díaz, G, Liberman, S and Ehrenfeld, A, 2020. A robust stochastic approach to mineral hyperspectral analysis for geometallurgy, *Minerals*, 10:1–32.

Ehrig, K, Liebezeit, V, Smith, M, Macmillan, E and Lower, C, 2014. Geologists and the value chain – how material characterisation by modern mineralogy can optimise design and operation of processing facilities, in *Proceedings of the Ninth International Mining Geology Conference*, pp 5–13 (The Australasian Institute of Mining and Metallurgy: Melbourne).

Ellis, A B, 1993. *Teaching General Chemistry: A Materials Science Companion* (ERIC).

Erten, G E, Mokdad, K, Nisenson, J, Brandao, G and Boisvert, J, 2025. Human-AI interaction: machine learning-based geostatistical hybrid models, *Applied Soft Computing*, 182.

Fallon, G N, Fullagar, P K and Sheard, S N, 1997. Application of geophysics in metalliferous mines, *Australian Journal of Earth Sciences*, 44:391–409.

Fullagar, P K and Fallon, G N, 1997. Geophysics in metalliferous mines for orebody delineation and rock mass characterisation, in *Proceedings of the Fourth Decennial International Conference on Mineral Exploration – Exploration '97*, pp 573–584.

Germiquet, J P and Minnitt, R C A, 2016. Rock Strength and Geometallurgical Modelling, Mogalakwena Mine, *Journal of the Southern African Institute of Mining and Metallurgy*, 116:247–250.

Godoy, M and Dimitrakopoulos, R, 2004. Managing risk and waste mining in long-term production scheduling of open pit mines, *SME Transactions*, 316.

Goodfellow, R and Dimitrakopoulos, R, 2016. Global optimization of open pit mining complexes with uncertainty, *Applied Soft Computing*, 40:292–304.

Govender, C N, 2025. Advancing geometallurgical modelling with machine learning: recovery and comminution hardness estimation at Mogalakwena Platinum Mine, in *Proceedings of the SAIMM Geometallurgy Conference 2025* (Southern African Institute of Mining and Metallurgy: Muldersdrift).

Gupta, A and Yan, D S, 2016. *Mineral Processing Design and Operations: an Introduction* (Elsevier).

Harraden, C L, Scher, S, Deyell-Wurst, C and Carey, R, 2019. Geometallurgical Considerations: Processing Mineralogy vs Alteration Footprints, in *Proceedings of the Procemin Geomet 2019* (Gecamin: Santiago).

Hatherly, P, 2013. Overview on the application of geophysics in coal mining, *International Journal of Coal Geology*, 114:74–84.

Jaimez-Salgado, E, Ortiz, J M and Rodríguez, C, 2021. Comparative analysis of machine learning techniques for the estimation of flotation response based on mineralogy, *Minerals Engineering*, 167.

Jang, H and Topal, E, 2020. Transformation of the Australian mining industry and future prospects, *Mining Technology*, 129:120–134.

Johnson, C L, Browning, D A and Pendock, N E, 2019. Hyperspectral Imaging Applications to Geometallurgy: Utilizing Blast Hole Mineralogy to Predict Au-Cu Recovery and Throughput at the Phoenix Mine, Nevada, *Economic Geology*, 114:1481–1494.

Kakani, S L and Kakani, A, 2004. *Material Science* (New Age International).

Kitzig, M C, Kepic, A and Grant, A, 2018. Near Real-Time Classification of Iron Ore Lithology by Applying Fuzzy Inference Systems to Petrophysical Downhole Data, *Minerals*, 8:276.

Lamberg, P, 2011. Particles – the bridge between geology and metallurgy, in *Proceedings of the Konferens I Mineralteknik 2011* (Luleå Tekniska Universitet).

Lechuti-Tlhalerwa, R, Coward, S and Field, M, 2019. Embracing step-changes in geoscientific information for effective implementation of geometallurgy, *Journal of the Southern African Institute of Mining and Metallurgy*, 119:355–360.

Linton, P, Montoya, P, Jahoda, R, Harris, P and Pendock, N, 2013. The application of hyperspectral core imaging to exploration and geometallurgy: a case study from the La Colosa gold porphyry, Colombia, in *Proceedings of the Second AusIMM International Geometallurgy Conference* (The Australasian Institute of Mining and Metallurgy: Brisbane).

Lishchuk, V, Koch, P H, Ghorbani, Y and Butcher, A R, 2020. Towards integrated geometallurgical approach: Critical review of current practices and future trends, *Minerals Engineering*, 145.

Matus, T C, 2020. Development of a methodology and validation of the Geopyörä breakage test, MSc thesis, The University of Oulu, Oulu.

McCoy, J T and Auret, L, 2019. Machine learning applications in minerals processing: A review, *Minerals Engineering*, 132:95–109.

McKay, N, Vann, J, Ware, W, Morley, C and Hodkiewicz, P, 2016. Strategic and tactical geometallurgy – a systemic process to add and sustain resource value, in *Proceedings of the Third AusIMM International Geometallurgy Conference*, pp 29–36 (The Australasian Institute of Mining and Metallurgy: Melbourne).

Merrill Cifuentes, J, 2024. Quantification of mineral textures for geometallurgical predictive modelling, PhD thesis, The University of Tasmania, Tasmania.

Morales, N, Seguel, S, Cáceres, A, Jélvez, E and Alarcón, M, 2019. Incorporation of geometallurgical attributes and geological uncertainty into long-term open pit mine planning, *Minerals*, 9(2):108.

Morley, C and Arvidson, H, 2017. Mine value chain reconciliation-demonstrating value through best practice, in *Proceedings of the Tenth International Mining Geology Conference*, pp 279–292 (The Australasian Institute of Mining and Metallurgy: Melbourne).

Morley, C, 2024. Ore control based on value – experiences in design and execution, in *Proceedings of the AusIMM International Mining Geology Conference 2024*, pp 209–219 (The Australasian Institute of Mining and Metallurgy: Melbourne).

Olson, G B, 1997. Computational Design of Hierarchically Structured Materials, *Science*, 277:1237–1242.

Ortiz, J M, Kracht, W, Townley, B, Lois, P, Cárdenas, E, Miranda, R and Álvarez, M, 2020. Workflows in geometallurgical prediction: Challenges and outlook, *Mining, Metallurgy and Exploration*, 37:1529–1540.

Parbhakar-Fox, A and Baumgartner, R, 2023. Action Versus Reaction: How Geometallurgy Can Improve Mine Waste Management Across the Life-Of-Mine, *Elements*, 19:371–376.

Parker, H M, 2012. Reconciliation principles for the mining industry, *Mining Technology*, 121:160–176.

Rogers, P and Freuler, R J, 2015. The "T-shaped" engineer, in *Proceedings of the 122nd ASEE Annual Conference and Exposition*, pp 26–1507 (American Society for Engineering Education: Seattle).

Rollinson, H R, 1993. A terrane interpretation of the Archaean Limpopo Belt, *Geological Magazine*, 130:755–765.

Schouwstra, R, Vaux, D D, Muzondo, T and Prins, C, 2013. A Geometallurgical Approach at Anglo American Platinum's Mogalakwena Operation, in *Proceedings of the Second AusIMM International Geometallurgy Conference (GeoMet)*, pp 85–92 (The Australasian Institute of Mining and Metallurgy: Melbourne).

Shirmard, H, Farahbakhsh, E, Müller, R D and Chandra, R, 2022. A review of machine learning in processing remote sensing data for mineral exploration, *Remote Sensing of Environment*, 268.

Silversides, K, Battalgazy, N and Dutch, R, 2025. Automated Coal Picking in MWD Logs, in *Proceedings of the SEG Conference* (Society of Economic Geology: Brisbane).

Stone, P, Froyland, G, Menabde, M, Law, B, Pasyar, R and Monkhouse, P, 2018. Blasor—Blended iron ore mine planning optimisation at Yandi, Western Australia, *Interfaces*, 48(1):33–48.

Tiu, G, 2017. Classification of Drill Core Textures for Process Simulation in Geometallurgy : Aitik Mine, MSc thesis, The Luleå University of Technology, Sweden.

Tiu, G, Ghorbani, Y, Jansson, N, Wanhainen, C and Bolin, N J, 2021. Texture-based classification in optical microscopy and FIB-SEM images through machine learning, *Minerals*, 11(7):738.

Tiu, G, Ghorbani, Y, Jansson, N, Wanhainen, C and Bolin, N-J, 2023. Quantifying the variability of a complex ore using geometallurgical domains, *Minerals Engineering*, 203:108323.

Tiu, G, Wanhainen, C, Jansson, N, Liwicki, F and Sand, A, 2024. Data fusion using machine learning: Towards real-time implementation of geometallurgical models for ore tracking, in *Proceedings of the Process Mineralogy'24*, South Africa.

Trogstad, I K, Kokkula, S and Van Der Aker, J, 2021. Application of T-shaped engineering skills in complex multidisciplinary projects, in *Proceedings of the INCOSE International Symp,* 31:1113–1129.

Tusa, L, Frenzel, M, Pereira, L, Thiele, S, Tolosana-Delgado, R and Gutzmer, J, 2025. Geometallurgy: Future Directions, *SEG Discovery*, 140:27–39.

Vatandoost, A, 2010. Petrophysical Characterization of Comminution Behavior, PhD thesis, The University of Tasmania, Tasmania.

Vezhapparambu, V S, 2016. Increased geometallurgical performance in industrial mineral operations through multivariate analysis of MWD-data, *Mineralproduksjon*, 7.

Wills, B A and Finch, J, 2015. *Wills' Mineral Processing Technology*: *An Introduction to the Practical Aspects of Ore Treatment and Mineral Recovery* (Butterworth-Heinemann).

Wyllie, P J, 1992. Experimental petrology: Earth materials science, in *Understanding the Earth*, pp 67–87 (Cambridge University Press, Cambridge).

Open to anything

Case studies

It's all about relationships – the use of ML in density imputation at Dugald River

M Angus[1] and D Kaeter[2]

1. Technical Consulting Director, ERM Australia, Brisbane Qld 4000. Email: maree.angus@erm.com
2. Managing Technical Consultant, ERM Canada, Vancouver BC V6E 4A6, Canada. Email: david.kaeter@erm.com

INTRODUCTION

A reliable bulk density (BD) data set is critical to the robustness of any Mineral Resource estimate (MRE) as it underpins tonnage and metal calculations. Representative coverage of BD measurements across a deposit can be difficult to achieve, as can consistency of measurement protocols. This is often the case when many phases of work have been completed, often by different companies.

Machine learning (ML) has been used at the Dugald River zinc-lead-silver mine to generate additional proxy density data and to review historical measurements for consistency. The process has been refined over several iterations, and a hybrid data set is now used for Mineral Resource estimation.

CASE STUDY

The Dugald River Zn-Pb-Ag mine is located approximately 65 km north-west of Cloncurry in Queensland. The main Dugald Lode is hosted within a major north–south striking, steeply west dipping shear zone which cross-cuts the strike of the Dugald River Slate stratigraphy from hanging wall to footwall at shallow depths and to the north.

The deposit was first drilled in the late 1960s, with sporadic drill programs up until 1988, at which time more intensive drilling programmes began. As at many early-stage exploration projects of that era (and still some today), BD measurements were taken on 10–15 cm pieces of drill core that were deemed representative, generally at a rate of one measurement per drill core tray. From around 2015/2016 onwards, BD measurements were generated on the full assay-sample length. Assessment of this more comprehensive data set demonstrated that BD is variable across the orebody and that the BD values for older drill holes often are inconsistent with those obtained from full-assay-length density measurements (Figure 1).

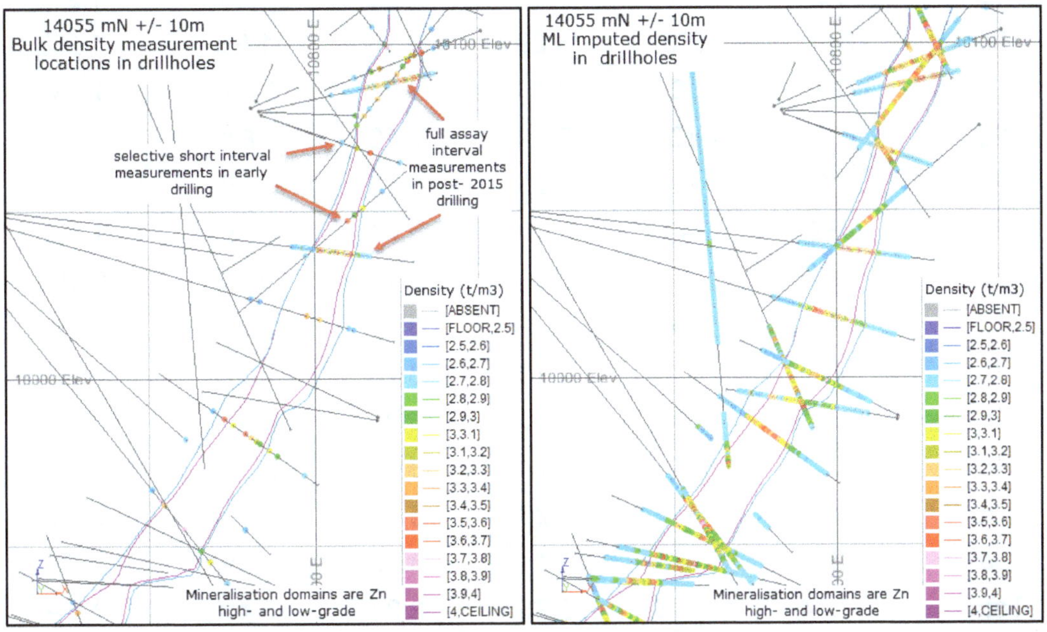

FIG 1 – Example of selective versus full length BD measurements and ML BD predictions.

The ability of ML-based tools to consider a larger number of variables when defining relationships was seen as an opportunity for improvement of the density estimation process with two main aims: 1) assess the local density changes in the estimate when both the primary and infill production drill fans have BD data (measurements only collected on primary drill fans); and 2) improve the granularity of BD data across the orebody at depth.

A hybrid data set comprising the historic point measurements as well as the more recent full-assay-length measurements was used for the initial ML density data imputation (Extreme Gradient Boosting). Input variables included zinc, lead, silver, copper, silver, iron, copper/zinc, zinc/iron and copper/iron. Three separate regression models were generated, for samples of zinc-lead-silver mineralisation, copper mineralisation and waste. The resulting ML-based density predictions were used to supplement the measured data to estimate density into the block model.

During the initial round of ML modelling by Datarock in 2023, issues were identified with the training data set, particularly for the modelling in the waste domain, where the holdout samples (subset of data withheld from the training data set) had a separate population in the predicted data with a mean of ~2.75 t/m^3 plotting against measured values from 2.0 t/m^3 up to almost 4.5 t/m^3. Site geologists suggested possible explanations that might contribute to the issue: differences in lithology and alteration or domaining issues. Nothing under consideration explained the larger deviations observed in the scatter plots. However, despite the issues, the initial ML-based models performed marginally better than the previous regression models, so the density predictions were added to the data set for the MRE. A comparison between the density estimate generated using the combined data set and the data set of measured values showed a global tonnage difference in the mined area of <0.5 per cent. On a mining area basis, local differences were variable (Table 1, Figure 2).

TABLE 1

Comparison of mined tonnage above cut-off, to end of December 2022.

Mining area	Tonnage (kt)		Bulk density (t/m^3)		Percentage difference	
	Measured	ML	Measured	ML	Tonnage	Bulk density
1a	254	253	3.37	3.36	-0.1%	-0.1%
1b	1083	1059	3.36	3.29	-2.2%	-2.2%
1c	527	529	3.18	3.19	0.2%	0.2%
1d	822	813	3.21	3.17	-1.1%	-1.1%
1e	509	503	3.25	3.22	-1.2%	-1.2%
2b	2611	2578	3.29	3.25	-1.2%	-1.2%
2c	279	280	3.19	3.21	0.5%	0.5%
2d	474	482	3.14	3.19	1.8%	1.8%
2e	111	112	3.20	3.23	0.7%	0.7%
3b	3732	3743	3.23	3.24	0.3%	0.3%
3c	137	137	3.18	3.17	-0.2%	-0.2%
3d	597	599	3.21	3.22	0.2%	0.2%
3e	7	7	3.15	3.20	1.5%	1.5%
4b	690	689	3.20	3.20	-0.1%	-0.1%
4c	12	12	3.24	3.25	0.3%	0.3%
4d	9	9	3.28	3.31	0.8%	0.8%
5d	0.35	0.37	3.14	3.25	3.7%	3.5%
Total	11 854	11 806	3.25	3.23	-0.4%	-0.1%

Reporting inside stope and development shapes, with 2023 net smelter return (NSR)>AU$161.

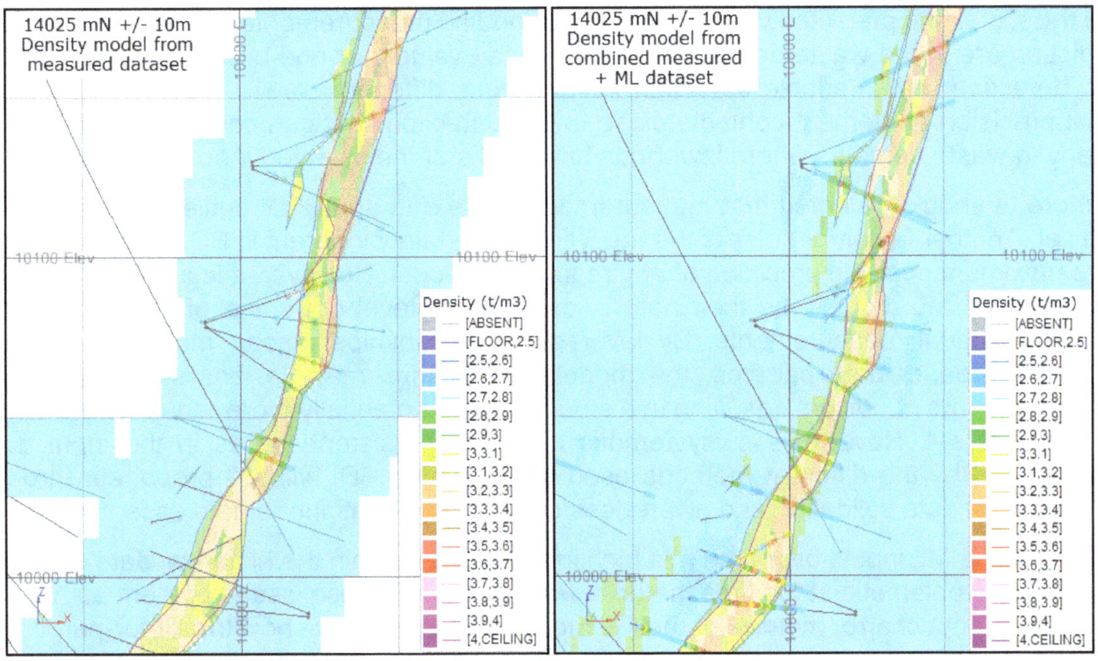

FIG 2 – Example section through estimates using measured data and combined measured + ML data set.

Subsequent investigations determined that there are likely to be two main contributors to the anomalous trends and noise observed for the holdout data and in validation plots for the whole of the imputed data set:

The historic point BD measurements are not always representative when the whole assay interval is considered. Consequently, the pre-2016 measurements were removed from the ML modelling data set for 2025 and this improved the model performance (Figure 3).

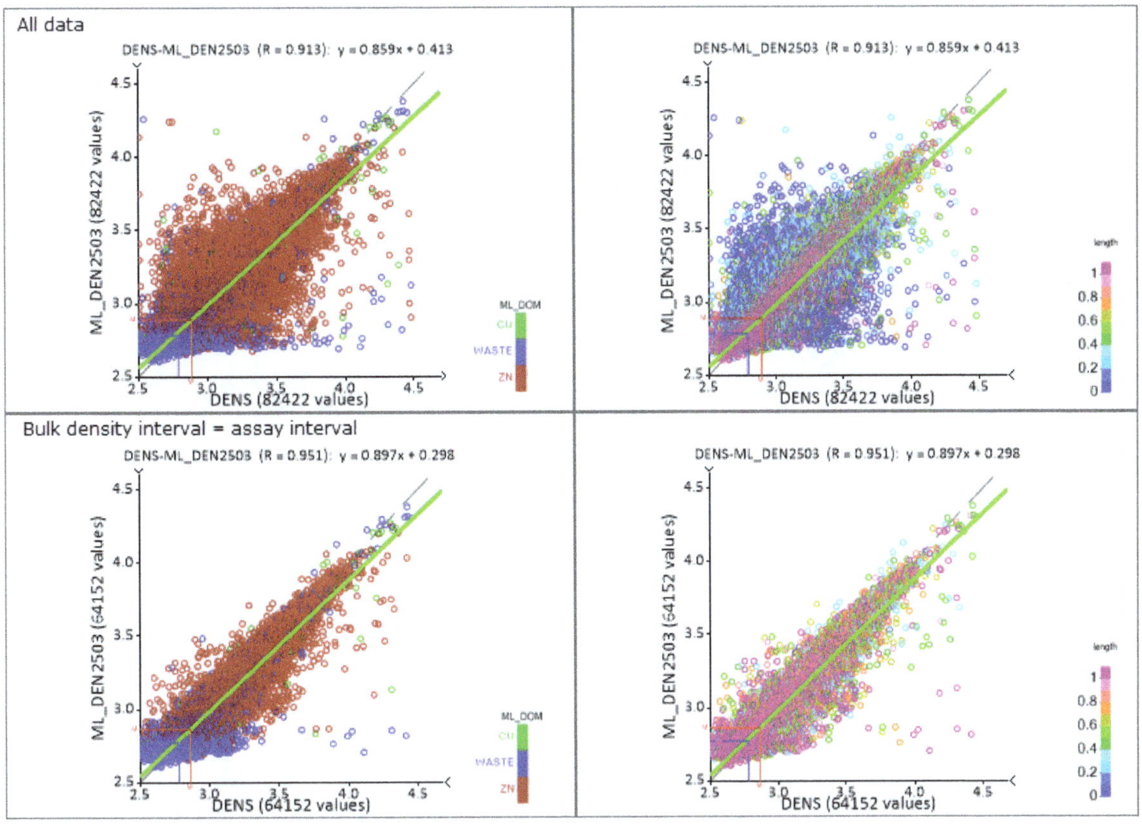

FIG 3 – Scatterplots of measured versus predicted BD for all data versus data where BD sample interval is equal to the assay interval.

Many of the remaining predicted values with corresponding measurements >10 per cent higher than the prediction are in the waste domain and have assay values for one or more of the metals at or close to detection limit. Variable detection limits across differing assay methodologies and poor analytical precision at element contents close to the detection limit can contribute to model noise, particularly in waste domains where lower concentrations of the selected input variables are seen.

Furthermore, it should be noted that measurement errors and analytical outliers can be present in the data set. Another factor for noise is the analytical uncertainty inherent in BD measurements. This limits the maximum possible precision of any regression model. However, it is also worth noting that regression-based ML models, by their nature, can help to level out some of the analytical noise present in individual measurements. By leveraging relationships among multiple variables and learning from broader data patterns, the model's predictions may be less sensitive to isolated measurement errors or outliers, resulting in a smoother and potentially more robust estimation of BD across the data set. Nevertheless, systematic errors or persistent biases in the input data, for example caused by the different methods used to determine BD, will still propagate through the model, so careful data validation and awareness of data set limitations remain essential.

Refinement of the ML inputs over time has highlighted issues within the historical data set that were not considered problematic at the time the data was collected. These issues present as noise in the validation plots, impact error metrics for the ML model, and underscore how the 'fit-for-purpose'-ness of a data set can shift as analytical methods and project requirements evolve. This phenomenon is likely common across many exploration data sets. Generation of ML models without a thorough understanding and ongoing validation of data set limitations can result in less reliable outcomes. Continuous scrutiny and refinement of both data acquisition and modelling approaches are, therefore, essential to ensure that resource estimates remain robust and defensible as new data and techniques become available.

Third time's the charm – a case study on the restart at Woodlawn Mine

H T Bannister[1] and D E Taylor[2]

1. Geology Superintendent, DEVELOP Global Ltd., Perth WA 6100.
   Email: hugh.bannister@develop.com.au
2. Senior Exploration Geologist, DEVELOP Global Ltd., Perth WA 6100.
   Email: daniel.taylor@develop.com.au

ABSTRACT

The restart of the Woodlawn volcanogenic massive sulfide (VMS) mine in New South Wales presented an uncommon geological and organisational challenge: rebuilding the full geology function while underground development, process plant recommissioning, and production ramp-up proceeded in parallel. In contrast to brownfields operations with mature technical systems, Woodlawn was effectively recommenced without an embedded geology operating model, integrated database architecture, or standardised workflows suitable for an active underground mine. This created a requirement not only to restore site-specific capability but to establish a modern geological framework that could be transferred across a growing multi-asset company.

This paper presents the development and implementation of a modular nine-pillar geological operating framework at Woodlawn. The framework comprised:

1. Mapping and digital data capture.
2. Sampling and QA/QC.
3. Geological modelling and domaining.
4. Drill planning and execution.
5. Estimation and reconciliation.
6. Material tracking and stockpile control.
7. Operational readiness and mine-technical integration.
8. Capability and workforce development.
9. Continuous improvement and governance.

Each pillar was designed to satisfy immediate operational needs while remaining site-neutral, interoperable and scalable across different commodities and mining methods.

Implementation required sequential rebuilding of core technical foundations, including the drill hole database, core handling and photography systems, logging and domaining standards, ore flow tracking, and reconciliation workflows aligned with Parker-style F1, F2 and F3 factors (Parker, 2012). Digital tools, including tablet-based mapping, photogrammetry and machine-assisted core image workflows, were selected for both operational fit and future redeployment potential.

By 2025, Woodlawn had transitioned from a restart project with fragmented legacy data to an operating underground mine supported by an integrated geology function. The principal outcome was not simply a set of site procedures, but a transferable geological operating model that linked orebody knowledge, operational control, reconciliation and governance. The Woodlawn case demonstrates that mine restarts can be used to establish scalable geological systems in parallel with production ramp-up, provided that system design is modular, disciplined and explicitly aligned with future multi-asset growth.

INTRODUCTION

The recommissioning of a legacy mine is commonly presented as a technical restart problem. In practice, it is equally a systems problem. Geological data may be abundant but inconsistent, legacy interpretations may be poorly documented, and workflows that once supported mining may no longer be compatible with contemporary operational, digital or corporate requirements. Where underground development and processing recommence before geological systems are fully rebuilt, there is a risk that geology becomes reactive: focused on immediate production support rather than on constructing the technical foundations required for consistent orebody knowledge, reconciliation and long-term planning.

The Woodlawn deposit in New South Wales represents a particularly instructive example (Large, 1992). Woodlawn is a polymetallic VMS deposit hosted in the eastern Lachlan Orogen and has been the subject of geological investigation since discovery in 1970 (Whiteley, 1981; Jones *et al,* 2017). Classical studies documented the stratiform sulfide lenses, copper-rich stockwork mineralisation, alteration zoning and subsequent tectonic overprint (McKay and Hazeldene, 1987; Petersen and Lambert, 1979; Glen *et al,* 1995). However, the existence of a well-known deposit does not guarantee the existence of a modern operating geology system. When DEVELOP acquired Woodlawn in 2022, the opportunity was therefore broader than simply restarting a mine: it was to rebuild the geology function from first principles.

The strategic context was also important. DEVELOP is a growing multi-asset mining company, and Woodlawn offered a practical test case for a geology operating model that could extend beyond a single site. The objective was not to create an overly customised set of local procedures, but to develop a modular framework that could be adapted across commodity types, levels of data maturity and mining methods. This design philosophy shaped both the order of implementation and the selection of technical systems.

This paper describes the architecture and implementation of that framework. It does not seek to reinterpret the genesis of the Woodlawn deposit, which has been addressed in previous academic work, but instead examines how geological systems were rebuilt to support mine restart and future organisational scalability. The paper focuses on the nine operational pillars, the sequencing of their implementation, and the practical lessons relevant to geology leaders working in brownfields restarts or expanding mining portfolios.

GEOLOGICAL SETTING

The Woodlawn deposit is located approximately 210 km south-west of Sydney and 50 km north-east of Canberra in New South Wales, Australia. It is hosted within a felsic volcanic-sedimentary sequence of Silurian age within the Lachlan Orogen. Woodlawn is widely described as a volcanogenic massive sulfide deposit containing zinc, copper, lead and silver, with associated gold. The mineralised sequence includes felsic volcanic rocks, volcaniclastic units and mudstones, later intruded by mafic sills and dykes and subsequently affected by deformation and low-grade metamorphism (McKay and Hazeldene, 1987; McKay, 1989).

Previous work established that mineralisation at Woodlawn comprises two principal components: stratiform to lensoidal massive sulfide bodies and an underlying copper-rich stockwork system (Petersen and Lambert, 1979; McKay and Hazeldene, 1987). The massive sulfides are dominated by pyrite, sphalerite, chalcopyrite and galena, while gangue and alteration assemblages include quartz, chlorite, sericite, talc and phlogopite (Jones *et al,* 2017). The stockwork zone occurs within a strongly altered footwall, with hydrothermal chlorite-sericite-silica assemblages reflecting proximity to the hydrothermal upflow system (Petersen and Lambert, 1979; McKay and Hazeldene, 1987).

The deposit has also been recognised as tectonically modified. Glen *et al* (1995) documented both syn- and post-tectonic mineralisation features and highlighted the importance of structural overprint in understanding the present geometry of the ore system. For operating geology, this has direct implications. It affects domain continuity, the relationship between geological and grade boundaries, and the scale at which observations from core, face mapping and stope performance can be reconciled into a coherent interpretation.

For the restart team, the geological setting therefore imposed three immediate technical requirements. First, the deposit needed to be approached as a structurally complex VMS system rather than a simple tabular orebody. Second, inherited models and logging conventions required critical review before they could be relied upon operationally. Third, the geology function needed systems capable of integrating multiple evidence streams – drilling, face observations, development mapping, reconciliation and material flow – into repeatable decision-making.

RESTART CONTEXT AND DESIGN PRINCIPLES

At the commencement of the restart, Woodlawn had underground infrastructure, a processing plant and an extensive historical data set, but it did not have the suite of integrated geological systems normally expected in a modern operating mine. Data existed in multiple locations and formats. Logging standards had evolved through time. Database structure and assay mappings required validation. Production tracking logic had not been built to the level of fidelity required for sustained underground operations. In addition, the geology team itself had to be assembled as production-facing expectations increased.

Under these conditions, the geology rebuild could not proceed as a set of independent improvements. It required an operating philosophy. Three design principles were adopted.

Modularity: Geological work was decomposed into functional pillars that could be designed, prioritised and improved without losing sight of their interdependence. This prevented the restart from becoming dominated by the loudest immediate issue while foundational gaps remained unresolved.

Transferability: Woodlawn-specific details were unavoidable, but workflows, decision logic, naming conventions, governance structures, and digital tools were selected with a future portfolio in mind. The question asked repeatedly was whether a given system would still make sense if deployed in a different deposit style, at a different stage of maturity, or at another DEVELOP operation.

Operational integration: Geological systems were not treated as purely technical repositories. They had to function within an active mine and influence daily decisions. A workflow that could not support planning meetings, ore movement control or reconciliation review was not considered complete, regardless of its conceptual merit.

These principles led to the development of the nine-pillar framework shown below. The framework provided both a communication tool and an implementation roadmap. It allowed work to be sequenced in a practical manner while preserving a whole-of-system view of the geology function.

THE NINE-PILLAR FRAMEWORK

The nine-pillar framework was developed as an operating architecture that links geological data capture, interpretation, operational control, and governance (Figure 1). The pillars were not conceived as isolated workstreams; each was designed to strengthen the others. Mapping standards improve domaining. Sampling and QA/QC underpin estimation. Material tracking and reconciliation create feedback to both geology and operations. Governance ensures that improvements are embedded rather than remaining person-dependent.

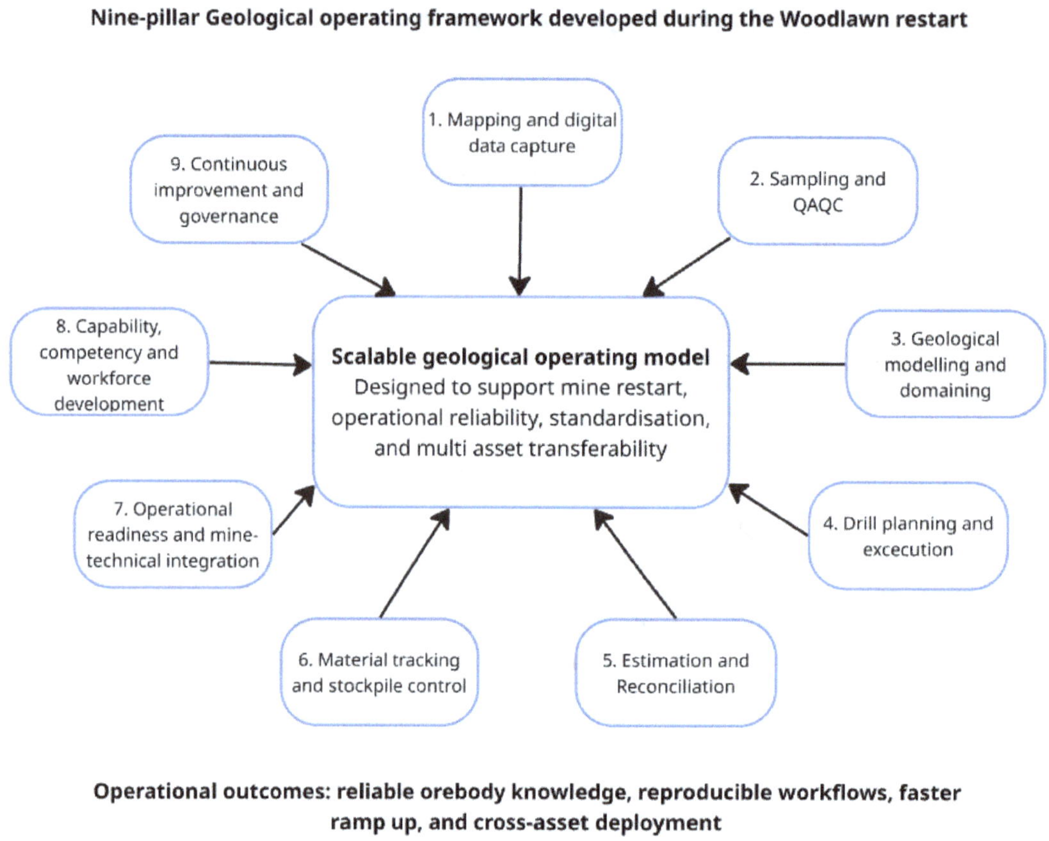

FIG 1 – Nine-pillar framework.

TABLE 1

Summarises the framework and its primary purpose.

Pillar	Primary purpose
Mapping and digital data capture	Standardise underground and drill-derived observations in forms that are auditable, spatially controlled and directly usable in downstream interpretation.
Sampling and QA/QC	Create representative and reproducible sample streams for grade control, resource definition and confidence in reported assays.
Geological modelling and domaining	Translate observations into explicit, repeatable domain rules and interpretations suitable for estimation and operational use.
Drill planning and execution	Coordinate drilling priorities, execution standards, core handling and turnaround between geology, contractors and technical services.
Estimation and reconciliation	Build fit-for-purpose model update workflows and use reconciliation as an active feedback mechanism for geological improvement.
Material tracking and stockpile control	Preserve source integrity from underground extraction to mill feed and enable reliable ore flow reporting and reconciliation.
Operational readiness and mine-technical integration	Embed geology outputs into the timing, language and routines of an active underground operation.
Capability, competency and workforce development	Define, train and maintain the practical geological skills required for current operations and future asset transferability.
Continuous improvement and governance	Control versions, review cycles, ownership and improvement pathways so systems remain durable rather than person-dependent.

Pillar 1 – Mapping and digital data capture

The first pillar addressed the need for reliable geological observations captured in a structured, accessible form. Legacy mine restarts often inherit a mixture of paper records, ad hoc spreadsheets and partially digitised mapping. This creates a lag between observation and use. At Woodlawn, the objective was to move field data capture toward standardised digital workflows, including development mapping, structural measurements, lithological logging support and photographic records.

Digital capture was valuable not simply because it reduced transcription. It imposed discipline on terminology, field selection and spatial control. In an environment where new geologists may join during ramp-up, standardised digital capture reduces interpretive drift and supports consistent import into modelling packages. It also provides a platform for later analytics, machine-assisted classification and auditability.

Pillar 2 – Sampling and QA/QC

The sampling strategy was reviewed because sampling is the critical interface between observed geology and reported grade. In structurally complex mineralisation, small changes in boundary interpretation can materially affect operational decisions. Sampling protocols were therefore redesigned to prioritise representativity, reproducibility and practical usability. This included review of sample intervals, increased inclusion of sub-economic and wing material where necessary to better characterise boundary conditions, and the move toward whole-core approaches in relevant resource definition and grade control contexts.

QA/QC was treated as a production-critical workflow rather than a compliance exercise. Reference materials, blanks, duplicates and laboratory checks only add value if they are visible within the decision cycle. Accordingly, QA/QC review needed to sit close to routine assay validation and database governance. At Woodlawn, the intent was to ensure that geological, estimation, and operational confidence were based on the same validated sample stream.

Pillar 3 – Geological modelling and domaining

Inherited geological interpretations were reviewed from the level of logging codes upward. Historical coding systems, while often comprehensive, can become operationally unwieldy when they contain overlapping, obsolete or highly deposit-specific categories. A practical subset of logging codes was therefore developed to preserve geological meaning while supporting repeatable interpretation. This simplification improved communication between geologists and reduced the risk of inconsistent coding being embedded into modelling.

Domaining standards were then established to support consistent wireframing across the progression from exploration interpretation to grade control. In the Woodlawn setting, this required balancing stratigraphic context, alteration intensity, sulfide style and structural overprint. The goal was not to impose artificial simplicity on a complex system, but to ensure that domain boundaries were based on explicit rules that could be tested against drilling, development exposures and reconciliation outcomes.

Pillar 4 – Drill planning and execution

Drilling was treated as a system rather than a sequence of individual campaigns. Resource targeting, resource definition and grade control drilling each serve different purposes, but at restart they compete for the same people, platforms, assay pathways and interpretation capacity. A structured workflow was therefore required to define drill priorities, design standards, execution expectations and handover points between geology, drilling contractors and technical services.

At Woodlawn, this pillar also included the review of core processing layout and photography standards. Logging workflows were designed first, and core yard infrastructure followed those workflows. The objective was to create a core processing environment that reduced manual handling, improved consistency and produced images of sufficient quality to retain long-term geological value. This reflected an explicit recognition that core photography is not merely archival; in mature underground operations it often becomes a durable data set for reinterpretation, training and machine-assisted analysis.

Pillar 5 – Estimation and reconciliation

Estimation workflows were progressively rebuilt as underground exposure and new data revealed limitations in inherited assumptions. The initial objective was to stabilise a fit-for-purpose workflow suitable for active mining, including interpretation timing, model update cadence, sample support considerations and practical controls around estimation inputs. As mining advanced, the system matured into a feedback loop in which reconciliation became an active driver of geological improvement.

Parker-style reconciliation factors were adopted to provide a common language around orebody knowledge, selectivity and process performance. In simple terms, these factors support comparison between reserve or long-range expectations, ore control estimates and plant outcomes. Their value at Woodlawn extended beyond technical diagnosis. They helped align multiple departments around a shared understanding that reconciliation is not owned solely by geology or metallurgy. It is a whole-of-value-chain measure (Parker, 2012).

Pillar 6 – Material tracking and stockpile control

Material tracking required redevelopment from first principles. Early tracking methods built during commissioning were sufficient for immediate needs but were not robust enough for a sustained underground operation with multiple ore sources, surface handling stages and reconciliation requirements. A new ore flow logic was therefore developed to track movements from underground extraction to transfer pad, ROM and mill feed in a manner suitable for later transcription into SQL architecture.

The importance of this pillar is often underestimated. In a multi-stage flow path, reconciliation can fail not because the geology model is poor, but because material movements are not controlled or recorded in a way that preserves source integrity. At Woodlawn, the rebuilt framework incorporated stockpile segregation logic, movement coding, source attribution and reporting structures capable of supporting both day-to-day ore flow control and end-of-month reconciliation. Development face sampling and ROM sampling were also integrated to provide redundancy where structural complexity limited the reliability of a single data source.

Pillar 7 – Operational readiness and mine-technical integration

Operational geology succeeds or fails at the interface with mining and technical services. This pillar focused on the routines, escalation pathways and meeting structures through which geological information influenced execution. It included mark-up protocols, communication of ore and lithological boundaries, integration of face observations into development decisions, and the timing of information release relative to production needs.

The key lesson at Woodlawn was that technical quality alone is insufficient. Geological outputs had to be available in a form and at a time that mine planning, underground operations and processing personnel could use. Operational readiness therefore meant turning geology from a specialist function into an embedded part of the mine operating system.

Pillar 8 – Capability, competency and workforce development

Because the restart required building the team in parallel with the systems, capability development was treated as a formal pillar rather than an assumed by-product of work. Competency expectations were defined around the actual tasks required in an underground polymetallic operation: mapping, sampling, mark-ups, logging, reconciliation interpretation, data handling and communication. Structured onboarding and deliberate cross-training reduced reliance on individual institutional knowledge.

This pillar also supported transferability. A company intending to scale across multiple assets needs not only common systems, but also a common language of capability. By framing training and performance around repeatable geological tasks, Woodlawn became a platform for developing geologists who could operate effectively in future DEVELOP environments.

Pillar 9 – Continuous improvement and governance

The final pillar ensured that the framework remained alive. Governance included document control, review cycles, sign-off pathways, version discipline and clarity around the system of record for geological and production data. Continuous improvement mechanisms were linked to reconciliation performance, audit findings, implementation experience and practical lessons from underground execution.

This pillar was essential because mine restart environments generate rapid local workarounds. Some are useful; others create hidden technical debt. Governance distinguishes adaptive improvement from uncontrolled drift. At Woodlawn, it provided the mechanism by which useful changes could be standardised and exported across the broader business.

IMPLEMENTATION SEQUENCE

Although the nine pillars were interdependent, they were not implemented simultaneously. The practical sequence began with data integrity and capture. Database stabilisation, logging standards and core handling workflows were early priorities because later decisions would only be as reliable as the inputs supporting them. At the same time, mapping and observation capture had to be standardised so that new data added during the restart would be more consistent than the data inherited.

Once these foundations were in place, attention shifted toward interpretation and ore control. Domaining standards, drilling workflows and estimation controls were progressively tightened as underground exposure increased. During this period, the operating model remained deliberately iterative. Early interpretations were not treated as fixed end points, but as hypotheses to be tested against development, operational performance and reconciliation.

Material tracking and reconciliation matured as the surface stockpile system evolved and deficiencies in initial tracking became visible. This phase of implementation was particularly important because it closed the feedback loop between geology and production. Without this loop, the site risked improving interpretation in isolation while losing confidence in what material was actually being delivered to the mill.

The final stage was organisational embedding. Competency frameworks, governance routines, and multi-departmental reconciliation communication ensured that the system no longer relied on a small number of individuals to drive it manually. This transition – from build phase to embedded operating rhythm – was arguably the most significant indicator that the geology function had moved from restart mode toward sustainable operation.

DIGITAL TRANSFORMATION AND SCALABILITY

A notable feature of the Woodlawn rebuild was the early adoption of selected digital tools. These included digital mapping platforms, photogrammetric methods and machine-assisted core image technologies. The rationale for implementation was broader than technology adoption for its own sake. Tools were assessed against four questions: did they suit the site, could they be maintained operationally, were they interoperable with broader data systems, and could they be redeployed across future assets?

This matters because digital projects in mining frequently fail at the transition from pilot to routine use. At Woodlawn, digital systems were selected that could directly strengthen one or more pillars of the framework. Tablet-based mapping improved data capture discipline. Photogrammetry supported geological context and visual documentation. Structured core imagery improved the long-term value of drilling data sets and created options for future machine learning applications. In each case, the test was whether the tool improved decision quality or workflow reliability, not whether it was novel.

The broader lesson is that digital transformation is most effective when anchored to an operating framework. Technology should reinforce geology, not distract from it. By linking digital tools to defined business and geological functions, Woodlawn avoided treating digital initiatives as standalone projects.

DISCUSSION

The principal contribution of the Woodlawn case is the demonstration that a mine restart can be used to build a geological operating model, rather than merely to restore historical practice. This distinction is important. Legacy sites commonly inherit habits, assumptions and fragmented data structures that are tolerated because they are familiar. Restart creates a short window in which those defaults can be questioned. If that opportunity is used well, the result can be a system better suited to both the site and the broader organisation than anything that existed previously.

Three practical lessons emerged.

First, geology must define its operating architecture early. Without a framework, restart work becomes dominated by urgent production requests, and technical debt accumulates in the background. The nine-pillar model was effective because it made dependencies visible and helped justify foundational work that might otherwise have been deferred.

Second, reconciliation and material tracking are not downstream reporting tasks. They are core geological functions because they determine whether orebody knowledge can be tested against reality. At Woodlawn, the redevelopment of ore flow tracking and the use of factor-based reconciliation materially improved the site's ability to identify where errors or losses entered the value chain.

Third, scalable systems depend as much on governance and capability as on geology. A technically strong workflow that cannot be taught, audited or reused at another asset has limited strategic value. The Woodlawn framework was intentionally designed so that its logic, rather than only its local details, could transfer.

There are, however, limits to transferability. No two deposits share the same geological complexity, data density or organisational context. The framework should therefore be viewed as modular rather than prescriptive. The pillars are broadly applicable, but the emphasis and implementation sequence may differ between operations. For example, a greenfields underground development may prioritise drill planning and data architecture differently from a brownfields restart with existing workings and stockpiles. The key transferable element is not the exact procedure set, but the disciplined linkage between data capture, interpretation, control, feedback and governance.

CONCLUSIONS

The restart of the Woodlawn VMS mine provided an opportunity to rebuild the geology function from first principles while supporting active underground and processing recommissioning. Rather than developing a collection of site-specific procedures, the work was structured as a modular nine-pillar operating framework that links geological data capture, sampling, modelling, drilling, estimation, material tracking, operational integration, workforce capability, and governance.

This framework allowed foundational systems to be rebuilt in a practical sequence without losing sight of their interdependence. It also ensured that key outputs, particularly material tracking and reconciliation, were integrated with the broader mining value chain. The result by 2025 was a functioning geology operation that supported production while also providing a transferable blueprint for future assets within DEVELOP.

For geology leaders, the Woodlawn case reinforces that a restart is not simply a time to recover inherited systems. It is a rare opportunity to design better ones. Where system architecture is modular, governance is explicit and geological workflows are built for operational use, the restart of a single mine can generate a durable technical advantage across a wider business.

ACKNOWLEDGEMENTS

The authors acknowledge the Woodlawn geology and technical services teams for their contribution and significant efforts to rebuild the site geology function during restart.

Particular acknowledgement is also made to colleagues and consultants who supported database consolidation (W Porter), modelling review (A van der Hayden), structural review (T Murphy) and implementation of new digital workflows (N Lui, D Baumann, J Hoskins).

The authors acknowledge DEVELOP Global Ltd for granting permission to publish this paper.

REFERENCES

Glen, R A, Walshe, J L, Bouffler, M, Ho, T and Dean, J A, 1995. Syn- and post-tectonic mineralization in the Woodlawn deposit, New South Wales, Australia. *Economic Geology,* 90:1857–1864.

Jones, S A, Downes, P M, Fitzherbert, J A and Hine, K, 2017. Woodlawn zinc-copper-lead-silver-(gold) deposit, *Discoveries in the Tasmanides 2017*, AIG Bulletin, 67:44–50.

Large, R R, 1992. Australian volcanic-hosted massive sulfide deposits: features, styles, and genetic models, *Economic Geology,* 87:471–510.

McKay, W J, 1989. A study of the geological setting, nature and genesis of the Woodlawn base metal deposit, New South Wales, Australia, PhD thesis, Australian National University, Canberra.

McKay, W J and Hazeldene, R K, 1987. Woodlawn Zn-Pb-Cu sulfide deposit, New South Wales, Australia: an interpretation of ore formation from field observations and metal zoning, *Economic Geology,* 82:141–164.

Parker, H M, 2012. Reconciliation principles for the mining industry, *Mining Technology,* 121(3):160–176.

Petersen, M D and Lambert, I B, 1979. Mineralogical and chemical zonation around the Woodlawn Cu-Pb-Zn ore deposit, southeastern New South Wales, *Journal of the Geological Society of Australia,* 26(3):169–186.

Whiteley, R J (ed), 1981. *Geophysical case study of the Woodlawn orebody, New South Wales, Australia* (Pergamon Press: Sydney).

Successful use of hyperspectral imaging to discriminate material types in an open cut lithium pegmatite mine

M Kamperman[1], T Paterson[2] and F Quach[3]

1. MAusIMM, Geology Manager, Covalent Lithium, Kwinana WA 6167. Email: marcel.kamperman@covalentlithium.com
2. MAusIMM, Head of Geology, Plotlogic Pty Ltd, Zillmere Qld 4034. Email: thomas.paterson@plotlogic.com
3. GM Solutions, Plotlogic Pty Ltd, Zillmere Qld 4034. Email: felice.quach@plotlogic.com

ABSTRACT

Critical to the profitability of lithium pegmatite mines is the correct truck-by-truck allocation of ore and waste material to waste dumps, stockpiles, and direct feed to the mill. Incorrect allocation of material results in, *inter alia*, valuable ore being incorrectly treated as waste material, and waste material, being processed through the plant. This results in poor ore recovery and lower profitability.

Mines attempt to ameliorate this risk of misallocation through resource drilling, grade control drilling, and in-mine ore spotting. However, this approach has an inherent lack of precision owing to the spacing of the grade control drill program, movement of material post-blasting, and the qualitative and subjective nature of ore-spotting practices. Absence of a timely feedback loop to confirm decision-making compounds the challenge. This lack of precision may cause a reduction in the mine profitability and accelerates the depletion of the orebody.

An innovative approach to address this lack of precision has been developed and implemented at Covalent Lithium's Mt Holland Mine. It builds on the existing grade control model by conducting daily assessments of mining areas, stockpiles, and waste dumps, utilising a digital imaging system that quantitatively predicts the proportion of pegmatite of any scanned rock in the mine allowing for precise allocation of each truck to correct locations. The imaging system uses a combination of LiDAR, hyperspectral imaging, and Artificial Intelligence to produce results. Using this approach correct allocations can be made in advance, during or post-mining, and results validated on a daily basis to ensure compliance to plan. Furthermore, the digital imaging provides consistent feedback, supporting ongoing, systemic improvement.

The results from the extensive testing and implementation of this new approach at Covalent Lithium's Mt Holland Mine are presented in this paper, along with recommendations for future research and development to further improve the value from ongoing implementation of this new approach.

INTRODUCTION

Parker (2011) illustrated the large economic impact on a mining asset of deviations from the plan for delivery of ore to the process plant, focusing on two broad causes for such deviations, firstly problems in the reserve or grade control models and secondly mining practices which lead to ore loss and or dilution. The problems and solutions at Mt Holland described in this paper relate to the second category.

The Mt Holland open cut mine commenced operations in 2022 (Wesfarmers Chemicals, Energy and Fertilisers (WesCEF), 2024). The deposit comprises several pegmatite bodies, broadly tabular and dipping 5–15° to the NW. The main body has thickness up to ~90 m. Country rock to the pegmatite intrusion is predominantly meta basalt. Minor, steep mafic dykes post-date the pegmatite.

In practice, ore control at Mt Holland is based on lithology such that all pegmatite is ore and all other units are waste. Variation in lithium grade and mineralogy within the pegmatite is not significant in the current production areas.

Despite the strong lithological contrast between ore and waste, mixing at the ore-waste contact occurs and this material is stockpiled separately (GeoInnova Consultores SPA, 2025). Consequences are typical for an open cut mine including:

- Excess allocation of ore to mixed ore-waste stockpiles (locally referred to as SORT).

- Requirements to mine additional waste to maintain ROM stock levels.
- Mill feed outside of specification.

The root causes of the problem include:

- The requirement to produce a 3 per cent waste cut-off is difficult to deliver with reliability. In operating conditions, low light, with intermixed materials and the presence of dust, the predominant colour observed is grey, rather than the ideal 'black and white' of the basalt and pegmatite.
- Complex geological contacts. The tabular pegmatite body exhibits splays and irregular terminations, variably sized xenoliths, and shallow dipping upper and lower undulating contacts and offsets. Post blast heave adds to the spatial complexity.
- The absence of timely feedback loops to Pit Technicians.

The Mt Holland Geology team approached Plotlogic over its proprietary OreSense® imaging solution with three hypotheses regarding the problem. Firstly, that pegmatite (alumino-silicate mineralogy) and basalt (ferromagnesian mineralogy) could be discriminated by hyperspectral methods. Secondly, that the technology was sufficiently mature to deliver useful results in an operational context (scale, speed, ambient conditions). Finally, that making the technology available to end-users at the mine would lead to best uptake and value impact. A staged approach to deployment, to work through these hypotheses in a logical sequence, was adopted.

Plotlogic developed OreSense® as a solution to mining industry imperatives for higher precision mining and decision support to operators using rapid and reliable data. OreSense® consists of three main technical components:

- Light vehicle mounted scanning hardware, comprising hyperspectral cameras, LiDAR and supporting components.
- AI processing using proprietary models and knowledge based accumulated by Plotlogic.
- Web based User Interface where client end-users can access processed results.

Headline results will be described further in this paper but include the following:

- Effective discrimination of ore and waste by OreSense®. As Covalent personnel had predicted, the dominantly aluminosilicate pegmatite and ferromagnesian dominant meta basalt were spectrally distinct materials. Scanning time, logistics and movements and format of results for end-users were found to meet the requirements of the Use Case.
- Early expansion of the scope to scan the SORT, ROM and Waste piles for checking of correct truck destination and quality control. Over an approximately three month evaluation period, 301 truckloads were found to be incorrectly placed, of which 49 loads were potentially recoverable ore placed on SORT, 134 waste loads place on SORT and 11 waste loads placed on ROM.
- Reduced variance in Pit Technicians' calls when using OreSense®.
- Opportunities for further improvements, features and value-add were identified and are currently the subject of joint planning between Covalent Lithium and Plotlogic.

MT HOLLAND MINE

Operation

Covalent Lithium is a 50:50 Joint Venture between Wesfarmers Chemicals, Energy and Fertilisers (WesCEF) and Sociedad Quimica y Minera de Chile (SQM) of Chile. Assets include the mine and concentrator at Mount Holland and a refinery at Kwinana, south of Perth. Mt Holland is located approximately 500 km East of Perth. Mining commenced in February 2022, and first lithium hydroxide was produced in August 2025 (Covalent, 2025). Operational capacity is ~2 Mt/a ROM, 380 kt/a spodumene concentrate and ~50 kt/a battery grade lithium hydroxide (WesCEF, 2024).

Mining is by conventional open cut methods. Horizontal benches of 5 m are mined in two passes. Load and haul operations are via 200 t hydraulic excavators and 160 t CAT785 trucks. Three destination categories are defined:

1. ROM Ore: pegmatite, ≤3 per cent basalt contamination.
2. SORT: pegmatite, >3–65 per cent basalt contamination.
3. Waste: >65 per cent basalt.

Ore loads are tipped at a set of ROM fingers adjacent to the primary crusher. SORT material is sent to one of several SORT stockpiles for future processing eg by an ore sorter. Layout of ore and waste zones is based primarily on infill drilling, supplemented by blasthole geology and pit floor mapping. Blastholes are not sampled.

Delineation of ore, waste and SORT material while loading and truck destination decision is the responsibility of a team of Pit Technicians, stationed in the excavator with excavator operators.

Geology

The Archean Earl Grey Pegmatite is the economically mineralised unit at the Mt Holland Mine. Country rock comprises steeply dipping mafic and ultramafic units of the NS trending Forrestania greenstone belt. Yilgarn granites bound and intrude the greenstone regionally (Phelps-Barber, Trench and Groves, 2022). The Early Grey pegmatite comprises a main tabular body up to ~90 m thickness with numerous hanging wall and footwall splays. The main body dips at 5–15° to the NW. Strike length is about one kilometre and known dip extent of approximately two kilometres.

The pegmatite is pale white to grey and brown and generally coarse grained, although texturally highly variable. Mineralogy of the pegmatite is predominantly a simple albite-quartz-microcline-spodumene-petalite assemblage. Minor biotite, muscovite and tourmaline occur. Other lithium bearing minerals are known but spodumene and lesser petalite predominate. Mining is focused on the spodumene dominant portions of the deposit. Minor mafic dykes cut the pegmatite. The mafic country rock is dark green-grey, medium grained and dominated by hornblende-plagioclase, with holmquistite observed in the mafic country rock in contact alteration zones with the pegmatite. The effects of weathering extend to approximately 30 m at Mt Holland and cause a general decrease in spodumene and lithium content. Both pegmatite and mafic waste exhibit variable orientation of mineral grains, are competent where fresh and generally tend to break to equidimensional blocks.

A variety of typical pegmatite geometry and contacts are observed at deposit and outcrop scale. While grossly tabular, the pegmatite bodies exhibit undulating surfaces, coalescing and splitting of units, occasional abrupt terminations, steep dyke like forms almost perpendicular to the main bodies and xenoliths of country rock at a range of scales (Figure 1).

FIG 1 – Geology exposed at Mt Holland lithium mine. Pale shallow dipping unit in second lowest bench is pegmatite ore. Note local offsets, mafic xenoliths, and lateral pinch outs.

Operational challenges

A primary mining problem encountered early in the mine life was inconsistent discrimination of ore, SORT and waste categories and incorrect destination of trucks with these loads.

These problems manifested in three forms:

1. Elevated allocation of trucks to SORT stockpiles, running above scheduled rates, and both ore and waste being incorrectly sent to SORT. This was the most unambiguous symptom of the problem.
2. Requirements to mine additional waste to maintain ROM stock levels.
3. Contaminated ore feed to the concentrator.

Three inherent factors drove the problem:

1. Difficulty of visual estimation of mixtures to a cut-off. Pegmatite ore and mafic waste lithologies while very distinct under favourable conditions tend to exhibit a range of grey colours in broken faces at a distance of metres to tens of metres.
2. Pegmatite geometry can produce complex boundaries, especially for thinner bodies with main contacts below top or base of the flitch, or where ore irregular contacts daylight or pinch out within a flitch.
3. Absence of an objective, rapid and reliable results feedback loop for the Pit Technicians.

INITIAL HYPOTHESES

Covalent approached Plotlogic with a structured and hypothesis-driven approach to deployment of OreSense®. This allowed for objective and stepwise testing of various critical requirements of any solution.

Ore and waste can be distinguished by hyperspectral approach

The distinct mineralogical difference between ore and waste lithologies should result in distinctive hyperspectral absorption features which can be readily distinguished.

Suitable technology might exist already

Hyperspectral applications in mineral deposits are well established for individual samples and drill core. Covalent needed to establish whether any provider in the market has already developed a solution for mine faces and stockpiles.

Scanning can be done at useful production rate in operational conditions

In addition to valid geoscientific results, any solution must deliver sufficiently rapid results, cope with the variation of ambient illumination, dust and required set-back distances from materials to be scanned.

Improvement in ore/waste discrimination will be value-adding

Objective site metrics will confirm material value-add for the operation.

Technology in technicians' hands will allow improved decision performance

By providing useful decision support data to technicians, they will be able to integrate into workflows and maximise the benefit of the improved information.

ORESENSE® TECHNOLOGY

OreSense® is a versatile imaging scanner, ruggedised for the mining environment which provides rapid and quantitative results delivered via a web interface. It employs various sensors including hyperspectral cameras. Hyperspectral absorption features allow for identification of a wide range of minerals. Progress has been made in recent decades in the analysis of geological materials, most notably in remote-sensed (satellite), airborne and drill sample applications. Fraser, Whitburn and Ramanaidou (2006) showed results from a hyperspectral imaging prototype in the mining environment for various deposit types. Maximum utility of a rapid imaging system in the mining environment requires the following capabilities, each with various levels of scientific and engineering challenge:

- Analysis of mixed samples, with mixed spectra. Rock types, ores and waste materials of variable composition are of greater practical relevance than individual mineral grains or end-member material types.
- Quantification of minerals present, for example constituent minerals are each allocated a percentage for the sample being analysed.
- Quantification of chemical makeup, based on mineral proportions.
- Generation of a geolocated 3D image covering an area relevant to mining dimensions, rather than spot analyses.
- Operation in typical open cut and underground mining conditions. Both environments are prone to extremes of dust, temperature, humidity and vibration. Open cut has variable solar illumination and stand-off distances ranging from 10–150 m while underground requires illumination to be supplied and stand-off distances around 7–8 m. Rapid provision of decision support outputs to end users, operators and managers. In the production environment there is little time and expertise to interpret or convert data layers into decisions.

OreSense® has substantially solved these challenges in the mining environment. A full description of the system is well beyond the scope of this paper and involves proprietary information. In summary form, the system is best presented as five components, each briefly described further.

1. Solution-Validation Framework.
2. Sensor Hardware.
3. AI processing.
4. User Interface.
5. Service and Support.

Solution framework

New technology is challenging to establish and embed in mining operations, see for example Ediriweera and Wiewiora (2021). Technical performance, establishment of a business case, compatibility with workflows, change management and leadership support are critical factors, amongst others. Plotlogic has established a framework (Figure 2) to maximise chances of success, provide logical stage gates and manage risk for Clients and asset teams. The Framework spans project stages from initial engagement to full deployment. In the middle stage, key workstreams of technical validation and integration planning are done.

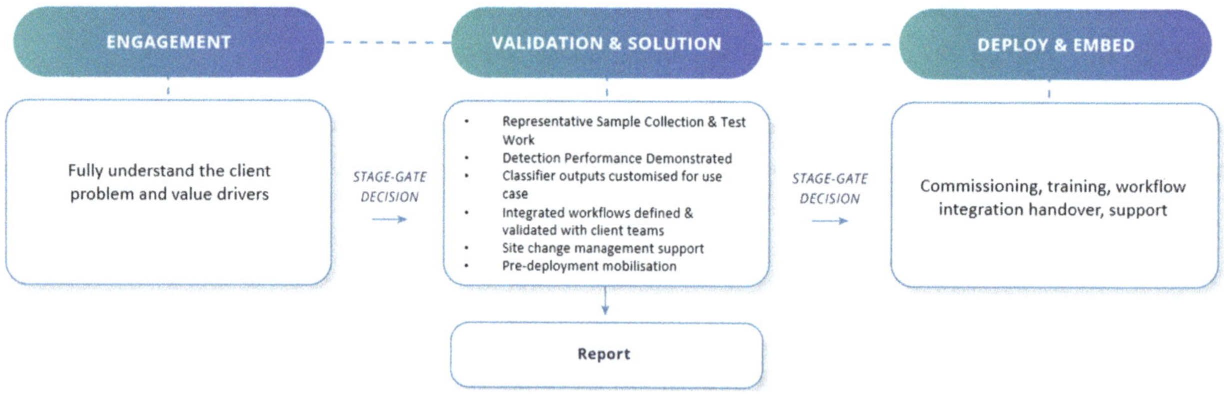

FIG 2 – Solution framework employed for client deployments.

Sensor hardware

Figure 3 shows a light-vehicle mounted system OreSense® scanner. This configuration is designed for open cut mines and is termed the Ranger system.

FIG 3 – OreSense® Ranger system hardware.

Hardware components include:

- Hyperspectral cameras.
- LiDAR.
- Supporting equipment including power supply, cooling and ruggedised enclosure.
- Ranger is typically mounted on a tray back light vehicle.

AI processing

The AI approach includes the following:

- Use of large training data sets, including traditional reference spectra for minerals species and rock samples from the deposit in question. The latter provide crucial training and test samples for the use case in question, eg discrimination of ore, waste, geometallurgical types and similar parameters.
- Preprocessing to fuse data from multiple sensors and which is subject to various artifacts of collection in the operating environment.
- A combination of intensive computing resources and experimentation with various machine learning approaches over some years of development have resulted in reliable models. These models are of a size and efficiency which can be run locally on the sensing unit. This reduces overall latency.

User interface

Near real time (approximately five mins) results are accessible via a web interface. In addition to any classifier(s) applied to a scan (eg Ore versus Waste in false colour highlights), users can access 2D and 3D rendering, RGB, location maps and catalogue search (Figures 4–6). Key workflow tools are user generated polygons for Regions of Interest (ROIs) for which statistics are available. Scans are downloadable for use in mining software including RGB and classified point clouds and user created ROIs as dxfs.

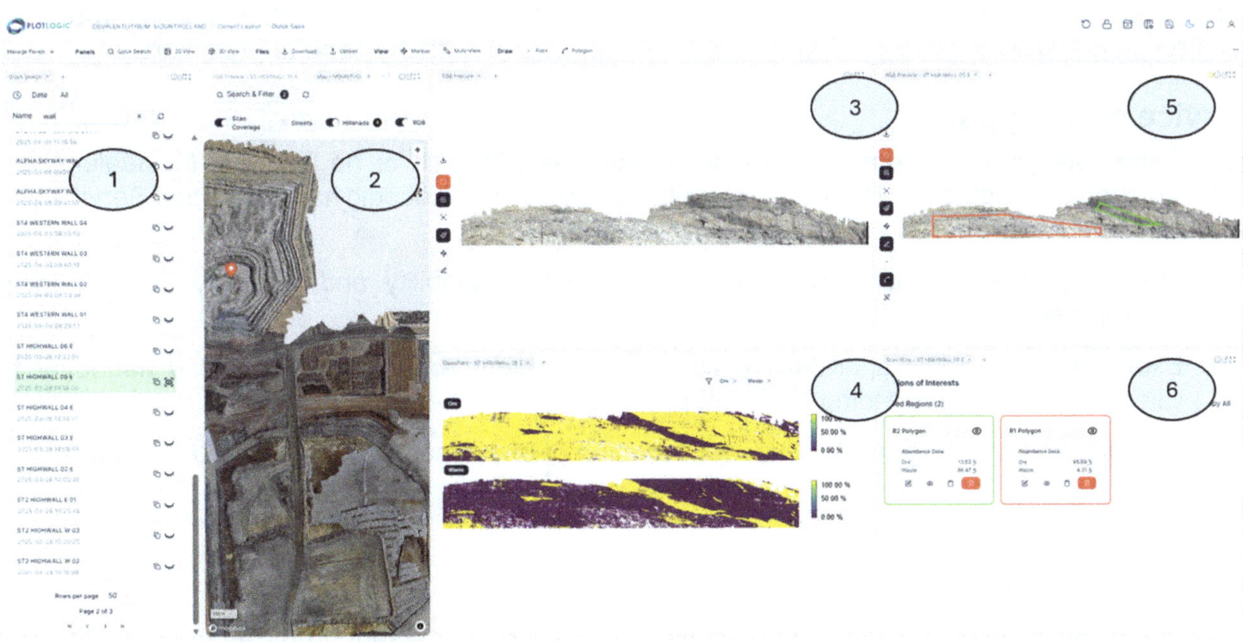

FIG 4 – Web-based User Interface. Illustration of working layout and tools, customisable by User. Panels as follows: (1) searchable scan catalogue, (2) map view, (3) RGB view of scanned face, (4) Classifier view of scanned face, (5) ROI digitised by user, (6) Classifier statistics for each ROI.

FIG 5 – More detailed view of the mining face shown above in Figure 4, panel 4. Upper image is RGB and lower image is ore-waste classifier. The classifier view of *in situ* geology assists in recognition of smaller waste bodies and defining the nature and orientation of contacts. Width of scene approximately 100 m.

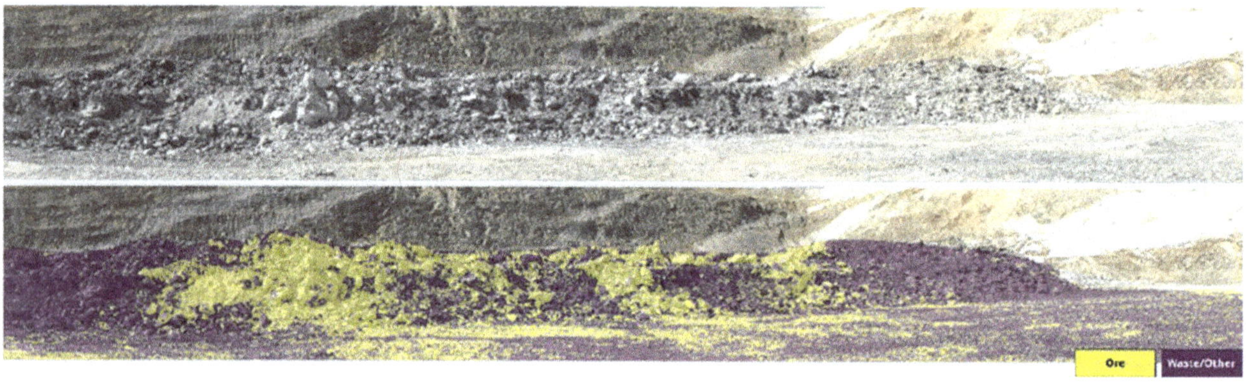

FIG 6 – Example of spectral distinction between pegmatite ore (yellow) and basalt waste (purple) in broken material. Blasted production bench, Mt Holland. Width of scene approximately 120 m.

Service and support

Mining sites are generally remote and have workforces with little or no flexibility for dealing with ad hoc technology problems. Through accumulated experience Plotlogic has developed Service and Support with a focus on:

- Inherently reliable machines engineered for high availability and reliability in the mining environment.
- Experienced Plotlogic operators to commence deployments with training and handover to Client technicians depending on Client preference.
- Ongoing technical support to deal with evolution of use cases and unexpected results.

SOLUTION AT MT HOLLAND

Phased deployment

Given the novel application in lithium pegmatite, Covalent management preferred to take a phased approach in which successive milestones could be demonstrated, in broad alignment with the hypotheses described earlier. This approach aligned well with Plotlogic's three-stage Solution Framework, described above.

- Engagement over problem to be solved.
- Site geology and representative sampling.

- Validation of scanning results and integration planning.
- Deployment.
- Staged reporting to client to confirm results and hypothesis testing.
- Ongoing: new workflows, improvements. Adjustment to focus areas of the operation as directed by the client.

Workflows prior to OreSense®

The system prior to OreSense® relied heavily on the technicians, both in the moment of loading and destination decisions, and in follow-up checks. For loading at the mine, the procedure was for a trained technician to spend the shift on the excavator to assist the operators in destination decision for a given truckload, ie ROM, Waste or SORT. Additionally, and especially following night shift, visual checks of stockpiles are made by the Geology team. If a SORT or ROM stockpile was observed to have an anomalous looking load or set of loads, an inspection would be made by the Geology team to ascertain if an error had occurred. This visual approach has a range of implications as described in Table 1.

TABLE 1

Attributes and Implications of visual ore spotting practices.

Attribute	Implication
Simple and effective when geological complexity is lower.	In the case of ore zones which are wide relative to SMU dimensions of the mining equipment, low levels of loss and dilution and high destination accuracy of truckloads can be achieved.
Breaks down with geological complexity.	Narrower or more complex ore zones result in mixed buckets, mixed truckloads and therefore more difficult destination decisions.
Estimation of 'grade' of a mixed material is more difficult than identification of binary types.	Ore and Waste at Mt Holland is based on lithology. The lithologies are relatively distinct. Where dilution of pegmatite ore occurs, technicians are required to make calls on, say, 3% dilution (Ore) versus 6% dilution (SORT). This is more difficult and produces variable results.
No feedback to Technicians.	Absence of any consistent sampling and analysis system means that technicians do not have a means of correcting or fine-tuning visual estimates.
Visual checks of tipped loads are challenging close to cut-offs.	Feed quality, loss and dilution can be improved (eg remove obvious waste from ROM) where there is a large deviation, but smaller deviations are prone to subjective visual spotting errors.
Traditional visual spotting type observations are not a robust and versatile data layer.	Visual spotting and checks of a locality are often verbal or paper records and relate to spot locations. They are not geospatial, quantitative and don't carry primary data.

The 'ore spotting challenge'

A quantitative illustration of the challenges associated with visual ore and waste control was made with a simple site exercise. There were 24 defined areas of various stockpiled material categorised by the site team. All available personnel from the Geology department, Geologists and Technicians, 10 in total, undertook the following:

- Each person in the test group individually completed the challenge without any OreSense® data. No time limit was set. Results were recorded. Required outputs were (1) Classification into ORE, SORT or WASTE (2) estimated percentage of waste.

- Each person in the test group individually repeated the challenge with OreSense® data. Results were recorded for the same required outputs.

Results from the challenge are shown in Figures 7 and 8. At the time of the challenge, the primary aspiration was that variance between individual technicians' results would be reduced.

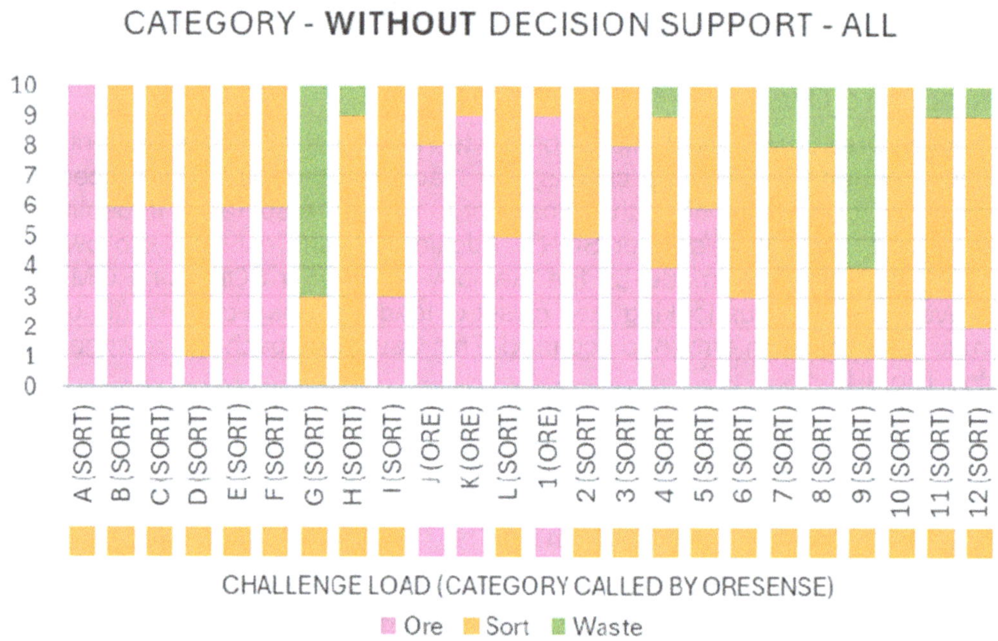

(a)

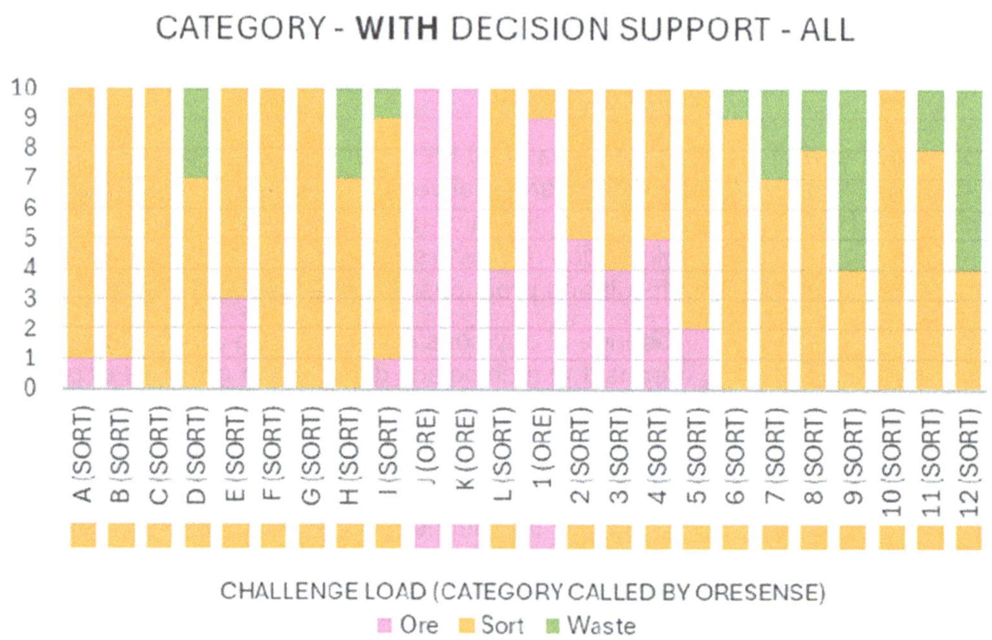

(b)

FIG 7 – (a) Material type classification, no OreSense® data. Each Y-axis increment represents a category determination by Geology department participants. (b) Material type classification, with OreSense® data.

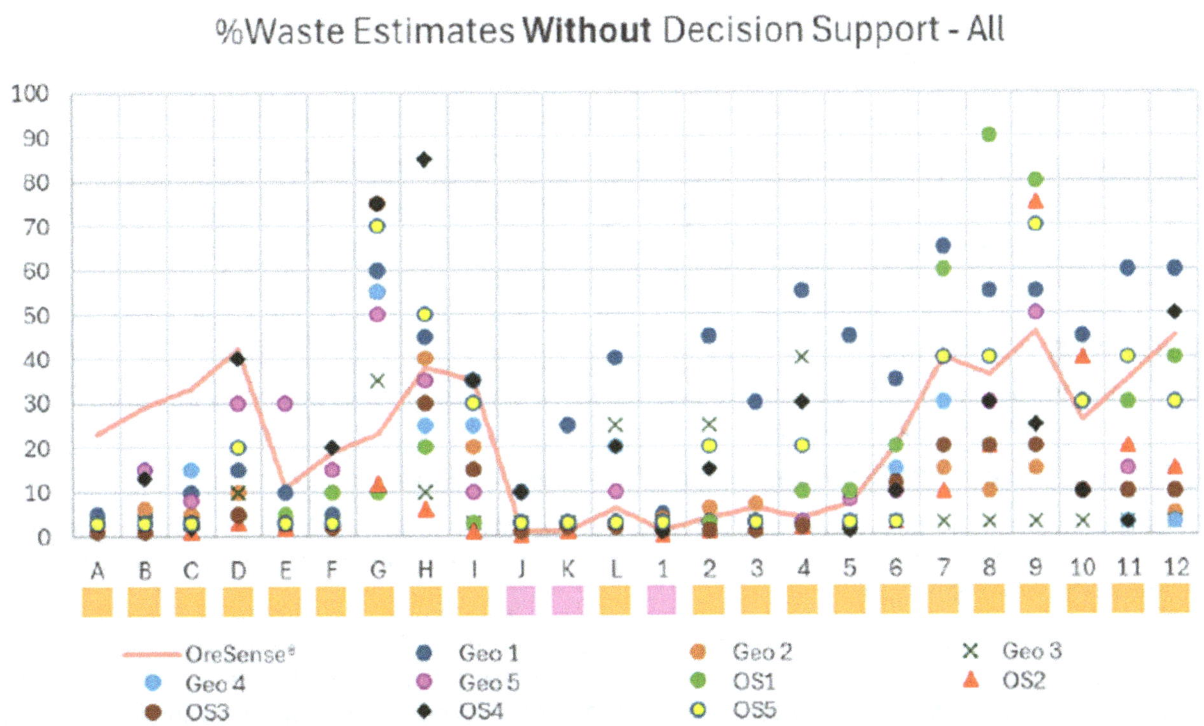

(a)

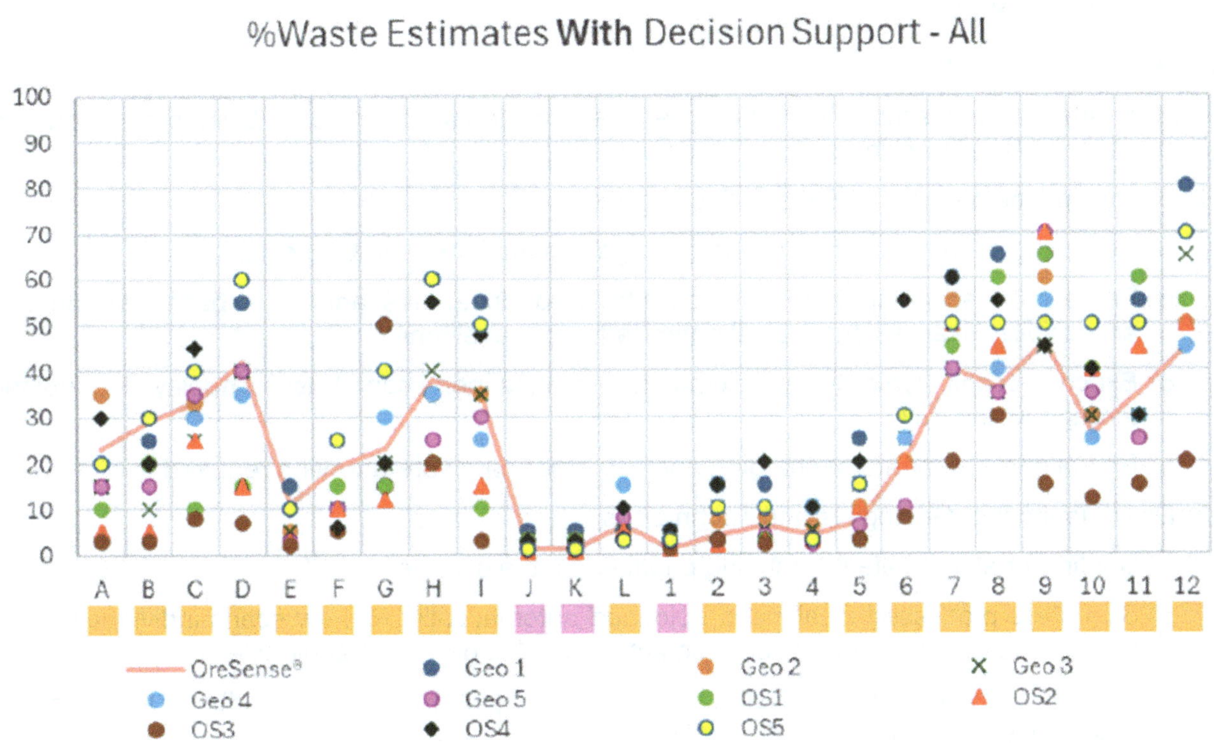

(b)

FIG 8 – (a) Waste% estimate, no OreSense® data. The Y-axis represents the percentage estimation range determined by Geology department participants. (b) Waste% estimate, with OreSense® data.

Variance of absolute %waste estimates improved significantly when using OreSense® data. Spreads in waste estimates were large without OreSense® data:

- Challenge area H – spread of 79 per cent (low 6 per cent, high 85 per cent).

- Challenge area 8 – spread of 87 per cent (low 3 per cent, high 90 per cent).

When compared to estimates with OreSense® decision support:

- Challenge area H – spread of 40 per cent (low 20 per cent, high 60 per cent).
- Challenge area 8 – spread of 35 per cent (low 30 per cent, high 65 per cent).

Without the OreSense® available, the average 1sd of collective waste estimates for loads J, K, L, 1, 2, 3 and 4 (in the area of sub 5 per cent waste in ore range. ie the critical feed range for the processing plant) was circa 9.5; with the OreSense® available this reduced to 3.5.

Results with OreSense®

Following the introduction of OreSense® there has been an evolution of practices and workflows at Mount Holland summarised in stages:

1. Initial use case of OreSense® at the mine face with trial application on stockpiles.
2. Stockpile scanning (ROM and SORT) became the focus of scanning. Scanning campaigns in the pit depending on conditions and requirements. Occasional scans of waste and trial scans of Crushed Ore Stockpile.
3. The addition of 'Rehandle Stockpiles' as a modification to ROM management. These short-term stockpiles are for mixed material with loads or zones of material considered to be 'clean SORT' which can be scanned and subsequently recovered by suitable machinery. Rehandle and ROM stockpiles are the most frequently scanned working areas at present. The degree of focus between mine face and stockpiles varies with time depending on geology exposed in the pit and other operational events.

Workflows

Workflows are briefly described here at three levels: Scanning operations, Daily monitoring and adjustments, and ROM tracking and management.

1. Scanning Operations (OreSense® Technicians):
    - Beginning of shift – confirm scanning target instructions from Geologists.
    - Proceed to scanning areas across site. Sequence of scanning is based on priority, accessibility of areas and logical optimisation of travel time.
    - Execute scans, typically approximately five mins per scan. Ensure quality and upload throughout the day.
    - End of shift – park up for charging and any other hardware attention required.
2. Daily Monitoring (Geologists and Technicians):
    - Confirm or adjust default scanning priorities for the scanner.
    - Scans are processed locally and uploaded automatically. Review scans individually or in logical sets as they are uploaded, eg active ROM fingers, Rehandle piles.
    - Assess if any area is out of specification and requires action, eg feedback to Mining, or recover clean material from Rehandle pile etc.
3. ROM tracking and management (Geology Supervision):
    - Monthly compilation of ROM quality deviations by location.
    - Feedback to teams involved.
    - Planning of priority areas for next period and potential changes to ROM practices, eg Rehandle pile utilisation.

Quantifying the impact

To quantify the general benefit of reducing loss and dilution, a pragmatic measure of impact on the operation is a count of loads sent to incorrect destination including identification of those potentially recoverable. For example, Ore on SORT pile, Waste on ROM etc. A systematic count was made over a defined period when the site focus had moved to stockpile scanning. Figure 9 shows the results over the three-month test period.

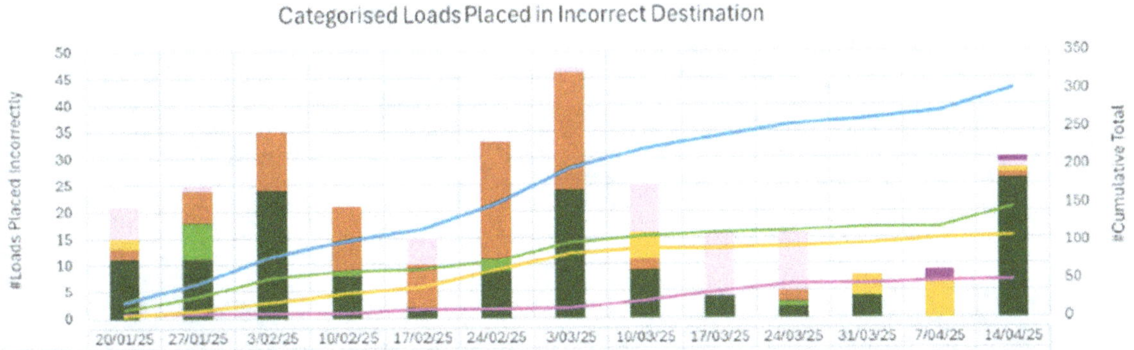

Week Starting	20/01/25	27/01/25	3/02/25	10/02/25	17/02/25	24/02/25	3/03/25	10/03/25	17/03/25	24/03/25	31/03/25	7/04/25	14/04/25
Ore on DUMP	0	0	0	0	0	0	0	0	0	0	0	2	1
Ore on SORT	6	1	0	0	5	0	1	9	12	11	0	0	1
Sort on DUMP	2	0	0	0	0	0	0	5	0	0	4	7	1
Sort on ROM	2	6	11	12	8	22	22	2	0	2	0	0	1
Waste on ROM	0	7	0	1	0	2	0	0	0	1	0	0	0
Waste on SORT	11	11	24	8	2	9	24	9	4	2	4	0	26
Cumulative Ore	6	7	7	7	12	12	13	22	34	45	45	47	49
Cumulative Sort	4	10	21	33	41	63	85	92	92	94	98	105	107
Cumulative Waste	11	29	53	62	64	76	99	108	112	115	119	119	145
Cumulative Total	21	46	81	102	117	150	197	222	238	254	262	271	301

FIG 9 – Count of mis-placed loads over a three-month period.

In total, 301 loads were placed in the incorrect destination during the period. Specifically:

- 49 potentially recoverable ore load were identified on the SORT stockpile and Dump.
- 134 waste loads (>55 per cent WST) were recorded on the SORT stockpiles. These loads contribute significantly to SORT volumes increasing ahead of schedule and are likely to materially impact ore sorting performance when the stockpiles are processed through available technologies.
- 11 waste (>65 per cent WST) + 88 SORT (8–65 per cent WST) loads were recorded on the ROM.

Future applications

Identified further value-adding potential applications of the system include:

- Optimised blending into plant feed. With better information about contained quality in mine faces and ROM and Rehandle stockpiles, it is possible to blend to meet plant requirements rather than meet strict cut-offs at each component stage.
- Identification of minor material types which can impact feed quality from time to time, eg high clay or highly weathered zones.
- Stockpile models – scanning during stockpile build and reclaim allows for a simple representation of contained material, as expressed by successively scanned faces (Figure 10).
- Scanner mounted on digging units rather than light vehicle leading to further improved resource extraction, reduced rehandling, and optimal stockpile blend building.

FIG 10 – Stockpile build view. Yellow represents classified pegmatite, dark colour classified waste (as seen in several zones to the left). Density of information can be increased by more frequent scans during stockpile build.

CONCLUSIONS

Traditional challenges of ore loss, dilution and ROM management in general remain a critical driver to operational and financial performance in mining. Rapid and reliable data to provide decision-support to technicians and operators in the production environment remains elusive. Drilling is costly and slow while face sampling exposes people to hazardous environments and struggles to cover areas at scale. Visual estimation is highly prone to error.

The deployment of an imaging scanner, in the form of OreSense® is to the best of our knowledge, a novel application for a hard rock lithium operation. To date it has been both technically and operationally successful at Mount Holland. As presented in this paper, new workflows have been established which provide rapid, reliable and geolocated representations of Ore and Waste for technicians and operators.

Good technical performance was based on:

- A good fit between the geology, material types and the technology employed.
- Effective engagement with the site technical team from the outset.
- Sufficient initial work including representative sampling and early discussion of results with site technical team.
- Understanding of integration requirements such that new workflows would be adopted and embedded.
- Stepwise validation of a set of key hypotheses.

Operational success is evidenced by:

- A material reduction in variance in ore spotting results.
- Identification of a significant number of mis-placed loads on stockpiles. Scanning avoids mis-placed loads in the first place and secondly helps recover or react to ore loss and dilution prior to ROM being fed to the process plant.
- A controlled expansion of use cases as directed by the site team. Specifically, the focus on stockpiles, including dedicated re-handle stockpiles has grown since the initial mine-focused engagement.

A broad observation is made that the solution has supported a progression from essentially an initial remedial phase, ie value-preservation, to eventual identification of optimisation opportunities, ie value-add (Table 2). This expanding set of applications is fundamentally enabled by the generation and utilisation of a rapid, reliable and geolocated primary data set within the ROM handling workflow.

TABLE 2

Progression of ROM management methods supported by OreSense®.

Remedial	Reactive	Tactical	Optimisation
Destination compliance problems	Identify and reclaim misallocated loads	New interim rehandle stockpiles	Ex-pit to ROM blending, new sensor configurations

ACKNOWLEDGEMENTS

Covalent Lithium and Plotlogic management are thanked for permission and support in publication of this work.

REFERENCES

Covalent, 2025. Covalent Lithium achieves first product at Kwinana lithium hydroxide refinery, Covalent Lithium. Available from: <https://www.covalentlithium.com/news> [Accessed: 20 November 2025].

Ediriweera, A and Wiewiora, A, 2021. Barriers and enablers of technology adoption in the mining industry, *Resources Policy*, 73:102188.

Fraser, S J, Whitburn, L B and Ramanaidou, E, 2006. Mineralogical face-mapping using hyperspectral scanning for mine mapping and control, in *Proceedings Sixth International Mining Geology Conference*, pp 227–232 (The Australasian Institute of Mining and Metallurgy: Melbourne).

GeoInnova Consultores SPA, 2025. Mount Holland Lithium Project Technical Report Summary. Available from: <https://www.sec.gov/Archives/edgar/data/909037/000090903725000012/exhibit965-technicalrepo.htm> [Accessed: 27 November 2025].

Parker, H, 2011. Reconciliation Principles for the Mining Industry, 35th APCOM Symposium (The Australasian Institute of Mining and Metallurgy: Melbourne).

Phelps-Barber, Z, Trench, A and Groves, D I, 2022. Recent pegmatite-hosted spodumene discoveries in Western Australia: insights for lithium exploration in Australia and globally, *Applied Earth Sciences (Trans Inst Metall B) 2022*, 131(2):100–113.

Wesfarmers Chemicals, Energy and Fertilisers (WesCEF), 2024. Covalent Lithium opens Mt Holland lithium mine and concentrator. Available from: <https://wescef.com.au/2024/03/07/covalent-lithium-opens-mt-holland-lithium-mine-and-concentrator/> [Accessed: 20 November 2025].

Appropriate certified reference materials or analytical method, a potential dilemma, headache or not important

S L Martin[1]

1. Senior Database Geologist, Aurelia Metals Limited, Brisbane Qld 4000.
   Email: shauna.martin@aureliametals.com.au

ABSTRACT

The use of certified reference materials (CRMs) to monitor the quality of analytical results as part of a wider quality assurance and quality control (QA/QC) programme is well understood. This is a requirement under the reporting codes for publicly listed companies involved with mineral exploration, resource estimation and mining grade control.

The focus on CRMs is around their performance to monitor accuracy and bias relative to the nominal values for the material. This provides a level of confidence in results generated for the unknown samples submitted to a laboratory. However, while CRMs are certified for a few different methods, the methods used to analyse these and the unknown samples needs to be considered in more detail.

The use of aqua regia method for base metals has an upper detection limit for key elements copper (Cu), lead (Pb) and zinc (Zn) of 1 per cent. If samples are tested that exceed these values, they will require further analysis by a different method to achieve absolute values. If the CRM also exceeds 1 per cent base metal, can the performance be assessed accurately across the various methods for the surrounding samples? Alternatively, if the CRM does not exceed 1 per cent base metals, and the surrounding samples do, is the CRM accurately assessing the surrounding samples? This issue, of CRM grades being unsuitable for the samples being tested (ie grades outside the detection ranges for method used), was identified at Aurelia Metals in 2022. In co-operation with the partner laboratory, a suitable solution was implemented that resulted in most samples undergoing the same analytical method which streamlined the analytical process. This had the added benefit of reducing turnaround time and costs, while maintaining the quality and confidence in the results.

INTRODUCTION

In 2022, prompted by a laboratory contract renewal, Aurelia Metals Limited (Aurelia) conducted a review, of the analytical services used across its business. Aurelia had three operational mines. These were the Peak Operation (including its Peak South and New Cobar mines) and Hera Mine (including then Federation Project) in the Cobar Basin region of western New South Wales (NSW) and Dargues Mine, near the town Braidwood in the NSW southern tablelands.

The key issues that needed to be addressed in this review included the following:

- Were the precision levels of the analytical methods in use appropriate for each deposit style and quality control (QC) samples in use?
- Was there an impact on turnaround times because of the choice of methods currently in use?
- Was the company getting the best quality and value for the money currently being spent?

The results of the review would determine what, if any, changes needed to be made that would deliver the best quality results suitable for each of the sites in a timely and cost-effective manner.

ANALYTICAL METHODS

The analytical methods for all three sites were reviewed. While gold analysis was included in the initial review, this report will focus solely on the base metal methods. Additional methods relating to sample preparation, sizing and weighing were excluded for the comparison. Table 1 shows that while the Peak Operation and Hera Mine were using the same base metal methods, the difference was the number of elements being analysed.

TABLE 1

Primary analytical methods for base metals used for each site in 2022.

Site	Department	Lab method	No. of elements	Comments
Dargues Mine	Exploration	ME-MS61	48	Full suite
Peak operation	Exploration/ mining	ME-ICP41	35	Full suite
		ME-OG46	4	Used when UL on ICP41 reached
		S-OG46	1	All samples
		ME-OG46h	4	Used when UL on OG46 reached
Hera Mine	Exploration	ME-ICP41	9	
		ME-OG46	4	Used when UL on ICP41 reached
		S-OG46	1	All samples
		ME-OG46h	4	Used when UL on OG46 reached
	Mining	ME-ICP41	8	
		ME-OG46	4	Used when UL on ICP41 reached
		S-OG46	1	All samples
		ME-OG46h	4	Used when UL on OG46 reached

The primary method ME-ICP41 is an aqua regia digest with an inductively coupled plasma (ICP) determination and is specifically referred to by ALS as a Targeted Exploration Analytical Suite. It is designed as a first pass exploration suite. The method has upper detection limit for copper, lead and zinc of 1 per cent and any sample reaching this is then analysed by ME-OG46, an ore grade method with upper limits of 50 per cent for copper, 20 per cent for lead and 30 per cent for zinc. This method also uses an aqua regia digest with ICP determination. For high-grade samples exceeding those limits, ME-OG46h was then used. Figure 1 shows a typical drill hole and which samples would undergo the different methods.

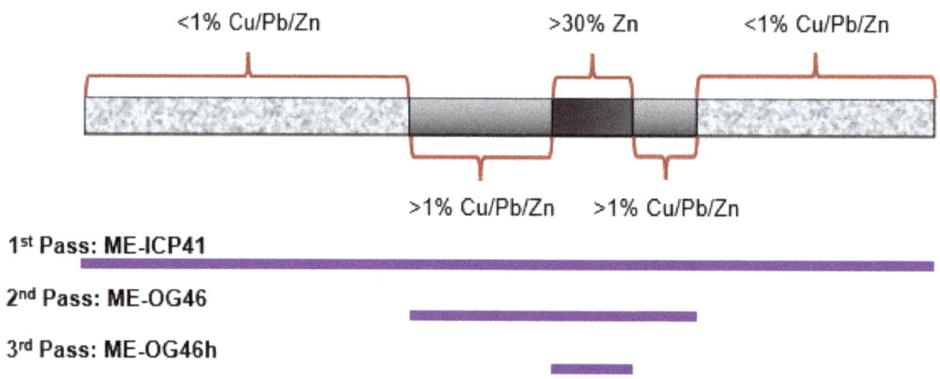

FIG 1 – The use of the analytical methods across a drill hole.

The sulfur method, S-OG46, was used to generate a more accurate sulfur value because at this time the specific gravity for each sample was calculated. This method pre-digests the samples in nitric and hydrogen bromide which is designed to help oxidise the sulfide to sulfate to improve sulfur recovery. While ME-ICP41 does produce a sulfur value the sulfur recovery is poor and therefore the results are not as reliable. This protocol would see all samples analysed by up to four different methods depending on their grades resulting in increased turnaround times. Furthermore, these methods have different method precisions which would give rise to varying degrees of accuracy and quality of the results within drill holes and across the deposit (Table 2).

TABLE 2

Method precisions for each method.

Analytical method	Precision
ME-ICP41	10%
ME-OG46	5%
S-OG46	5%
ME-OG46h	5%

CRM REVIEW

After examining the analytical methods, it became clear that the CRMs in use also needed to be examined. The review focused on the Peak Operation, specifically looking at the CRMs used in 2022. Table 3 shows the CRMs used were primarily certified for either gold or base metals. Upon checking how the CRMs were used, there were often drill holes that had only base metals CRMs present despite the presence of high-grade gold or *vice versa*. This would mean there were holes that had no quality control for either gold or base metals.

TABLE 3

CRMs used in 2022 at the peak operation.

Standard ID	Certified methods	Element#	Nominal value
G300-7	50 g Fire Assay	Au	1.00 ppm
G312-2	50 g Fire Assay	Au	1.51 ppm
G312-4	50 g Fire Assay	Au	5.30 ppm
G915-2	50 g Fire Assay	Au	4.98 ppm
GLG901-2	50 g Fire Assay	Au	1.76 ppm
GBM315-13	Not Specified	Cu	12565 ppm
	Not Specified	Pb	34135 ppm
	Not Specified	Zn	37270 ppm
GBM319-14	Not Specified	Cu	29546 ppm
	Not Specified	Pb	7331 ppm
	Not Specified	Zn	22491 ppm
OREAS-620*	Aqua Regia Digestion	Cu	0.175%
	Aqua Regia Digestion	Pb	0.774%
	Aqua Regia Digestion	Zn	3.12%

The table is focused on the main base metal elements of interest but there are other elements certified for these CRMs.

* Introduced in November 2022.

The grades of the CRM's particularly for base metals used prior to November 2022, all exceed the 1 per cent copper, lead or zinc upper limit for the primary analytical method ME-ICP41. This raised the question about how the quality of the routine samples could be assessed when the QC samples used may be subjected to a different method or not certified for the elements of interest. Using the same drill hole shown in Figure 1 and adding in CRMs as shown in Figure 2 the issue was highlighted.

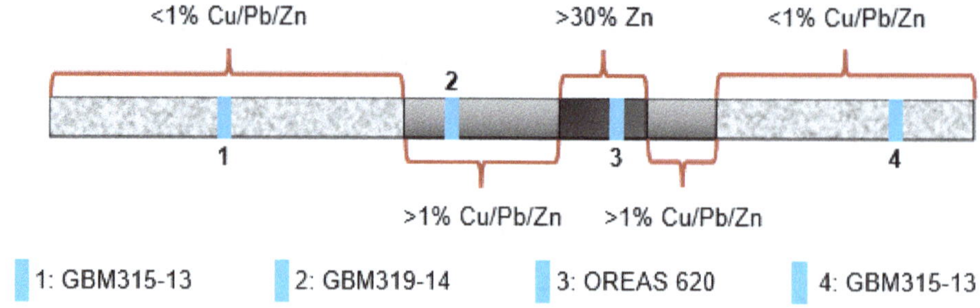

FIG 2 – A drill hole with CRMs inserted and how these relate to the routine samples.

The drill hole has been split into four sections with each section containing a CRM:

1. The first and fourth sections have routine samples all reporting <1 per cent Cu, Pb and Zn but both have a CRM whose certified values for Cu, Pb and Zn all exceed 1 per cent for these elements. All routine samples would have been analysed by ME-ICP41 and as the CRM is the only sample that would have reached the upper limits for this method, the values for these elements would have been generated by method ME-OG46. Therefore, the quality of the routine samples cannot be assessed on the performance of the CRM as they underwent different analytical methods.
2. The second section shows the routine samples and the CRM to report over 1 per cent Cu, Pb and Zn and the CRM also has certified values exceeding 1 per cent for Cu and Zn. In this case, both the routine and CRM samples will have undergone ME-ICP41 followed by ME-OG46 to generate definite results for these elements. Therefore, the quality of the routine sample can be assessed based on the CRMs performance as they underwent the same methods.
3. The third section show the routine samples exceed 30 per cent for Zn and CRM submitted has a certified value of 3.12 per cent. This means the performance of the CRM cannot give any indication of the quality of the result for the routine samples as these would have been analysed by the ME-OG46h methods to generate definitive results and the CRM would have been analysed using ME-OG46 only.

Across this hole, section 2 is the one area where the performance of the CRM can be used to assess the quality of the results for the surrounding routine samples having underwent the same analytical method. If this example contained CRMs that were only certified for gold (G312-4 or G915-2), there would be no way to assess the performance of the routine samples for base metals.

PROPOSED SOLUTION

It was clear from the review that the analytical methods were not suitable and a change was needed. With the aqua regia digest being the primary method, it was decided to continue with this but look to an intermediate level suite. In this case the most appropriate method was ME-ICP41a. This is a modified method of the ore grade method ME-OG46. It has the same digestion process but ME-ICP41a is designed to offer a more widespread range of reportable elements in the intermediate ranges with higher upper limits for Cu, Pb and Zn of 5 per cent. However, this still required the high-grade zones in the deposits be followed up with another ore grade method.

After consultation with the laboratory, it was confirmed that the method has defined absolute upper limits. Although these limits are not routinely applied, particularly when a more suitable method exists for samples that consistently exceed them, these thresholds can be used to generate a quantitative result rather than reporting an 'over-range' value, while still maintaining the required level of accuracy and quality (Table 4). For any samples then exceeding the absolute upper limits would be followed up with method ME-OG46h.

TABLE 4

Detection limits for ME-ICP41a with absolute upper limits shown.

ME-ICP41a				
Analyte	**Unit**	**Lower limit**	**Upper limit**	**Absolute upper limit**
Ag	PPM	1	200	1500
Cu	PPM	5	50 000	400 000
Pb	PPM	3	50 000	200 000
Zn	PPM	10	50 000	370 000

Changing to this method would result in a cost increase. To quantify the cost, 66 batches that were returned between September to November 2022 from the Peak Operation were examined. Using the costs at the time, the total cost by the original methods was compared to the costs of the new methods (Table 5). Over the time period assessed, had the new method been implemented (ME-ICP41a) the cost saving is estimated to be $9240.48. Annually, this was projected to save the company approximately $42 000 a year for 300 despatches.

TABLE 5

Cost comparison of 66 batches from the peak operation.

	No. despatches	**Total spend**
Original analysis	66	$116 018.04
Propose new analysis	66	$106 777.56

The added benefit of moving to this method would be a reduction in the turnaround time because of the 66 despatches, only four would have required the overlimit method ME-OG46h.

Moving to ME-ICP41a would also help to reduce the issue of having CRMs analysed by a different method to the method used for the routine samples. If the base metal CRMs in use at the time remained unchanged, taking the example shown in Figure 2, then potentially all routine samples and CRMs would be subjected to the one method ME-ICP41a. Only Zn samples in the high-grade ore zones exceeded 37 per cent, would they then undergo analysis by ME-OG46h. However, in addition to the proposed method change, recommendations were made to source more suitable CRMs. As the Peak Operation, Hera Mine and Federation Project are all classed as polymetallic deposits, CRMs certified in the elements of interest Ag, Au, Cu, Pb and Zn produced by Ore Research and Exploration were identified as being the most suitable on the market at the time. Not only would they provide quality assessment of the gold and base metals they ranged in grade from low, to medium, to high-grade options that would all return results by the new method ME-ICP41a.

DISCUSSION

A fundamental component that underpins the mining value chain is the geological data, specifically the assay data. The collection of assay data needs to be at a level that provides quality and accuracy from which a level of confidence can be gained to optimise business decisions. Typically, the QC of a QA/QC programme includes the insertions of field blanks, CRMs and duplicates in each submission to the laboratory. However, while it is important to include the QC samples, the analytical methods planned to be used need to be considered in the early stages of a project. Not only does the method used need to provide results that are meaningful but the QC samples included should also be analysed by the same method to allow them to do their job in assessing the quality of their performance and the surrounding routine samples. When selecting CRMs, the element concentrations should be at levels where decisions are made which will also be dependent upon the stage of the project. For example, greenfields exploration would use CRMs with low level expected values to allow for clear discrimination between background and anomalous values, while delineation drilling in a mine may have CRMs that are at mine cut-off grade, an average grade and higher grade (Smee *et al,* 2024).

It can be argued that a similar process needs to be undertaken when selecting the analytical methods to be used. In greenfields exploration, a trace element analysis is ideal for identifying pathfinder elements, unless specifically targeting elements that are expected to show low-grade mineralisation. In areas known to have significant mineralisation, intermediate methods would be more economic while specific deposits including sulfide would benefit from using ore grade methods.

Exploration and mining activities have been ongoing in the Cobar region for over 100 years. Aurelia has been operating in the Cobar Basin region since 2007 and have taken many samples during this time. For Aurelia, the primary focus for many years at the Peak Operation was on gold mineralisation but then the company's focus shifted to include base metals in 2018. This review of analytical methods and CRMs identified that neither had been considered in detail since the company's focus shifted and that the QA/QC was being performed without real consideration of what it was designed to do.

The recommendations of the review were approved and implemented in 2023, with the Peak Operation, Hera Mine and Federation Project being standardised to using ME-ICP41a as the primary method and ME-OG46h as the overlimit method for base metals. During 2023, 234 despatches were sent to the laboratory for analysis of which only five despatches had samples that exceeded the upper limits for Pb and Zn by ME-ICP41a and required the ME-OG46h method. The average turnaround time in 2022 was 54 days and in 2023 this reduced to an average of 30 days, therefore showing this change to the analytical method had a significant impact on the reduction in turnaround time.

CONCLUSIONS

The review into the analytical methods and CRMs used at Aurelia resulted in changing to a more deposit/mineralisation appropriate analytical method and introducing CRMs that were certified for all the elements of interest within the deposits. The benefits of these changes included:

- Standardising and simplifying the analytical methods used across the company.
- Ensuring the maximum data was provided for each site ie all sites getting results for the full analytical suite and not a limited number of elements.
- A reduction in turnaround times.
- QA/QC analysis that is more meaningful, as the comparison between QC and routine samples is like for like.
- The accuracy and overall quality of the base metal analysis significantly improved as the method precision tightened from 10 per cent to 7 per cent, reflecting reduced variability, enhanced consistency, and greater confidence in the analytical results.
- The reduction in assay costs as the new method is more cost-effective.

ACKNOWLEDGEMENTS

The author wishes to thank Aurelia Metals Ltd for their approval in allowing this paper to be published and all that provided comments and suggestions during the review period.

REFERENCES

Smee, B W, Bloom, L, Arne, D and Heberlein, D, 2024. Practical applications of quality assurance and quality control mineral exploration, resource estimation and mining programmes: a review of recommended international practices, *Geochemistry: Exploration, Environment, Analysis*, 24(2):2023-046. <https://doi.org/10.1144/geochem2023-046> [Accessed: 8 May 2024].

New geological interpretation and grade control strategy at the Federation Pb-Zn Deposit, Cobar Basin

J Mole[1] and C Cavill[2]

1. MAusIMM, Senior Mine Geologist, Aurelia Metals Limited, Nymagee NSW 2831.
   Email: jeremy.mole@aureliametals.com.au
2. Principal Geologist, Aurelia Metals Limited, Brisbane Qld 4000.
   Email: chloe.cavill@aureliametals.com.au

ABSTRACT

The Federation Pb-Zn deposit, located approximately 15 km south of Nymagee, New South Wales, lies on the south-eastern margin of the Palaeozoic Cobar Basin. Discovered by Aurelia Metals Limited in 2019, the deposit was initially characterised in the feasibility study as an epigenetic, structurally controlled system, with massive sulfides hosted in steeply plunging, short-strike shoots surrounded by breccia zones.

Subsequent infill drilling during the early stages of the mine development phase quickly led to a revision of the geological model. Within the broadly east-north-east-trending mineralisation envelope, the deposit hosts discrete, steeply dipping, high-grade lenses of massive sphalerite and galena, typically striking north-north-east. These lenses generally exhibit sharp contacts with the generally undeformed, barren siltstone host rocks, highlighting the importance of understanding the structural and stratigraphic controls.

This revised understanding in orientation and orebody geometry prompted a change in both drilling strategy and grade control methods. Tightly spaced infill drilling, conducted in a new orientation from revised drill platforms to better define the short-strike lenses was implemented. Additional, detailed mapping and the introduction of face sampling was required to sufficiently define the economic zones for production. The implementation of rapid-turnaround grade control block modelling further supported efficient, reactive mine planning, effective stope design and minimised dilution between discrete ore zones.

Unlike most Cobar-style deposits, Federation is distinguished by its overall east-north-east structural orientation and structurally controlled, narrow lens style. Mineralisation is hosted by gently folded, south-east-verging sedimentary host rocks. The structural control is evident, with mineralisation textures, tenor and structural measurements key to defining the individual lenses. The integration of geological reinterpretation using modern techniques with adaptive operational practices has been critical to reacting and optimising early-stage mine planning and maintaining the project development schedule.

INTRODUCTION

The Federation Lead-Zinc (Pb-Zn) deposit was discovered by Aurelia Metals Limited in 2019 upon drilling an induced polarisation and Pb soil anomaly proximal to Aurelia's Dominion prospect (Aurelia Metals Limited, 2019). It is located approximately 15 km south of Nymagee in New South Wales. The Maiden Resource Estimate for Federation was released in June the following year (Aurelia Metals Limited, 2020), and the feasibility study in June 2022 (Aurelia Metals Limited, 2022a). Resource infill drilling from underground commenced in early 2024 with first development ore mined in July of that year.

The process from Resource interpretation to grade control model can be a varied one, where the importance of close scale drilling can play a large part in determining the mine plan. Variations from feasibility study designs to monthly scale mine designs are expected and the consistent and detailed logging of infill drilling and follow on modelling has set-up Federation to manage this variation smoothly.

BACKGROUND

Regional geology

The Cobar Superbasin is a large Palaeozoic sedimentary–volcanic basin system in central New South Wales developed within the Central Lachlan Orogen. Basin development occurred predominantly during the Silurian to Devonian as a series of interconnected troughs formed in response to crustal extension and strike-slip faulting. The basin fill comprises thick successions of marine turbiditic sandstones and siltstones with lesser shale, chert and limestone, interlayered with mafic to felsic volcanic and volcaniclastic rocks. Basin inversion at approximately 400 Ma reactivated earlier extensional structures and generated new contractional faults, providing pathways for hydrothermal fluids during mineralisation (Glen, 1991).

The Cobar Superbasin hosts numerous structurally controlled hydrothermal Cu–Au–Pb–Zn–Ag deposits, including CSA, Peak, Endeavor New Cobar and Hera deposits. These deposits are typically pipe-like to lensoidal and spatially associated with major fault zones and fold hinges, typically anticlines. Mineralisation consists of massive to disseminated sulfides dominated by chalcopyrite, sphalerite, galena, pyrrhotite and pyrite. Ore formation is interpreted to have occurred primarily during Late Silurian to Devonian basin-scale fluid circulation, with subsequent deformation locally upgrading and remobilising mineralisation (Fitzherbert, Downes and Blevin, 2022).

Federation geology

The Federation deposit is located near the south-eastern margin of the Cobar Basin, approximately 5 km north of the non-conformable contact with the Silurian Erimeran Granite. Mineralisation is hosted within gently folded siltstones and fine-grained sandstones of the lower Amphitheatre Group (Thomas, Smith and McKinnon, 2022). Regional metamorphism reached lower greenschist facies, and a moderately developed, near-vertical cleavage is present, although deformation intensity in the Federation area is weaker than at Hera and Nymagee to the north (Thomas, Smith and McKinnon).

At feasibility stage, Federation was interpreted as a structurally controlled, epigenetic Zn–Pb–Au–Ag system developed within a 070° striking structural corridor along the south-eastern basin margin (Thomas, Smith and McKinnon, 2022). This orientation is sub-orthogonal to nearby deposits such as Hera, Nymagee and Peak, which trend parallel to the Rookery Fault system. District-scale gravity and magnetic interpretations indicate that both structural orientations are significant, with Federation-trending structures commonly offsetting Rookery-parallel features (Thomas, Smith and McKinnon). The geological model recognised three principal massive sulfide–breccia lenses (North-east, Western and Hesperides) extending to depths greater than 500 m. High-grade mineralisation was interpreted to be hosted within steep breccia zones associated with south-east-verging parasitic anticlines and north-east–south-west shears, with additional control from north-west-dipping bedding-parallel faults localised at rheological contrasts between chlorite–pyrrhotite and silica–sericite–pyrrhotite altered sedimentary packages. The Main Thrust at ~450–500 m depth was considered an important structural control and fluid pathway (Thomas, Smith and McKinnon).

Mineralisation comprises a relatively simple sulfide assemblage dominated by sphalerite (commonly low-Fe) and galena, with minor chalcopyrite and pyrrhotite, local native gold, and rare pyrite and magnetite (Thomas, Smith and McKinnon, 2022). High-grade zones were interpreted as massive sulfide and breccia bodies with relatively low pyrrhotite compared to peripheral alteration. Alteration within the mineralised corridor is characterised by silica–sericite–pyrrhotite ± chlorite assemblages, grading outward to weaker chlorite–silica alteration (Thomas, Smith and McKinnon).

NEW GEOLOGICAL INTERPRETATION

The 2022 feasibility study identified areas for further investigation, such as the late bedding parallel faults, that were identified but not yet understood in how they interact with the mineralised system. The 2022 Mineral Resource and Ore Reserve statement identifies that the available drilling reasonably supports the orientation of the mineralisation, with closer spaced drilling expected to improve confidence in the Mineral Resource Estimate. The underground drilling at Federation has allowed the further investigation into both the faulting and the orientation of the mineralisation. Progression of the data from Inferred and Indicated to a mine planning-suitable grade control model level of spacing (12.5 × 12.5 m), as well as the progression of drill collars from surface to underground has allowed for a new interpretation.

Underground drilling at Federation commenced in January 2024 from the decline stockpile, initially targeting the bulk sample area. Early drilling in the top north-east of the deposit returned several high-grade massive sulfide intersections, confirming the tenor of the system. However, the geometry and internal character of mineralisation observed in this new drilling were inconsistent with the prevailing breccia-dominated geological model. Intersections were narrow, discrete and sharply bounded, commonly with little to no visible alteration or economic mineralisation in adjacent host rock. These characteristics contrasted with expectations for a massive sulfide and sulfide breccia system and varied from the existing geological interpretation. The implications for mine design and extraction strategy prompted a comprehensive review of structural controls, orebody geometry and grade control methodology.

To investigate this further, detailed structural core logging and systematic orientation data collection were prioritised. Measurements of bedding, cleavage and fold geometry were broadly consistent with previous work. Bedding typically dips gently to moderately north to north-west, axial planar cleavage trends north-north-east, and folds are open, non-cylindrical and south-east verging (Figure 1). Anticlines within the mining area are parasitic in character. While no reliable younging indicators were identified in the initial infill drill-out area, exploration drilling had identified indicators present lower in the sequence indicating that the stratigraphy is upright. Macrostructural observations in core support this interpretation, with open, parasitic folds exhibiting north-north-east-trending axial planes (Figure 2). Bedding and cleavage data sets are internally consistent and provide little evidence of broad brecciation, suggesting an overall low-strain deformational history relative to nearby deposits.

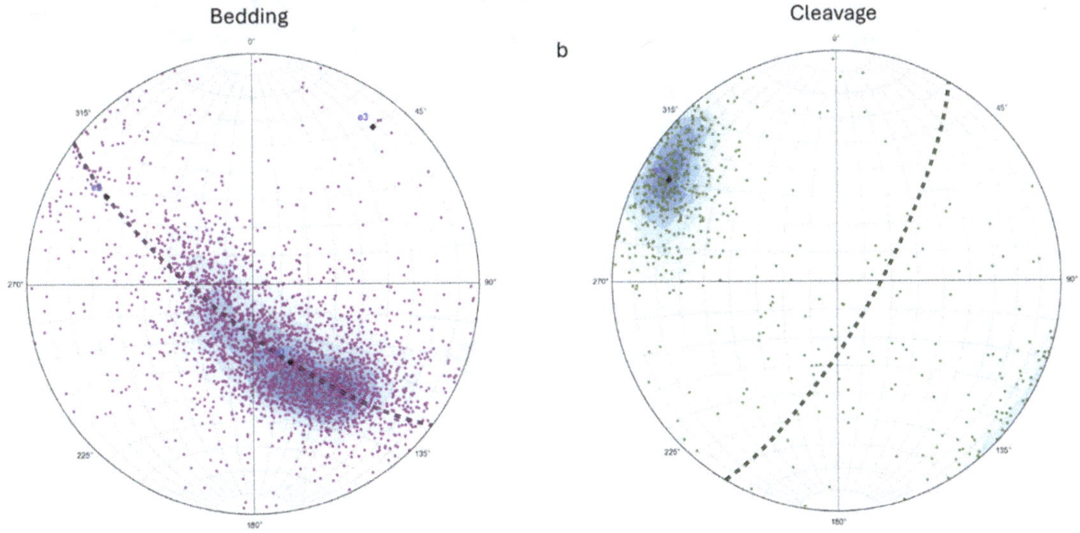

FIG 1 – Bedding (a) and cleavage (b) stereonets illustrating gently east-verging, north-east plunging asymmetric open folds with an axial plane trending to ~030°. Measurements plotted as poles to planes.

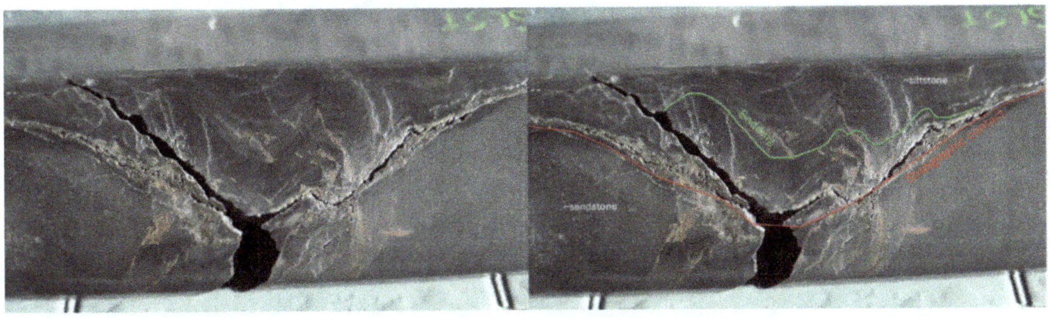

FIG 2 – Macrostructures in core – stratigraphic contact an important control on the location and tenor of PbZn mineralisation? Core photo courtesy Tucker (2024).

Structural measurements of mineralisation contacts, alteration contacts and sulfide foliation, with specific codes, were introduced early in the underground drill-out (Figure 3). These data sets consistently indicate that mineralisation, and associated alteration, trend north-north-east to north-east and dip steeply to the east-south-east (Figure 4), sub-parallel to the dominant cleavage and fold axial planes. Structurally controlled breccias also share this orientation. Vein analysis further supports this interpretation: sphalerite-bearing veins are predominantly steeply dipping to the south-east to south-south-east, defining a well-clustered structural population (Figure 4). Chalcopyrite-bearing veins are less common but define two orientations: one comparable to the sphalerite-bearing set (subvertical SE to NW dipping) and a subordinate, gently west-dipping population (Figure 5). Collectively, these data sets demonstrate that mineralisation is structurally aligned with the host rock architecture.

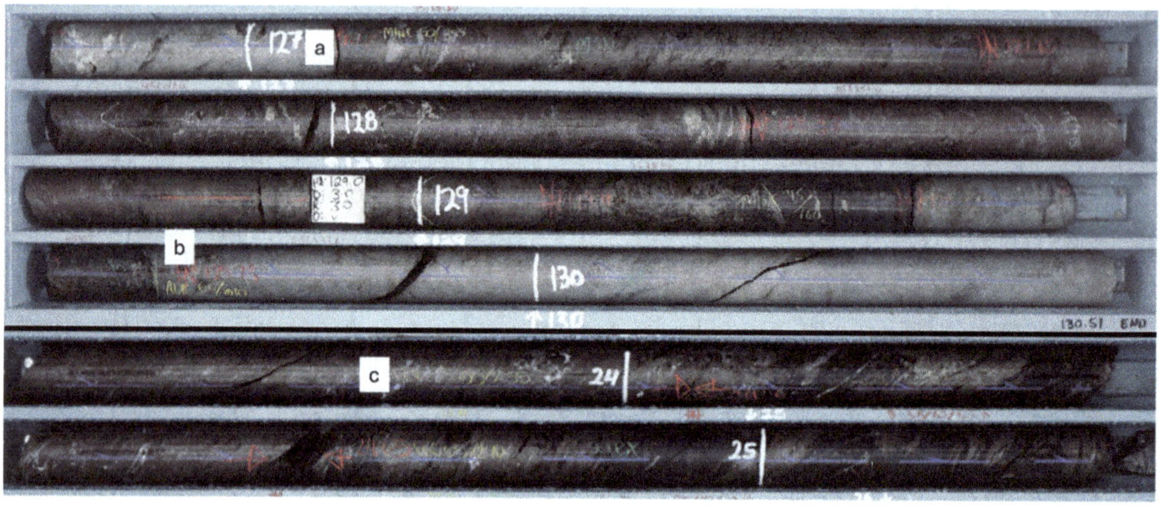

FIG 3 – Photos of drill core with geology mark-up. Examples of mineralisation contacts ('MINC', a), alteration contacts ('ALTC', b) and foliation ('FOL', c) marked in yellow chinagraph on drill core.

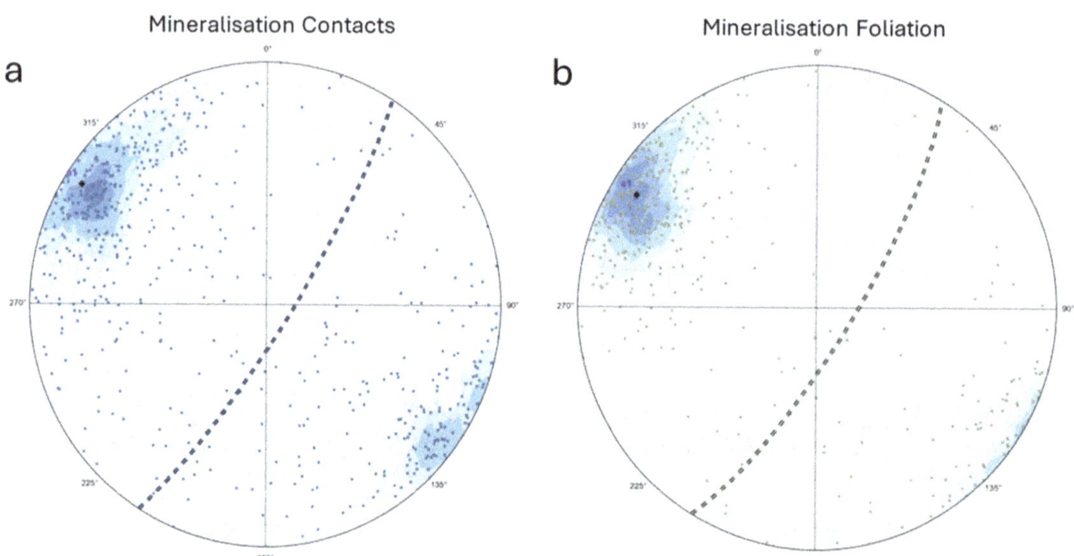

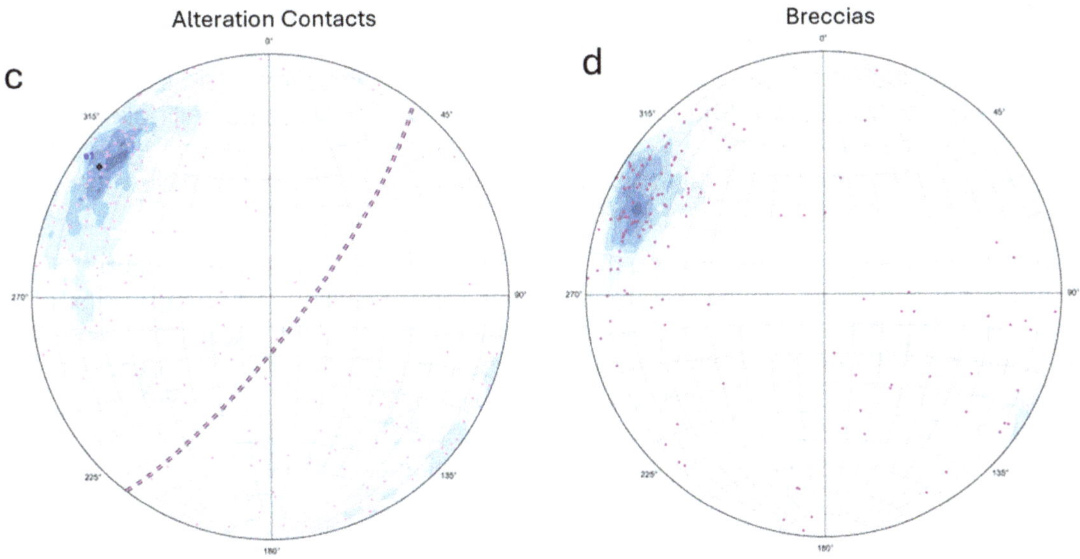

FIG 4 – Stereonets with structural measurements from drill core plotted as poles to planes: (a) mineralisation contacts, (b) foliation of mineralisation, (c) alteration contacts and (d) breccias.

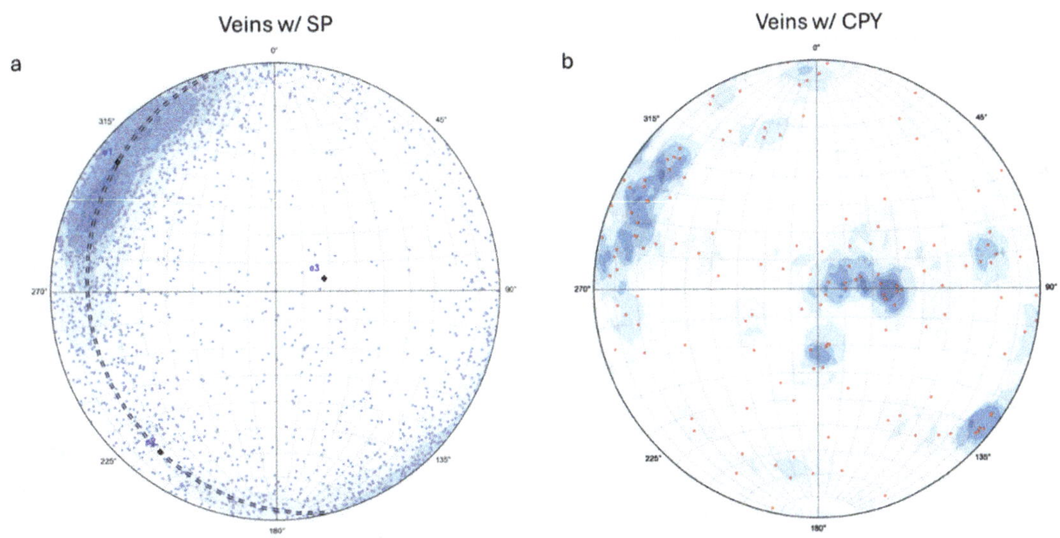

FIG 5 – Stereonets with vein structural measurements plotted as poles to planes, taken in drill core and categorised by logged contained minerals. (a) Sphalerite-bearing veins, (b) chalcopyrite-bearing veins.

This revised structural understanding formed the basis for development of a new local geological model focused on the initial production areas. Implicit modelling (Leapfrog Geo) was used to interpret mineralisation as a series of *en echelon*, steeply plunging massive sulfide lenses trending approximately 030°, sub-parallel to cleavage. In addition, several low-grade, bedding-parallel stratiform lenses were recognised. Existing underground workings were repurposed as drill platforms to improve intersection angles to the newly interpreted lenses.

Historically, drilling at Federation started on an east–west orientation before progressing to majority south-east oriented holes with some north-west (scissor holes) for orientation assessment at a broader (25 to 50 m) spacing aligned with the broader mineralisation envelope. While appropriate for testing the moderate confidence model, these orientations resulted in suboptimal intersection angles to the newly recognised north-north-east-trending *en echelon* lenses. Under the revised structural interpretation, drilling is now oriented west to north-west from the south of the orebody (or east to south-east from the north), maximising intersection angles to the 030° trending, steeply east-south-east-dipping lenses, improving true-thickness representation and lens discrimination.

The new interpretation was first applied to the upper-east (Feas: 'Northeast Breccia') mining area, where resource drilling prior to infill had limited structural information due to the use of RC drill holes. It was subsequently tested in the upper-west (Feas: 'Western Breccia'). As suitable underground drill platforms were not available for the west area, a single surface diamond drill hole was completed to test the 030° trending *en echelon* lens model and guide the justification and establishment of a purpose-mined drill platform. Structural logging from this hole confirmed mineralisation orientations consistent with those defined in the upper-east area.

Approval was subsequently granted for 40 m of dedicated drill-platform development south of the orebody, enabling drilling in a westerly to north-westerly direction and improving intersection geometry to the interpreted north-north-east trending lenses. The first underground drill hole from this platform intersected four discrete massive sulfide-bearing lenses – two previously unrecognised – at near-true thickness and high angles to the core axis: 5.3 m @ 22.8 per cent Pb+Zn, 3.3 m @ 46.4 per cent Pb+Zn, 3.0 m @ 14.4 per cent Pb+Zn and 3.1 m @ 12.4 per cent Pb+Zn (Figure 6). These results provided strong validation of the revised geological model.

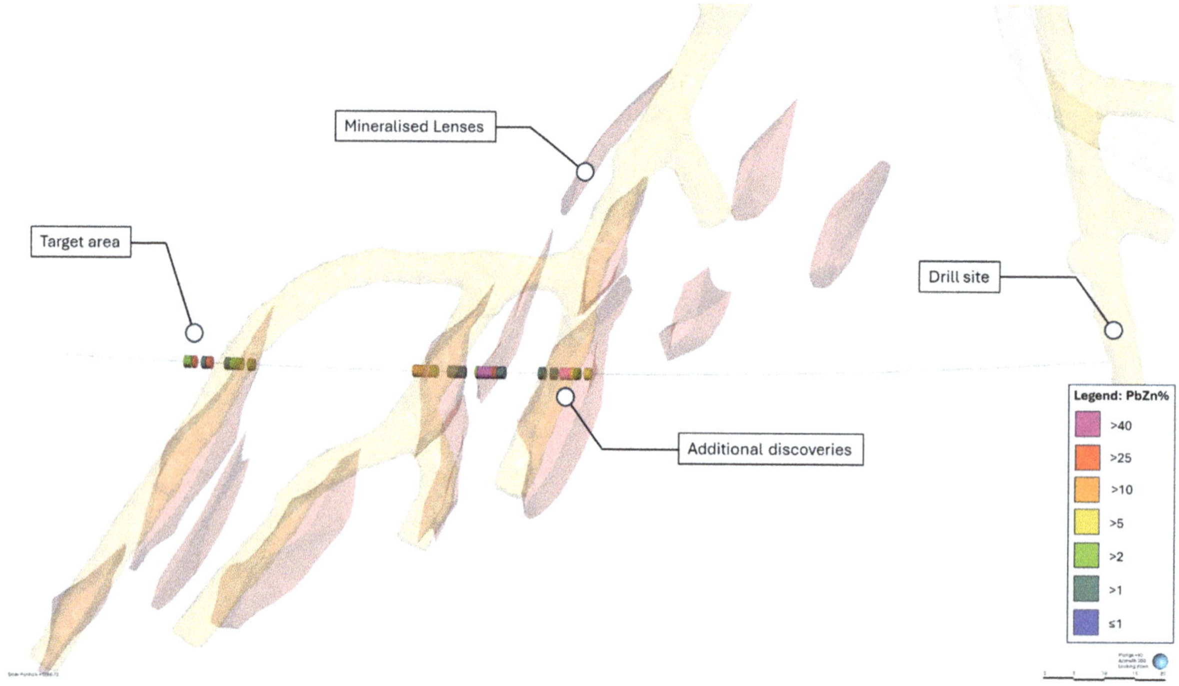

FIG 6 – First underground diamond drill hole on the western lenses at Federation utilising a new drilling orientation.

The reinterpretation represents a shift in understanding of the Federation orebody. Rather than a broad breccia-dominated system, mineralisation is now recognised as a structurally partitioned, cleavage-parallel *en echelon*-lens system. The highest grades are hosted within narrow, steeply east-south-east-dipping, north-north-east-trending lenses that are strike-limited but plunge-extensive, occasionally extending over 300 m. Dilation zones of enhanced grades appear to occur where these steep lodes intersect low-grade, discontinuous stratiform sulfide horizons. Mineralisation textures range from blebby and disseminated to semi-massive and foliated massive sulfides (Tucker, 2024), and individual lens geometry is structurally controlled and reasonably predictable down plunge.

Two dominant mineralised orientations are recognised within the upper western limbs of open, non-cylindrical, south-east-verging parasitic folds:

1. Stratiform lenses: gently north-west-dipping, bedding-parallel and generally lower grade.
2. Steep lenses: north-north-east-trending, cleavage-parallel, narrow, high-grade and plunge-extensive, sometimes extending over 300 m. Some lenses appear interrupted along plunge, with the economic ore being discontinuous, however the geological controls on this are yet to be completely understood.

Given the implications of this interpretive shift (Figures 7 and 8), underground grade control activities were ramped up during development toward the orebody. Detailed geological mapping utilising LiDAR, systematic channel sampling, stockpile sampling and targeted sludge sampling were implemented to validate orebody knowledge at mining scale. Mapping in the first two ore drives observed narrow (0.5 to 4 m) massive sulfides over short strike lengths, with strong continuity down plunge, confirming the revised interpretation. Both mapping and the channel sampling confirmed that there was minimal economic mineralised material between lenses in most cases.

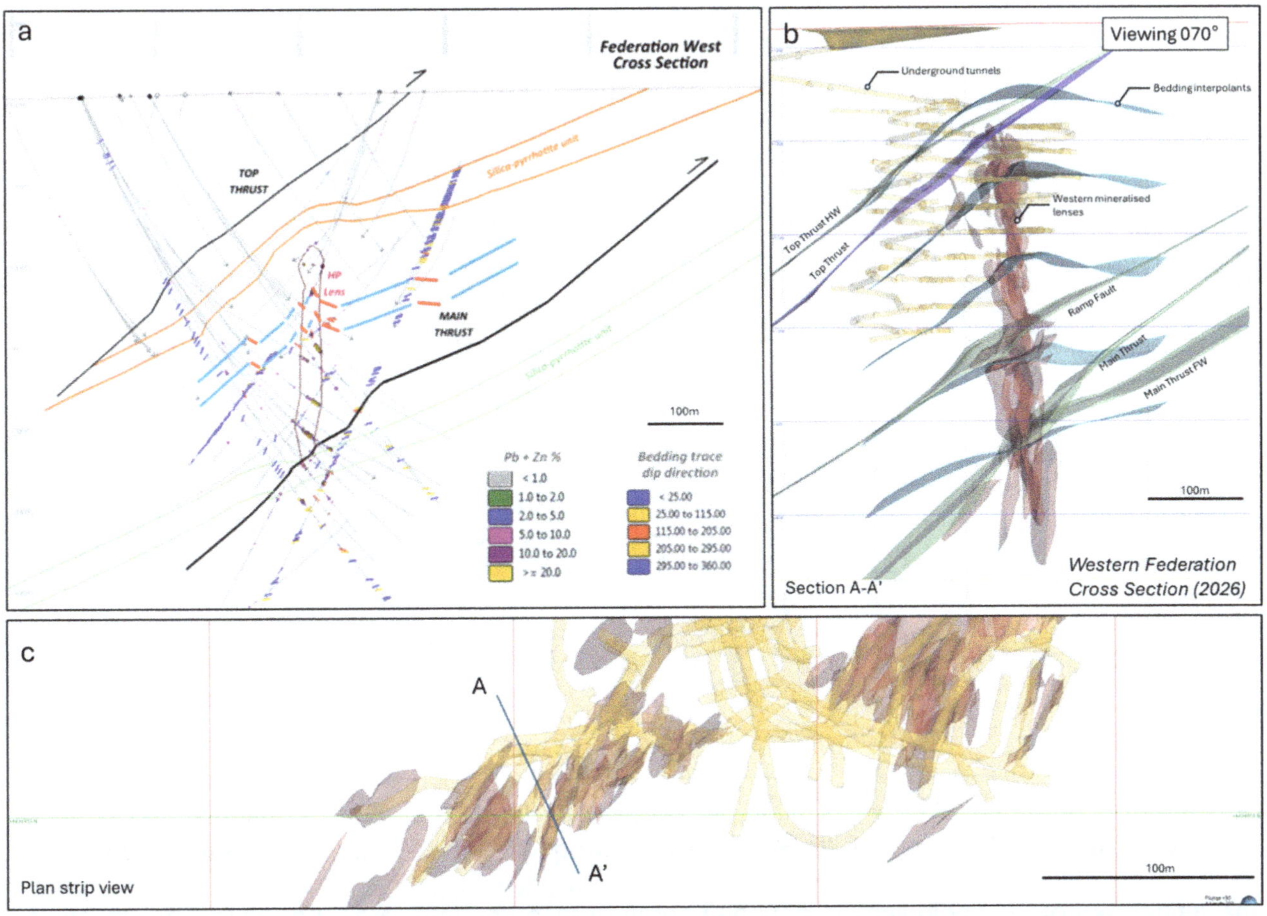

FIG 7 – Cross-sections of the western side of the Federation orebody. (a) Interpretation of the Hesperides (HP) lens viewing approximately 070° from 2022 (modified after Thomas, Smith and McKinnon, 2022). (b) Current interpretation of the western side of the main Federation orebody maintaining a 070° viewing angle in approximately the same location as (a). (c) Plan view showing location of cross-section on right.

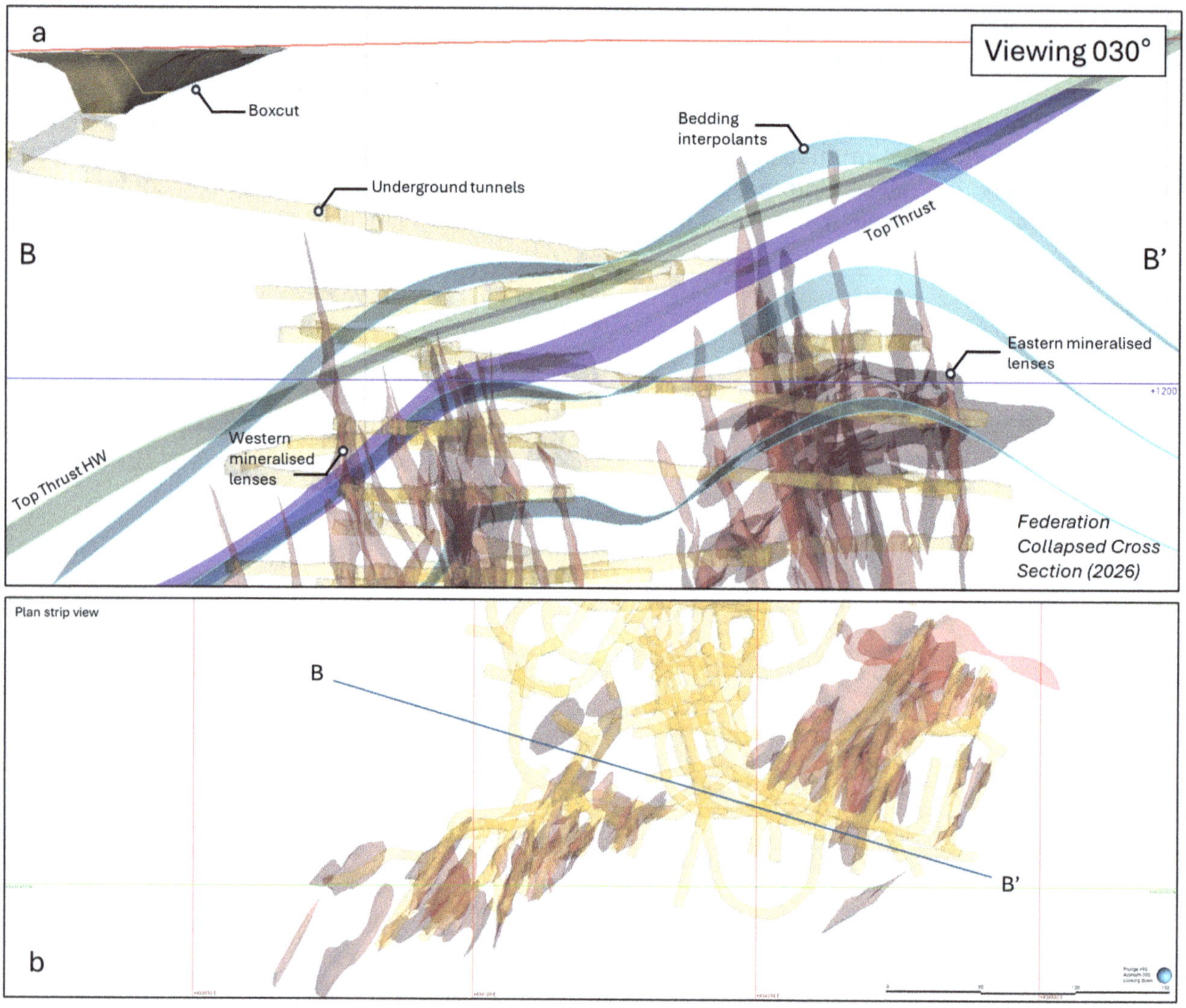

FIG 8 – (a) Current (2026) interpretation of the upper Federation orebody. Cross-section at 120°, viewing in a 030° heading. (b) Strip view showing location of collapsed cross-section.

The refinement of the geological model has improved resource infill targeting, drilling efficiency and underground grade control strategy, enabling increased mining selectivity and clearer distinction between high-grade ore shoots and low-grade, subeconomic sulfide mineralisation.

Modelling and domain estimation

The feasibility study model was built using sulfide domains based on lead-zinc (PbZn) grades greater than 3 per cent combined PbZn and refined using lithology (Figure 9a). Three domains were used for estimation of Pb, Zn, Cu, As, Bi, Fe and S. The domains followed the east north-east trend of the mineralisation, with a slight (50 m) southerly offset of the eastern domain. There is no obvious faulting between these domains however, folding through the region shows two north-east trending anticlines as the primary hosts of the mineralisation. The third sulfide domain is to the south towards the Dominion prospect. Pb, Zn, Cu, As, Bi, Fe and S were estimated from 1 m composites using ordinary kriging (Aurelia Metals Limited, 2022b).

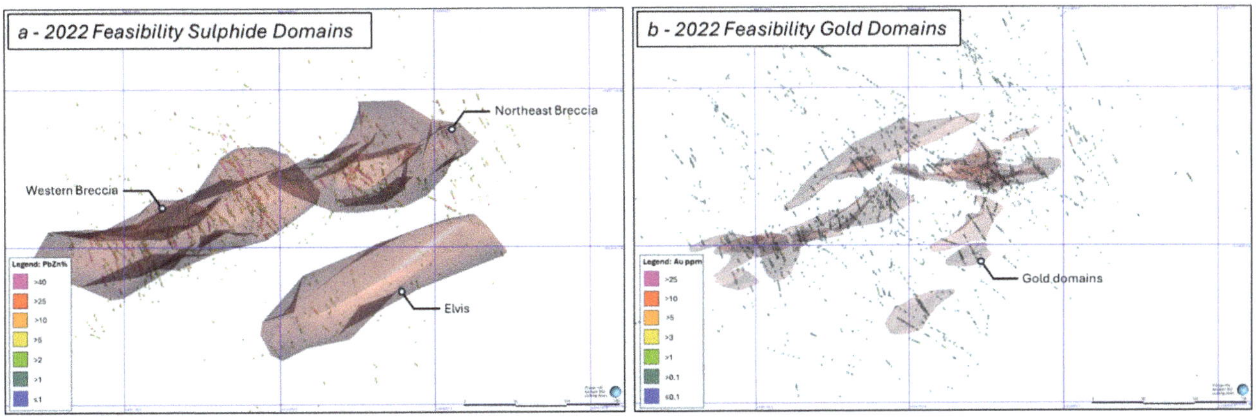

FIG 9 – Estimation domains used in the 2022 Federation feasibility study. (a) Sulfide domains with drill hole assays >3 per cent PbZn displayed, (b) Gold domains with drill hole assays >0.1 ppm displayed.

In addition to sulfide domains, gold estimated for the feasibility study was hard bounded into 14 gold domains, or narrow plunging shoots based on a 0.2 g/t gold grade shell (Figure 9b). The gold was estimated inside and outside these domains using multiple indicator kriging (MIK).

Blocks were constructed into 2 × 10 × 10 (X, Y, Z) metres sizing and rotated along a bearing of 18° to suit the ENE trend of the broad mineralisation. Search ellipses and variograms were also oriented along this trend.

RESULTS OF NEW INTERPRETATION

The new interpretation resulted in an increase from three broad domains to more than 90 discrete domains, each estimated independently using ordinary kriging with hard boundaries. Drill spacing (aimed at a 12.5 × 12.5 m interval prior to development) is now essential for both geological modelling and grade estimation. Revised interpretation, reflecting the more narrow, lens-like nature of the orebody, has substantially increased Pb, Zn, and Au grades, while overall tonnage has decreased due to an improved understanding of orebody geometry. Infill drilling continues to identify additional lenses within the existing footprint, that were not encountered due to the previous spacing (minimum 25 m for Indicated) and orientation of drilling.

The increase in assay data as well as improved geological understanding has allowed the change from MIK to ordinary kriging for gold estimation. Gold is estimated using the same domains as the sulfides, often occurring along the edges of the domains. Elevated (greater than 2 g/t) gold is seen in only eight domains out of the 90. Due to the small size of most domains, there are limited data points available, therefore variography has been conducted using all the domains of each nature (Pluto for Eastern domains, Venus for Western domains, Neptune for stratigraphic domains), and the direction of the variogram forced in the orientation of the domain (north-north-east) to prevent reverting to an east-north-east orientation. Grade control estimation is hard bounded for individual domains, preventing smearing. This estimation relies on accurate and precise domaining of the lenses to get the best results and requires infill drilling at close spacing as well as geological mapping and face sampling.

The application of improved geological knowledge to the interpretation of domains outside the infill drilling area has been conducted, using orientations of mineralised contacts, texture, assay information and faulting. However, infilling these areas does generally result in changes to the lens geometry as well as the identification of additional lenses as new drill platforms become available, allowing more appropriate drilling orientations.

ONGOING IMPLICATIONS

The revised orebody geometry has implications for mine design, most notably an increase in required development metres. Longitudinal and transverse stopes designed in the feasibility study have been replaced with sub-level open stope longitudinal retreats mined along strike on development levels. The proliferation of narrow lenses and the need for faster infill drilling have placed additional pressure on mine scheduling, with stope drilling commonly commencing shortly after level development completion. However, this has often been offset by the availability of additional mining fronts – a result of the *en echelon* arrangement of mineralised lenses. Figure 10 illustrates the contrast between the 2022 feasibility study model and mine plan, and the 2025 grade control model and actual mined development (displaying model blocks greater than 5 per cent Pb+Zn).

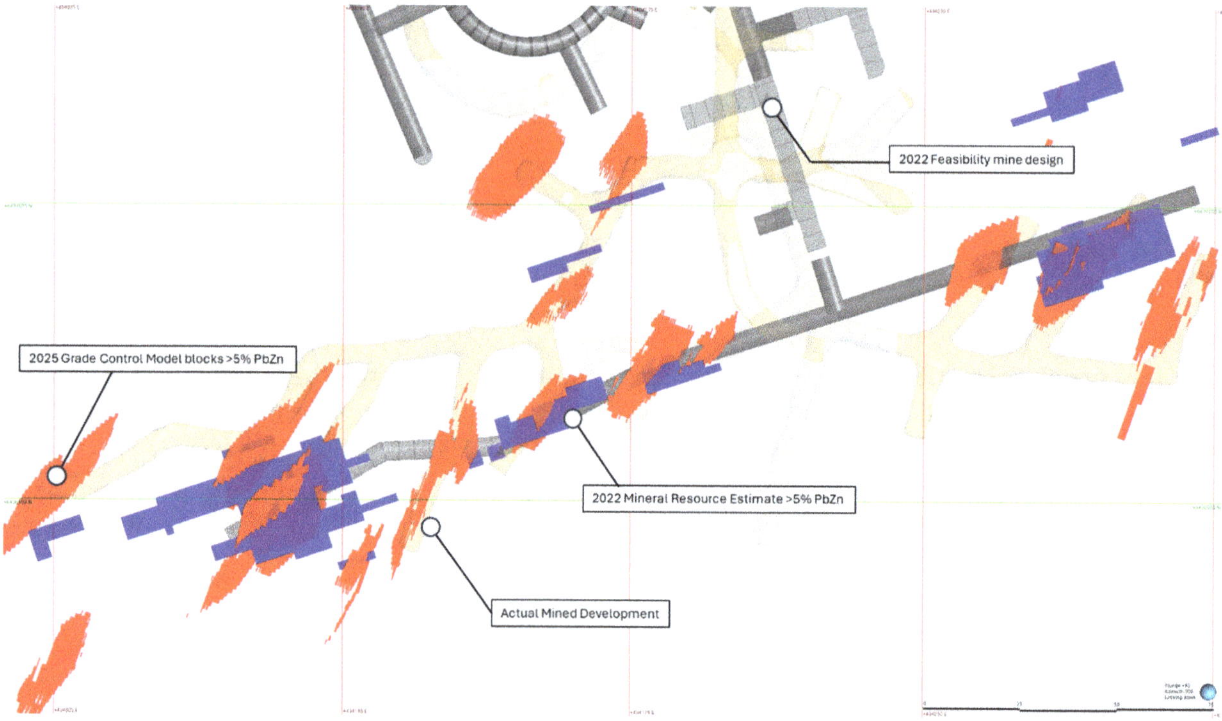

FIG 10 – 2022 Feasibility Resource Model blocks (blue) and Mine Plan (grey) versus Actual Mined (orange) and 2025 Grade Control Model blocks (red). Both models are filtered to display blocks >5 per cent PbZn.

In addition to ongoing infill diamond drilling, face mapping and channel sampling are routinely undertaken to refine and confirm local geological understanding. Underground face samples are processed at the Peak Mine laboratory to enable rapid result turnaround, and drill core is 'quick logged' (rough log of mineralisation) as it comes out from underground, to enable earlier interpretation to drive quick reaction to unexpected intercepts (further diamond drilling, drill plan alteration, mine design alteration). Incorporating this continuous flow of geological information into grade control models supported by implicit modelling techniques has been critical for reliable production forecasts. As mining advances, infill drilling and iterative grade control model updates remain imperative components of mine-planning. The developed geological understanding within the mine footprint continues to positively contribute to other aspects of both geology and the broader business – reconciliation, mine design optimisation, resource extension and exploration targeting.

CONCLUSIONS

The new geological interpretation of the Federation Pb–Zn deposit represented a shift in understanding of detailed orebody geometry, structural controls and grade distribution. What was previously modelled at the feasibility stage as a broad, breccia-dominated system has been demonstrated, through systematic underground drilling and structural analysis, to comprise a structurally partitioned, cleavage-parallel *en echelon*-lens system. Detailed structural logging of mineralisation contacts, alteration boundaries, sulfide foliation and vein orientations consistently indicates that high-grade mineralisation is sub-parallel to cleavage and fold axial planes, trending north-north-east (~030°) and dipping steeply to the east-south-east. Rather than forming large, continuous breccia bodies, high-grade mineralisation occurs as narrow, strike-limited but plunge-extensive massive sulfide lenses, accompanied by subordinate bedding-parallel, lower-grade stratiform horizons. Enhanced grades are commonly developed where steep lenses intersect these stratiform units.

The implications of this reinterpretation have been widespread. The feasibility study model utilised three broad sulfide domains, whereas the current grade control model comprises more than 90 well-informed, implicit, discrete, hard-bounded domains estimated independently. This refinement has substantially reduced grade smearing and improved estimation accuracy locally. Locally, grades have increased, contained Pb and Zn have slightly decreased while contained Au has increased. The improved geological understanding also enabled the transition from multiple indicator kriging for gold to ordinary kriging within structurally constrained sulfide domains.

A key driver of this new interpretation was a detailed structural analysis that resulted in a more targeted drilling orientation. Earlier surface resource drilling provided the orientation of the broader mineralisation, though the detailed orientation and full extent of the continuity and geometry as *en echelon* lenses became clearer with subsequent work. Reorientation of underground drilling to intersect mineralisation at high angles validated the new model, with individual drill holes intersecting multiple discrete high-grade lenses at near-true thickness. Infill drilling at approximately 12.5 m spacing has proven essential for reliable domaining and estimation in this structurally controlled, Cobar style *en echelon*-lens system, with ongoing drilling continuing to identify previously unrecognised lenses. Underground observations in the form of detailed ore-drive mapping and sampling supported the new interpretation.

The revised geological model has influenced mine design. The majority of transverse stoping designed at feasibility has been replaced by longitudinal retreat stoping along strike, reflecting the narrow, *en echelon* configuration of mineralised lenses. Although this has increased development metres and intensified drilling requirements, it has also enabled improved mining selectivity, clearer separation of high-grade ore shoots from subeconomic sulfide material, and more reliable production forecasting. The availability of multiple *en echelon* lenses has, in many cases, provided additional mining fronts that has partially offset increased development demands.

The Federation case study demonstrates the importance of early acquisition and rigorous interpretation of structural data in structurally controlled systems. Where mineralisation geometry is closely aligned with host-rock architecture, drill orientation, structural measurement and precise geological domaining are critical to reducing geological risk. Importantly, the case highlights the need for geological models to remain dynamic during early production. Underground exposure and close-spaced drilling provide opportunities to test and, where necessary, challenge established interpretations.

Ultimately, the Federation experience reinforces several key principles applicable across structurally complex deposits: drill orientation must be considered to test structural hypotheses; structural data must guide modelling decisions; and tighter infill drill spacing is often significantly less costly than the consequences of mining misinterpreted geometry. By prioritising geological accuracy and iterative model refinement, geological risk can be materially reduced, with direct benefits to resource estimation confidence, mine design optimisation and ultimately business success.

ACKNOWLEDGEMENTS

This paper is published with the permission of Aurelia Metals Ltd. Acknowledgement is made to all past and present staff of Aurelia Metals, the Exploration team including Todd McGilvray and Owen Thomas and especially the Federation Mine Geology team of Emma White, James Hill, Omar Cortes-Manzo, Karina Williams, Dylan Watson, Stephanie Bender and Elizabeth Hemlin for their past and ongoing contributions to orebody knowledge.

The early contributions of both Patrick Tucker and Kenneth Bush of Core Prospecting Geological Solutions would also like to be acknowledged, both contributing to, and peer reviewing early ideas and/or models in relation to the Federation deposit.

REFERENCES

Aurelia Metals Limited, 2019. Discovery of High-Grade Mineralisation at the Federation Prospect, South of Hera, Company release to the ASX dated 6 May 2019.

Aurelia Metals Limited, 2020. Maiden Federation Resource Estimate, Company release to the ASX dated 9 June 2020.

Aurelia Metals Limited, 2022a. Federation Mine Feasibility Study, Company release to the ASX dated 10 October 2022.

Aurelia Metals Limited, 2022b. Group Mineral Resource and Ore Reserve Statement, Company release to the ASX dated 10 October 2022.

Fitzherbert, J, Downes, P and Blevin, P, 2022. District-scale characteristics of mineralisation in the Cobar superbasin: attempting to understand timing and genesis, *AIG Bulletin 70 – Discoveries in the Tasmanides*, pp 115–124.

Glen, R, 1991. Inverted transtensional basin setting for gold and copper and base metal deposits at Cobar, New South Wales, *BMR Journal of Australian Geology and Geophysics,* 12(1):13–24.

Thomas, O, Smith, P and McKinnon, A, 2022. Federation Zn-Pb-Au-Cu-Ag Deposit, Nymagee NSW, *AIG Bulletin 70 – Discoveries in the Tasmanides*, pp 249–262.

Tucker, P, 2024. Federation Ore Textures Relogging – internal report to company.

Living with Schrödinger's kittens – using moving point of origin (MPO) analysis to look at resource estimation uncertainty, model to mine reconciliation and infill drilling efficacy

J Moore[1], M Grant[2], D Corley[3], W Randa[4] and V Leal[5]

1. Head of Resource Development, Oceanagold, Dunedin 9016, New Zealand.
   Email: jonathan.moore@oceanagold.com
2. Senior Geologist Resource Development, Macraes, Oceanagold, New Zealand.
   Email: matthew.grant@oceanagold.com
3. Principal Geologist Resource Development, Oceanagold, South Brisbane Qld 4101.
   Email: doug.corley@oceanagold.com
4. Principal Geologist Resource Development, Oceanagold, South Brisbane Qld 4101.
   Email: wesly.randa@oceanagold.com
5. Superintendent Geology, Didipio, Oceanagold, Nueva Vizcaya, Philippines.
   Email: vyron.leal@oceanagold.com

ABSTRACT

In May 2023, the authors' paper 'Schrödinger's kittens – lifting the lid on resource drill hole data after mining' (Moore *et al*, 2023), revealed an important aspect of resource estimation uncertainty that, although previously recognised, is typically overlooked. The paper presented a case study using closely spaced production gold assay data resurrected from OceanaGold's mined-out Globe Progress deposit in New Zealand. The Globe Progress deposit was 'redrilled' 49 times at 35 m × 35 m spacings with an iterated moving point of origin (MPO) algorithm. There was found to be considerable spread across the 49 extracted drill hole sets (the Schrödinger effect), with eight of the 49 drill hole data sets not adequately representing the histogram of the in-ground mineralisation they were being used to estimate.

Since the 2023 paper, four additional gold case studies from OceanaGold's Waihi (New Zealand), Didipio (Philippines), Macraes (New Zealand) and Haile (USA) operating mines have been completed, confirming the original Globe Progress analysis. Given the reproducibility of the Schrödinger effect for the five case studies, the Schrödinger effect needs to be accepted as a fundamental component of resource estimation uncertainty, albeit not knowable until after mining.

- Being an inherent property of the data, for a given drill spacing, the 'Schrödinger effect' represents the fundamental lower limit of estimation uncertainty.
- For many estimates, the Schrödinger effect is the main driver of estimation uncertainty and therefore resource model performance.

This data-related uncertainty, or 'Schrödinger effect', is a product of chance (luck of the draw) and because it is not knowable until after mining, presents the resource geologist with an invisible layer of uncertainty, prior to estimation.

- This is problematic when attempting to quantify forward-looking estimation ranges.
- The implications for resource classification are more layered and discussed in this paper.

Surprisingly, the magnitude of the effect is similar across the five case studies despite diverse geological settings and mineralisation styles. This observation may help develop some 'rules of thumb'.

Building on the findings presented in the 2023 paper, this paper works through five case studies to explore the potential consequences stemming from the Schrödinger-effect; forward-looking estimation uncertainty, resource classification, model to mine reconciliation, as well as typical infill drill efficacy versus the surprisingly erratic outcomes of infill drilling on a campaign-by-campaign basis.

MPO analysis is a powerful tool, operating in data-rich production environments, but is labour intensive and time-consuming. This provides fertile ground for existing and emerging technologies to streamline workflows.

INTRODUCTION

Moore *et al* (2023) presented a case study using data resurrected from OceanaGold's mined-out Globe Progress deposit in New Zealand, now in the final stages of restoration. The retrospective study was based on the closely spaced (5 m × 5 m), high quality reverse circulation GC sample data from the mined-out Globe Progress open pit. Utilising this closely spaced GC data removed the need for assumptions regarding short range continuity, which are necessary with forward-looking analyses based upon broader spaced resource drilling (eg conditional simulations).

The exhaustive Globe Progress grade control data was used to repeatedly 'redrill' the deposit by extracting 35 m × 35 m spaced subsets from the original 5 m × 5 m spaced grade control data with a moving point of origin (MPO) algorithm, iterated by incrementing the grid origin in 5 mE, and 5 mN steps. Given that there are 7 × 5 m steps per 35 m resource spacing in both east and north directions, there are 49 unique drill hole sets (ie 7 × 7 permutations) that can be extracted. Figure 1 shows two of the 49 extracted drill hole sets superimposed over a grade control-derived, cumulative gold map for the deposit.

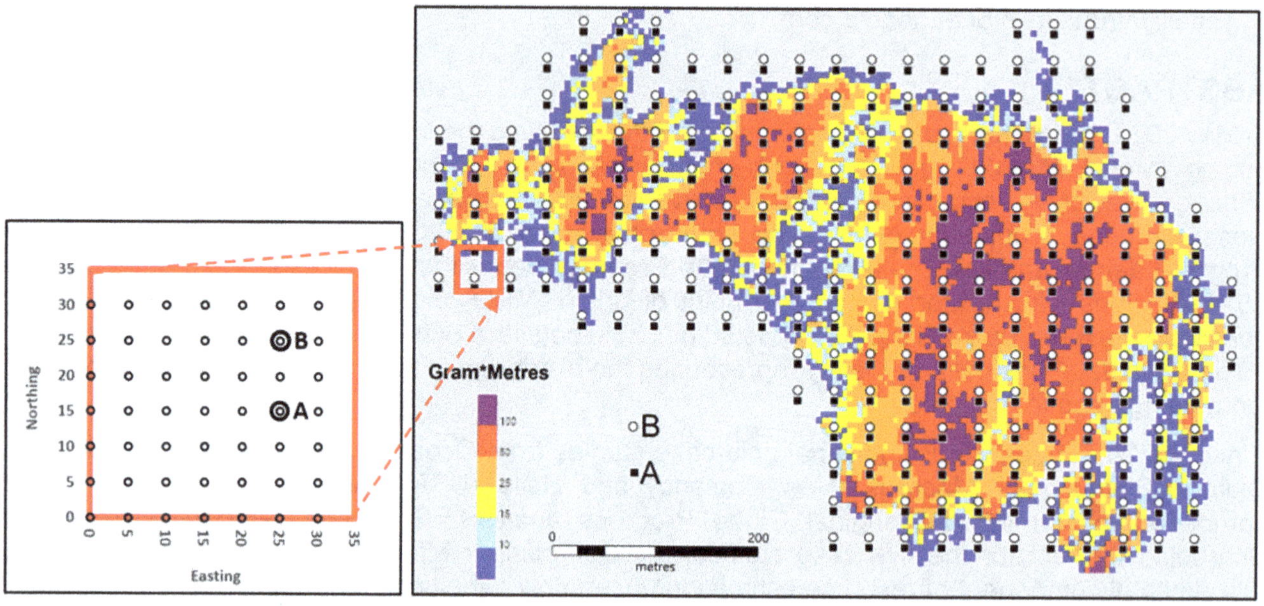

FIG 1 – Plan view of the Globe Progress Deposit. The right-hand image shows two of 49 possible drilling grids, sets A and B, over RC Grade Control Gram-metres. The grid of possible origin points in the left-hand image shows the relative location each set.

Individual resource estimates were then completed for each of the 49 extracted drill hole data sets as well as a grade control estimate based upon the exhaustive 5 m × 5 m data set. The 49 estimates were then compared against each other and the grade control estimate (Figure 2). Whilst the data unpinning each of the 49 estimates changed for each estimate, the geological assumptions, variography, and modelling parameters remained constant. This approach was taken to isolate the impact of changing the input data.

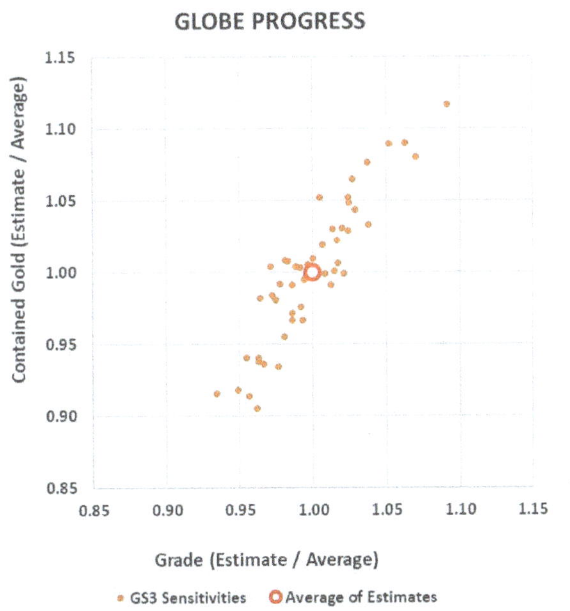

FIG 2 – Scatterplot of comparisons between estimates and grade control for grade and contained-gold. Both contained gold and grade are normalised by dividing by model average contained gold and grade respectively.

Note that combining all 49 drill holes sets exactly reproduces the grade control data set.

That the mean contained gold, tonnes and grade of all 49 sensitivity estimates (10.7 Mt @ 1.63 g/t and 561 koz gold) is close to that of the GC estimate (10.4 Mt @ 1.69 g/t and 564 koz gold), suggests the estimation methodology is reasonable and that the drill spacing is sufficient to support reasonable estimates. The focus of this study however, is on the component of estimation uncertainty related to the underlying data, which is reflected in the spread across the estimates.

- For example, the resource estimate based on the drilling set A in Figure 1 is 11.0 Mt @ 1.78 g/t for 627 koz gold, yet the resource estimate based on the drilling set B, only 10 m offset from set A, is 10.1 Mt @ 1.57 g/t for 509 koz gold.

The spread across the 49 global estimates, from highest to lowest, for this particular case study was found to be approximately 20 per cent in grade and metal which is attributable solely to the underlying data.

The Globe Progress case study quantified a component of the estimation uncertainty, coined the 'Schrödinger effect', that is inherent in all drill hole data and is distinct from the uncertainties associated with sample and subsample quality, and drill hole spacing-related interpolation uncertainty. Importantly, this data-related uncertainty, that is, whether or not your resource drill hole data is representative of the mineralisation being estimated, is a product of chance (luck of the draw) and is unknowable prior to mining. The resource drill hole data can only be compared against GC data after mining has taken place.

CASE STUDIES

Since the 2023 study, four more case studies using grade control data from OceanaGold's New Zealand, Philippine and USA operations have been completed.

Figure 3 provides plan view slices for each study area with grade control sample grades, 0.5 g/t grade contours and 35 m × 35 m grids. The previous Globe Progress area is included.

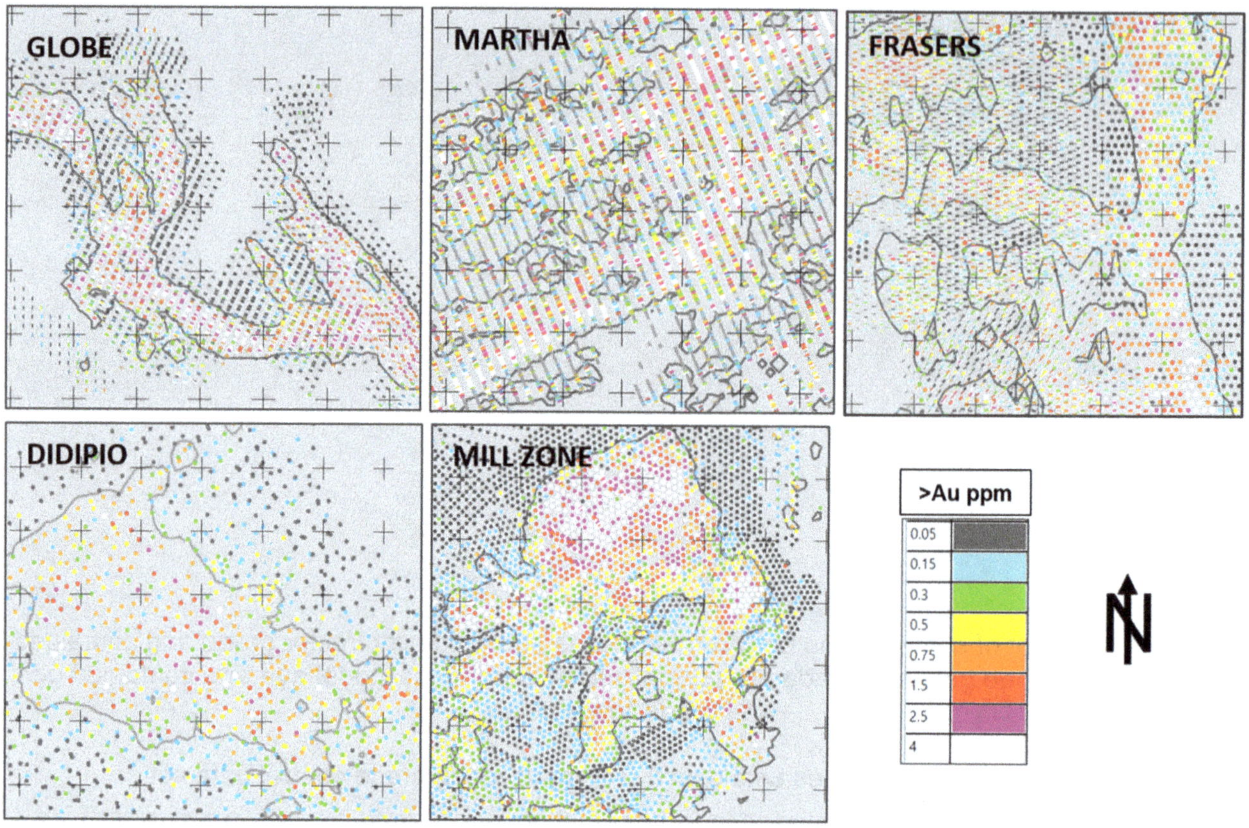

FIG 3 – Plan view slices (± 1.25 m) of grade control sample grades for (clockwise from top left) Globe Progress, Martha, Frasers, Didipio and Mill Zone. Shown with 0.5 g/t grade shell (grey) and 35 mE × 35 mN grids as crosses. Colour legend is same for all figures.

Clockwise from top left these are:

- Globe Progress open pit (Reefton, South Island, New Zealand), representing structurally controlled, metasediment-hosted gold.
- Martha open pit (Waihi, North Island, New Zealand), representing epithermal vein hosted gold-silver.
- Frasers open pit (Macraes, South Island, New Zealand), representing shear-zone hosted gold-tungsten, both mineralised shears and quartz vein arrays.
- Mill Zone open pit (Haile, South Carolina, USA), representing structurally controlled, metasediment/metavolcanic-hosted gold.
- Didipio open pit (Luzon, Philippines), representing alkalic porphyry copper-gold mineralisation.

In each case the same GC spacings (5 m × 5 m), extracted resource drill hole spacings (35 m × 35 m), MPO methodology and estimation method were used. Note that OceanaGold open pit resources that are drilled to 35 m × 35 m spacing would typically be classified as Indicated.

RETROSPECTIVE ANALYSIS – USING MPO TO QUANTIFY ESTIMATION UNCERTAINTY

The results of MPO analysis for the subsequent four case studies confirm the original Globe Progress analysis (see scatterplots in Figure 4). Given the reproducibility of the Schrödinger effect across the five case studies, the effect needs to be accepted as a fundamental component of resource estimation uncertainty. Because the Schrödinger effect is an inherent property of the underlying data, for a given drilling spacing, this represents the lower limit to estimation uncertainty.

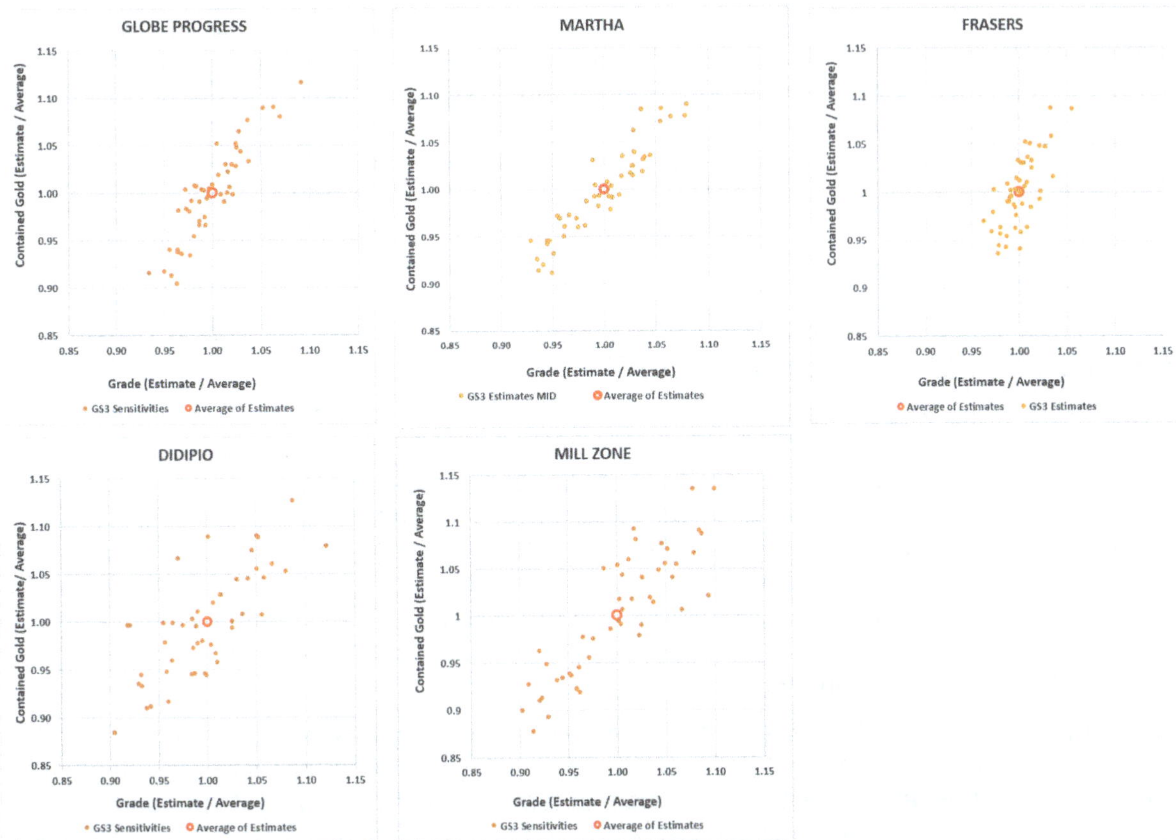

FIG 4 – Scatterplot of comparisons for Globe Progress, Martha, Frasers, Didipio and Mill Zone between estimates and average of all 49 estimates for grade and contained gold.

Surprisingly, the magnitude of the Schrödinger effect is similar across the five case studies in terms of contained gold, despite diverse geological settings and mineralisation styles.

Figure 5 consolidates the scatterplot results (LHS) of all five studies, given that the scatterplots have been normalised as ratios. These are summarised as a cumulative frequency distribution (RHS). The consolidated scatterplot and cumulative frequency distribution are based on almost 250 points, allowing meaningful analysis.

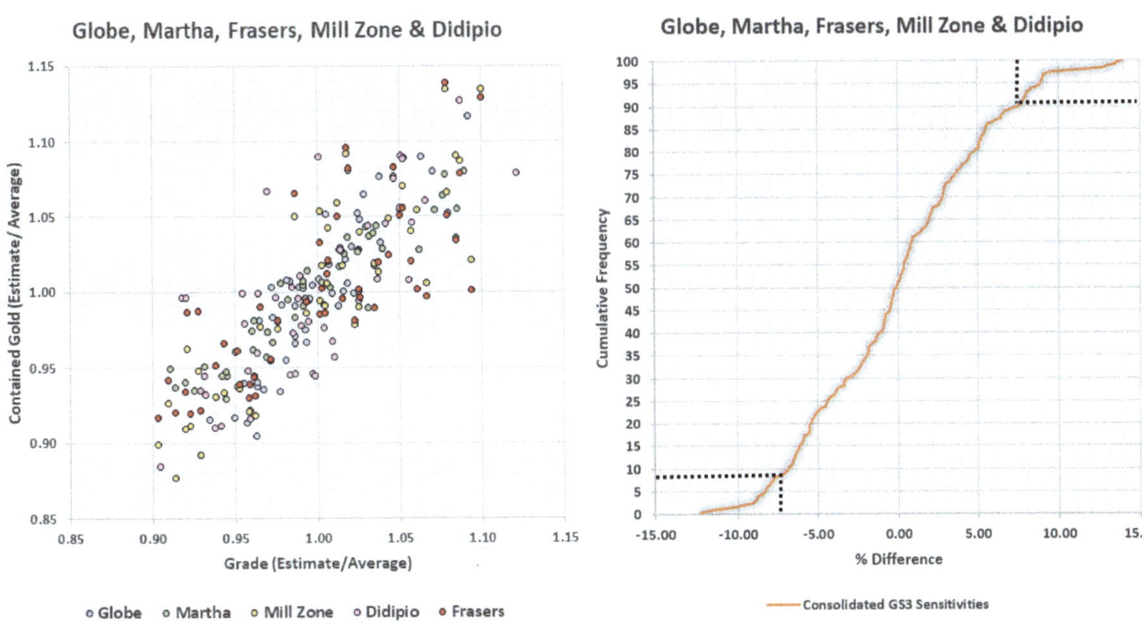

FIG 5 – LHS: Consolidated scatterplot of comparisons for Globe Progress, Martha, Frasers, Didipio and Mill Zone between estimates and GC for grade and contained-gold. RHS: Cumulative frequency of contained gold deviation from mean.

Across the five case studies (Globe Progress, Martha, Frasers, Didipio and Mill Zone), approximately 50 per cent of the sensitivity estimates fell within 4 per cent of the GC contained-gold estimate, suggesting that for many projects the histogram of drill hole data is unlikely to differ significantly from that of the in-ground resource. However, about 18 per cent of the sensitivity estimates across the five case studies differed by more than 7.5 per cent, suggesting that a not-insignificant proportion of drill hole date sets will materially misrepresent the *in situ* resource. Whether or not the available drill hole data is representative, comes down to the 'luck of the draw' and cannot be known at the time of resource estimation.

FORWARD-LOOKING ANALYSIS – USING CONDITIONAL SIMULATION TO QUANTIFY ESTIMATION UNCERTAINTY

Following estimation, but prior to mining, conditional simulation is often used to bracket forward-looking resource estimation uncertainty. Because of the typically broad drill hole spacing available (eg for these case studies 35 m × 35 m), the conditional simulations are based on a modelled, but largely assumed, variogram. It is also assumed that the histogram of the available (ie drawn) drill hole sample data is representative of the in-ground mineralisation. As demonstrated above however, for a not-insignificant number of cases (18 per cent of drill sets for five case studies > 7.5 per cent divergence), the available drill hole data will not be representative. So it is only after mining that it can be known if the preconditions for simulation (ie a representative sample histogram and variogram), were met. This is problematic.

To better understand the implications of the flawed representativity assumption, resource estimation uncertainty was bracketed for four of the case study areas. Two scenarios were used in each study by selecting the highest and lowest drill hole sets in terms of contained gold. One hundred Turning Band conditional simulation realisations were generated for each drill hole set. The realisations were then re-blocked to reflect open pit mining selectivity and their frequency distributions plotted. Each frequency distribution reflects an equally possible evaluation of the estimation uncertainty, based entirely upon the drill data available for that evaluation.

Figure 6 superimposes the highest and lowest scenario frequency distributions of contained gold for each case study area. A median case is included for Globe as a reference point, that is a reasonably representative drill hole set. Notably, the highest and lowest scenario distributions barely overlap, providing conflicting forward-looking uncertainty ranges. Furthermore, generally the highest case (red) has a larger spread than the minimum case (green).

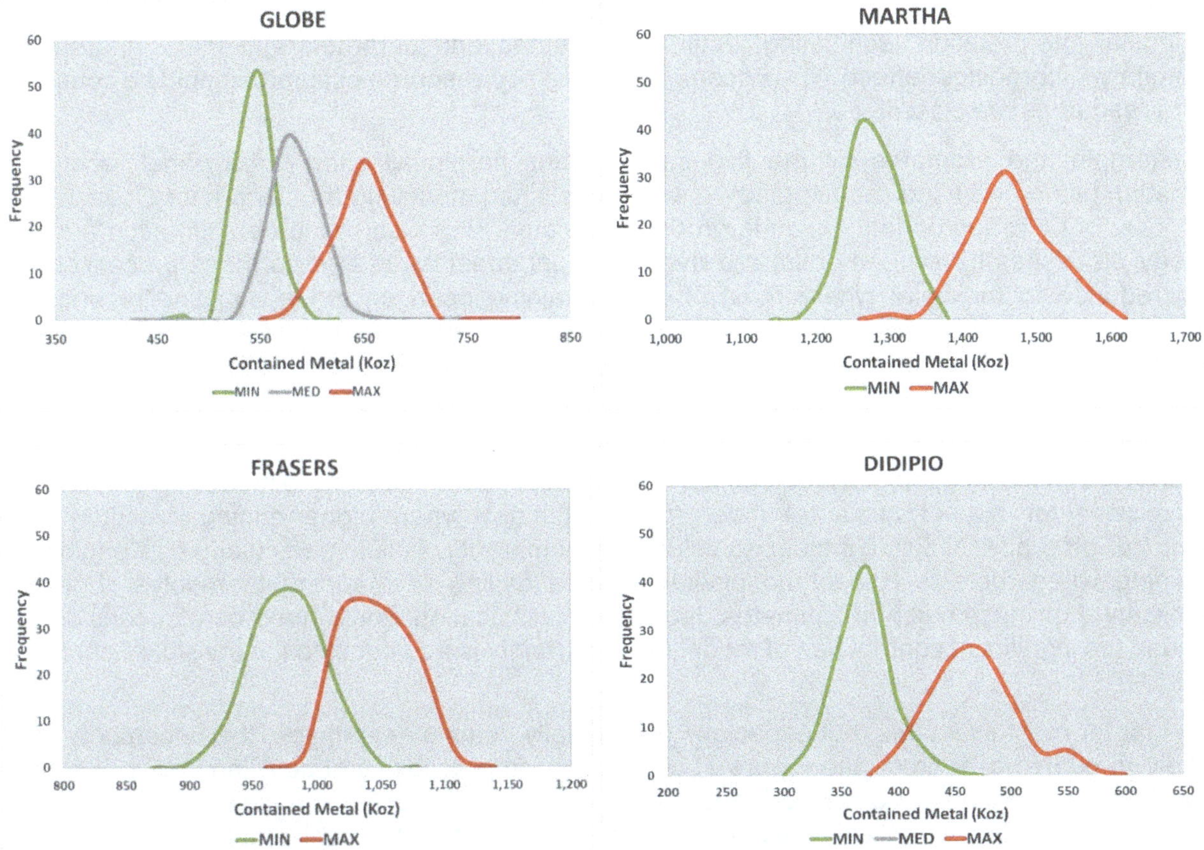

FIG 6 – Superposition of conditionally simulated frequency distributions for minimum, median and maximum drill hole sets from Globe Progress, Martha, Frasers, and Didipio.

At the time of evaluating forward-looking uncertainty, it isn't possible to know whether the available data is a high-case, low-case or something in between, meaning that the actual forward-looking uncertainty is significantly larger than can be quantified from the data at hand.

Can a rule of thumb be applied? The five case studies completed across diverse geological settings reveal a similar magnitude of the Schrödinger effect in each case. In the absence of other information, might it be reasonable to assume the consolidated probability distribution in Figure 5 (approximately ± 8 per cent at 90 per cent confidence) is a reasonable indicator of data uncertainty for most gold estimates drilled to a similar spacing? Additional case studies would be helpful.

RESOURCE CLASSIFICATION

The JORC Code, 2012 Edition provides guidance on resource classification but is non-prescriptive in terms of long-term, annual, quarterly or monthly performance metrics for Measured, Indicated or Inferred resource categories.

For example:

> *'An Indicated Mineral Resource' is that part of a Mineral Resource for which quantity, grade (or quality), densities, shape and physical characteristics are estimated with sufficient confidence to allow the application of Modifying Factors in sufficient detail to support mine planning and evaluation of the economic viability of the deposit.'*

At OceanaGold's operations, resource model performance factors, based upon those discussed by Parker (2011), underpin reconciliation analysis. Indicated Resources are expected to reconcile ±15 per cent on an annual basis, at 90 per cent confidence (ie better than -15 per cent, 19 out of 20 years), albeit the exclusion of Inferred Resources from the reconciliation process tends to skew the metric towards +17.5 per cent / -12.5 per cent. Because MPO analysis is retrospective, in hindsight it is tempting to view the scatter in Figure 5 as being within the range of uncertainty considered acceptable for Indicated Resources and therefore, as being fully captured by the resource classification process. But is this really the case?

Different resource estimates have different uncertainty profiles in terms of process uncertainty (geological interpretation, domaining, estimation method and parameters) versus inherent data uncertainty (Schrödinger effect). The uncertainty profile of a resource estimate should be considered before resources are classified.

For example, the estimates for the five case studies herein all used large panel recoverable estimation (via multiple indicator kriging), a well-established modelling framework for OceanaGold's open pits, but not appropriate for underground estimates. Because resource estimation and mine geology processes have been optimised over time, uncertainties related to these processes have been reduced to the point where most of the remaining open pit model to mine reconciliation divergence is related to the underlying data. For these estimates, the Indicated classification would accommodate most of the scatter seen in Figure 5 because the contribution of the modelling processes to the overall estimation uncertainty is relatively low.

Some other types of estimates however, are materially compounded by geological complexity, introducing significant geological (domaining/volumetric) uncertainty into the modelling process. This is more usual for (but not unique to) underground estimates, where higher mining selectivity and/or mining cut-off grades result in increased volumetric complexity. This is particularly challenging prior to mining when there is little or no closely spaced drilling or mapping to resolve short-scale complexity. For these types of estimates, Indicated classification would need careful consideration because the significant contribution of modelling uncertainty is additional to the invisible Schrödinger effect.

Note that for most estimates, but especially geologically complex estimates, the uncertainty profile is likely to evolve over time (see Figure 7). As mining progresses, orebody knowledge increases, modelling assumptions are optimised, operational geology risks better managed, the contributions of these processes to overall estimation uncertainty are reduced. Meanwhile, unless the drill spacing is reduced, the Schrödinger component stays constant, meaning that the uncertainty for mature resource estimation processes moves towards a more data-dominated profile over time. So, the classification of the resources may evolve as mining progresses.

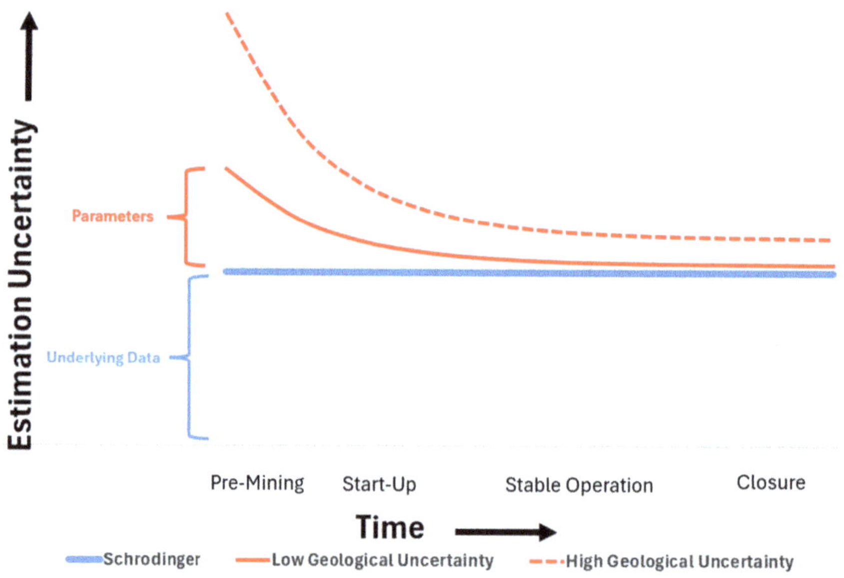

FIG 7 – Conceptual uncertainty profiles over time for high and low geologically complexity estimates.

MODEL TO MINE RECONCILIATION

Typically, when a sustained deterioration in F1 model to mine performance is experienced, this is attributed to poor geological interpretation, poor estimation methodology or parameter choices. In some instances, steps might be taken to 'fix' the model before the root cause of poor F1 performance has been properly established. The case studies below use MPO sensitivity estimates to isolate a contribution to poor model performance that is not typically considered: the Schrödinger effect.

For four of the five case studies, the highest and lowest contained gold estimates of the 49 MPO sensitivity estimates were reconciled against grade control estimates using a conceptual mine schedule (see Figure 8). The schedule was constrained within an actual as-mined volume, aggregating benches into approximate 30 koz packages to broadly represent quarterly (three-monthly) periods. The red line represents the highest estimate of the 49 estimates and the green line, the lowest. Both are ratioed against the grade control estimate for contained gold (black line), where a ratio of 1.0 is a neutral reconciliation.

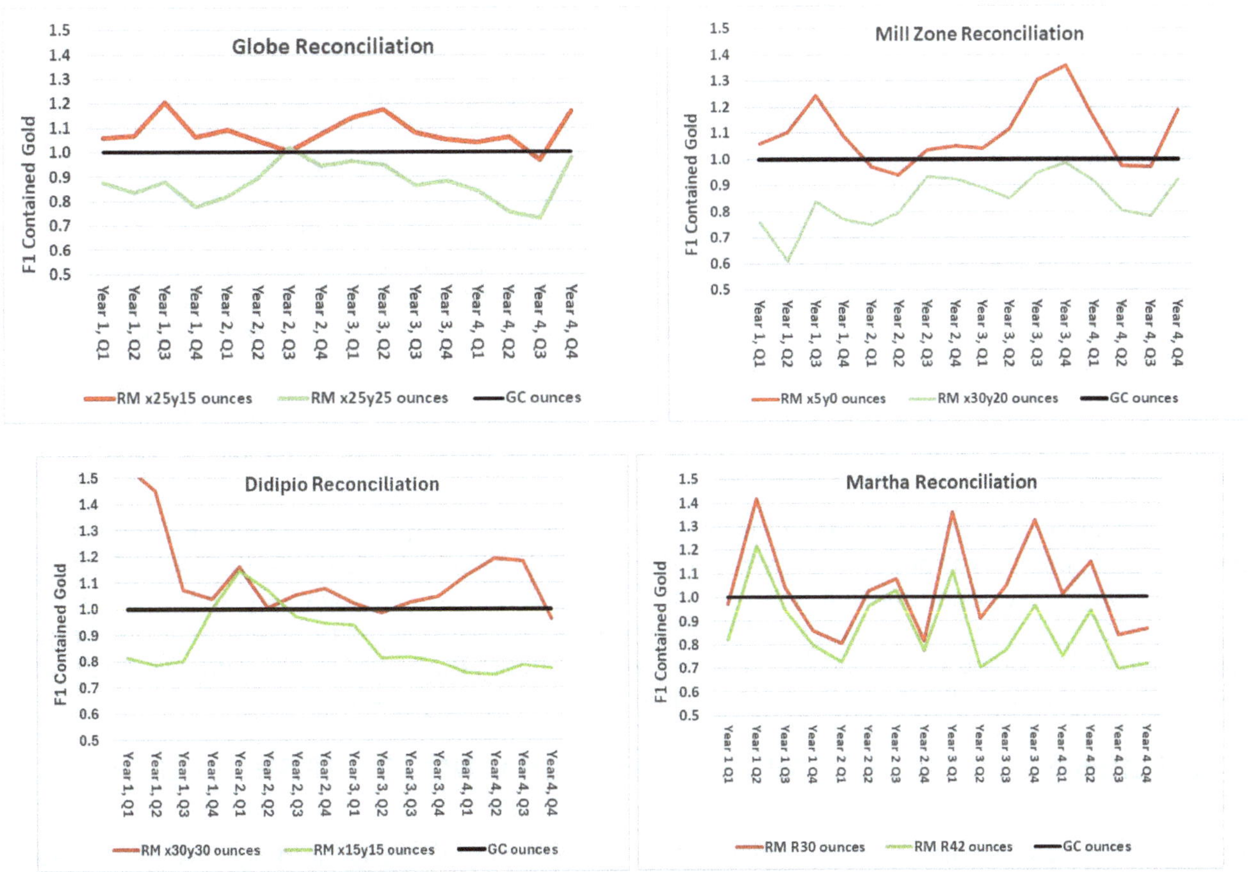

FIG 8 – Model to grade control reconciliation for approximate mining quarters (clockwise from top left) Globe Progress, Martha, Mill Zone and Didipio. Shown with 0.5 g/t grade shell (grey) and 35 mE × 35 mN grid as crosses. Colour legend is same for all figures.

In each case the high and low scenario sensitivity estimates persistently overestimate and underestimate respectively, yet the modelling methodology and parameters were the same for both high and low scenarios. In operational environments, either high or low scenario might lead to a review of the modelling parameters and potentially result in a 'recalibrated' estimate despite the root cause being data, not the modelling.

The Schrödinger effect can only be determined after mining, that is, retrospectively. Model to mine reconciliation is similarly a retrospective process and is normally updated on a monthly-basis as mining progresses. The monthly mined tonnages of mineralisation are aggregated until statistically meaningful model to mine reconciliation can be undertaken. If poor F1 model performance is observed (either positively or negatively), it may be possible to determine whether the Schrödinger effect is a contributing factor by comparing the declustered means of the grade control versus resource sample gold grades in conjunction with nearest-neighbour quantile-quantile (QQ) analysis to first eliminate any possibility of sampling bias.

For example, Table 1 summarises comparisons between the resource drill hole data and the grade control data, for each of the case studies. In each case the ratioed declustered means for high scenario models (left hand column) show positive bias while the low scenario models (right hand column) show a negative bias. The direction of the reconciliation correlates with the Schrödinger effect.

TABLE 1

Declustered gold grade means estimate/GC. Left column high-cases, right column low cases.

Globe	x25y15 = 1.11	**Globe**	x25y25 = 0.94
Mill Zone	x05y00 = 1.08	**Mill Zone**	x30y20 = 0.87
Martha	R30 = 1.05	**Martha**	R42 = 0.89
Didipio	x30y30 = 1.11	**Didipio**	x15y15 = 0.95

Implicating the Schrödinger effect is relatively straightforward for the MPO case studies, but not necessarily so in active operational environments. For the case studies, the resource data sets were extracted directly from the grade control data. Given the common parent data source, the divergence between the mean grades of the resource and grade control data can be attributed to the Schrödinger effect. However, in active operational environments, resource and grade control drilling, sampling, sample preparation and assaying processes are often independent. Divergence might instead be related to differences between the grade control and resource data collection processes. Fortunately, this is testable if a reasonable number of closely spaced sample pairs (resource drilling versus grade control drilling) are available for nearest neighbour QQ analysis.

So, caution is recommended when 'calibrating' your resource estimate to improve model to mine reconciliation unless you understand the underlying cause of the poor reconciliation. Otherwise, you may be 'fixing' something that isn't broken.

Any investigation into poor F1 model performance should include checks for disparities between the resource and grade control sample data via declustered mean comparisons (a feature built into GS3M software, developed by Neil Schofield of FSS International Consultants (Australia), since its inception. GS3M software has been used by OceanaGold for large panel recoverable estimation via multiple indicator kriging for 25 years). Where practical, nearest neighbour QQ analysis could also be used to test for potential sampling, sample preparation or assay biases.

ESTABLISHING RESOURCE MODEL PERFORMANCE THRESHOLDS USING MPO

The five MPO case studies in this paper revealed similar magnitude Schrödinger effects, despite diverse geological settings and mineralisation styles. In the absence of other data, it seems reasonable to use production data from similar previously mined areas of mineralisation, either from the same or other operations, to establish tolerance limits for resource model performance. An example is presented below from Frasers open pit at Macraes gold mine where quartz vein array mineralisation predominates.

Using the same MPO methodology as for the five case studies above, multiple drill hole sets were extracted from closely spaced open pit blasthole sample data successively at 12 m × 12 m, 17 m × 17 m, 25 m × 25 m, 35 m × 35 m and 50 m × 50 m spacings. Note that each spacing increase requires approximately half the drilling of the previous spacing. Individual resource estimates were then completed for each of these extracted drill hole data sets as well as a grade control estimate based upon the exhaustive 5 m × 5 m data set. Due to time-constraints, approximately half of the extracted drill hole sets were completed. Completion of all extracted drill hole sets would improve the analysis, but the general outcomes and probabilities would be similar. The annual performance analysis is the most impacted because there are only one quarter of the number of data points that exist for quarterly analysis. The estimates were compared against the grade control estimate within volumes approximating annual and quarterly mining volumes and the outputs used to produce the following figures.

Figure 9 presents stacked charts based upon, from top to bottom, monthly, quarterly and annual performance periods. Each chart compares resource model performance estimates based upon 12 m × 12 m, 17 m × 17 m, 25 m × 25 m, 35 m × 35 m and 50 m × 50 m spacings respectively. Resource model performance is measured against the grade control estimate and deteriorates as drill spacing is increased. Additionally, as would be expected the resource model performance improves significantly by increasing the comparison period from monthly to quarterly to annual, reflecting the volume variance effect.

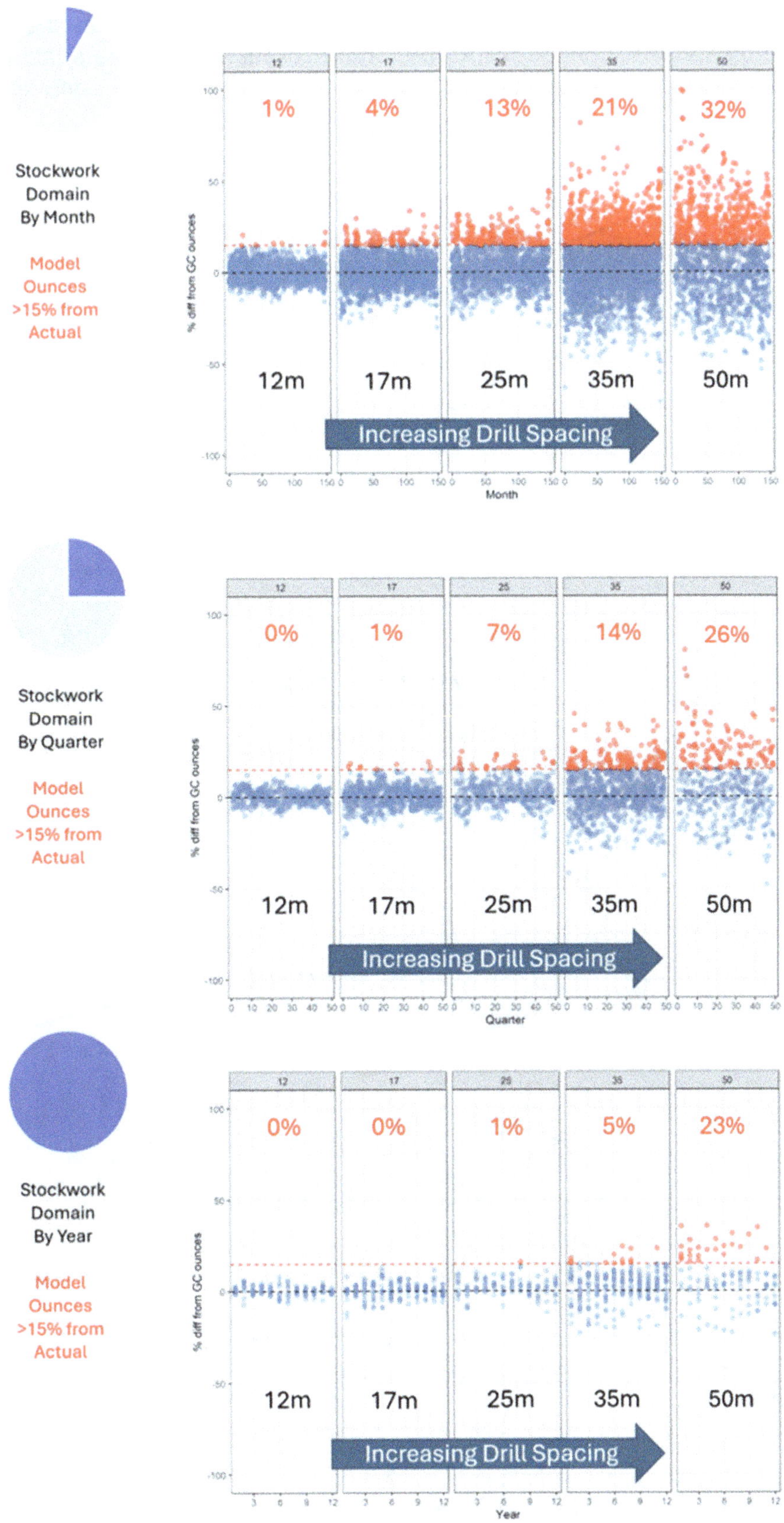

FIG 9 – Top to bottom, Monthly, Quarterly and Annual resource model performance versus drill spacing.

Figures 10 and 11 summarise resource model performance for quarterly and annual periods respectively by plotting probability/resource model performance contours for each drill spacing. This plot, which has been constructed for Frasers quartz vein array mineralisation, maps desired confidence (probability) to quarterly model variance for any of the drill hole spacings. So, not only can this be used to establish model performance thresholds for quarterly and annual periods at any of the presented drill spacings, but this also provides a basis for cost-benefit evaluation for infill drilling to improve model performance, or for performance-based drill hole spacing studies. Other considerations such as lost production and rig/mining fleet interaction also need to be factored in.

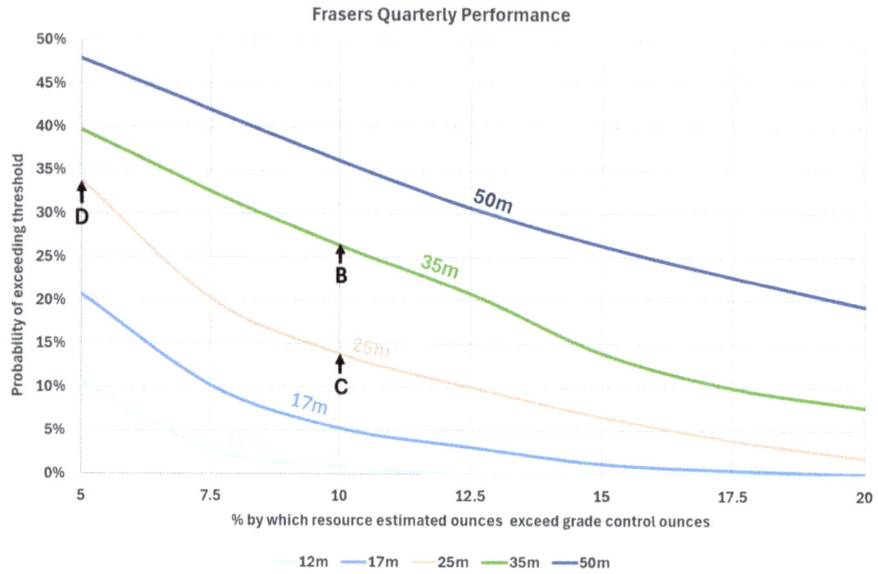

FIG 10 – Probability contours of quarterly resource model performance versus drill spacing.

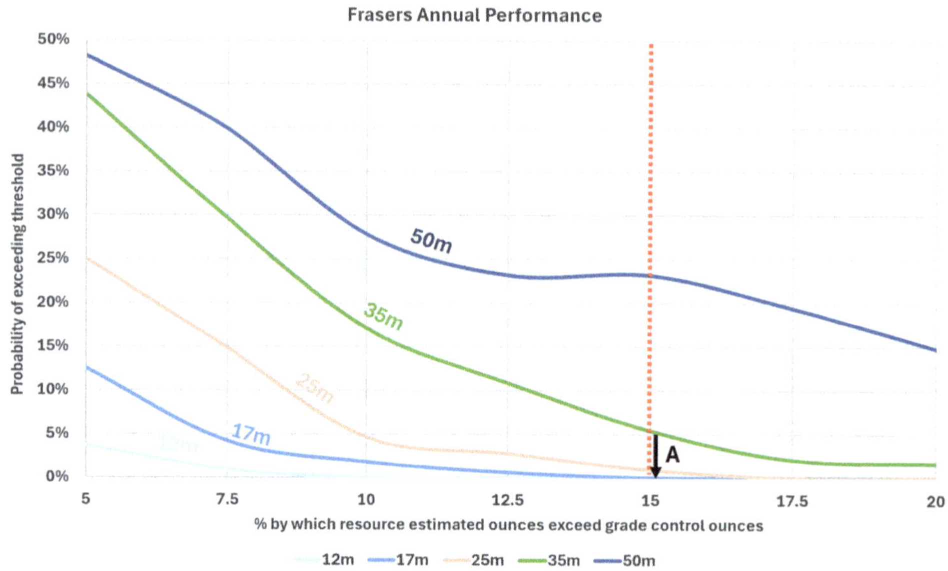

FIG 11 – Probability contours of annual resource model performance versus drill spacing.

Two examples of practical MPO application are provided below. The examples are provided to illustrate how Figures 10 and 11 can be used. Following the process is more important than outcomes themselves.

Example 1

A hypothetical scenario. It is the fifth month of a budget year, and the operation (open pit only) is 10 per cent behind budget in terms of gold production due an unexpected processing plant shutdown. The mine schedule has June to September primarily stripping waste with Q4 delivering the remaining 40 per cent of the full-year's gold production (60 per cent was mined January to May). The resources are drilled to 35 m × 35 m, and for the purposes of this example it is assumed that the mineralisation is analogous to Frasers Quartz Vein Array mineralisation in Figure 10. In order to meet the lower range of external guidance, 90 per cent of the gold scheduled for Q4 must materialise. There is a desire to infill drill the Q4 resources to increase the confidence in the resource estimate and therefore that the ounces will be delivered. Only one pit stage is being mined and the working area in the pit is very tight. Given the constrained pit area, mining fleet interruptions during drilling (to 25 m × 25 m, doubling the drilling density) are likely to reduce mining productivity by 5 per cent and produced gold by 5 per cent. How much would the infill drilling improve confidence in achieving annual guidance?

Referring to Figure 10, we can see that with 35 m × 35 m drilling there is a 27 per cent chance (point B) of the resource model over-estimating by 10 per cent. This is a significant risk. Infill drilling the Q4 volume to 25 m × 25 m would reduce this to a 14 per cent chance (point C) of the resource model over-estimating by 10 per cent. However, if the Q4 scheduled volume was drilled, gold delivery would be reduced by 5 per cent due to drilling activity impacting mining fleet productivity. This means that if the infill drilling to 25 m × 25 m did proceed, over-estimating by more than 5 per cent, not 10 per cent (10 per cent previously, minus 5 per cent productivity loss) would result in missing guidance. At 25 m × 25 m spacing, there is a 33 per cent (point D) chance of the resource model over-estimating by 5 per cent. Infill drilling would therefore increase the likelihood of missing guidance from 27 per cent to 33 per cent, as well as introduce additional interactions between drilling personnel and mining fleet within a limited working area.

Example 2

At the time of drilling Frasers, MPO was not available as a tool. The 35 m × 35 m spacing for Indicated Resources was selected as a 'harmonic' of the existing 50 m × 50 m. With the benefit of hindsight, was 35 m × 35 m a reasonable choice of spacing?

Referring to Figure 11, at a 35 m × 35 m spacing, there is 5 per cent chance of over-estimating (y axis) by more than 15 per cent (x axis) on an annual basis (arrow A). This so happens to align with the OceanaGold definition that 'Indicated Resources are expected to reconcile ± 15 per cent on an annual basis, at 90 per cent confidence (ie better than -15 per cent, 19 out of 20 years)'. It seems that the choice was reasonable.

Note that the MPO analysis is not the sole consideration in establishing operational resource model performance tolerances. Allowances may need to made for downstream operational uncertainties; operational model reconciliation performance is measured against grade control (normally mill-adjusted), introducing reconciliation uncertainties related to, for example, grade control sampling, grade control modelling, mining extraction, material tracking, weightometer measurements, in-circuit inventories and tails sampling. These uncertainties would further increase for operations with multiple mill-feed sources (eg confluent underground, open pit and stockpile mill feed).

THE SURPRISINGLY ERRATIC OUTCOMES OF INFILL DRILLING

Figure 11 demonstrates a large improvement in estimation outcomes as drill spacing is reduced from 50 m × 50 m to 35 m × 35 m to 25 m × 25 m to 17 m × 17 m to 12 m × 12 m drilling. Each infill step approximately doubles the drilling, with the aggregate from 50 m × 50 m to 12 m × 12 m representing a 17-fold increase in drilling.

These generalised outcomes, based upon averages of many estimation outcomes are not unexpected. However, as will become evident through the following MPO analysis, the averages conceal some surprisingly erratic individual outcomes related to 'divergent' drill hole sets and may initially seem counterintuitive. But if viewed within the MPO framework, these can be seen as inevitable consequences of the Schrödinger effect.

Within the MPO framework, infill drilling can be achieved by combining two unique extracted drill hole sets. For example, in Figure 12, a 35 m × 35 m drill hole set (black squares on left grid) can be 'infilled' to 25 m × 25 m by combining it with another MPO 35 m × 35 m drill hole set, selected from the 49 available sets to be offset by 17.5 m in both east and north directions (red circles on centre grid).

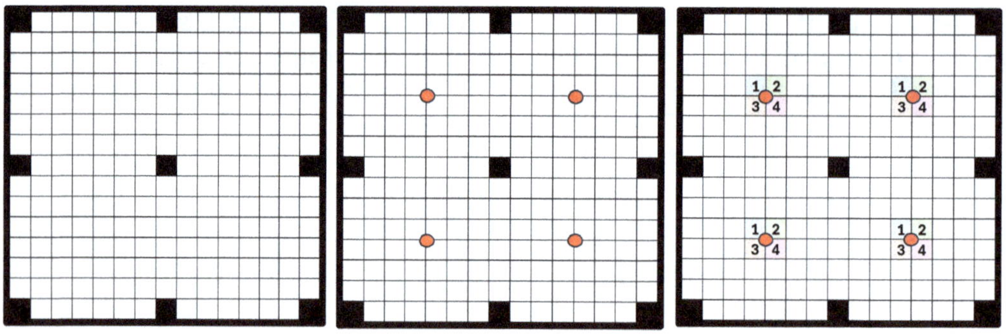

FIG 12 – From left to right, 35 m × 35 m grid, 25 m × 25 m grid, and close approximations of 25 m × 25 m grids.

However, the 5 m × 5 m resolution of the exhaustive grade control data limits drill hole spacings to exact multiples of 5 m, and the17.5 m offset is not a multiple of 5 m. The closest available offsets are either 15 m or 20 m. For this reason, rather than the +17.5 mE/+17.5 mN offsets, each 35 m × 35 m was 'infilled' using all four available offset options (see coloured squares 1, 2, 3 and 4 in right grid):

- +15 m E/+20 mN, square 1.
- +20 m E/+20 mN, square 2.
- +15 m E/+15 mN, square 3.
- +20 m E/+15 mN, square 4.

Each of the four options is a reasonable approximation to 17.5 m offsets, given the target 25 m × 25 m spacing but by testing and comparing outcomes for each of the four infill options, the efficacy of infill options and their sensitivity to small divergences in pattern spacing can be observed.

Due to time constraints, it was not possible to 'infill' each of the 49 drill hole sets. Only the eight most divergent 35 m × 35 m drill hole sets, that is the four highest and four lowest, were chosen for infill drilling (see red and blue circles respectively in Figure 13). It was reasoned that the eight most divergent (four highest and lowest) 35 m × 35 m drill hole sets had the most potential to be improved by infill drilling. This required 32 infill estimates for each case study area (four infill options for each of the eight most divergent drill sets).

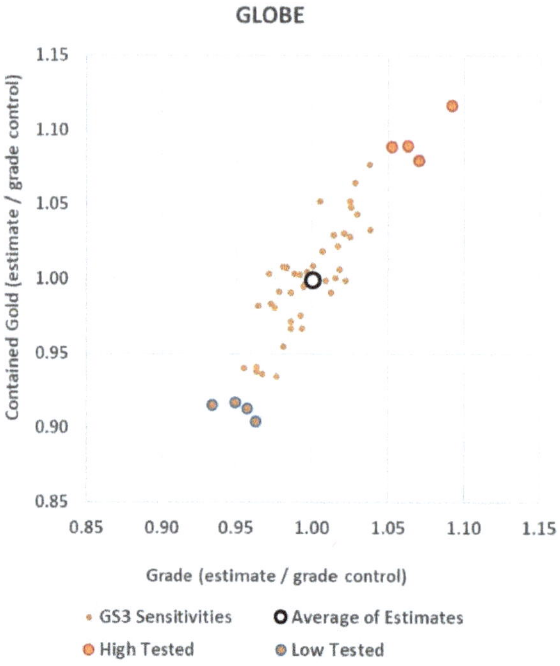

FIG 13 – Scatterplot of comparisons between estimates and grade control for grade and contained-gold (small circles). Eight most divergent estimates as bold circles (red high, blue low).

To-date, time has only allowed Globe Progress, Frasers and Mill Zone infill studies to be completed. Figure 14 compares the outcomes of the estimates based upon the infill drilled 25 m × 25 m pattern against the estimates based upon the parent 35 m × 35 m drilling.

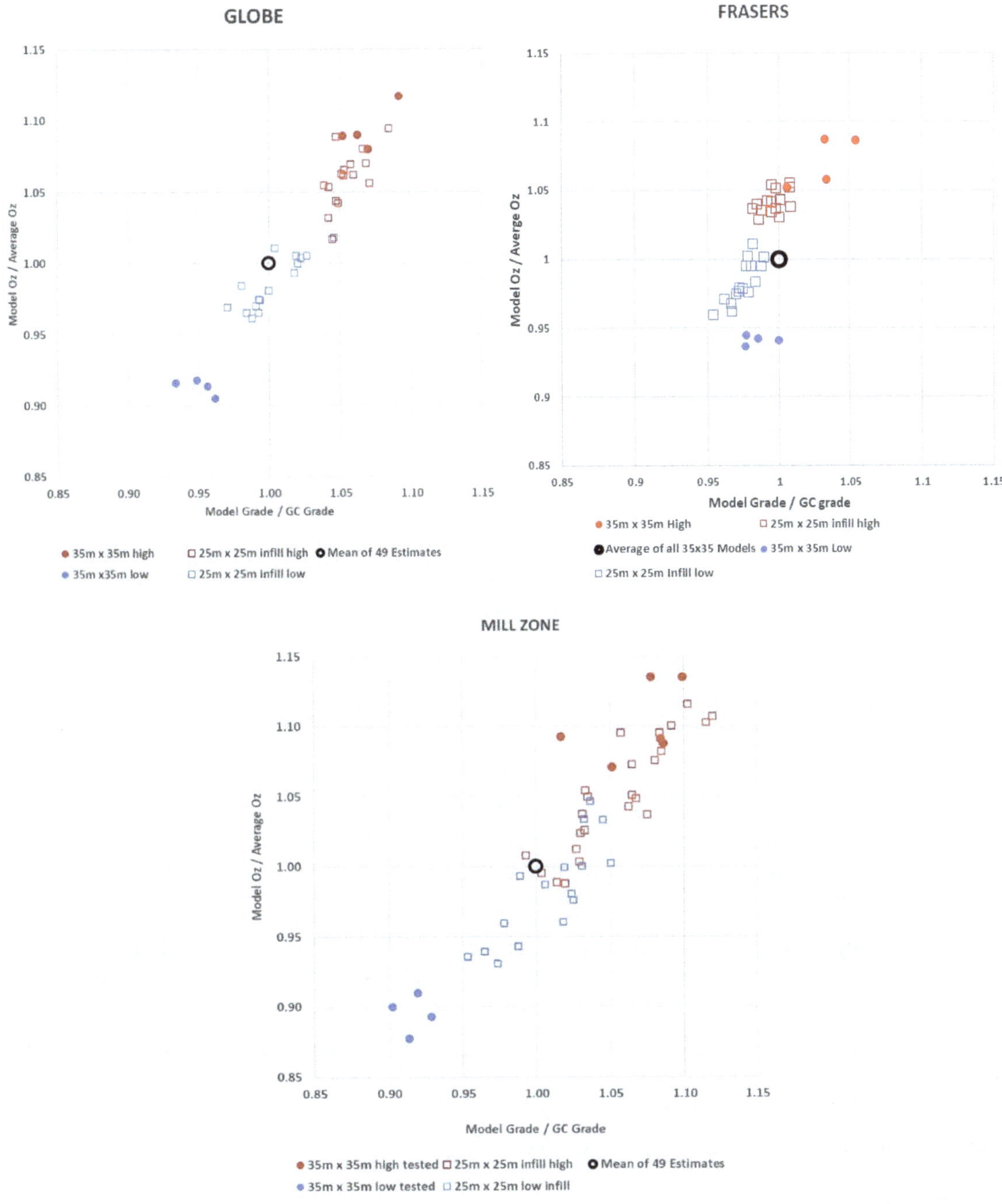

FIG 14 – 35 m × 35 m parent and 25 m × 25 m infill estimates (clockwise from top left); Globe, Frasers and Mill Zone. High parent sets red dots, low parent sets blue dots. Hollow squares for high (red) and low (blue) 25 m × 25 m infill.

Infill drilling of the 35 m × 35 m low-cases significantly improves estimation outcomes as expected (particularly for Globe and Mill Zone). The blue squares in Figure 14 are located much closer to the origin than the parent low-case sets (blue dots). Infill drilling of the high-cases is not as effective. The red squares in Figure 14 generally move closer to the origin than the parent high-case sets (red dots) but much less so than the low-cases.

Figure 15 compares the outcomes of estimates based upon high-case parent drill sets (large dots) versus their four infill daughters (hollow squares). Colour coding associates specific parents with their daughter outcomes so we can look at outcomes on a case-by-case basis. Infill drilling typically moves outcomes (daughters) towards the origin but not nearly to the extent seen for the low-cases and the outcomes vary markedly for each parent, ranging from significant improvement to actually worsening of outcomes.

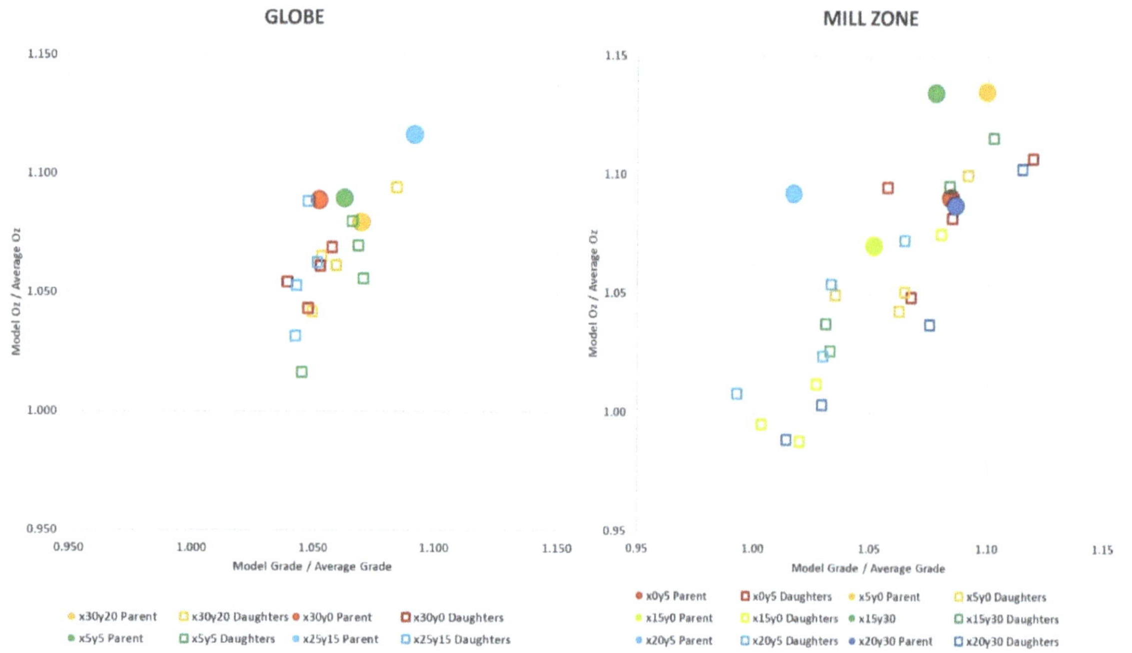

FIG 15 – 35 m × 35 m parent and 25 m × 25 m infill estimates for High-cases. Globe and Mill Zone examples.

It is the high outliers that would normally present the greatest risk to a project, so more research is required. Although the positive skewness of the grade distributions is likely the key to this, exactly how, remains unclear.

Using the MPO framework, allows us to see that infill drilling (the combination of two unique drill hole data sets), can be likened to two consecutive 'visits to the casino'; the luck of the draw for the initial 35 m × 35 m drill set and subsequently, the luck of the draw for the infill 35 m × 35 m drill set. Whilst likely in general to improve estimation outcomes, it is clear that infill drilling has the potential to produce erratic outcomes.

Based upon the analysis to-date, it is tempting to ask whether the impact of infill drilling is overvalued if your current drill hole spacing already supports Indicated resource classification. However, more immediate questions are:

- Why some drill hole sets seem more resistant than others to improvement by infill drilling.
- How infill drilling could be more effectively deployed.

These questions deserve considerably more attention.

CONCLUSIONS

As the saying goes, 'personal experience is statistically insignificant'. Instead of a single personal experience, the MPO framework offers access to a spectrum of equiprobable outcomes within a controlled environment, allowing statistical analysis, and providing a lens through which to gain valuable insights. Unlike forward-looking simulation analysis, which is based upon broadly spaced drilling (itself inherently uncertain), the use of MPO analysis on closely spaced grade control data avoids the need to make assumptions about short range grade continuity or the representativity of the mineralisation being estimated.

Work to-date has revealed that the Schrödinger effect is present in all five gold case studies and that for a given drilling spacing, the 'Schrödinger effect' represents the fundamental lower limit for estimation uncertainty, because it is inherent to the data underlying the estimates. Without reducing the drill spacing, the total estimation uncertainty cannot be reduced below this threshold. The magnitude of the effect is similar across all five case studies despite varied geological settings and mineralisation styles.

The actual impact of the Schrödinger effect can only be determined after mining, which means that conditional simulation-based evaluations of forward-looking uncertainty will significantly under-call estimation uncertainty. The similarity in magnitude across all case-studies however, provides some justification for using rules of thumb to evaluate resource estimation uncertainty via generalised MPO analysis.

The uncertainty profile (modelling processes versus data-inherent Schrödinger effect) of a resource estimate should be considered before resources are classified. For many estimates, the Schrödinger effect is the main driver of estimation uncertainty and therefore resource model performance. So, in operational environments, the Schrödinger effect can manifest as poor reconciliation outcomes which may be mistaken for poor resource modelling/estimation methodology. This is testable via comparison of declustered data means and nearest neighbour QQ analysis, and may prevent 'fixing' something that isn't broken.

Unrealistic expectations of resource model performance result in operational self-harm, distracting attention from useful activities by futilely chasing noise rather than signal. MPO analysis can be used to establish realistic operational thresholds, as well as to estimate the cost of improving resource performance via infill drilling and for performance-based drill hole spacing studies. In many cases, allowances would need to made for downstream operational uncertainties.

More research into infill drilling efficacy is required. Whilst in general, infill drilling is expected to improve estimation outcomes in terms of local and global accuracy, on a case-by-case basis, outcomes can be unpredictable. MPO analysis to-date suggests that the efficacy of infill drilling is asymmetric; mitigating over-estimation appears to be more challenging than mitigating under-estimation (the former generally poses a larger risk to operations). This needs to be better understood so that infill drilling might be more effectively deployed.

MPO analysis is a powerful tool, operating in data-rich production environments, but is labour intensive and time-consuming. Over 450 resource estimates and 10 simulation studies across five case studies and a separate performance threshold study were completed to support analysis in this paper. This provides fertile ground for existing and emerging technologies to streamline workflows.

To conclude, each year vast troves of closely spaced production data from mined out projects are retired to company databases, becoming sunken treasures. These abyssal hoards need to be recovered. Retrospective analyses based on these high-resolution data sets will unlock answers to many problems currently confronting resource and mine geologists.

ACKNOWLEDGEMENTS

Kate McKercher and Kristine Colaljo for assisting with resource estimations. Greg Hollett, Mike Stewart and the anonymous AusIMM reviewer for valuable feedback. OceanaGold for allowing publication. Neil Schofield as always for providing the spark for many of the ideas. Thank you all.

REFERENCES

JORC, 2012. Australasian Code for Reporting of Exploration Results, Mineral Resources and Ore Reserves (The JORC Code) [online]. Available from: <http://www.jorc.org> (The Joint Ore Reserves Committee of The Australasian Institute of Mining and Metallurgy, Australian Institute of Geoscientists and Minerals Council of Australia).

Moore, J, Grant, M, Corley, D and Randa, W, 2023. Schrödinger's kittens – lifting the lid on resource drill hole data after mining, in *Proceedings of the Mineral Resource Estimation Conference 2023*, pp 141–150 (The Australasian Institute of Mining and Metallurgy: Melbourne).

Parker, H M, 2011. Reconciliation principles for the mining industry, in *Proceedings of the 35th APCOM Symposium 2011*, pp 47–67 (The Australasian Institute of Mining and Metallurgy: Melbourne).

Reconciliation

Data to decisions

Enhancing underground mine reconciliation – a unified F- and R-series framework

R Hargreaves[1], T Elkington[2], I Glacken[3] and A Jewbali[4]

1. Principal Consultant, Datamine, Perth WA 6000.
   Email: rayleen.hargreaves@dataminesoftware.com
2. General Manager, Snowden Optiro, Perth WA 6000.
   Email: tarrant.elkington@snowdenoptiro.com
3. Executive Consultant, Snowden Optiro, Perth WA 6000.
   Email: ian.glacken@snowdenoptiro.com
4. Head, Resource Management, Global, Newmont Corporation, Denver CO, USA.
   Email: arja.jewbali@newmont.com

ABSTRACT

Reconciliation is a critical governance control in mining operations, providing transparency over the relationship between planned and actual performance across the resource, reserve, mining, and processing value chain. The F-series factors, originally proposed by Parker (2012), are widely used to assess the accuracy of mine planning and execution by comparing reserve, production, and plant outcomes. Building on this framework, the R-series factors (Hargreaves *et al,* 2022) extend reconciliation upstream by tracking how modifying factors such as dilution, ore loss, and information effects are applied from the resource model through to production.

While these frameworks are increasingly being adopted, their application in underground mining remains challenging. Selective extraction, irregular geometries, limited visibility, and inconsistent depletion definitions frequently result in spatial misalignment between models, leading to reconciliation outcomes that are difficult to interpret and, in some cases, misleading. Without a consistent depletion basis, apparent variances may reflect differences in reporting volumes rather than true geological or operational performance.

This paper presents a standardised approach to underground reconciliation based on a 'common volume' depletion methodology. The common volume represents the unified spatial extent of material mined or sterilised during a reporting period and can be applied consistently across resource, reserve, grade control, design, and production models. This approach enables spatially valid calculation of F- and R-series factors and ensures that reconciliation results reflect genuine modifying factors and execution outcomes rather than artefacts of inconsistent geometry.

The methodology is demonstrated through a case study from Newmont's Tanami underground gold operation. The results show how integrating F- and R-series factors within a common spatial–temporal framework improves insight into dilution, ore loss, and information effects, supports calibration of planning assumptions, and strengthens confidence in reconciliation performance. The approach is framed within the evolving industry context, including proposed updates to the JORC Code, reinforcing the role of reconciliation as a foundation for transparent, defensible reporting in underground mining operations.

INTRODUCTION

Reconciliation is a critical governance function within mining operations, serving as a key mechanism for comparing planned and actual performance across the value chain. Historically, reconciliation has informed month-end metal accounting, operational forecasting, and auditability by comparing estimates of tonnage, grade, and contained metal at multiple stages; resource, reserve, grade control, mined, plant received and processed (Wild, 1998; Shaw, 1991; Sides, 1992; Gover and Assibey-Bonsu, 1996; Schofield, 2001; Morley, 2003; Morley and Moller, 2005; Fouet *et al,* 2009; Helm, Hargreaves and Morley, 2013). Numerous case studies have demonstrated its value in identifying underperformance and supporting continuous improvement (Coles, Hadlow and Levy, 1993; Westley, 1986; Morley and Arvidson, 2017).

In response to the need for more structured evaluation, Parker (2012) introduced the F-series factors, which provide a framework for assessing the predictive accuracy of estimation and mine planning at various stages, such as how well reserves forecast mined tonnes or how accurately mine production translates to plant feed. Recognising the need to extend this framework, Hargreaves *et al* (2021) proposed the R-series factors to complement the F-series by examining how modifying factors are applied through the planning pipeline, from the resource model through to actual production, enabling deeper analysis of ore loss, dilution, and planning changes over time.

While these frameworks have gained traction, their application in underground mining presents unique challenges. Unlike open pit operations, where spatial depletion volumes can be clearly defined, underground mines are almost always characterised by selective extraction, leading to inconsistencies in depletion definitions and unreliable model comparisons. These complexities reduce the effectiveness of traditional reconciliation approaches and create barriers to transparency and consistency (Hargreaves and Booth, 2019).

This paper proposes a standardised approach to underground reconciliation through the definition of a 'common volume' depletion methodology. This spatially valid technique enables consistent calculation of F- and R-series factors, enhancing reconciliation insight and reliability. The method is demonstrated through a case study at Newmont's Tanami operation and is framed within the evolving context of industry reporting standards, including proposed updates to the JORC Code. In parallel, this approach aligns with guidance from broader industry initiatives such as the AMIRA P745 Reconciliation Code of Practice (AMIRA, 2007), the draft JORC Code (2024) and the Reconciliation Standard (Newmont Corporation, 2021), which promote consistent terminology, model classification, and depletion strategies. These standards reinforce the need for spatial and temporal clarity in reconciliation processes, which are particularly relevant in underground environments where variability in depletion definitions can lead to ambiguity in model comparisons and undermine reporting reliability.

It is important to note that this paper does not address reconciliation in caving operations (eg block caving or sub-level caving), as these require flow-based modelling approaches outside the scope of the common volume methodology proposed here.

BACKGROUND AND RECONCILIATION FRAMEWORKS

Reconciliation is more than a reporting requirement; it is a vital control mechanism that supports transparency, planning confidence, and measures operational performance across the mine value chain. Standardisation of reconciliation factors has been an ongoing industry objective (Fouet *et al*, 2009). To systematise reconciliation analysis, two complementary frameworks have emerged: the F-series and R-series factors.

The F-series factors

Early recognition procedures for open pit operations were formalised in a number of studies (Parker, 2006). The F-series factors, developed by Parker (2012), focus on measuring how well key planning and operational estimates align with actual outcomes. These factors allow practitioners to identify tonnage, grade and metal subdivisions between mine planning, production, and processing stages:

- **F1** compares mine production (ASM) against the reserve model (RSV), answering: *How well does the reserve predict what was mined?*
- **F2** compares plant feed or processed material (PLT) with as-delivered tonnes, grade and metal (ASD), assessing: *How consistent are the mine and plant measurement systems?*
- **F3** (often calculated as F1 × F2) reflects how well the reserve predicts actual processed outcomes.

These metrics are widely used in monthly and quarterly reporting cycles and are fundamental for evaluating short-term plan performance.

The R-series factors

To complement the F-series, Hargreaves *et al* (2021) introduced the R-series framework (Figure 1) to enable granular tracking of modifying factors throughout the planning and execution pipeline. Rather than simply comparing outputs, the R-series captures how planning inputs evolve, from resource estimation through to physical production and processing, with a focus on understanding where and how modifying factors are applied.

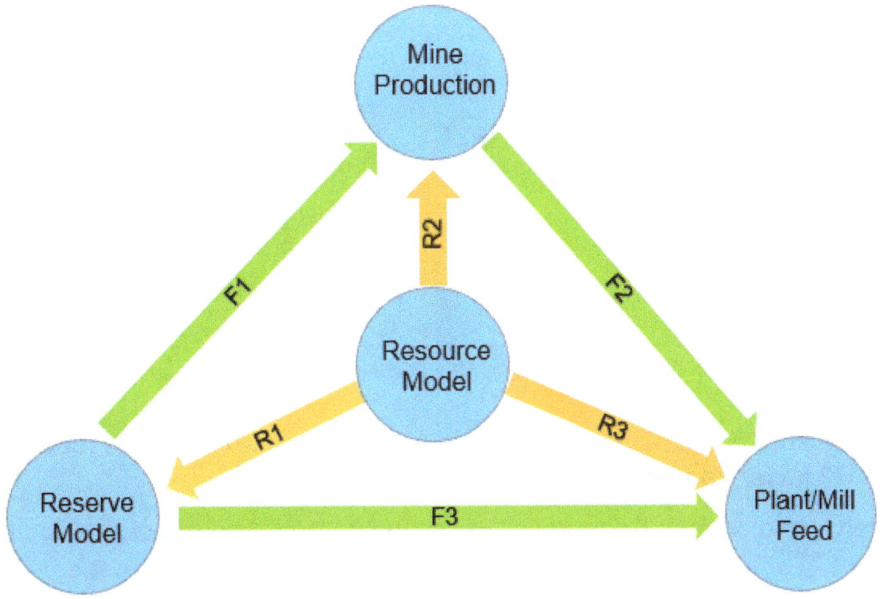

FIG 1 – F- and R-series reconciliation framework.

Each R-factor traces a specific stage in this transformation, namely:

- **R1:** Reserve versus Resource – Tracks the application of modifying factors from the resource to reserve model.
- **R2:** Mine Production versus Resource – Measures the cumulative effect of all modifying factors, including information effect, dilution, and ore loss.
- **R3:** Plant Feed versus Resource – Provides an overall end-to-end performance factor from estimation to processing.

Importantly, Hargreaves *et al* (2022) emphasise that these comparisons must be conducted within a common spatial depletion volume to ensure validity, particularly in underground environments, where selective mining means not all models cover the same physical extent.

The F- and R-series frameworks are designed to work in tandem. While F-factors address *how close reality is to the plan*, R-factors track *how the plan was shaped in the first place*. When used together, they enable operations to move beyond reporting variances to understanding the root causes.

Each of the purposes listed in Table 1 are tools to identify potential improvement in an operation, but only if the measurements are available, consistent, reliable, and relevant. Without *all* of these items, the value of the reconciliation system erodes quickly.

TABLE 1

Summary of F- and R-factors – numerators, denominators, and purpose.

Factor	Numerator	Denominator	Purpose
F1	Mine Production (ASM)	Reserve Model (RSV)	How well is the Reserve predicting mine production?
F2	Plant Feed (PLT)	As-Delivered (ASD)	How consistent are the mine and plant system measurements?
F3[1]	PLT	RSV	How well is the Reserve predicting the actuals?
R1	RSV	Resource Model (RSC)	What are the trends in modifying factors applied for mine planning?
R2	ASM	RSC	What are the trends modifying the resource to mine production?
R3[1]	PLT	RSC	What is the trend in resource recovery to actuals?

Note: 1. Due to stockpiles F3 and R3 should never be directly calculated. Rather F3 = F1 × F2 and R3 = R2 × F2 = R1 × F3.

All of the F- and R-factors can be further broken down into intermediate steps for additional insights. For example, the R2 factor can be subdivided into the following:

- **R2a:** Grade Control versus Resource – Highlights information effect due to short-range data differences.
- **R2b:** Mine Production versus Grade Control – Captures actual dilution and ore loss during mining.
- **R2c:** Mine Production versus Design – Represents execution fidelity – unplanned dilution, misclassification, and physical loss.

UNDERGROUND RECONCILIATION – CHALLENGES AND OPPORTUNITIES

While reconciliation is a critical control in both open pit and underground operations, the two environments diverge significantly in how they operate, collect data, and measure outcomes. Underground mining is inherently more selective, constrained by development access and often defined by irregular geometries. Unlike open pits, where each bench represents a relatively uniform horizontal slice of the orebody, underground operations follow narrow veins, stope outlines, or panel shapes that may vary considerably in orientation, thickness, and continuity.

This introduces a fundamental challenge: defining what has been mined is less straightforward underground. Visual inspection is limited, multiple development headings may be active concurrently, and survey practices can vary based on access. Furthermore, development activities may sterilise nearby resources unintentionally, but these effects are rarely accounted for in reconciliation workflows.

Where open pit mines can apply simple volumetric depletion to resource and grade control models by bench or pit shell, underground operations require more nuanced and site-specific definitions of 'depletion volume'. Without a standard approach, model-to-model comparisons can become invalid, leading to poor governance visibility and data misinterpretation.

Depletion inconsistencies and model misalignment

The central challenge in underground reconciliation lies in how depletion volumes are constructed and applied. In many operations, each model – resource, reserve, grade control, design, and actual production – is measured using different spatial volumes, filtering criteria, or levels of granularity and accuracy in how well they represent actual mined conditions. For example:

- The resource model may cover an expansive geological interpretation with no accounting for current access.

- The reserve model, if spatially coded at all, may exist only in a spreadsheet summarising tonnage and grade, disconnected from a physical block model.
- The grade control model may be applied in parts of the mine where recent drilling has occurred but may be absent in legacy or development headings.
- Production data often lacks reliable survey volumes or may be based on truck counts with approximate tonnages and no reliable grade allocation at all.

When these data sets are compared, the resulting reconciliation factors are often distorted. For instance, R2c comparisons (actual production versus planning model) may show high ore loss or dilution, which may not be due to operational issues but to inconsistent depletion shapes between the planning model and survey solids. In some cases, the reported resource model depletion grades are lower than the grades in undeveloped areas, which is not only misleading, but statistically impossible.

This is the result of comparing apples and oranges; in other words, depletion volumes drawn from different shapes, time periods, or modelling standards. The outcome is a reconciliation system that produces numbers, but not insight.

Sampling constraints and dilution

The sampling regime in underground mines is often constrained by access, safety, and logistics. Development headings are typically sampled via face, channel sampling or visual grade estimation, while production stope grades may be inferred from limited drilling, historical stope data, or simply estimated. In contrast, open pit grade control benefits from regular drilling patterns and predictable bench advance, providing more reliable and consistent data sets. As a result:

- Ore loss may be underestimated when post-mining voids are poorly surveyed or when remnant stopes are not captured in reconciliation solids.
- Dilution may be overestimated if based on design assumptions rather than actual mined volumes, especially where designs are conservative or not updated regularly.
- Stockpiled material moved from underground to surface may be rehandled multiple times, with its source and grade becoming uncertain by the time it reaches the plant.

These issues introduce ambiguity into the reconciliation data chain, making it difficult to reliably track metal from the resource model through to final plant feed. In particular, stockpile rehandling can distort delivered grade comparisons and undermine confidence in overall material balances.

Gaps in spatial reserve and design models

In many underground operations, the reserve and short-term design models often do not exist as spatially coded entities within the geological software environment. Instead, they may live in Excel sheets, databases, or other planning outputs that summarise tonnes and grades but cannot be intersected with reconciliation solids. This makes five of the nine F- and R-series comparisons either impossible or unreliable, as they require volume-based alignment to be meaningful. For example:

- R1 (Resource versus Reserve) and R2b (Grade Control versus Design) are spatial comparisons by nature. If one of the models does not exist in spatial form, the factor has little or no significance.
- This also prevents validation of cut-off application, modifying factor transparency, accuracy and granularity between plan and actual.

If reconciliation is to be a serious governance tool (Shaw *et al*, 2013), then each input must be spatially and temporally aligned – not just tabulated.

THE COMMON VOLUME METHODOLOGY

Accurately defining depletion volumes has long been recognised as one of the most persistent challenges in underground reconciliation. Unlike open pit operations, where surveyed pit shells typically provide a clear and repeatable boundary for model reporting (Sides, 1992; Shaw, 1991), underground mines, by their very nature, are characterised by selective stoping, variable design execution, remnant zones, and inconsistent survey coverage. This complexity frequently results in misleading or non-comparable reconciliation outcomes, particularly when different models are assessed over different spatial footprints. Historical studies highlight how such inconsistencies can distort grade and tonnage comparisons and obscure the true sources of variance (Wild, 1998; Schofield, 2001).

These issues become more pronounced when reserve, design, grade control, and production models rely on different depletion shapes or are only partially captured in reporting volumes. When the reporting geometries do not align, apparent discrepancies may reflect differences in spatial definition rather than meaningful operational performance (Morley, 2003; Fouet *et al,* 2009). The absence of a standardised depletion approach therefore introduces material risks into the reconciliation process, impacting not only technical interpretation but also corporate reporting and governance. As industry expectations evolve, particularly with the proposed revisions to the JORC Code, the need for a defensible, consistent, and spatially rigorous reconciliation methodology has become increasingly critical.

To address these challenges, this paper proposes the common volume methodology, a unified spatial framework designed to ensure that all models are assessed against the same physical volume during a reconciliation period. By capturing all material sterilised through mining or design decisions, whether mined, partially mined, or rendered inaccessible, the common volume provides a consistent foundation for calculating F- and R-series factors in underground environments. This standardisation allows reconciliation outcomes to reflect genuine differences in orebody interpretation, modifying factors, and operational performance rather than artefacts of inconsistent depletion logic.

Defining the common volume

The common volume is defined as a unified depletion geometry that represents all material *rendered inaccessible or removed during a reconciliation period*, regardless of whether it was physically mined, planned, or sterilised by proximity. This approach diverges from traditional reconciliation methods that rely solely on surveyed voids or mine production data, which often underestimate the true spatial extent affected by underground extraction decisions.

Instead, the common volume is constructed as the union of four critical domains:

1. Surveyed as-built, CMS or drone-derived shapes representing actual mined material.
2. Planned or designed stopes, including any material left behind.
3. Blocks from resource or grade control models intersected by the mining envelope.
4. Sterilised regions such as pillars or remnant zones rendered unrecoverable by surrounding extraction.

By combining these components, the common volume ensures that each model (resource, reserve, design, grade control, mine production) is depleted over an equivalent spatial extent (Figure 2), enabling meaningful comparison of modifying factors and accurate calculation of R-series factors such as R1, R2a, R2b, and R2c. It also allows flexibility for each model to apply its own grade filters (eg cut-off, ore blockout, reserve adjustments), preserving internal logic while ensuring external consistency.

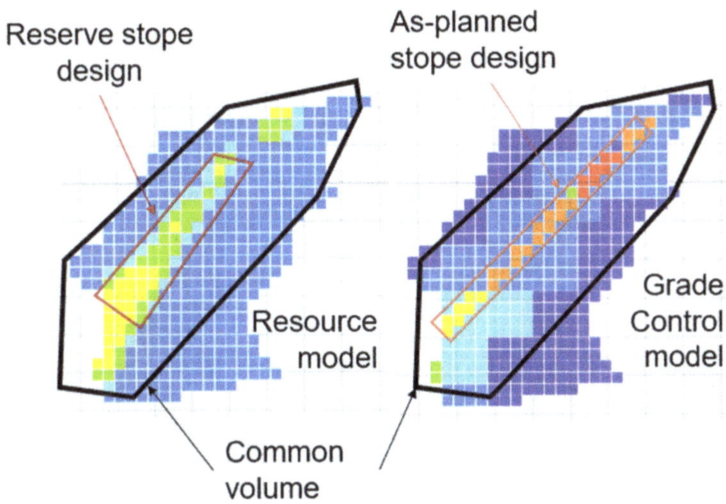

FIG 2 – Common volume concept.

This methodology addresses long-standing issues observed at underground operations, where low or misleading resource model grades were reported due to narrow depletion volumes that failed to capture the full planning envelope (as previously noted by Morley (2003) and Hargreaves *et al* (2021)). It eliminates 'apples-to-oranges' comparisons between planning and production and instead supports a spatially valid and defensible basis for model evaluation, ore loss tracking, and reconciliation reporting.

Link to R-series factors

The common volume methodology is essential for the reliable calculation of R- and F-factors in underground mines. Because each relevant factor compares two planning or production models, the result is only meaningful when both models are depleted over the same spatial footprint. Without a common volume, differences may reflect inconsistent depletion shapes rather than true changes in modifying factors, information effect, dilution, or mining performance. Using a unified depletion volume ensures that each R- and F-factor reflects real operational or geological drivers, rather than artefacts of shape definition.

Reconciliation in the F-/R-series framework relies on seven quality measurements. A summary of the recommended basis is in Table 2 (for open pit mines) and Table 3 (for underground mines) incorporating the common volume concept.

The common volume approach is most important for measuring the RSC and GCM, which are only used in R-Series factors. Consequently, because this is only a new concept and most reporting has been using only the F-Series factors, underground operators have been able to work around this issue. Therefore, the utility of the common volume is largely linked to the value of the R-Series factors.

TABLE 2

Measurement point definitions for open pit mines.

ID	Measurement	Grade source	Depletion shape	Shape filter	Grade filter	Factoring
RSC	Resource model	Resource model	Mined pit	Nil	Cut-off	Nil
RSV	Reserve model	Resource model	Mined pit	Reserve shapes	Cut-off	Reserve factors
GCM	Grade control model	Grade Control model	Mined pit	Nil	Cut-off	Nil
ASP/DES	As-planned/ Design	Grade Control model	Mined pit	Ore blockout	Nil	Nil
ASM	Mine production	Grade Control model	NA	NA	NA	Survey factor
ASD	Mine delivered	Stockpile model	NA	NA	NA	Survey factor
PLT	Plant Mill Feed	Belt Samplers	NA	NA	NA	NA
PLR	Plant reconciled	Plant reconciled	NA	NA	NA	NA

TABLE 3

Measurement point definitions for underground mines.

ID	Measurement	Grade source	Depletion shape	Shape filter	Grade filter	Factoring
RSC	Resource model	Resource model	Common volume	Nil	Cut-off	Nil
RSV	Reserve model	Resource model	Common volume	Reserve shapes	Nil	Reserve factors
GCM	Grade control model	Grade Control model	Common volume	Nil	Cut-off	Nil
ASP/DES	As-planned/Design	Grade Control model	Common volume	Final design	Nil	Nil
ASM	Mine production	Grade Control model	NA	NA	NA	Survey factor
ASD	Mine delivered	Stockpile model	NA	NA	NA	Survey factor
PLT	Plant Mill Feed	Belt Samplers	NA	NA	NA	NA
PLR	Plant reconciled	Plant reconciled	NA	NA	NA	NA

Scope considerations – stoping versus caving methods

The common volume methodology proposed in this paper has been specifically developed for underground mining operations that use selective stoping methods, such as room and pillar, longhole open stoping or cut-and-fill. These methods are generally characterised by defined stope geometries, discrete extraction units, and spatially mappable depletion boundaries, making them well-suited to a standardised spatial reconciliation framework.

However, not all underground mining methods operate within this paradigm. Sub-level caving (SLC) and block caving (BC) rely on fundamentally different extraction principles (Table 4). Instead of selectively removing predefined shapes, caving methods use gravity-assisted material flow to recover ore through a network of drawpoints. As such, the depletion volume in caving is dynamic, three-dimensional, and strongly influenced by flow mechanics rather than stope boundaries.

TABLE 4

Stoping versus caving features.

Feature	Stoping methods	Caving methods
Extraction style	Selective mining of designed shapes	Continuous draw from cave front
Geometry control	Discrete, surveyed voids (CMS, as-builts)	Flow-based draw zones; no defined void boundaries
Reconciliation volumes	Based on design and CMS solids	Inferred from material flow modelling
Dilution mechanisms	Often design-related or due to stop overbreak	Internal to cave, driven by flow and fragmentation
Tracking methods	Model comparisons using spatial solids	Flow-based tracking, drawpoint sampling, simulations

Because caving methods involve ongoing material flow, internal mixing, and time-delayed draw responses, they require reconciliation frameworks that incorporate geomechanical flow models, drawpoint-based material tracking, and recovery/dilution simulation tools. The spatial reconciliation logic underpinning the common volume method is not directly transferrable to these settings without significant adaptation.

As such, the scope of this paper is limited to underground operations using stoping-based mining methods.

CASE STUDY – APPLYING THE F- AND R-SERIES FRAMEWORK AT TANAMI

To demonstrate the practical application of the reconciliation framework and common volume methodology discussed in this paper, a case study from Newmont's Tanami Gold Operation (Tanami) is presented.

Tanami overview

The Tanami Gold Operation is a large-scale underground gold mine located in the Tanami Desert of Australia's Northern Territory, approximately 550 km north-west of Alice Springs (Figure 3). The operation is owned and operated by Newmont Corporation and comprises underground mining at the Callie deposit, with ore processed at The Granites processing plant located approximately 40 km from the mine.

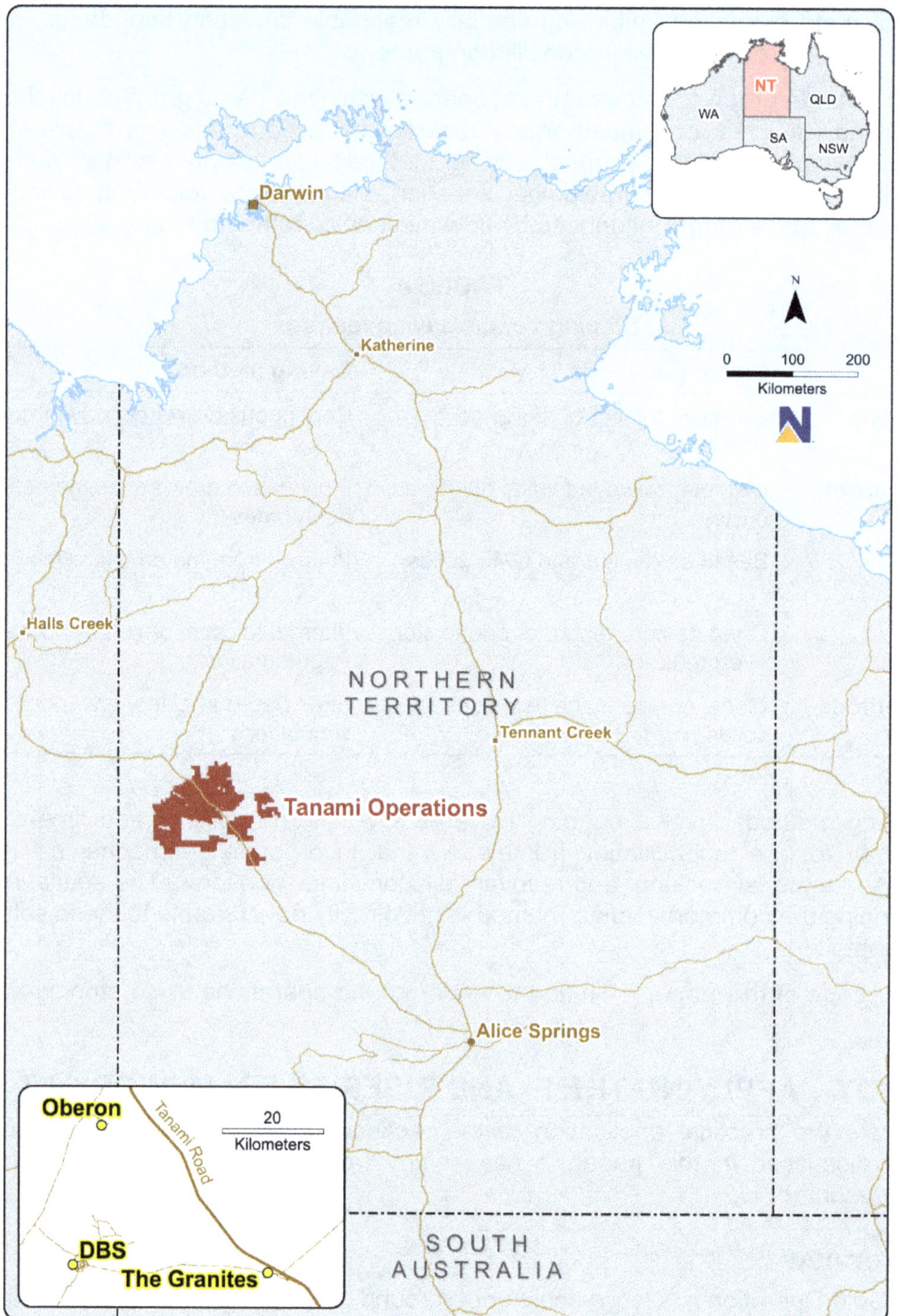

FIG 3 – Newmont's Tanami Gold Mine, Northern Territory, Australia.

Mining at Tanami targets narrow, high-grade orogenic gold mineralisation hosted within structurally complex shear zones. Mineralisation is characterised by extreme grade variability and the presence of coarse gold, which has implications for both mining dilution and metallurgical reconciliation. *In situ* resource grades are high relative to many underground gold operations; however, the mining method results in significant dilution to enable safe and productive extraction at scale.

Underground production is undertaken using bulk stoping methods, primarily longhole open stoping, supported by development. These methods allow high throughput but require the inclusion of low-grade or waste material around high-grade structures. As a result, planned dilution and ore loss are material components of the mine plan and are explicitly incorporated into reserve modifying factors.

A long section of the mine is shown in Figure 4.

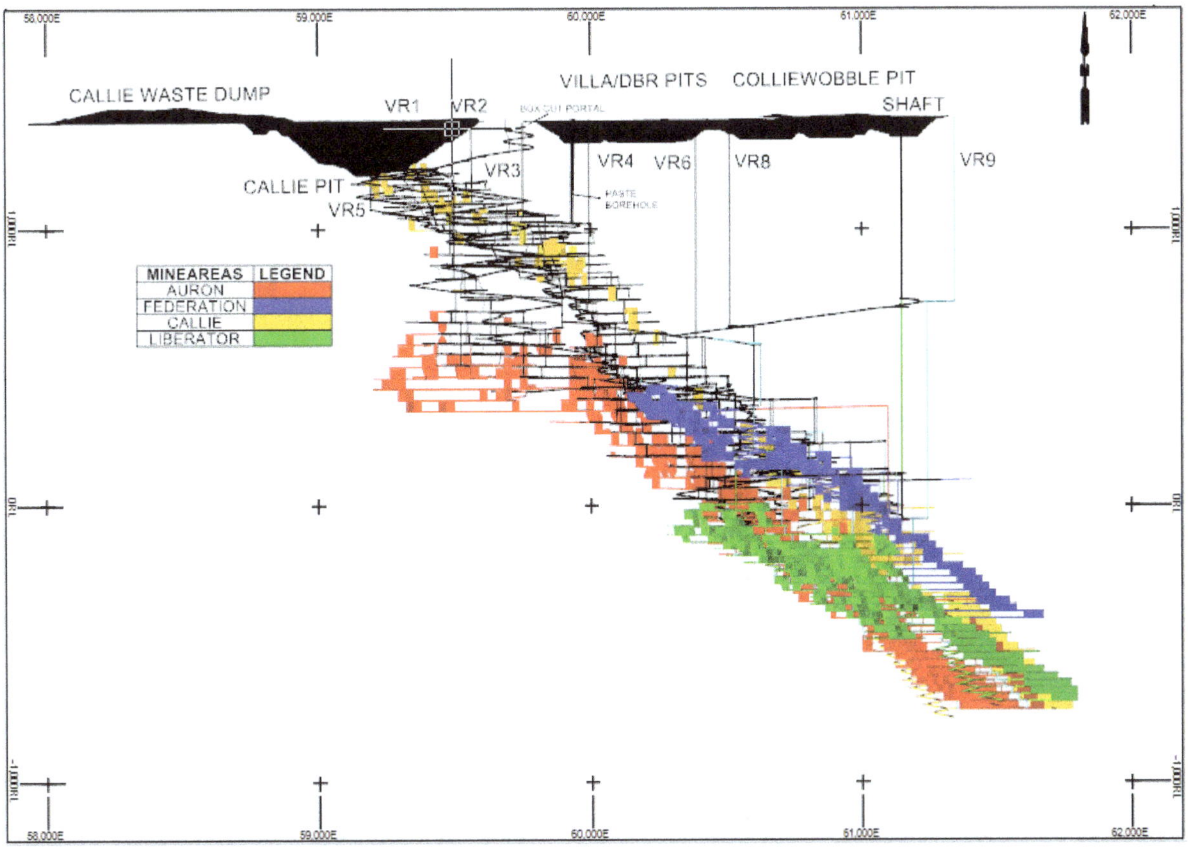

FIG 4 – Tanami underground long-section.

During the period considered in this study (2021–2024), Tanami operated at an average production rate of approximately 2.3–2.7 Mt/a, with a mill feed grade typically in the range of 5–7 g/t Au, resulting in annual gold production of the order of 500 koz. Monthly production tonnes and grades exhibit variability due to stope sequencing, development access, and the selective nature of underground mining, with additional short-term variability introduced by coarse gold behaviour during processing.

These characteristics make Tanami a technically demanding environment for reconciliation. High dilution, strong grade variability and complex material movement through stockpiles challenge traditional planned-versus-actual reconciliation metrics and can obscure the distinction between inherent mining method effects and unanticipated performance issues. Consequently, Tanami provides a suitable case study for demonstrating how a spatially consistent reconciliation framework can be applied to expose and interpret modifying factor behaviour in an underground mining context.

The following sections present the resulting F- and R-series reconciliation trends.

F-series reporting

Newmont's primary measurements are used to derive F-series factors as well as a range of other useful intermediate metrics. For the calculations used in this case study, the F-Series at Tanami are defined as:

F1 = Final mine design plan/reserve plan

F2 = Received at mill/delivered to mill

F3 = F1 × F2

The trends in F-series between 2021 and 2024 are shown in Figure 5.

(a) F1 = Final mine design plan/reserve plan.

(b) F2 = Received at mill/delivered to mill.

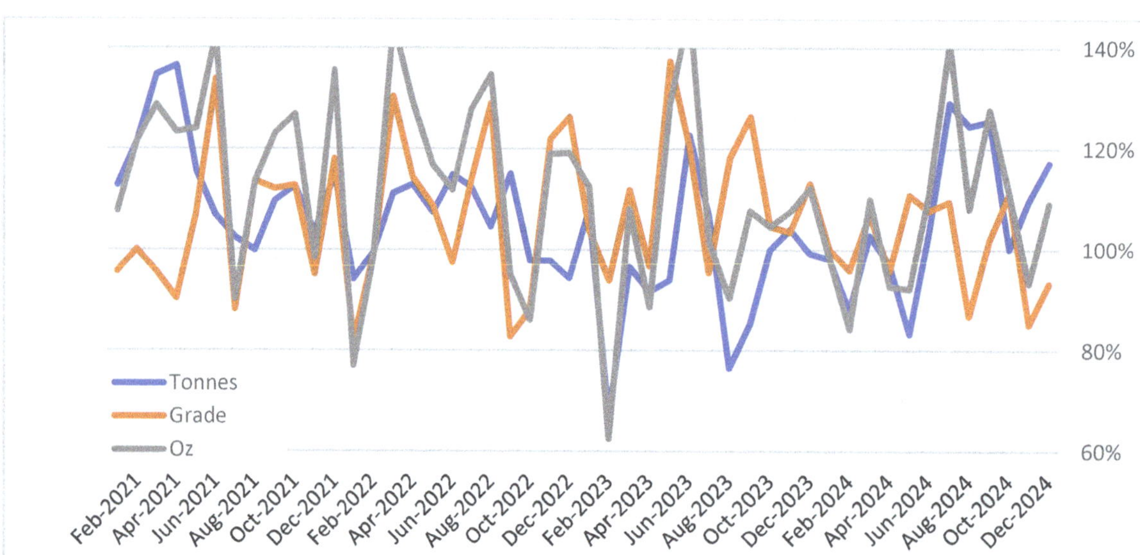

(c) F3 = F1 × F2.

FIG 5 – Tanami F-series reporting (2021–2024).

The F1 reconciliation reflects the relationship between reserve planned and final design mining tonnes, grade and contained metal. Over the period reviewed, F1 exhibits moderate month-to-month variability, with tonnes and ounces showing greater dispersion than grade.

Short-term underperformance observed in parts of 2022 and early 2023 is primarily attributable to operational factors, rather than deficiencies in the underlying resource or reserve models. These include restricted access to planned stopes, deferral of lower-confidence stopes, and sequencing changes driven by grade control drilling availability and underground development constraints.

Grade reconciliation under F1 is comparatively stable, generally remaining close to unity Improvements in F1 stability observed from late 2023 into 2024 are consistent with increased infill drilling, improved stope definition and enhanced mine execution at Liberator and Federation.

Overall, F1 performance supports the appropriateness of the mining modifying factors applied in the reserve estimate, with no evidence of systematic grade bias.

The F2 reconciliation displays higher short-term variability than F1, particularly for grade and ounces, while tonnes remain relatively stable. This behaviour is characteristic of Tanami-style coarse gold mineralisation, particularly Callie-style vein-hosted mineralisation, where gravity recovery performance and gold distribution can vary significantly over short time frames.

Monthly spikes and troughs in F2 are driven by a combination of factors, including gravity circuit performance, batch processing of high-grade development ore, stockpile movements, and timing differences between mine production, mill feed and gold pouring.

Importantly, no sustained positive or negative bias is evident in F2 when assessed as longer-term rolling averages, indicating that metallurgical recovery models and processing assumptions remain appropriate at an annualised scale.

The combined F3 reconciliation, which compares the reserve prediction with what was received at the mill, incorporates both mining delivery and processing effects and therefore exhibits amplified short-term variability relative to F1 and F2 individually. Months where deviations in F1 and F2 coincide result in larger departures from unity, which is expected for a combined metric at monthly resolution.

Despite this volatility, F3 does not demonstrate any persistent bias above or below 100 per cent over the review period. Rolling averages converge toward unity, consistent with reconciliation performance reported in the 2023 and 2024 period. This behaviour supports the conclusion that observed short-term variability is driven by operational and metallurgical timing effects rather than systematic issues with the mineral resource or mineral reserve estimates.

The reconciliation performance across F1, F2 and F3 is consistent with expectations for an underground gold operation characterised by coarse gold mineralisation and complex stope sequencing. Short-term variability is evident at a monthly scale; however, longer-term reconciliation trends demonstrate convergence toward planned outcomes.

The reconciliation results support the continued use of the current geological models, modifying factors, recovery assumptions and reserve estimation methodologies. There is no evidence to suggest material bias in grade, tonnes or contained metal at an annual or life-of-mine scale.

R-series reporting

To extend the Tanami reconciliation to include R-series reporting, an additional metric, 'Resource Model', was introduced. This measurement represents the depletion of the resource model above cut-off that was either physically mined or rendered inaccessible (sterilised) during a reporting period. This metric forms the reference point for R-series comparisons. As this metric was not routinely generated within Newmont's standard reconciliation workflow, it was constructed specifically for the purposes of this study. For the calculations used in this case study, the R-Series at Tanami are defined as:

R1 = Reserve plan/resource model

R2 = Final mine design plan/resource model

R3 = R2 × R2

Monthly common volume depletion solids were not available to directly support this calculation. Consequently, the common volume for each month was approximated using the best available spatial and production data, namely final stope CMS solids and monthly stope-level tonnage and grade reports. A series of simplifying but reasonable assumptions were applied, including bottom-up extraction within individual stopes, to reconstruct the temporal progression of depletion.

The procedure used to derive monthly resource model depletion volumes was as follows:

- Establish the chronological order of stope extraction to ensure that overlapping stopes were coded from the time of first mining.
- Code stope identifiers onto the resource block model using final stope CMS solids, including all stopes mined to date and future reserve stopes.
- Expand each stope solid by 10 m to define a sterilisation buffer ('halo') representing material rendered inaccessible by mining.
- Assign an extraction month to blocks within each stope based on monthly stope production records.
- Assign a sterilisation month to blocks within the halo based on the extraction month of the nearest mined block.
- Apply the appropriate resource model density and grade corresponding to the extraction month, noting that resource models at Tanami are updated annually.
- Apply the reserve cut-off grade to the depleted resource model for each month and report tonnes and grade.

The resultant R-series factors are shown in Figure 6.

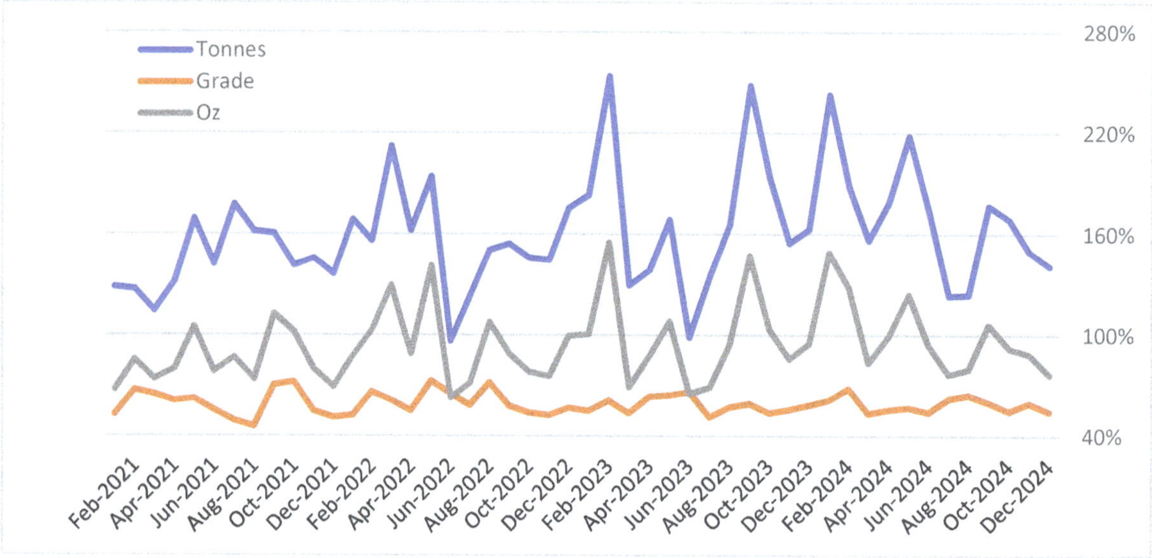

(a) R1 = Reserve plan/resource model.

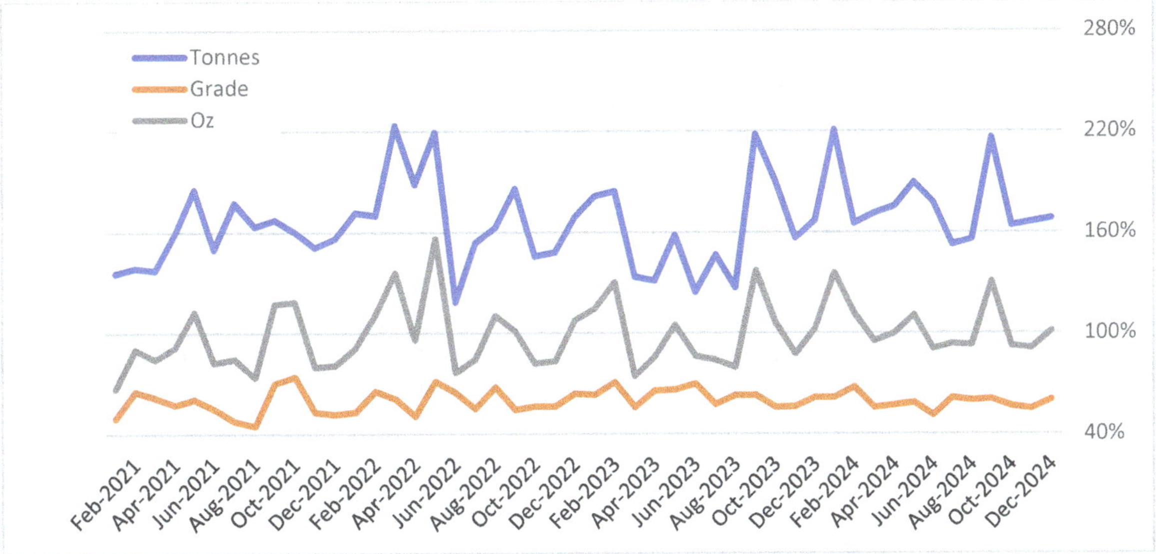

(b) R2 = Final mine design plan/resource model.

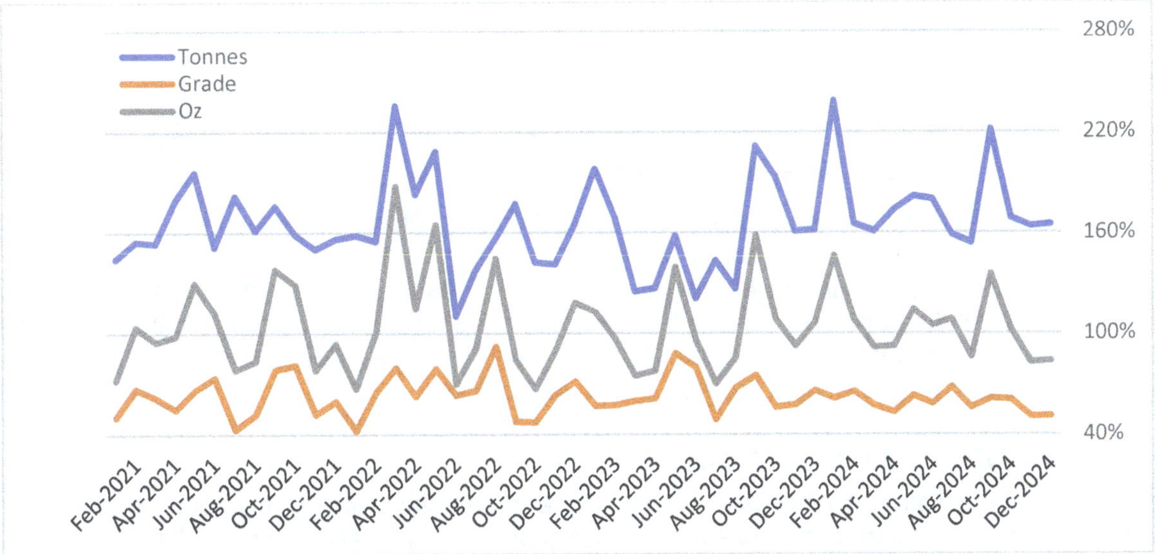

(c) R3 = R2 × F2.

FIG 6 – Tanami R-series reporting (2021–2024).

R1 reflects the conversion of the resource model to the reserve within the common volume. The trend is characterised by a pronounced increase in tonnes accompanied by a corresponding reduction in grade, consistent with the application of bulk underground mining methods to a narrow, high-grade orebody. Planned dilution and ore loss are material and are embedded within the reserve definition to enable safe and productive extraction at scale.

Despite the magnitude of these modifying factors, R1 remains stable over the analysis period, indicating that dilution and loss assumptions have been applied consistently and transparently during reserve generation. When assessed in terms of contained metal, the reduction in grade is largely offset by the increase in tonnes, resulting in metal outcomes that align with reserve expectations. This behaviour reflects a mature planning framework in which dilution is an inherent and well-quantified component of the mining strategy.

R2 captures the cumulative effect of all modifying factors realised in mining, including information effects, planned dilution, and unplanned dilution. The R2 trends mirror those observed in R1, with elevated tonnage factors and reduced grade factors, demonstrating that the level of dilution anticipated in the plan is largely realised in practice.

Short-term variability in R2 reflects localised operational and geological effects, including stope sequencing, access constraints and short-range geological uncertainty. However, over longer periods, R2 tracks closely with R1, particularly when considered in terms of contained metal. This indicates that while dilution is significant, it is neither unexpected nor uncontrolled, and that actual mining outcomes are broadly consistent with the modifying factors assumed at the reserve stage.

R3 represents the end-to-end recovery of metal from the resource model through to plant feed and incorporates both mining and processing effects. As expected, R3 exhibits greater short-term variability in tonnes and grade due to metallurgical timing effects, gravity recovery behaviour and stockpile movements. However, longer-term trends show that metal recovery remains stable and consistent with expectations once these effects are averaged.

The R3 response demonstrates that the additional tonnes introduced through dilution do not translate into proportionate metal loss. Instead, the operation consistently captures most of the metal predicted once modifying factors are applied, confirming that dilution, while substantial, is effectively anticipated and managed through the planning and execution chain.

The R-series trends shown in Figure 6 highlight the extent of dilution inherent in both the mine plan and the operating reality at Tanami and provide context for understanding how that dilution propagates through tonnes, grade and metal. Elevated tonnage factors and reduced grade factors are a defining characteristic of the operation, reflecting the chosen mining method rather than deficiencies in geological modelling or execution.

Viewed through a metal lens, the reconciliation performance indicates that the operation delivers outcomes that are consistent with plan over the long-term. The close alignment between predicted and realised dilution and ore loss supports confidence in the modifying factors applied during resource-to-reserve conversion and demonstrates that high dilution, in this context, is a controlled and well-understood feature of the mining system.

The R-series methodology is central to revealing these relationships, ensuring that changes in tonnes, grade and metal reflect genuine planning and operational effects rather than artefacts of inconsistent depletion geometry. In combination, the R-series factors provide transparency around the true impact of dilution and ore loss in underground mining and enable reconciliation performance to be evaluated in a manner that is both spatially valid and operationally meaningful. They also provide critical data for determining reserve modifying factors.

CONCLUSIONS AND RECOMMENDATIONS

Underground reconciliation is inherently complex due to selective mining, irregular geometries, and inconsistent spatial definitions of depletion. Conventional reconciliation approaches often lack the spatial consistency required to reliably distinguish between geological uncertainty, modifying factor application, and operational performance.

This paper demonstrates that integrating the F- and R-series reconciliation frameworks within a common volume depletion methodology provides a practical and defensible approach to underground reconciliation. The Tanami case study shows that reconciling all models over an equivalent spatial footprint enables meaningful interpretation of changes in tonnes, grade, and metal, allowing dilution, ore loss, and information effects to be clearly identified.

The R-series factors, in particular, add significant value by revealing the magnitude and consistency of dilution inherent in both the mine plan and operating reality. At Tanami, elevated tonnage factors and reduced grade factors reflect the deliberate application of bulk mining methods to a relatively narrow, high-grade orebody. Long-term metal R-series trends demonstrate that this dilution is well understood, consistently predicted, and broadly realised in practice, supporting confidence in the modifying factors applied during resource-to-reserve conversion.

Recommendations for operators

To realise the full value of F- and R-series reconciliation in underground operations, the following practices are recommended:

- Adopt a common volume depletion methodology to ensure that all reconciliation comparisons are spatially valid and are based upon an equivalent physical footprint.

- Maintain actual spatially coded resource, reserve, design, and grade control models to support meaningful calculation of R-series factors.
- Implement regular and systematic capture of mined voids and development geometry to result in realistic depletion volumes and reduce reconciliation uncertainty.
- Interpret reconciliation results over appropriate and different temporal scales, recognising that short-term variability may reflect mining sequence, dilution behaviour or metallurgical timing effects, rather than model bias.
- Embed reconciliation outcomes into planning and review cycles so that modifying factors, dilution assumptions, and execution performance are routinely tested and calibrated.

Areas for further work

Several areas of further work would strengthen the methodology and support broader industry adoption, namely:

- Automation of common volume generation through integration of routine CMS/drone capture, development surveys, and design solids into repeatable monthly or quarterly workflows.
- Extension of the framework to caving-based mining methods through incorporation of flow-based depletion, drawpoint tracking, and reconciliation logic consistent with R-series principles.
- Development of industry guidance on minimum spatial data standards for reserve, design, and grade control models, aligned with emerging JORC and other reconciliation standards.
- Application of the common volume and R-series framework across multiple underground operations to enable benchmarking of dilution, ore loss, and information effects and to support calibration of modifying factors at a portfolio scale.
- Deeper integration of R-series reconciliation outputs into reserve governance, life-of-mine planning updates, and risk assessment processes.

ACKNOWLEDGEMENTS

The authors would like to express their sincere appreciation to Newmont Corporation for their openness and collaboration in developing this paper. In particular, we acknowledge the support and contributions of Arja Jewbali and Sam Vine, whose insights and cooperation were instrumental in shaping the methodology and case study presented. Their commitment to continuous improvement and industry knowledge sharing has been invaluable in advancing underground reconciliation practices.

REFERENCES

AMIRA, 2007. *Metal Accounting; Code of Practice and Guidelines*, P745, Release 3. AMIRA International, South Africa.

Coles, A J, Hadlow, H and Levy, I, 1993. Reconciliations at the Kundana Gold Mine, Western Australia, Open Pit to Underground: Making the Transition, Perth, Australia, Australian Institute of Geoscientists, Bulletin 14.

Fouet, T, Hargreaves (née Riske), R, Morley, C, Cook, A and Centofanti, J, 2009. Standardising the Reconciliation Factors Required in Governance Reporting, in *Proceedings of the Seventh International Mining Geology Conference*, pp 127–139 (The Australasian Institute of Mining and Metallurgy: Melbourne).

Gover, C G and Assibey-Bonsu, W, 1996. The GORS reconciliation system: a case study, Mining Geostatistics Conference Proceedings, Geostatistical Association of South Africa.

Hargreaves, R and Booth, G W, 2019. Transparency and Standardisation in Metal Reconciliation Reporting, in *Proceedings of the Mining Geology Conference*, pp 36–42 (The Australasian Institute of Mining and Metallurgy: Melbourne).

Hargreaves, R, Elkington, T, Booth, G W and Shaw, W J, 2022. Mine reconciliation standardisation – R Factor Series, in *Proceedings of the International Mining Geology Conference 2022*, pp 366–374 (The Australasian Institute of Mining and Metallurgy: Melbourne).

Helm, S, Hargreaves, R and Morley, C, 2013. Technology drives improvements in mine reconciliation, *The AusIMM Bulletin*, April (The Australasian Institute of Mining and Metallurgy: Melbourne).

Joint Ore Reserves Committee (JORC), 2024. *Draft Australasian Code for Reporting of Exploration Results, Mineral Resources and Ore Reserves.* Section 10 – Reconciliation.

Morley, C and Arvidson, H, 2017. Mine value chain reconciliation - demonstrating value through best practice, in *Proceedings of the Tenth International Mining Geology Conference*, pp 279–292 (The Australasian Institute of Mining and Metallurgy: Melbourne).

Morley, C and Moller, R, 2005. Iron Ore Mine Reconciliation – A Case Study from Sishen Iron Ore Mine, South Africa, in *Proceedings Iron Ore Conference*, pp 311–318 (The Australasian Institute of Mining and Metallurgy: Melbourne).

Morley, C, 2003. Beyond Reconciliation – A Proactive Approach to Using Mining Data, in *Proceedings Fifth Large Open Pit Conference*, pp 185–191 (The Australasian Institute of Mining and Metallurgy: Melbourne).

Newmont Corporation, 2021. Reconciliation Standard, Internal guideline (Document No: NEM-TES-STA-001), Newmont Corporation.

Parker, H M, 2006. Resource and reserve reconciliation procedures for open-pit mines, *AMEC Report*, p 39.

Parker, H M, 2012. Reconciliation principles for the mining industry, *Mining Technology*, 121(3):160–176.

Schofield, N A, 2001. The myth of mine reconciliation, in *Mineral Resource and Ore Reserve Estimation – The AusIMM Guide to Good Practice* (ed: A C Edwards), pp 601–610 (The Australasian Institute of Mining and Metallurgy: Melbourne).

Shaw, W J, 1991. Mining Geology, Grade Control and Reconciliation, unpublished course notes, Perth WA, unpaginated.

Shaw, W, Weeks, A, Khosrowshahi, S and Godoy, M, 2013. Reconciliation – Delivering On Promises, 36th APCOM (Application of Computers and Operations Research in The Mineral Industry).

Sides, E J, 1992. Reconciliation studies and reserve estimation, Case Histories and Methods, in *Mineral Resource Evaluation, AE Annels*, The Geological Society, London, Special Publications, 63:197–218, Torex Gold. Available from: <https://torexgold.com/site/assets/files/7852/torex_gold_-_corporate_presentation_-_august_2021-1.pdf>

Westley, J F H, 1986. Product monitoring, grade control and ore: waste reconciliation at Bougainville Copper Limited, Large Open Pit Mining Conference (The Australasian Institute of Mining and Metallurgy: Melbourne).

Wild, M W, 1998. Reconciliation in the Gold-Mining Industry: Problems and Practice, Geology and Geophysics Department, Centre for Strategic Mineral Deposits and UWA Extension, The University of Western Australia, Publication 29, p 66.

GLOSSARY OF TERMS

Term	Definition
As-Mined (ASM)	The actual material extracted during a reporting period, typically measured via production records, truck counts, CMS scans, or surveyed voids. Used as the denominator in R2c to compare with planned extraction.
As-Planned (ASP)	A general term referring to the intended mining volume or shape prior to execution, often aligned with the Design (DES). Includes all mining blocks/areas targeted for extraction, both ore and planned dilution.
Design (DES)	A short-term stope or extraction shape derived from the grade control model. Represents the planned extent of mining, including planned dilution. Used in R2b and R2c.
Common Volume Methodology	A standardised spatial framework for underground reconciliation that defines a consistent depletion volume across models (resource, reserve, grade control, production) to allow valid comparison.
Depletion Volume	The physical volume of material mined or sterilised during a reporting period, used as the basis for comparing model estimates.
Design Dilution	Waste material included with ore during mining, which lowers overall ore grade. Can be planned (R2b) or unplanned (R2c).
F-Series Factors	A set of standardised reconciliation factors (Parker, 2012) that quantify the accuracy of reserve estimation (F1), mine-to-plant delivery (F2), and reserve-to-plant conversion (F3 = F1 × F2 in most, but not all, cases).

Term	Definition
GCM (Grade Control Model)	A short-term model based on infill drilling and mapping, used for operational ore delineation.
Information Effect	The difference in grade between the resource and grade control models, reflecting additional data collected closer to mining.
JORC Code	The Australasian Code for Reporting of Exploration Results, Mineral Resources and Ore Reserves; proposed updates include emphasis on reconciliation practices.
Material Tracking	The process of monitoring ore movement through the mining and processing systems to maintain data continuity and reconciliation accuracy.
Mill Variance	A site-specific term referring to the difference between the plant feed tonnage (measured at the plant) and the reconciled plant output (after metallurgical balancing). It can result from stockpile changes, belt scale errors, or recovery assumptions. While related, it is not synonymous with the formal F2 reconciliation factor, which compares mine delivery to plant received.
Mine Production (ASM)	Actual or as-mined tonnage and grade, based on truck counts, sampling, or survey data.
Modifying Factors	Adjustments applied to resource models (eg dilution, recovery, geotechnical factors) to convert to reserves or mineable material.
Ore Loss	Material planned as ore that is not recovered during mining, often due to selective extraction or geotechnical constraints.
Plant Feed (PLT)	Material received by the processing plant, recorded via belt scales or weightometers.
Plant Reconciled (PLR)	The final metallurgically balanced amount of material recorded at the plant after processing, often reconciled against head assays and metallurgical recovery.
Planned Dilution	Waste material intentionally included in mine designs to enable safe or practical extraction. In the R-Series framework, this is represented by R2b, which compares the short-term design model (DES) to the grade control model (GCM), capturing the expected dilution resulting from design decisions.
R-Series Factors	An extension of the F-Series (Hargreaves *et al,* 2021) that track how modifying factors are applied from the resource model to production, enabling insight into dilution, ore loss, and information effects.
Reconciliation	The process of comparing estimated and actual performance across the mining value chain, often in terms of tonnes, grade, and metal content, to evaluate accuracy and identify areas for improvement.
Reconciliation Standard/Code	Several industry initiatives, including the AMIRA P754 reconciliation project, have promoted standardised terminology and reporting frameworks for mine reconciliation (AMIRA International, 2007) promoting consistent definitions, measurement points, and transparency in reconciliation reporting.
Reserve Model (RSV)	A planning model incorporating modifying factors, representing the portion of the resource considered economically mineable.
Resource Model (RSC)	A geostatistical estimate of *in situ* mineral content, typically classified as Measured, Indicated, or Inferred.
Stockpile	A temporary storage location for ore or waste material before processing or further transport. Can be Long-Term (ROM) or Short-Term (mill feed).

Term	Definition
STM (Short-Term Model)	A model used for near-term production forecasting or scheduling, often derived from grade control or design data with limited forecast horizon.
Survey Factor	A correction applied to mine production tonnage based on volumetric survey data.
Tanami Operation	A remote underground gold mine in the Northern Territory of Australia, operated by Newmont Corporation; used as a case study in this paper.
Unplanned Dilution	Waste material unintentionally extracted during mining due to factors such as misclassification, poor execution, or blast overbreak. In the R-Series framework, this is measured by R2c, comparing the as-mined production (ASM) to the design model (DES), highlighting discrepancies between planned and actual performance.

Is real time reconciliation just a dream – using representative moisture, multi-element and mass flow measurement on conveyed flows

H Kurth[1] and A Brodie[2]

1. Chief Marketing Officer and Minerals Consultant, Scantech International Pty Ltd, Camden Park SA 5038. Email: h.kurth@scantech.com.au
2. General Manager Marketing, Scantech International Pty Ltd, Camden Park SA 5038. Email: a.brodie@scantech.com.au

ABSTRACT

Mining processes rarely allow a well-defined orebody to be efficiently extracted. Unexpected changes in ore boundaries, complex ore and waste interfaces, blasting underbreak, overbreak, unexpected internal dilution or grade variation, mixing ore and waste during loading, and misallocation of material during haulage can all contribute to tonnage and grade varying from plan. When are these detected: when the mine geologist or grade controller sees unexpected material in an underground drawpoint or blasted pile, when the mine produces fewer or more tonnes than expected from a blast, when the run-of-mine (ROM) stockpile colour doesn't look right, when the process samplers show unexpected grade, or at the end of month metal accounting meeting? Measurement technologies have been successfully applied in many commodities to address these ore quality measurement requirements. Penetrative, continuous microwave-based free moisture analysis allows dry tonnes to be determined at 0.5 per cent moisture precision. High performance prompt gamma neutron activation analysis (PGNAA) provides representative, precise, and timely multi-elemental analysis of ore quality on primary crushed conveyed flows on parcels equivalent to every 30 seconds of flow. High performance camera-based volume and belt speed measurement systems can also utilise particle size distribution (PSD) data to adjust bulk density for accurate mass flow measurement to within one per cent of nucleonic weigh scale performance. These measurements provide unmatched quantification of material on a conveyor underground or on surface. Any variations from expected quality can be identified in minutes and actions taken to prevent waste parcels reporting to the crushed ore stockpile (COS) and being unnecessarily processed. The paper discusses case studies in multiple commodities.

INTRODUCTION

Previous papers on the technologies discussed herein have focused on grade control or process benefits which are well proven and accepted in many commodities in the resources sector. This paper is focused on the value of the data generated from real time measurement technologies for ore reconciliation and metal reconciliation (accounting) applications. Ore reconciliation is a process to define the tonnage, grade and metal content transferred between key operations in the mining value chain, to identify and manage variances with the aim of improving mine planning, grade control and processing.

Digitalisation of conveyed flow quality provides representative measurement with high precisions over short time frames or parcels of flow; usually much smaller than the scale at which tonnes and quality are predicted from orebody block models and mining schedules. This paper looks at the practicalities associated with using such data for real time reconciliation exercises.

Extracting orebodies for processing often involves mining to stable shapes that don't necessarily correspond to orebody boundaries resulting in planned and unplanned dilution and ore losses and overbreak. Reconciling the actual mined material with the expected mined material may be challenging as quantifying the removed material tonnage and quality usually involves assumptions as well as various measurement errors.

These errors are due to:

- Actual ore or grade boundaries being different to those expected, particularly relevant for mining near ore/waste boundaries or structurally complex zones, where drilling and mining exposure data are limited.

- Material being misallocated due to unexpected variations in quality, or poor measurement or subjective visual and surface-only assessment methods.
- Belt scale measurements of mass which can contain significant errors; 10–30+ per cent.
- Determination of dry tonnes due to application of a constant moisture factor; a few per cent, when moisture content varies due to several factors: weather, particle size, material handling, dust management methods.
- Estimation of material composition by sampling or observation, rarely representative.

It seems logical that the best measurements available will provide the most confidence, however, errors such as the examples above need to be considered. The usual approach is to perform a reconciliation process over a longer term to minimise the effect of individual measurement errors. This may be complicated by having multiple ore sources, covering pre- and post-calibration of specific equipment such as belt scales, or having (or not having) rain for part of an accounting period.

The traceability of ore and waste through materials handling and stockpiles can be complicated by segregation occurring on stockpiles, blending and mixing, material misallocation or incorrect placement. A typical Mine to Mill process may involve run-of-mine stockpiles for blending ores, a crusher which supplies a crushed ore stockpile that may have multiple feeders to blend ore into a mill. Sometimes additional intermediate stockpiles are used for multiple crushing stages as buffers to ensure processing can continue when parts of the comminution circuit are under maintenance. There are key measurement locations where the mass and quality can be measured to create a documented mine production history. The measurement points can be simultaneously used as a definitive mill feed quality record for metal accounting purposes and performance evaluation and improvement. Measuring the parameters in real time enables responses in real time so that operational performance can be continuously improved.

This paper focuses on some measurement techniques providing real time data on key parameters: material composition, mass flow and moisture content. The paper discusses applications on operating sites in iron ore, copper and gold using these technologies. Two of these technologies are penetrative and continuous measurement and one is surface measurement, however, the parameters are measured to high confidence levels and successfully used in ore reconciliation and metal accounting and also for grade control and process control (as described in related papers).

TECHNOLOGIES

The penetrative technologies are representative because they allow measurement without bias due to material composition or flow characteristics and measure continuously to provide average analysis over time increments to minimise inherent technology errors, such a variability in received response and effects of mass flow changes. Customised calibrations account for these variables. The technology has been described in a previous International Mining Geology Conference paper (Kurth, 2024) but included here again for convenience. They are accepted process analytical technologies (PATs) in many commodities.

Prompt gamma neutron activation analysis (PGNAA)

PGNAA utilises a radiation source located beneath the conveyed flow which emits a concentrated zone of neutrons into the tunnel section of the analyser. Many elemental nuclei in the material flow capture neutrons and release gamma energy spectral responses unique to each element. Spectral responses are accumulated over a measurement period using an array of high specification detectors enabling the proportion of each element in a conveyed parcel to be determined proportional to the mass flow over that period, corrected for moisture content, and reported (Figure 1). High performance PGNAA is the most representative analysis technique available measuring many elements from carbon onwards in the periodic table at high precisions over short measurement times to low levels of detection. Balzan *et al* (2022) discusses the recently developed capability to directly measure gold in conveyed ore using GEOSCAN-GOLD.

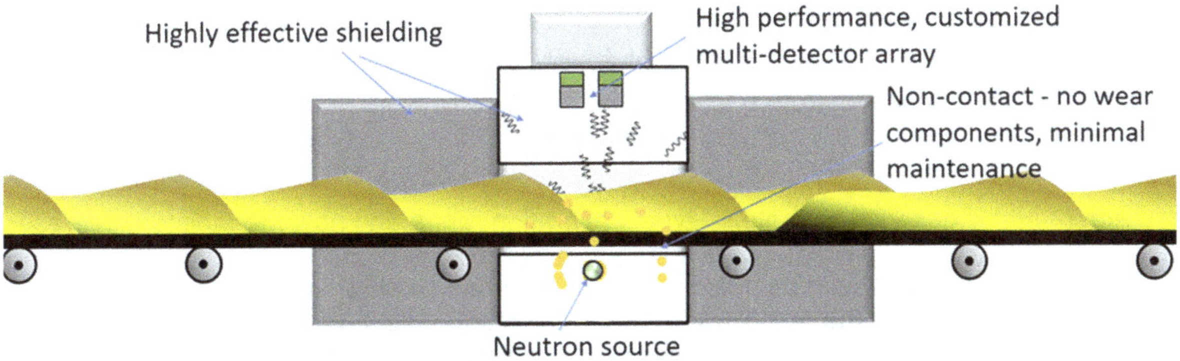

Au ppm	Cu %	S %	Fe %	Cr ppm	Ni ppm
1.28	0.43	0.74	11.64	69	44

FIG 1 – Cross-section through a PGNAA analyser showing main components and an example of elemental results for a typical measurement period (source: Scantech International Pty Ltd).

PGNAA is representative, continuous, real-time measurement of primary conveyed flows for analysis for each 30 sec to two mins of flow for many elements and for 10 mins analysis for average direct measurement of gold and silver content. It is typically utilised on crushed rock to determine composition of run-of-mine (ROM) material flowing from a primary crusher to a crushed ore stockpile (COS). It can also be used in underground locations to measure material to be hoisted. PGNAA has been used successfully for over 20 years in many mineral commodities (iron ore, manganese, copper, gold, zinc, lead, nickel, lithium, phosphate rock, bauxite, diamonds, chromium etc) to measure and control material quality in real time. The technique is unaffected by particle size, belt speed, mineralogy, dust, layering and moisture content.

Microwave transmission moisture measurement

Microwave transmission technology has been used successfully in conveyed material flows since the 1980s in many commodities with many hundreds in operation. Microwaves at frequencies in the high megahertz to low gigahertz range are transmitted from an antenna beneath the conveyor belt and pass through the belt and conveyed material (see Figure 2).

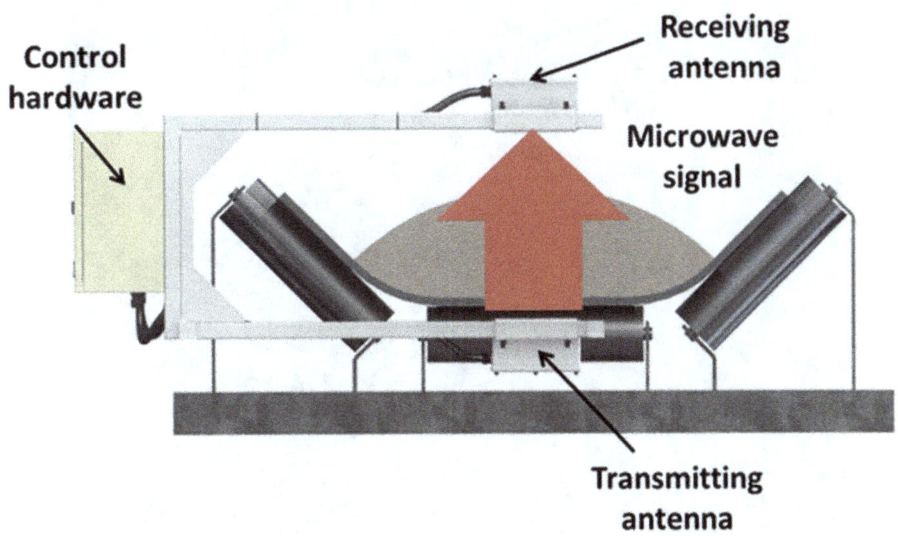

FIG 2 – A schematic section through a microwave transmission moisture analyser showing major components and illustrating the technique utilised (source: Scantech International Pty Ltd).

The microwave field is non-hazardous due to the extremely low power output of 0.18 mW/cm^2. The response related to the water is isolated due to the difference in di-electric constant of water relative to that of the rest of the conveyed material. Attenuation and group delay of the microwaves are measured and used to indicate the proportion of free moisture present. These parameters are correlated with known free moisture content during the calibration phase lasting one to two days.

Calibration performance varies from 0.05 per cent moisture to 1 per cent moisture in installed units but are commonly around 0.5 per cent moisture on primary crushed mill feed material and concentrates.

Measurement precisions vary by material type, calibration range and how the moisture is present in the material. Measurement precision can be affected by the presence of minerals containing loosely bound water in their matrix, such as goethite. Other minerals, such as magnetite, if present at high concentrations may prevent the technique working effectively as magnetite has a similar di-electric constant to water and is also highly magnetic, which detrimentally affects the microwave transmission. Belt scale inputs are needed to perform tonnage weighted moisture measurements for the average of each measurement period, which can be second by second or for periods corresponding with elemental analysis results if integrated with a PGNAA analyser. Standalone systems can report results over any required periods such as a shift or day or specific to a quantity of conveyed material such as a parcel of ore. This helps significantly in determining dry tonnes for an accounting period or batch.

3D infrared camera and proprietary algorithms

This paper discusses a solution for particle size distribution (PSD) that also accurately measures conveyor speed and conveyed volume. It was developed by COREM in Quebec City applying a 3D Infrared (3D IR) camera and advanced imaging algorithm system (see Figure 3) to overcome known performance limitations of typical PSD systems. It has also been successfully adopted for use in foreign object detection in conveyed flows.

FIG 3 – 3D Infrared camera PSD analyser which measures belt speed and volume and calculates tonnes per hour using a bulk density estimate (source: Scantech International Pty Ltd).

The 3D IR system measures belt speed and volume and calculates mass flow using assumed bulk density, with results updated every two seconds. A single calibration is performed, and PSD results compare to the full volume sieve analysis with an RMSE (root mean square error) of approximately 5 per cent. Recalibration is only performed if crushing parameters change. Mass flow correlates well

(R^2 = 0.964, tonnage to within 1 per cent) with a nuclear scale (Faucher *et al*, 2015). The capacity to measure flow volume and belt speed provides a useful verification of mass flow where a bulk density of the material is well known and not highly variable. The PSD results can be used to adjust the bulk density factor used in determining mass flow. It has particular benefit where there are no belt scales available on a conveyed flow feeding a plant or belt scales have a poor calibration. Other volume detection systems require belt speed input data from external sensors to enable volume and mass determination. Integrating these parameters into one system has significant advantages in measurement error reduction and control.

REAL TIME RECONCILIATION?

The abovementioned technologies are all able to provide data useful in conveyed flow tonnage, quality and contained metal reconciliation. However, the capability to do so is limited by the availability of real time expected mass flow and quality data for a given parcel or period of mine production and process feed. The data from the mine plan that is available is based on block model grade data which contains averaged quality for each block (often of thousands of tonnes), when there is much higher quality variability evident in downhole assays from diamond drill core and measured conveyed flows over short intervals, eg 30 seconds data. The minimum reconcilable quantity tends to be a stope, or defined volume extracted over a given period much longer than a measurable interval from any sensor or combination of sensors.

The mining effects also often confound the ability to match expected material quality with measured quality. Tracking ore through the mining and material handling processes (such as depicted in Figures 4 and 5) can be difficult considering:

- Blasting large tonnages can cause mixing across block boundaries and hence the ability to track a known volume of *in situ* rock is complicated.
- Stockpiles at various locations through the process may represent a large proportion of production over a short time frame.
- Shovel loading haul trucks, with subsequent dumping into a crusher and then measuring the conveyed flow cannot easily track material back to a source.
- Trackers are usually reliable for larger volumes where multiple tags are placed in blastholes.
- The scale of mining and material tracking is performed on much larger parcels than the resolution of material quality that is possible to measure using sensors.

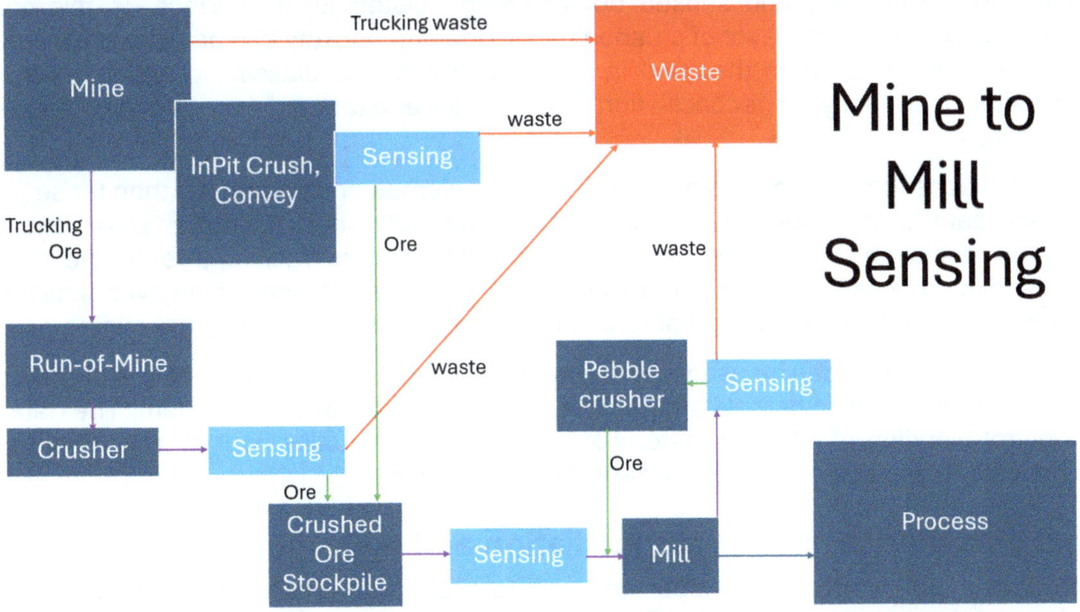

FIG 4 – Ore and waste flows in a typical open cut mine flow sheet and the opportunities to use sensors to measure material quality for ore and waste reconciliation (source: Scantech International Pty Ltd).

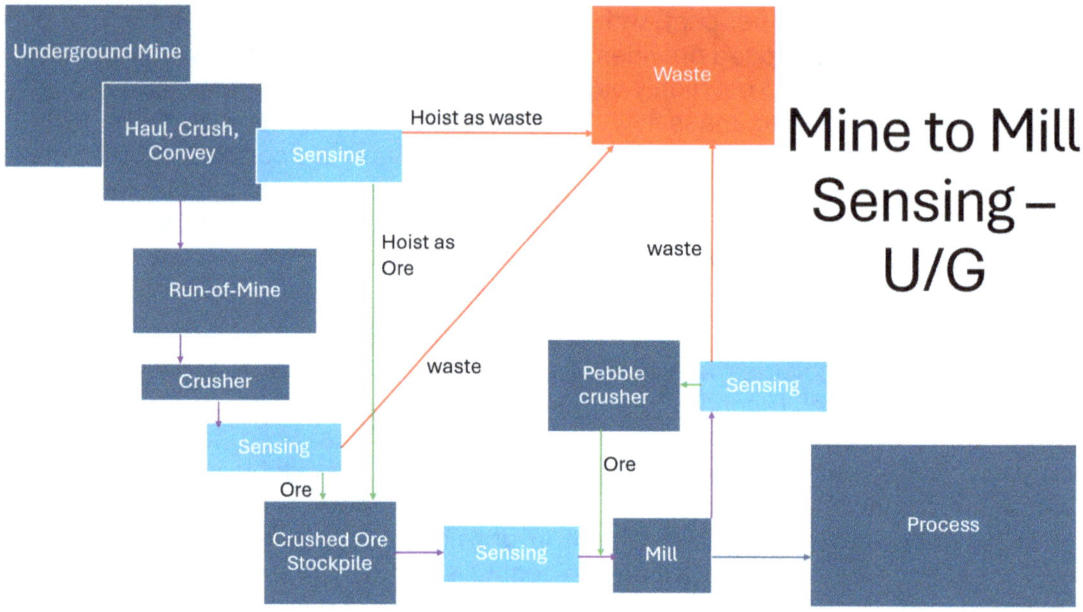

FIG 5 – Ore and waste flows in a typical underground mine flow sheet and the opportunities to use sensors to measure material quality for ore and waste reconciliation (source: Scantech International Pty Ltd).

Trucking material as ore or waste using grade allocation at the mine through normal grade control or superficial surface measurement technologies results in significant misallocation of material, particularly near a planned cut-off grade. Sensing material quality with accuracy does not normally occur until after primary crushing of material classified as ore. That sensing step can allow waste parcels to be removed from the 'ore' flow when combined with a diversion system. This enables an absolute measurement of quality at that point in the flow sheet, enabling confidence in reconciliation using that known data. The ability to measure quality representatively and at high accuracies where in-pit crushing and conveying is utilised results in far better 'automated' grade control. Reconciliation becomes problematic where there are diversions of a proportion of the flow and these are not measured and accounted for. Often these need to be estimated, or an average quality assumed as they cannot be reliably sampled.

The depiction of ore and waste flows in Figure 5 replicates an underground source of material which benefits from being crushed and sensed before being hoisted to the surface. In this case, the placement of sensors on a conveyor of crushed rock to loading pockets can be utilised for measuring the quality being extracted from the mine in a definitive way and classifying ore and waste. This creates a better opportunity for reconciliation of the material extracted from a mined sequence or individual source.

Reconciliation needs to be performed over much larger volumes or time periods than those for which measurement data is available. The question of whether real time reconciliation is possible is a function of the resolution of the 'expected to be mined data' and its tracking through the mining and materials handling process. Sensors can define quality and mass of increments even smaller than a loader shovel and hence the challenge of real time reconciliation falls back to the mining processes.

Multi-sensor analyser data can be used in real time for many advantages, including automated grade control, to divert uneconomic parcels and remove them from the plant feed stream. The data can be used to control blending of different materials to improve feed quality consistency which also has process benefits. blended from multiple run-of-mine stockpiles.

How real time data is used in reconciliation and metal accounting

Real time data is used in iron ore, copper and gold to assist with reconciliation and few sites have expected production data at the same scale as sensor data. Analysers help with reconciliation where they are implemented for bulk ore sorting applications as they record the mass and quality of each measured increment and total of the diverted and undiverted tonnes and average grade over customised time frames. The main applications are between mine and process plant. Real time data is often composited to produce weighted averages of tonnes and grade for direct comparisons.

Iron ore

Assmang Khumani operations in South Africa iron ore perform elemental balance continuously by measuring in up to 23 conveyed flow locations within their operations. The measurement starts at multiple remote mines where mined ore quality is measured at each site. All conveyors utilise belt scales for tonnage measurement. The ore is transferred via overland conveyors between the mines and a distant beneficiation plant, with ore quality measured on three overland belts, and parcels diverted from the flow if already product quality (Matthews and du Toit, 2011). Diverted ore quality and quantity is separately determined and recorded to undiverted material flow. Analysers also measure the feed and product streams to and from each jig and the discard stream quality. Product flows from the jigs are stockpiled and the train loading conveyor has an analyser to determine the flow and quality removed from each stockpile. The combination of measurement locations allows the mass and quality to be continuously determined and balanced to allow reconciliation to occur back to each mine for not only produced mass and quality but also recoveries and discard rates. The analysers on the jig feed conveyors can be used to monitor jig feed performance as each of the numerous ore types characterised by the measured elemental composition also has expected upgrade factors. This operation has the closest to real time reconciliation possible as all key flows are well measured and frequently monitored.

Copper

High variability in copper ore grade at the Sepon operation in Laos resulted in the purchase of an analyser to control ore blending in plant feed to eliminate the likelihood of metal tonnes to a leach circuit exceeding its capacity (Arena and McTiernan, 2011; Balzan *et al,* 2016). High-grade variability within each blasted block from multiple open pits did not allow reconciliation back to the geological model or mine schedule as the analyser measured ore blended from multiple run-of-mine stockpiles. The benefit of measuring the copper was extended to the ability to reconcile the mass and quality in real time using extensive datalogging of all process plant tags and reconciling product and discard quality with plant feed quality. The result was that copper content could be tracked to the nearest tonne across the processing circuit.

Gold

In the last few years, the capability to measure gold directly in ore to as low as 0.2 ppm and to precisions to below 0.15 ppm has enabled gold mines to measure ore and waste quality in plant feed. This has become very useful where non-sulfide gold is present and is difficult to measure representatively through sampling. At a USA (top ten global gold miner) gold operation, sampling of plant feed accounts for about half the contained gold in the ore with plant recoveries typically double that expected from sample assay data collected. The coarse gold is very challenging to quantify and hence the interest in direct measurement of the contained gold using an analyser. They are currently unable to reconcile process feed to the mine model due to the large discrepancies in gold sample and geological model (from drill hole assays) data.

Another operating gold mine in the USA (also a top ten global gold miner) has recently installed a PGNAA analyser and is undergoing calibration at time of writing the paper. Their project involves the use of analysers for reconciliation across multiple mining and processing sites. Currently mined ore and waste (as indicated from geological models) is sampled at high frequency (multiple times per hour) to ascertain gold content. The waste, which contains significant quality variation, is intermittently produced from the mine and stockpiled over the course of each production shift. A decision is made at the end of each shift that determines if that stockpile is trucked away as waste or sent as process feed. A separate ore stockpile is assumed to be high-grade and trucked continuously to the process plant. The site is unable to perform reconciliation with any reliability as ore and waste quality are not determined with high confidence. The analyser will be used to accurately measure the composition of the mine output to enable correct allocation of ore and waste stockpiles (grade control) to maximise ore recovery and minimise dilution and improve mine planning and materials handling strategies.

At the Telfer gold mine in Australia, an analyser is used to measure grade of ore sourced from the underground component of the plant feed to determine the contribution of gold from that source. The balance of the metal content determined by the process team is assigned to the open pit operations. Sampling and assaying underground ore had been unable to provide any confidence in its average quality and direct measurement was found to be the most reliable solution.

Moisture measurement is utilised in conjunction with elemental measurement at each of the mentioned sites so that dry tonnes can be determined for each of the conveyed flows. The dry tonnage previously determined using historical moisture factors proved ineffective as moisture content was more variable than expected. This resulted in expected stockpile quantities being very different to those observed and often led to unrealistic expectations of metal recovery. Additional moisture tonnes assumed to be ore reduced the back-calculated average grade in the ore.

SUMMARY

Effective reconciliation requires measurement of the actual mass and quality of material supplied by the mine to the process operations. Measurements of diverted parcels are tracked so that tonnage weighted averages of feed and reject material are captured. Ore reconciliation is performed to identify variances from planned ore quality, mass and metal content so that improvements can be made to mine planning, materials handling and processing. Representative and precise elemental, moisture and mass flow measurement data is generated by applying proven technologies to primary crushed conveyed flows. It is the most effective way to measure mined material quality once available on a conveyor. Data is provided in real time and used in many sites to improve grade control and resultant process performance. Data is only used periodically to reconcile expected to actual tonnage and grade in most applications due to the batch nature of performance comparisons. These batches represent fixed mining volumes or fixed time periods for which expected tonnes and grade are generated.

Gold measurement allows a previously unattainable level of ore and metal reconciliation to be performed. High performance prompt gamma neutron activation analysis (PGNAA) is proven in many mineral commodities to measure and control material quality in real time. Transmission microwave moisture has proven effective for free moisture determination and 3D IR camera-based technology has allowed mass flows to be accurately measured in the absence of belt scales. Digitalisation of the conveyed flow provides useful data for real time ore quality and feed forward process control and ideal data for tonnage, grade and metal reconciliation.

The biggest challenge for real time ore reconciliation is that sensor data is available at much higher resolution than block model or mine schedule data and therefore, unfortunately, real time reconciliation solutions remain elusive. Mine planning software that accommodates actual mining data should allow comparison with actual measured output from the mine on comparable mass increment for reconciliation to be more effective, assuming the effects of mining and material handling can be determined. Currently reconciliation over a complete mine block or longer mining period appears to be a practical limitation. Significant improvements to mine planning, grade control and processing are expected in future as real time ore reconciliation practices develop.

ACKNOWLEDGEMENTS

The authors wish to acknowledge permission of Scantech International Pty Ltd to publish this paper and acknowledge customers who have published their own papers on the successful implementation of the technology.

REFERENCES

Arena, T and McTiernan, J, 2011. On-belt analysis at Sepon copper operation, in *Proceedings MetPlant 2011*, pp 527–535 (The Australasian Institute of Mining and Metallurgy: Melbourne).

Balzan, L, Jolly, T, Harris, A and Bauk, Z, 2016. Greater use of Geoscan on-belt analysis for process control at Sepon copper operation, in *Proceedings XXVIII International Mineral Processing Congress* (Canadian Institute of Mining, Metallurgy and Petroleum: Quebec).

Balzan, L A, de Paor, A, Doorgapershad, A and Futcher, W, 2022. The end of the rainbow: real time direct gold analysis in run of mine ore at Newcrest's Telfer mine using GEOSCAN analysis, in *Proceedings of International Mineral*

Processing Conference – Asia Pacific 2022, pp 1140–1149 (The Australasian Institute of Mining and Metallurgy: Melbourne).

Faucher, A, Makni, S, Gagnon, G, Lavoie, F and Roberge, E, 2015. Ag Feed particle size distribution measurement of iron ore, in *Proceedings 47th Annual Canadian Mineral Processors Operators Conference,* pp 296–304 (Canadian Institute of Mining, Metallurgy and Petroleum: Montreal).

Kurth, H, 2024. Using representative elemental analysis on conveyed flows to complement grade control: a win-win for geologists and processors, in *Proceedings International Mining Geology Conference 2024,* pp 347–353 (The Australasian Institute of Mining and Metallurgy: Melbourne).

Matthews, D and du Toit, T, 2011. Validation of material stockpiles and roll out for overall elemental balance as observed in the Khumani iron ore mine, South Africa, in *Proceedings Iron Ore Conference*, pp 297–305 (The Australasian Institute of Mining and Metallurgy: Melbourne).

Carrapateena – a journey of reconciliation discovery in sub level cave mining

S Light[1], L Klingberg[2] and B Forster[3]

1. Principal Geologist, BHP Copper South Australia, Adelaide SA 5000.
   Email: shaun.light@bhp.com
2. Lead Geologist, BHP Copper South Australia, Adelaide SA 5000.
   Email: laura.klingberg@bhp.com
3. Geology Superintendent, BHP Copper South Australia, Adelaide SA 5000.
   Email: ben.forster@bhp.com

ABSTRACT

Carrapateena, a breccia-hosted iron oxide copper-gold (IOCG) deposit in South Australia, commenced sub level cave (SLC) mining in 2019. Establishing robust reconciliation processes during the initial years of production was critical for aligning geological models, cave flow predictions, and operational performance.

This case study outlines the development of a systematic end-of-month reconciliation framework integrating material tracking, sensitivity analyses, and consistent terminology across geology and mining teams. Challenges presented include discrepancies in tonnes and grade reconciliation, particularly gold overperformance linked to unmodelled structural features, and variance in tonnage attributed to stockpile density assumptions and weightometer calibration.

Investigations using radio frequency identification (RFID) tracking and calibration reviews improved understanding of material flow and measurement accuracy. Results demonstrate that reconciliation at Carrapateena requires balancing precision with practicality, emphasizing trend analysis over short-term anomalies. The approach has enhanced transparency in performance metrics and informed continuous improvement in estimation and cave management practices. This case study provides insights into reconciliation complexities in SLC operations and highlights strategies for managing uncertainty in high-variability underground mining environments.

INTRODUCTION

Carrapateena is a multicommodity copper (Cu), gold (Au) and silver (Ag) mine located in the highly prospective Gawler Craton in South Australia, approximately 160 km north of Port Augusta (Figure 1). This case study is aimed at outlining development of an end of month reconciliation process associated with the first two years of mine production for Carrapateena.

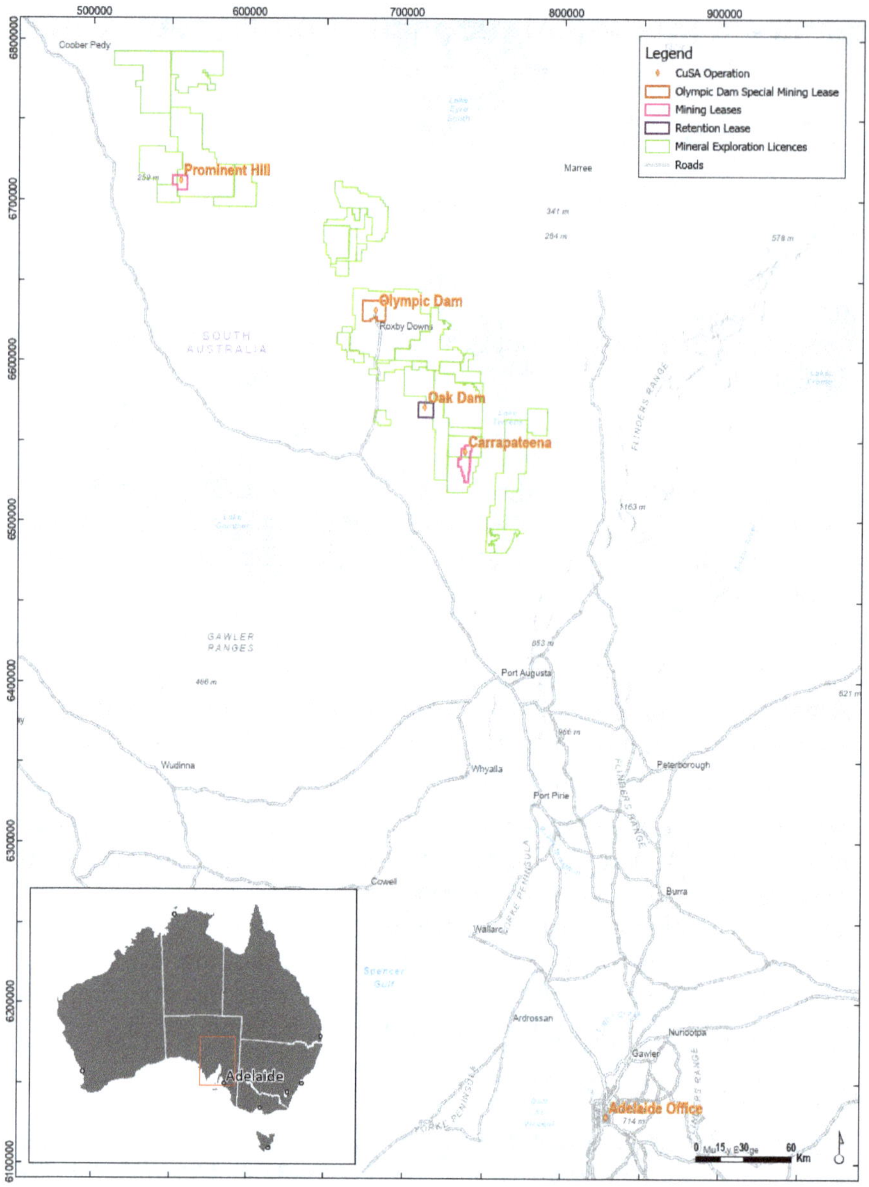

FIG 1 – Carrapateena location map.

Geology

Carrapateena is an iron oxide copper gold (IOCG) deposit. Mineralisation is hosted within the Carrapateena Breccia Complex (CBC), which is interpreted to be Mesoproterozoic in age. The CBC is a polymictic hematite-granite breccia with hematite-sericite-chlorite-carbonate alteration hosting disseminated Cu sulfides. The CBC is surrounded laterally by altered Palaeoproterozoic Donington suite granitoid (Figure 2) and is overlain above an unconformity by Neoproterozoic Stuart Shelf sediments. Mafic and felsic dykes interpreted to be Mesoproterozoic in age cross-cut the surrounding granite and to some extent cross-cut the outer parts of the CBC.

Copper sulfide mineralisation within Carrapateena consists of zones of bornite and chalcopyrite. Pyrite generally occurs with chalcopyrite in the lower grade areas of the deposit. Copper, gold and silver are the revenue metals.

Within the basement, at the basement-cover unconformity contact, there is a barren hematite zone that can range from 10–40 m thick. Underneath this zone is a gold enriched cap with no copper sulfides, ranging from less than one metre up to 10 m thick. This zone sits directly above the CBC copper bearing domains; it is not observed in association with unmineralised domains.

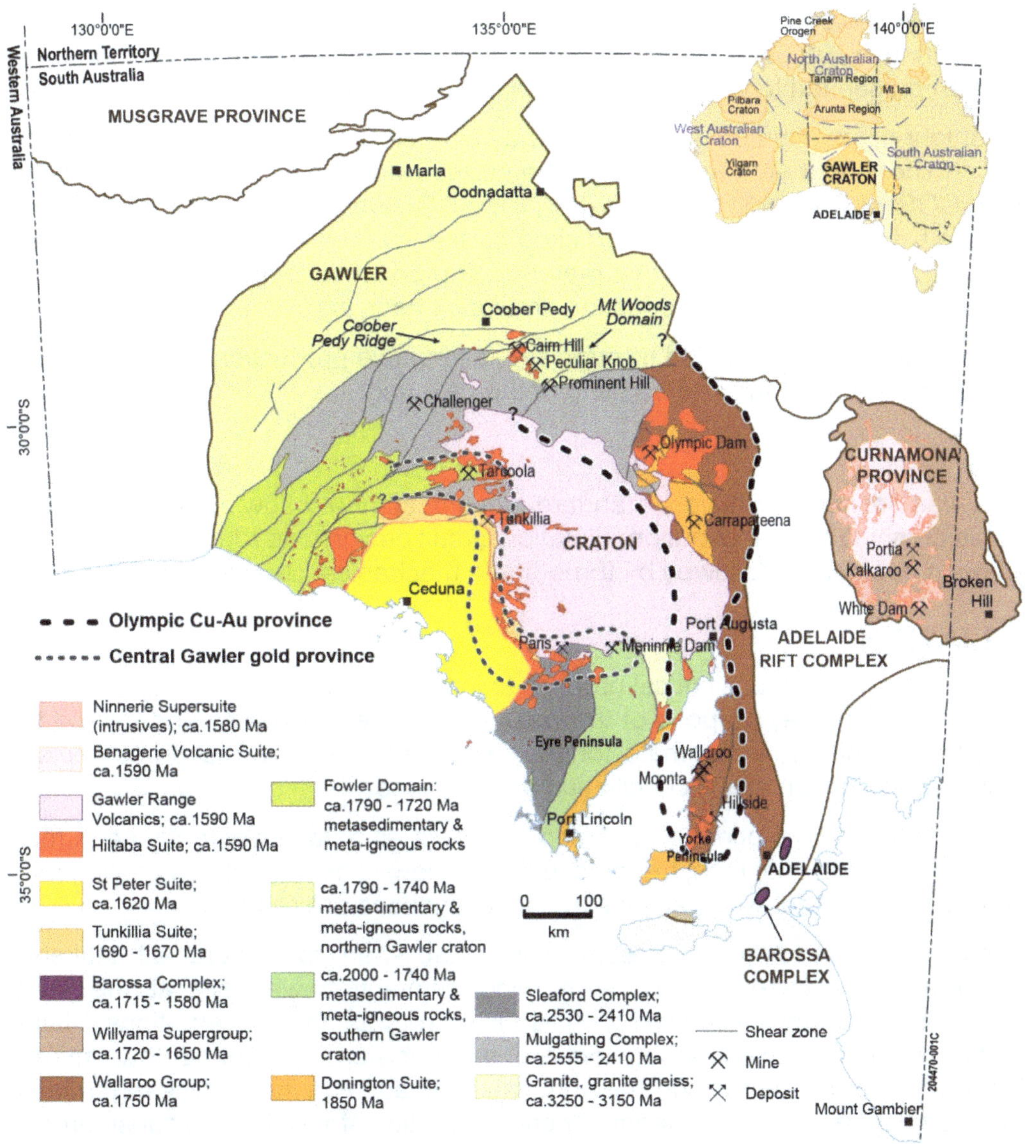

FIG 2 – Regional geology map showing Carrapateena deposit location (Reid, Wade and Jagodzinski, 2022).

Discovery

Carrapateena was discovered in 2005 by RMG Services which was followed by a surface diamond sub-vertical drill program to a 200 × 200 m drill pattern. Infill drilling to 100 x100 m spacing in areas of interest was completed as part of a joint venture with Teck Cominco (Light, 2025). OZ Minerals purchased the deposit in 2011 and followed up with angled and wedged surface holes through the highest grade zones of the deposit. OZ Minerals made the decision to commence early access with the construction of a decline in 2016 (Light, Neumann and Ehrig, 2025). In 2017 OZ Minerals announced board approval for the development of an SLC mine, aiming to produce 4.25 Mt/a (OZ Minerals, 2017). First concentrate was produced in late 2019 from development ore (OZ Minerals, 2019). Underground (UG) drilling with sub-horizontal holes commenced in late 2019 when decline and capital infrastructure UG development was sufficiently established in the competent basement granite.

Mine

Carrapateena is a sub level cave mine with a starting depth 500 m below the surface and an initial SLC planned mine depth of 4,585 m RL (OZ Minerals, 2017). The planned SLC had a level spacing of 25 m for the top half of the mine and 30 m in the bottom section with an underground crusher station at the mid-point (Hocking *et al,* 2018).

The top three levels of the sub level cave operated fixed draw of bogging 40 per cent of fired tonnes on first production level, 60 per cent on second and 90 per cent on third production level (Hronsky *et al,* 2020), then proceeded to variable draw for all following lower levels.

During the first four years of production, the majority of underground ore material was trucked to the underground crusher and then transported to the surface via conveyor (MHS) to either the Coarse Ore Stockpile (COS) or Rehandle Stockpile (RH) if the COS was at capacity. Before the crusher was commissioned all development ore was trucked to the surface and stored on a Pre-production Stockpile. Ore passes were used to move material off production levels to lower levels to allow for truck loading to be decoupled from production drawpoint bogging.

Production from the SLC is conducted through single longhole production ring firing and bogging to the design tonnes and then firing and bogging of the next ring.

Processing

Carrapateena was designed to have a standard industry standard flow sheet, with a SAG and ball mill with pebble crushing for comminution. Flotation circuit design was rougher flotation, concentrate regrind, Jameson cell cleaner followed by three-stage mechanical cleaners (OZ Minerals, 2017).

RECONCILIATION

End of month reconciliation was a critical process to implement and keep consistent in the set-up of the geology function. The key purpose of the reconciliation set-up was to understand the ability of the different geology models to estimate tonnes, grade and metal to the process plant and take actions to improve estimation and reconciliation performance.

The designed reconciliation process was aimed to collect all data consistently each month so that comparisons can be made over time, trends could be reviewed on monthly and rolling three-monthly bases, with a clear system for when actions and investigations are required.

Consistent naming for each data stream was established to ensure all stakeholders and reconciliation customers were aligned. A naming convention was set at the start, remains unchanged, and is consistently listed with definitions on reports and presentations. Material 'mined' is an example of aligning the language where the decision was made that 'mined' material classification is only applied once it exits the portal and unloaded on the surface regardless of surface location. Whether it lands on COS, rehandle or pre-production stockpiles it is considered mined. The second crucial point was to define the handover point for data between geology and processing; this was decided to be after the COS and before the mill. Material is delivered to the ball mill from the COS via a conveyor belt with a weightometer determining tonnes, grade and metal allocated through metal accounting systems from the processing team. It was also established by the site that all reconciliation would present data as dry metric tonnes.

This case study is looking at the geology and mining material delivery and reconciliation to the COS only. Therefore, it is assumed that Carrapateena Metallurgy processes are implicitly accurate. This is not the case and reconciliation is a joint process between Mining and Processing as the systems are interconnected and will impact on each other.

The initial step for defining the reconciliation process was mapping out the material flow for the mining method (Figure 3).

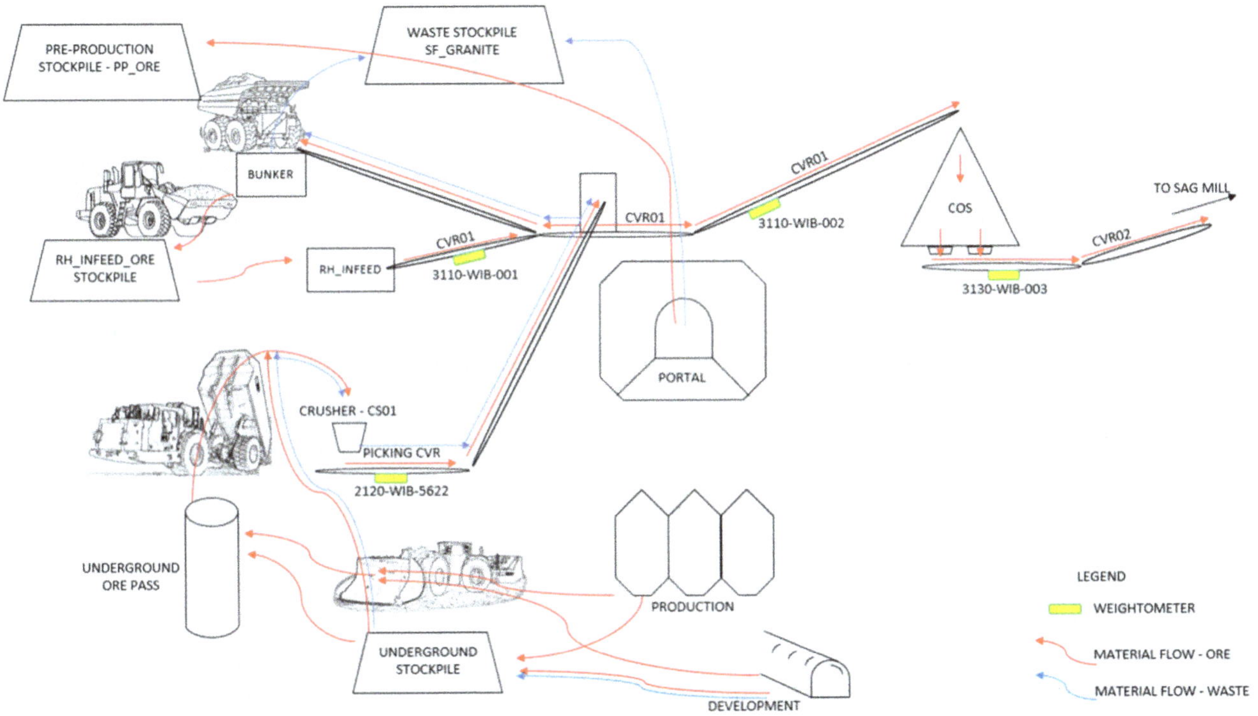

FIG 3 – Carrapateena material flow with weightometer positions.

Claimed ore

Claimed ore is the geology estimation of what has been mined utilising Datamine MineMarket software. Tracking of daily material movements occur in two stages: tonnes tracked for loader and truck movements through Micromine PITRAM through radio call ups from underground operators; and MHS Conveyor tonnes recorded from weightometers into a PI system (process control system) then manually transferred by control room operators into PITRAM. All tonnage movements are then imported from PITRAM via an automated macro process on a shift basis into the geology MineMarket material management software.

Grades are allocated to all primary source locations from the mid-term engineering schedule grades on a monthly basis for the month ahead. Development is calculated as all development for a drive planned for the month, evaluated through the latest grade control block model. Production tonnes (defined as tonnes from the cave) are assigned grades to each planned ring, from a cave flow engineer modelled evaluation, completed on a monthly basis for the month ahead. This is done with Power Geotechnical Cellular Automata (PGCA) particle-based flow modelling software.

Surface stockpiles

COS grade is calculated at EOM using a live and dead COS logic (Zamorano, 2006). The logic assumed for this time period was 30 per cent of the COS volume was live and 70 per cent was dead. The COS tonnes were tracked and reviewed at daily scale. If the COS remained above the total dead COS tonnes for the month, the dead COS grade remained consistent month to month. If the dead COS dropped below the total dead COS tonnes, the previous EOM dead COS grades were combined with tonnes added to the COS when rebuilding to the new dead COS level grades to set a new dead COS grade. The live COS grade was calculated as an average of the last three days COS feed of the month. The expected life on the COS is in the order of 12 hrs but can vary depending on shape and mill throughput; three days was used to provide a less noisy data point and smooth over any errors that may have a bigger impact if using only a single shift.

All surface stockpiles have a survey scan completed using a high-tech drone at or near end of month date. This volume is then used to calculate the final stockpile tonnes:

$$\begin{aligned} Stockpile\ Tonnes \\ &= Survey\ Volume\ \times stockpile\ density\ factor\ \times stockpile\ moisture\ factor \\ &+ \Delta tonnes\ from\ EOM\ to\ stockpile\ scan\ time \end{aligned}$$

Calculated plant feed

Calculated plant feed is not the daily MineMarket number as this does not use live and dead COS values. It is calculated at EOM using the following calculation:

$$Calculated\ Plant\ Feed = Claimed\ Ore + \Delta COS + \Delta\ Surface\ Stockpile$$

Underground stockpiles

Underground stockpiles are treated in MineMarket logic as 'last in, first out', orepasses are treated with 'first in, first out' logic, surface stockpiles are not reclaimed in specific order so these have a rolling weighted-average grade. tonnes for all underground stockpiles are visually audited on the last shift of the month and any obvious errors investigated and rectified as required.

The term 'underground stockpiles' at Carrapateena includes lateral development stockpiles, crusher stockpiles and orepasses. These stockpiles are not factored into any calculations for reconciliation as the data is of poor accuracy, introducing noise and error into the process. The numbers are recorded at the end of each month so are available to understand how they have changed and can be investigated to help explain differences.

Balanced actual ore

The decision was made that any attempt to back allocate material to primary sources would imply a level of precision that is not suitable for the mining type and material tracking systems available at Carrapateena. It was instead decided to only calculate a 'Balanced Actual Ore' to material out of the portal (Figure 4). This is calculated as a simple calculation:

$$Balanced\ Actual\ Ore = Plant\ Feed\ Balanced\ Actual + \Delta COS + \Delta\ Surface\ Stockpile$$

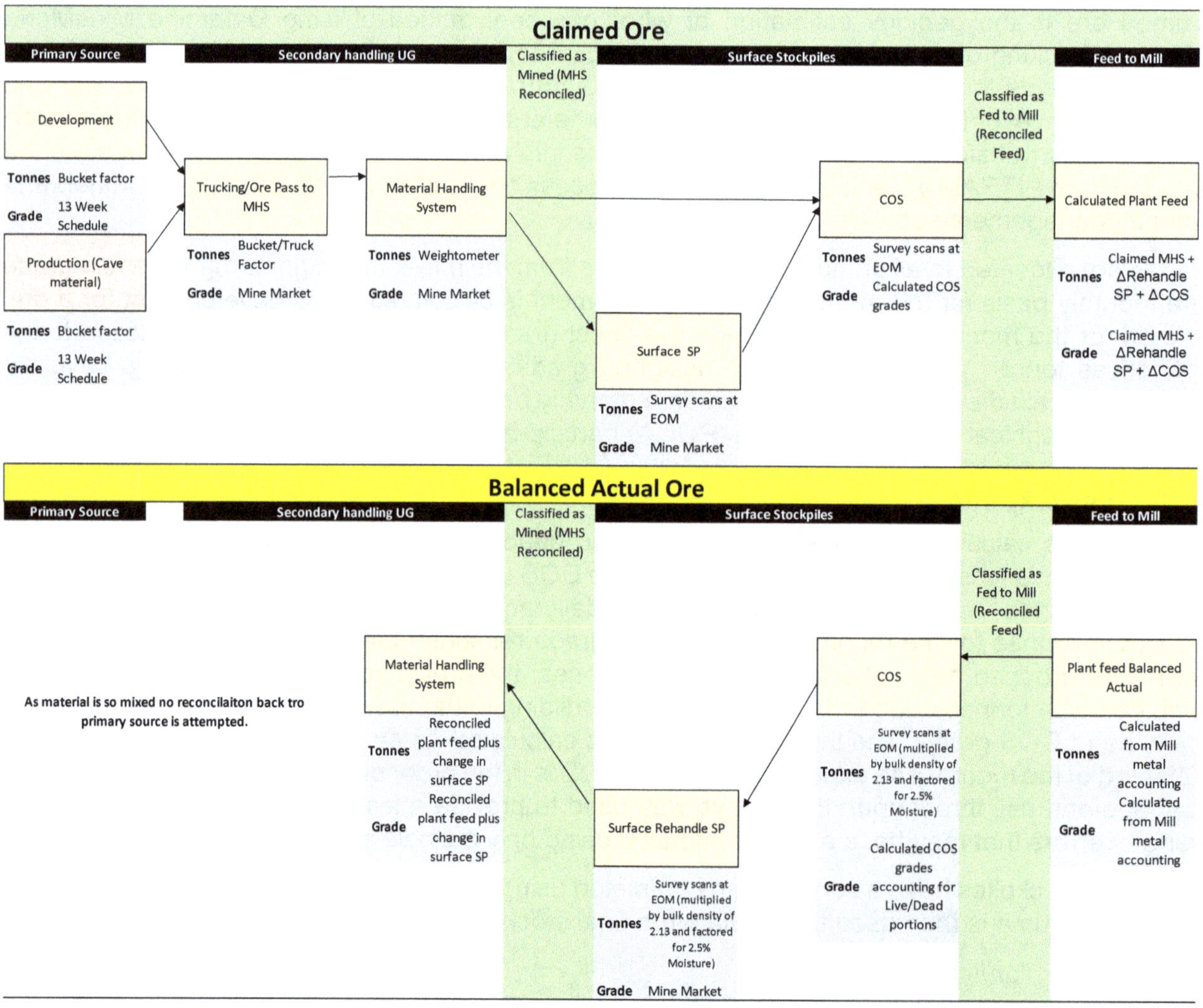

FIG 4 – Source and flow data for claimed and balanced actual.

Geology model mined

To calculate tonnes and grade that the geology model estimates is difficult in an SLC mine as there is no opportunity for survey scans or pick-ups to provide volumes of production voids. Ore development is calculated every month by cutting survey pickup triangulations at face position for start of month and end of month then running through the latest Grade Control Block Model (GCBM) and Resource Block Model (Figure 5).

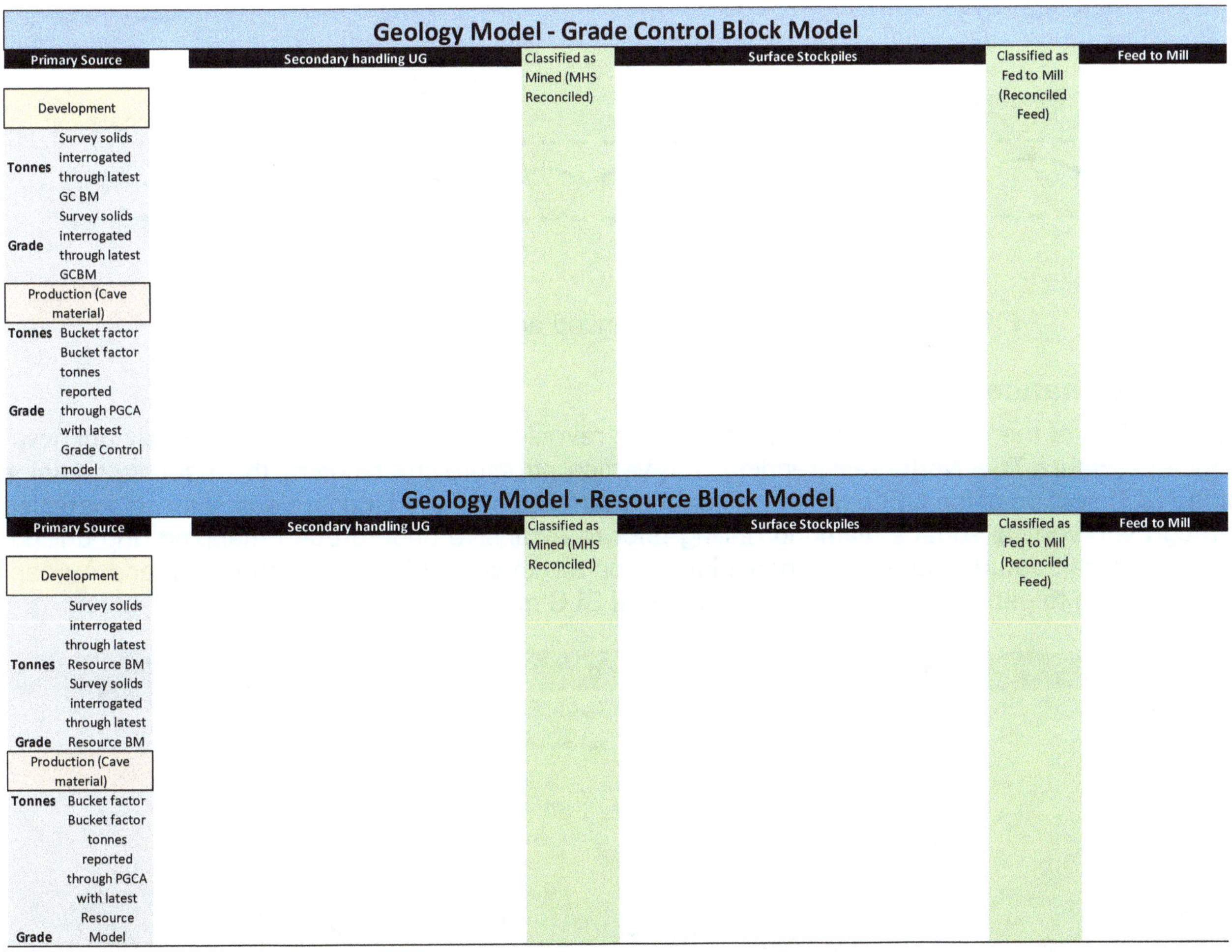

FIG 5 – Source and flow data for the geology model.

For cave production the loader bucket count reported by operators is reprocessed through PGCA to calculate an updated grade for the actual tonnes bogged from each ring throughout the month.

These development and production tonnes and grades are added together to make the Geology Model Mined.

The point in the mining process that the geology model mined numbers are is bogged from primary source not ex-portal, so underground stockpiles and orepasses that needs to be considered when comparing data.

3MRP (3 month rolling plan) forecast

Collation of the monthly forecast for tonnes, grade and metal is recorded and presented with end of month numbers, to allow for comparison with Compliance To Plan maps. This helps to frame the discussion of where variations have occurred.

RESULTS

At the end of each month the initial review is Claimed versus Balanced Actual to understand the monthly performance. Figure 6 shows tonnes, copper grade, gold grade, copper metal and gold metal. Between geology and management a decision was made to set the performance metrics as

±10 per cent based on a conditional simulation study, past experience from similar mining methods, and expectations of the broader business. There is a pattern of Au overperformance with three clear peaks decreasing with time. The other trend of note is the tonnes overperformance in 2021.

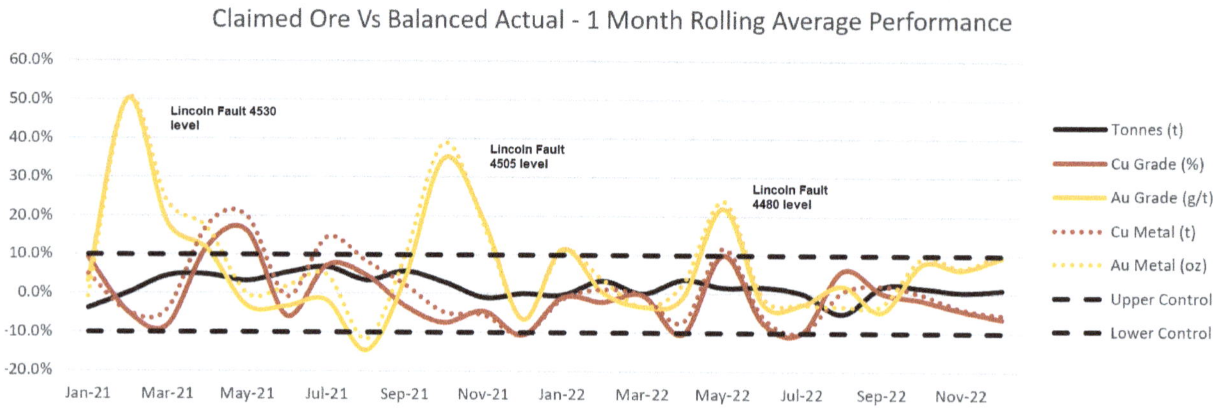

FIG 6 – Claimed ore versus balanced actual for 2021 and 2022.

Au performance

Investigation of the Au reconciliation performance revealed the presence of a previously unknown structure (Figure 7). A North-east trending, sub-vertical structure cross-cutting the main ore domains of the SLC section of the CBC, which was subsequently named the Lincoln Fault, was not identified through sub-vertical surface diamond drilling prior to extensive underground mapping and drilling. This structural feature had a topographic impact on the overlying barren leached cap and Au cap, causing them to penetrate deeper into the planned SLC mine area.

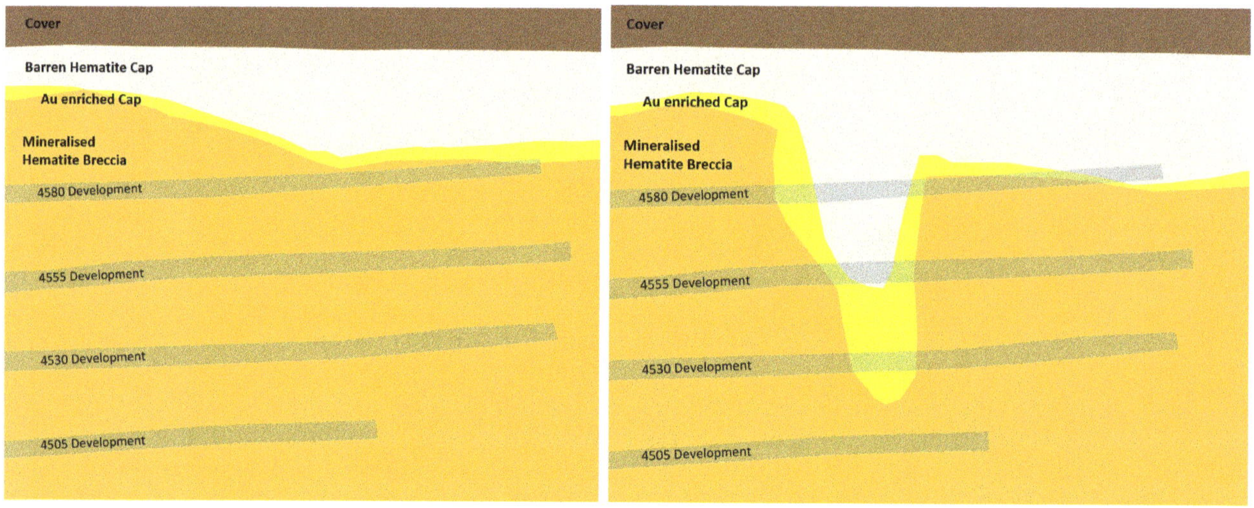

FIG 7 – Left shows original interpretation with surface only data, right image shows interpretation with wall mapping and limited sub-horizontal drilling for these levels.

This information was refined as each level mined through the zone and provided more detailed data and information. The operation at the time made the decision to accept the poor reconciliation and collect the information as mining occurred as it was in the middle of the SLC footprint and would not change the size of the cave. The first three levels were fixed draw and updated wireframes from development data available before production for any potential overdraw decisions from the wall mapping process.

Tonnes

The tonnes discrepancy in 2021 was of concern as there was 8 months in a row with an average of +5 per cent tonnes reconciliation. It was expected the tonnes would reconcile better as this was in essence a comparison between a weightometer on the underground conveyor and the weightometer on the COS to mill conveyor. The only other source of error is surveyed stockpile volume scans or the constant density being applied.

A review of rehandle stockpile density was conducted by completing scans before and after single movement periods and comparing to weightometer data. This data was noisy but did not indicate a clear issue with stockpile density.

A review was also completed on several months of data, tracking what the stockpile density and moisture needed to be to align Claimed with Balanced Actual. Table 1 is an example of a month looking at the change in stockpile volume and the required density and moisture needed to balance the tonnes. This example of 3.82 SG is above the average *in situ* density of 3.4 so is considered unrealistic. It is acknowledged that the density of the SP varies but the changes are generally small so are not a large source of error.

TABLE 1

Example sensitivity analysis modifying stockpile density and moisture content to calculate the Balanced Actual Ore. Moisture of 2.5 per cent and Broken Density of 2.13. A Broken Density of 3.82 for the SP volume change would account for the tonnes difference highlighted by the orange cell aligning with the Claimed Ore.

Claimed Ore (t)	371,402
Balanced Actual Mill Feed (t)	432,071
Surveyed Volume Change (m^3)	16,270

		Moisture %				
		1.0	**1.5**	**2.0**	**2.5**	**3.0**
Broken Density	**1.93**	400 984	401 141	401 298	401 455	401 612
	2.13	397 762	397 936	398 109	**398 282**	398 456
	2.25	395 830	396 013	396 196	396 379	396 562
	2.33	394 541	394 731	394 920	395 110	395 299
	3.00	383 749	383 993	384 237	384 481	384 725
	3.82	370 469	370 780	371 091	371 402	371 713

A review of weightometer calibration and set-up was completed by an external subject matter expert (SME). This review highlighted upgrades to the weightometer style, design and positioning, as well as improvements that could be made to the method of calibration for the weightometers. In September and October 2021 calibrations were undertaken by the external SME using their recommended upgrades, on the two main weightometers utilised in reconciliation process; this aligns with improvements in tonnes reconciliation in Figure 8. This problem of inconsistency and error observed in weighing devices is not unique to Carrapateena and is outlined as a source of error in Parker (2012).

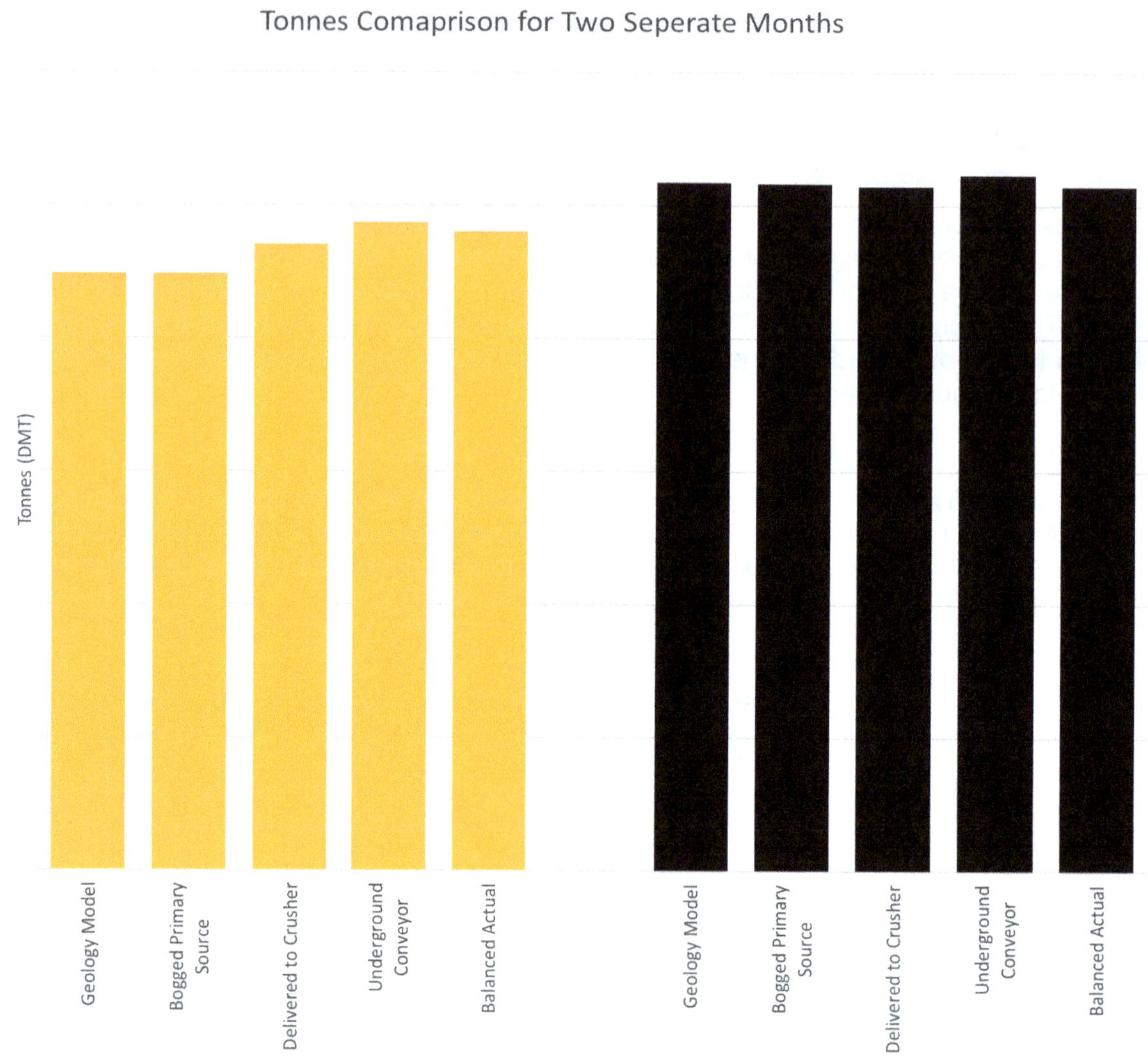

FIG 8 – Chart showing tonnes from different measurements accounting for change in stockpiles through the data flow, orange = month 1, black = month 2. Geology Model, Dev = Survey Voids Prod = Bucket Factor. Bogged primary = Bucket Factor, Delivered to Crusher = bucket/truck factor + Δ underground stockpiles. Underground conveyor = weightometer on underground conveyor.

Balanced actual ore versus geology model

A comparison of Balanced Actual Ore (exit portal) versus Geology Model (primary source) is considered difficult as they are measured at different points. A secondary source of error exists as there are no surveys of voids to run through the model for a comparison of *in situ* tonnes. Instead, the cave flow model relies on tonnes derived from bucket factors, known to be unreliable and variable. The decision was made for the Geology Model comparisons to be reviewed on one-month and rolling three-month windows, using the one-month data to aid in locating potential sources of variation.

To better understand the potential variables and influence in timing between the Balanced Actual Ore and Geology Model, a project was completed by the Carrapateena geology team using RFID tags. These tags were distributed underground across a representative population of primary source locations. These were then recorded by antenna on the mine to surface conveyor after the underground crusher. The data from this project shows that there is variable residence time in the system: of 179 tags distributed, 124 tags were recovered, 35 were within the first 24 hrs, 72 returned between one and seven days and 17 took longer than seven days. This variability is aligned with expectations for how material handling is observed, with development generally having greater variance than routine production. It also supports the view that the underground stockpile residence time is not material when presenting data over a rolling three-month window.

Bucket factors and tonnes drawn are key inputs to understanding the model performance, so consistent tonnes tracking was established to compare all the data sources available. To track performance of this data the geology team compile data each month, tracking sources of tonnes measurements for comparison. An example for two months is shown in Figure 8. This chart highlights the issue with bogged primary sources under-calling tonnes compared to Underground Conveyor and Balanced Actual for month 1. This resulted in an increase in the bucket factor utilised in month 2 and a more aligned and consistent data set for ore tonnes.

CONCLUSIONS

The aim of this case study is to present how Carrapateena approached the requirements to consistently measure a range of performance metrics at the end of each month, allowing for comparisons to understand performance and implement actions when performance is not aligned with expectations. This paper shows that an important step for Carrapateena has been understanding the accuracy and precision of the different measurement points and being able to communicate them to other stakeholders.

ACKNOWLEDGEMENTS

The authors would like to acknowledge the current and past members of the Carrapateena Geology team who all contribute to the reconciliation process with their efforts every day. We would also like to thank Kathy Ehrig for her support and review.

REFERENCES

Hocking, R, Balog, G, Ormerod, T and Pearce, H, 2018. Early cave management at the Carrapateena sublevel cave, in *Proceedings of the Fourth International Symposium on Block and Sublevel Caving – Caving 2018* (eds: Y Potvin and J Jakubec), pp 421–432 (Australian Centre for Geomechanics: Perth).

Hronsky, D, Mariager, S, Donaldson, S, Bobbermen, L and Eustace, C, 2020. Simulation modelling of Carrapateena Sub Level Cave production level performance, in *Proceedings of the Eighth International Conference & Exhibition on Mass Mining – MassMin 2020* (eds: R Castro, F Báez and K Suzuki), pp 658–667 (University of Chile, Santiago). https://doi.org/10.36487/ACG_repo/2063_44

Light, S, 2025. Carrapateena – Evolution of data collection and storage from exploration to operating Sub Level Cave, paper presented to South Australian Exploration and Mining Conference (SAEMC), Adelaide, 28 November 2025. Available from: <https://saemc.com.au/wp-content/uploads/2026/02/1605-Shaun-Light.pdf>

Light, S, Neumann, M, and Ehrig, K, 2025. Carrapateena IOCG Deposit: discovery to operating Sub Level Cave mine and beyond, paper presented to SEG 2025 Conference, Brisbane, 26–29 September 2025.

OZ Minerals, 2017. Development green light for OZ minerals Carrapateena copper gold mine. Available from: <https://minedocs.com/17/ozminerals_08242017_ASX_Release_Carrapateena_Feasibility_Study_Update.pdf>

OZ Minerals, 2019. Carrapateena produces first copper concentrate; operational ramp up outlined. Available from: <https://announcements.asx.com.au/asxpdf/20191220/pdf/44crw1zp78vvy9.pdf>

Parker, H M, 2012. Reconciliation principles for the mining industry, *Mining Technology*, 121(3):160–176.

Reid, A, Wade, C and Jagodzinski, E, 2022. Mafic dykes of the southeastern Gawler Craton: ca 1564 Ma magmatism with an enriched mantle source, *Australian Journal of Earth Sciences*, 69:1–22.

Zamorano, S, 2006. Coarse ore stockpiles, *Bulk Solids Handling*, 26(4):246–251.

Trying to fix what isn't broken – the value of mine chain reconciliation

W Randa[1], J Moore[2], W Vigour-Brown[3] and L Farmer[4]

1. Principal Geologist – Resources, Oceanagold, South Brisbane Qld 4101. Email: wesly.randa@oceanagold.com
2. Head of Resource Development, Oceanagold, Dunedin 9016, NZ. Email: jonathan.moore@oceanagold.com
3. Superintendent – Mine Geology, Oceanagold, Waihi 3610, NZ. Email: william.vigour-brown@oceanagold.com
4. Senior Geologist – Resource Development, Oceanagold, Kershaw SC 29067, USA. Email: lucian.farmer@oceanagold.com

ABSTRACT

Well-managed mine reconciliation processes provide the foundation of resource and reserve governance, as well as facilitating problem solving and continuous improvement. However, due to the complexity of the reconciliation chain, isolating the root cause(s) during periods of poor reconciliation can be challenging. As a consequence, failure to deliver to budget expectations is often conflated with poor resource model performance, despite resource performance being only one of several contributors to delivering on budget expectations. Permitting, mine planning, compliance to plan, resource model performance, ore extraction performance (dilution and ore loss), stockpile management, and mill throughput and recovery, all impact mine plan outcomes.

In response to these challenges, OceanaGold has developed an integrated mine chain reconciliation system to assist each of its operations in tracking production performance across the mine chain from: compliance to mine plan –> resource/reserve model performance –> extraction efficiency –> net stockpile transaction –> Mine to Mill feed –> received at mill –> metal produced/recovery. The system is based on performance factors documented by the late Harry Parker (Parker, 2011) and uses waterfall charts to summarise performance across the mine chain. Interactive dashboards and tables allow a visual and more detailed technical analysis.

The mine chain reconciliation (MCR) analysis approach has proven effective at identifying the link (or links) in the mining chain that are contributing to poor performance, avoiding misdirected, time-consuming investigations, and preventing reactive management, particularly to short-term reconciliation disparities that reflect normal operational variability rather than underlying model performance issues.

Initially piloted as an Excel-based process, OceanaGold's MCR has now been migrated to an interactive SharePoint platform accessible to site personnel, technical centres, and senior management. Establishing and maintaining good reconciliation processes, reporting processes, and communications has resulted in improved teamwork, focus, and cultural change across the business. The recent introduction of structured quarterly reconciliation meetings which are formally minuted, has increased transparency and strengthened internal controls processes. Although OceanaGold's operations comprise both open pit (OP) and underground (UG) mines, this paper mainly concentrates on UG reconciliation processes. OP reconciliation methodologies are already well established in industry practice and comprehensively documented across numerous publications (ie Morley and Arvidson, 2017).

This paper presents operational case studies using the OceanaGold Corporation's (OceanaGold) reconciliation process to identify some common reconciliation challenges. This paper is not intended as a comprehensive overview of mine reconciliation.

INTRODUCTION

Effective and timely reconciliation is a critical operational and corporate function that tracks and compares predicted versus actual outcomes across key geological, mine planning, mining execution, and processing domains, each which contributes to the overall operational performance.

OceanaGold's reconciliation process is sequential and begins with tracking mine compliance to plan, then resource model performance, mining extraction efficiency, Mine to Mill variance and mill throughput and recovery. Each link in this intricate chain carries intrinsic and potentially human errors.

Beyond technical validations, reconciliation plays a pivotal role in strategic decision-making. It informs budgeting and reforecasting and supports long-term planning by providing a closed-loop system of performance measurement. Advanced operations are increasingly integrating reconciliation data into a single framework (rather than in isolation) to enable predictive analytics and near real-time optimisation. As such, mining company performance can be assessed, corrected and continuously improved (Fouet *et al*, 2009; Macfarlane, 2013; Hargraves and Morley, 2014).

In practice, achieving budgeted mining targets can be hindered by any number of factors, including geological complexity not captured in the model, unexpected geotechnical/geometallurgical issues, operational disruptions, deviations from mine plans, and processing constraints such as plant downtime or sub-optimal throughput or recovery. However, when variances between budget and actual metal production occur, resource model performance is often the first suspect.

In response to this, OceanaGold has developed an integrated Mine Chain Reconciliation (MCR) system that tracks key metrics at along the value chain. This includes:

- Compliance to plan – comparing actual mining against the budget mine plan, both spatially and in terms of total tonnes and grade mined.
- Resource/Reserve model performance (F1) – comparing the resource model with mining modifying factors applied (referred to as the Reserve Model) against the grade control within detailed mine designs (GCD).
- Extraction performance (F2) – comparing GCD versus actual material extracted (CMS), identifying unplanned dilution or ore loss.
- Net stockpile transactions – tracking additions, withdrawals, and inventory changes across long- and short-term stockpiles.
- Mine-to-mill reconciliation (F3) – the resultant ore delivered to the mill, which may include ROM and stockpile transactions, is then tested against mill weightometer tonnages and back-calculated grades, based upon mass balancing including in-circuit inventory, recovered metal and rejected tailings. This can be challenging for operations will multiple, confluent mill feed sources.
- Processing recovery performance – comparing budget against actual recovery (both absolute recovery and recovery models based on actual plant feed).

OceanaGold's MCR metrics are based on the late Harry Parker's reconciliation performance factors. OceanaGold track performance across the full mining chain as waterfall charts, which offer clearer visibility than summary tables with lists of discrete values (Table 1).

TABLE 1

Example of summary table of MCR.

Mine chain	Tonnes	Au g/t	Au Oz	Delta
Budget RSV	906 674	2.44	71 195	
RSV Mined	960 562	2.25	69 629	-1566
GCD	1 022 882	2.46	80 930	11 301
Trucked	1 038 753	2.39	79 963	-967
Trucked to Stockpile	182 969	0.66	3907	-3907
Trucked to ROM	855 784	2.76	76 056	
Budget Rehandle	655 825	1.51	31 910	
Actual Rehandle	359 921	1.61	18 666	-13 244
Delta ROM	76 209	5.03	12 324	
Mill versus Mine Feed	1 291 912	2.58	107 046	
Received at Mill	1 317 888	2.66	112 612	5566
Budget Recovery			84%	
Actual Recovery			88%	4416
Budget Gold Produced			86 836	
Actual Gold Produced			99 259	12 423

The waterfall chart example in Figure 1 visualises the cumulative impact (from Table 1) of each step on final metal production, supported by interactive dashboards and drill-down tables for further investigation when required. The system allows users to monitor performance for each step of the mine value chain, avoiding the diversion of scarce operational resources to misdirected investigations and reducing reactive management in response to short-term variability.

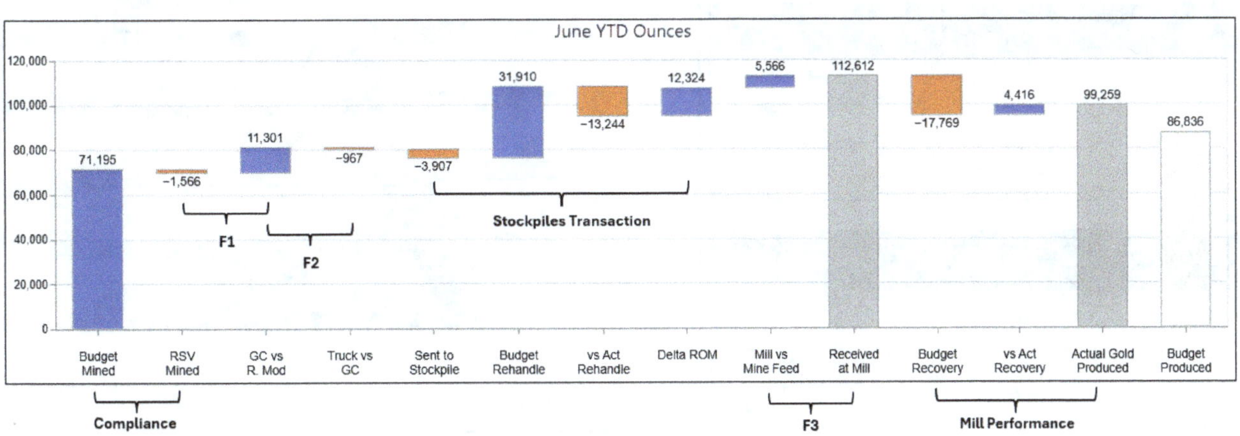

FIG 1 – OceanaGold's MCR waterfall chart, nesting the Compliance to plan, F1 to F3 metrics, and mill performance.

OceanaGold internally recommends three months as being the minimum performance period for investigation, in order to discourage reactive management. Discerning signal from noise is fundamental to rigorous reconciliation analysis, as it enables practitioners to determine whether variances are substantive and meaningful (signal) or short-term variation (noise) (Schofield, Moore and Carswell, 2012).

This paper presents case studies demonstrating applications of MCR at two OceanaGold operations. The analyses demonstrate the use of the reconciliation logic to isolate root-causes, including production shortfalls (Waihi in 2024) and complex multisource scheduling (Haile in 2025).

The strength of MCR is not as a remote, desk-top problem-solving tool, but more importantly as a means for operational personnel to isolate and improve areas of sub-optimal performance.

MCR has driven a significant cultural transformation across OceanaGold and improved communication by standardising terminology, creating a shared understanding ('speaking the same language'), and improving transparency through regular quarterly reconciliation meetings.

Compliance to plan

In compliance-to-plan assessments, OceanaGold benchmarks actual reserve extraction against the budget rather than the forecast. This approach reflects OceanaGold's position that the budget serves as the primary reference for market guidance and therefore represents the most appropriate basis for variance analysis.

The compliance-to-plan assessment is critical because even a highly accurate resource model cannot deliver budgeted metal if the mined ore tonnes and grades are not extracted in alignment with the mine plan. It is important to note that the compliance-to-plan metric applied in this context is quantitative rather than spatial. As highlighted by Morley and Arvidson (2017), comparisons based solely on tonnes and grade do not capture the spatial or temporal dimensions of mining performance. OceanaGold's MCR focus is typically on the isolation of compliance to plan issues by comparing actual pit floor, stope and development surveys (Figure 2) against the budget surfaces and solids. More detailed spatial and temporal non-compliance is generally left for the site teams to resolve given their deeper understanding of the site realities.

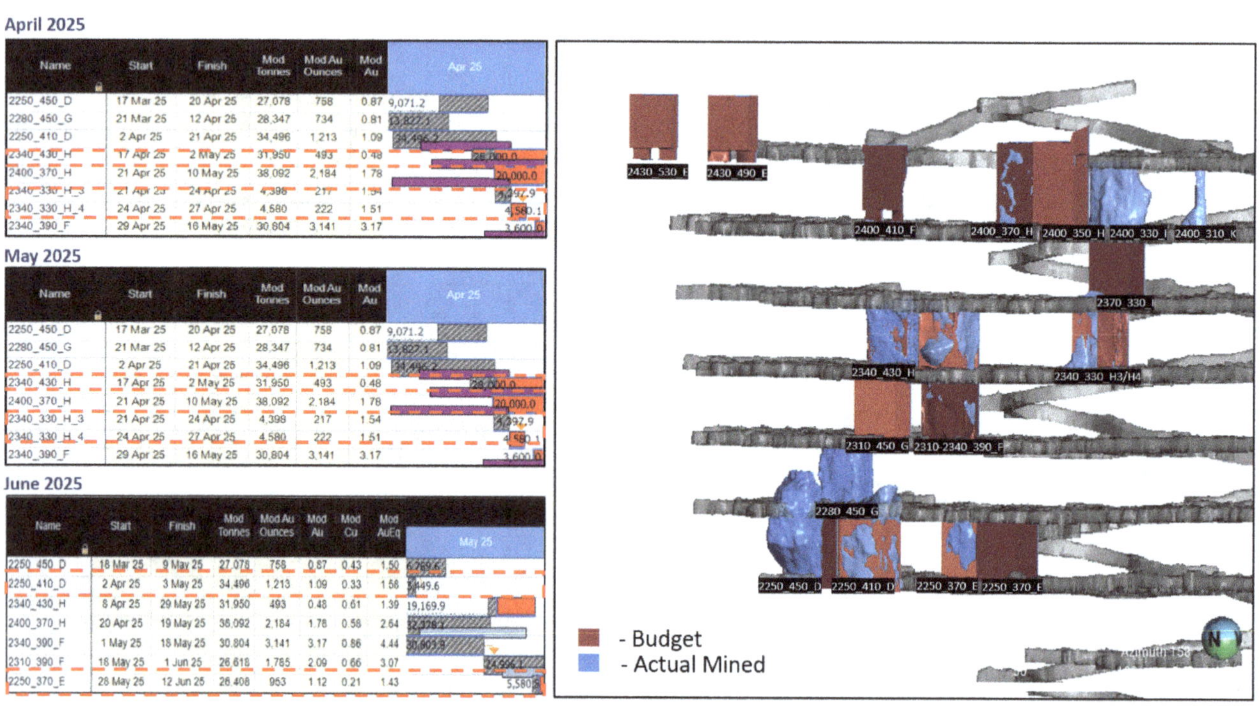

FIG 2 – Example of compliance to plan reconciliation review.

Common causes of non-compliance in mining operations could be any of a number of factors: Underperformance of the reserve model due to poor drilling coverage or inappropriate geological, sub-optimal estimation domaining or poor parameters choices. Unrealistic mine plans or designs. Geotechnical constraints such as wall failures or unexpected ground conditions. High rainfall or other adverse weather events. Unplanned maintenance, poor availability or labour availability.

Short-term Key Performance Indicators (KPIs) can result in reactive decision-making and erode value. For example, a short-range planning may prioritise easier and higher-grade areas to meet the month's target, at the expense of Life-of-mine (LOM) sequencing. This short-term gain can erode long-term value due to misalignment between the mine plan and mining actuals.

F1 – Reserve (RSV) model versus grade control design (GCD)

The F1 reconciliation metric compares the resource model with budget mine plan assumptions applied (RSV model) versus GCD. It is calculated as follows:

- RSV – Resource model Measured and Indicated blocks that fall within the specified budget 3D solids mined (RSV mined), then adjusted by applying any mining assumptions.
- GCD – GC model blocks that fall within the specified GCD 3D solids, then adjusted by applying mining assumptions (if required).

Resource model performance (F1) is measured as the ratio of tonnes, grade and metal between GCD/RSV (Table 2).

TABLE 2

Example of F1 reconciliation for an underground mine.

	Month			3 Month Trailing			Year to Date		
	Tonnes	**Au g/t**	**Au Oz**	**Tonnes**	**Au g/t**	**Au Oz**	**Tonnes**	**Au g/t**	**Au Oz**
Reserve Model	65 356	1.96	4123	169 203	2.37	12 906	310 258	2.76	27 573
Grade Control Design	54 874	2.04	3604	154 138	2.32	11 504	296 766	2.89	27 553
UF1	**0.84**	**1.04**	**0.87**	**0.91**	**0.98**	**0.89**	**0.96**	**1.05**	**1.00**

The 3D volume discrepancy between RSV 3D solids versus GCD 3D solids is typically small when there is a 12–18 month advance window of GC drilling (Figure 3). However, unexpected changes to the mine plan can result in incomplete grade control drilling coverage.

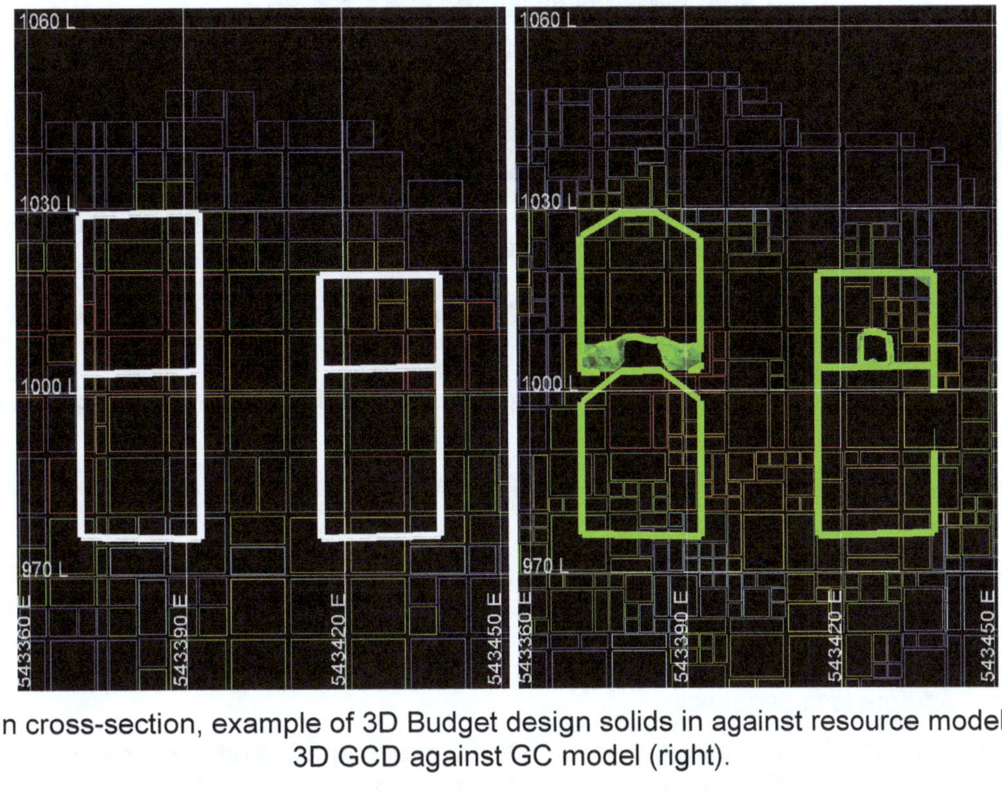

FIG 3 – In cross-section, example of 3D Budget design solids in against resource model (left) and 3D GCD against GC model (right).

A common source of F1 metric discrepancies is the differing data coverages supporting the respective resource and grade control estimates: resource models are typically based on wider-spaced drilling, whereas GC models rely on closely spaced data collected closer to the time of mining and are therefore expected to provide more accurate local estimates (Journel and Huijbregts, 1978). The F1 reconciliation metric quantifies the medium to long-term performance of the resource model.

F2 – Grade control design (GCD) versus cavity measurement system (CMS)

In general, the mine plan includes mining extraction assumptions such us overbreak, underbreak, dilution etc (planned). Therefore, the F2 metric tracks mining performance relative to these extraction assumptions (unplanned).

By reconciling grade control design (GCD) versus actual extraction (CMS), both visual and numerical evaluation of the extraction efficacy are possible (Figure 4), keeping in mind that the mine plan may include additional mining allowances, typically applied as line item factors to the mine schedule. Spatial analysis of ore loss and dilution sources enables continuous improvement; targeted training, improved blast control, and optimised equipment operation, translating into higher extraction recovery and reduced waste movement.

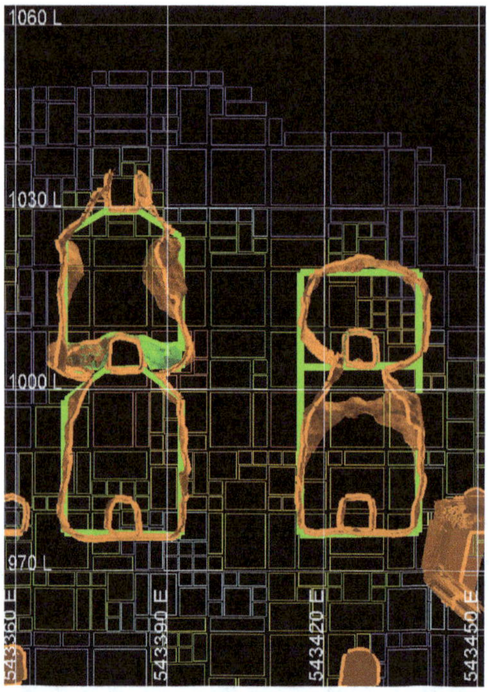

FIG 4 – In cross-section, an example of GCD solids (green) and CMS scan (orange) against GC model.

The F2 reconciliation metric compares actual extraction against expectation and is calculated as follows:

- GCD – GC model blocks that fall within the specified GCD 3D solids, adjusted by any additional mining assumptions.
- Trucked – GC model blocks that fall within the actual 3D CMS volume. The CMS captures actual ore loss and dilution.

CMS scans are not always available at the time of reconciliation, particularly when mining spans multiple months. In these situations, truck counts are typically used to approximate stope/development depletion for periods between CMS surveys. It is important that efforts are made to 'stitch' successive CMS volumes together carefully, to accurately reflect actual overbreak, recovered ore and ore loss sequencing.

Extraction performance (F2) is measured as the ratio of tonnes, grade and metal between CMS/GCD and interpreted as either unplanned dilution or unplanned ore loss (Table 3).

TABLE 3
Example of F2 reconciliation for an underground mine.

	Month			3 Month Trailing			Year to Date		
	Tonnes	**Au g/t**	**Au Oz**	**Tonnes**	**Au g/t**	**Au Oz**	**Tonnes**	**Au g/t**	**Au Oz**
Grade Control Design	54 874	2.04	3604	154 138	2.32	11 504	296 766	2.89	27 553
Trucked	54 981	1.97	3474	158 266	2.25	11 467	314 218	2.83	28 599
UF2	**1.00**	**0.97**	**0.96**	**1.03**	**0.97**	**1.00**	**1.06**	**0.98**	**1.04**

Stockpiles type

OceanaGold's stockpile inventory comprises both long-term and short-term stockpiles. The long-term stockpiles, typically composed of lower-grade material, contribute a significant proportion in Life-of-Mine (LOM) plan. These long-term stockpiles often carry an inherent uncertainty in their reconciled tonnes (assumed bulk density) and grade, adding variability to the mill feed stream. Given their potentially significant contribution as a feed source, any non-compliance has the potential to manifest later in the mine plan (Figure 5). The use of non-spatial grade modelling (ie using numerical average grades rather than 3D stockpile block models) can be problematic.

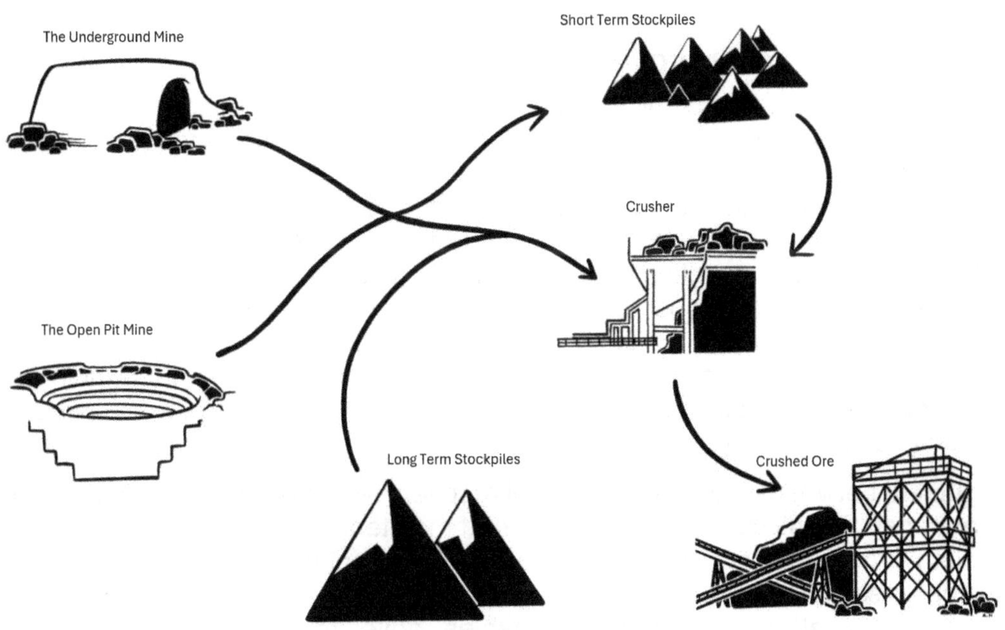

FIG 5 – Common Mine to Mill material movement.

Stockpile variability feeds directly into Mine to Mill reconciliation, as uncertainties in stockpile propagate through reclaim, rehandle, and blending processes. When multiple sources are fed, stockpile discrepancies can compound mine claim versus mill-received (ie the F3 reconciliation metric).

F3 – Mine to Mill versus mill reconciled

Material tracking from the Mine to Mill is not always straightforward. Errors associated with fleet dispatch and tracking, truck factors (partial loading, carry-back and spillage), ROM/stockpiles (management, surveying, accounting and ore blending) as well as mill weightometer calibration and moisture estimates can all compound.

The total Mine to Mill feed is estimated by aggregating the OP feed, net stockpile/ROM transactions, and the UG feed. The mill reconciled is based on end of month (EOM) mass-balancing, including in-circuit inventory, recovered metal and rejected tailings summarised.

The ratio of tonnes, grade and metal of mill reconciled/total Mine to Mill feed defines the F3 metric (Table 4).

TABLE 4

Example F3 reconciliation.

	Month			3 Month Trailing			Year to Date		
	Tonnes	**Au g/t**	**Au Oz**	**Tonnes**	**Au g/t**	**Au Oz**	**Tonnes**	**Au g/t**	**Au Oz**
Open Pit	5775	0.69	129	132 859	1.34	5703	582 669	2.58	48 344
Rehandle	154 280	1.34	6671	349 228	1.62	18 237	514 202	1.53	25 337
Underground	38 007	1.84	2250	140 461	2.40	10 858	316 897	2.95	30 091
Change in ROM	-14 996	3.29	-1586	-50 595	3.48	-5667	-91 205	4.74	-13 910
Total Mine to Mill Feed	213 058	1.55	10 636	673 143	1.87	40 464	1 504 972	2.43	117 683
Mill Reconciled	212 716	1.70	11 626	677 640	2.14	46 531	1 530 604	2.52	124 239
F3	**1.00**	**1.10**	**1.09**	**1.01**	**1.14**	**1.15**	**1.02**	**1.04**	**1.06**

Budget versus actual mill recovery

Budgeted gold recovery represents the planned metallurgical performance of a processing plant based on metallurgical test work, historical performance and metallurgical modelling. Actual recovery is derived from the proportion of gold that is recovered compared to the total contained gold fed to the mill. Differences between the two are expected because forecasts are based upon a limited number of test samples, and metallurgical models, out of necessity, are simplifications of reality, including assumptions about liberation grind size, reagent consumption, as well as flotation and leach kinetics. Operational reality sees ore variability – such as shifts in grade, hardness, sulfide content, clay levels, or refractory minerals. The gap between budget and actual performance can be further widened by operational issues like fluctuating throughput, suboptimal grinding or flotation, equipment downtime, and process-control limitations, along with sampling errors and poor mine-to-mil reconciliation (Dominy, 2018).

Improving alignment between budgeted and actual recovery requires adequate data support, diligent modelling, plant operating discipline and well-planned maintenance. Enhanced ore characterisation through geometallurgical testing, domain modelling, and real-time sensing allows metallurgical models to be optimised. Strengthened metallurgical surveillance, using frequent sampling, online analysers, and statistical process control, enables agile responses to real-time variability. Maintaining target grind size, optimising reagent regimes, stabilising throughput, and applying robust maintenance practices help sustain recovery. Clear communication and improved reconciliation between geology, mining and metallurgy ensure that feed quality changes are identified and communicated early, allowing proactive management rather than reactive correction (Hunt and Berry, 2017).

Table 5 presents a comparison of budgeted versus actual gold recovery for the year-to-date (YTD) period ending July 2025. The higher actual gold recovery primarily reflects the increased head grade delivered to the plant relative to budget (there is a positive correlation between recovery and gold grade). Although the actual recovery is higher than budgeted, it may be that actual recovery performance sits on the budget recovery/grade curve. Checking for this may validate (or invalidate) the recovery grade model.

TABLE 5

YTD July budget versus actual mined.

	Budget	Actual
Tonnes (Ore)	1 842 371	1 530 604
Au (g/t)	2.07	2.52
Au (Oz)	122 479	124 239
Au Recovery (%)	84%	87%
Gold Produced (Oz)	103 267	108 255

Signal versus noise

The distinction between *signal* and *noise* is fundamental to rigorous reconciliation processes and analysis. Distinguishing between signal and noise enables practitioners to determine whether deviations between planned and actual performance reflect substantive geological or operational issues or merely inherent system variability. Meaningful signal cannot be detected until large volumes of mineralisation have been mined, typically more than 3 months mining and may manifest as persistent tonnage, grade and/or contained gold disparities at any point along the reconciliation chain. Typical causes are sampling or locational biases in resource drilling (Moore *et al*, 2023), unresolved geological complexity, sub-optimal estimation, suboptimal mining extraction (ore loss/dilution) and/or mill weightometer biases. Conversely, noise comprises random, short-term fluctuations arising from natural grade variability, operational delays, blast performance heterogeneity, or plant-level recovery variation, none of which necessarily signal structural problems.

Effective reconciliation therefore requires applying appropriate statistical and spatial analysis techniques to filter out noise and isolate true signals that warrant operational or strategic intervention, a principle reinforced across Mine to Mill case studies demonstrating how misinterpretation of noise can obscure underlying performance issues (Schofield, Moore and Carswell, 2012).

Case study – Waihi, 2024

The Martha underground mine in Waihi includes the mining of areas that were previously mined. Mining in remnant areas is very challenging due to difficulties in accurately predicting mined out void volumes, the extent to which these were backfilled, the grade of the backfill and the geotechnical conditions in remnant areas.

By May 2024, the OceanaGold Waihi operation had fallen behind the YTD budget expectations, and the shortfall was initially attributed to poor resource-model performance (Table 6). At that stage, OceanaGold's MCR system was not yet fully implemented across all sites, limiting the ability to isolate discrepancies and determine their root-causes.

TABLE 6

YTD May 2024, Waihi budget versus actual.

	Budget	Actual
Tonnes (Ore)	216 378	233 752
Au (g/t)	4.67	2.81
Au (Oz)	32 488	21 118
Au Recovery (%)	94%	91%
Gold Produced (Oz)	30 466	19 330

As the investigation progressed, OceanaGold began using the reconciliation system to examine specific points along the value chain where variations in tonnes, grade, and ounces were occurring.

As illustrated in Figure 6, the waterfall charts reveal a significant reduction in tonnes mined relative to budget, at slightly lower grade (point a). In contrast, the GCD shows a substantial increase in tonnes compared with the reserve model mined, but significantly lower grade (point b). The waterfall charts also indicate that additional tonnes were extracted during mining, again at lower grade (point c).

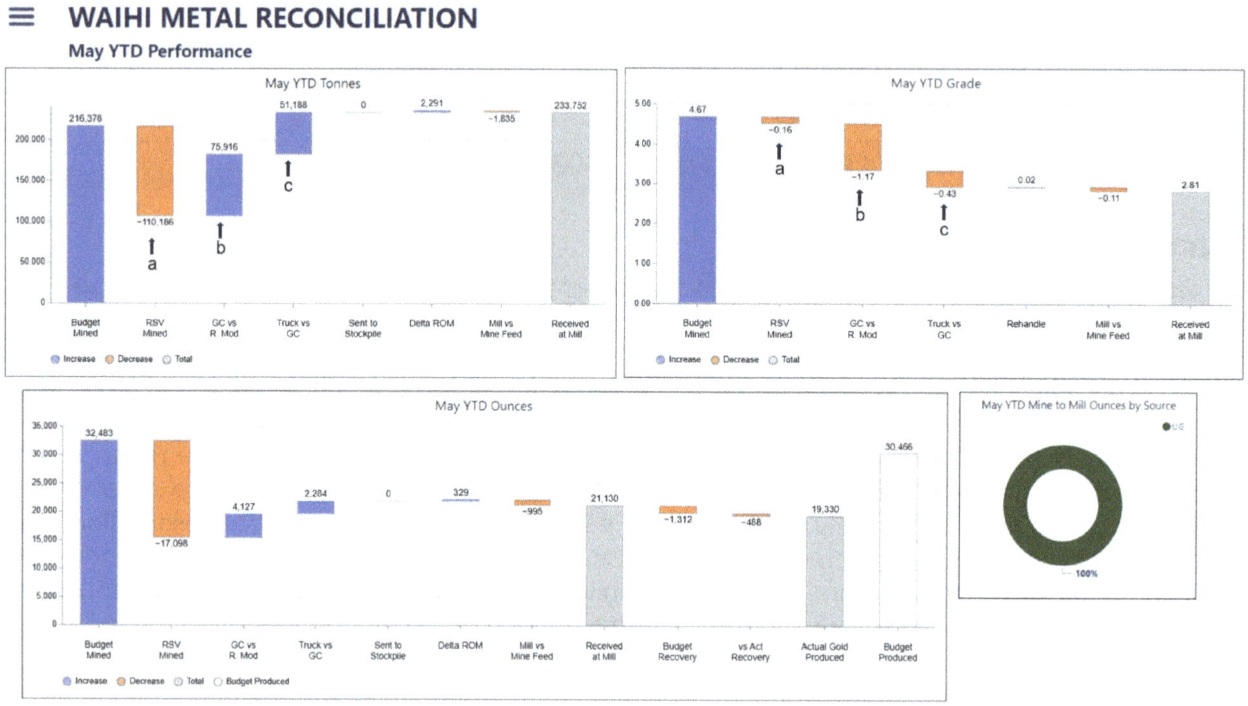

FIG 6 – Full chain mine reconciliation for Waihi YTD May 2024.

Comparisons of actual mining versus the budget mine plan showed that mining performance fell short of expectations due to a number of factors but primarily scheduled spatial compliance to plan. Actual mining outcomes were further impacted by additional low-grade remnant backfill being included in the GCD but not in the budget designs. Additionally, sub-economic overbreak, due to challenging blasting control during extraction, and underbreak, where high-grade skins within GCD were not extracted, resulted in higher tonnages and lower grades than budgeted (Figure 7).

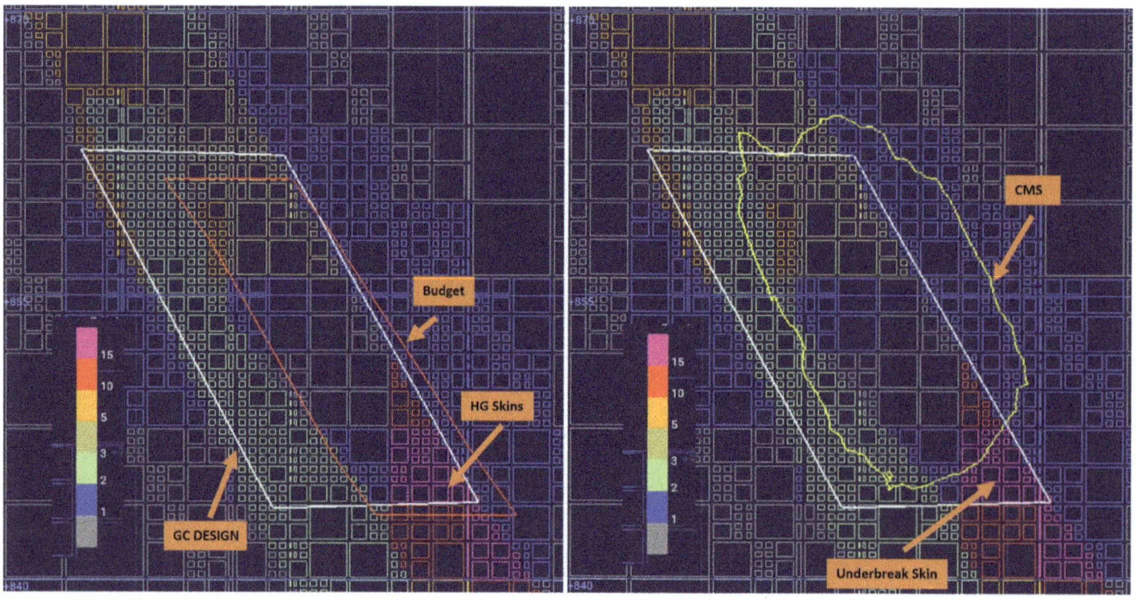

FIG 7 – A visual comparison between budget 3D solid (red) versus GCD 3D solid (white) on the left and GCD 3D solids (white) versus CMS (yellow) on the right.

Note that stopes targeting remnant skins tend to be severely impacted by underbreak (cf overbreak) because a large proportion of the stopes total gold content is located along the margins of the stope as high-grade skins. A small amount of underbreak therefore has the potential to result in a disproportionately large loss of contained gold.

OceanaGold successfully implemented several targeted operational improvements for 2025, focusing primarily on spatial compliance to plan by allocating additional resources to overcome the challenges associated with remnant mining. Sampling processes of remnant material were also improved.

Although the resource model occasionally contributes to reconciliation disparities, these impacts are generally minimal. Table 7 presents Waihi's performance against budget in 2025, showing a positive variance.

TABLE 7

YTD September 2025, Waihi budget versus actual.

	Budget	Actual
Tonnes (Ore)	404 939	509 997
Au (g/t)	3.72	3.47
Au (Oz)	48 397	56 503
Au Recovery (%)	94%	94%
Gold Produced (Oz)	45 570	52 921

Despite ongoing efforts to improve Mine to Mill reconciliation, remnant mining continues to present challenges. The reconciliation process provides clarity towards control strategies to meet sustainable outcomes.

Case study – Haile, 2025

By 2025, the OceanaGold's MCR system had been widely adopted across operational and planning teams, providing near real-time visibility into ore movements from the OP, UG and stockpile to the mill. Table 8 presents the 2025 YTD (as at June 2025) production profile, with actual gold production ahead of budget.

TABLE 8

YTD June 2025, Haile budget versus actual.

	Budget	Actual
Tonnes (Ore)	1 562 499	1 317 888
Au (g/t)	2.05	2.66
Au (Oz)	103 105	112 612
Au Recovery (%)	84%	88%
Gold Produced (Oz)	86 836	99 259

However, the aggregated YTD comparison masks underlying performance issues. The waterfall chart (Figure 8) highlights that the combined open pit and underground RSV tonnage mined exceeded budget, albeit at lower grade (point d). This is because, although open pit mining was ahead of budget, less high-grade underground ore than budgeted was mined (Figure 9).

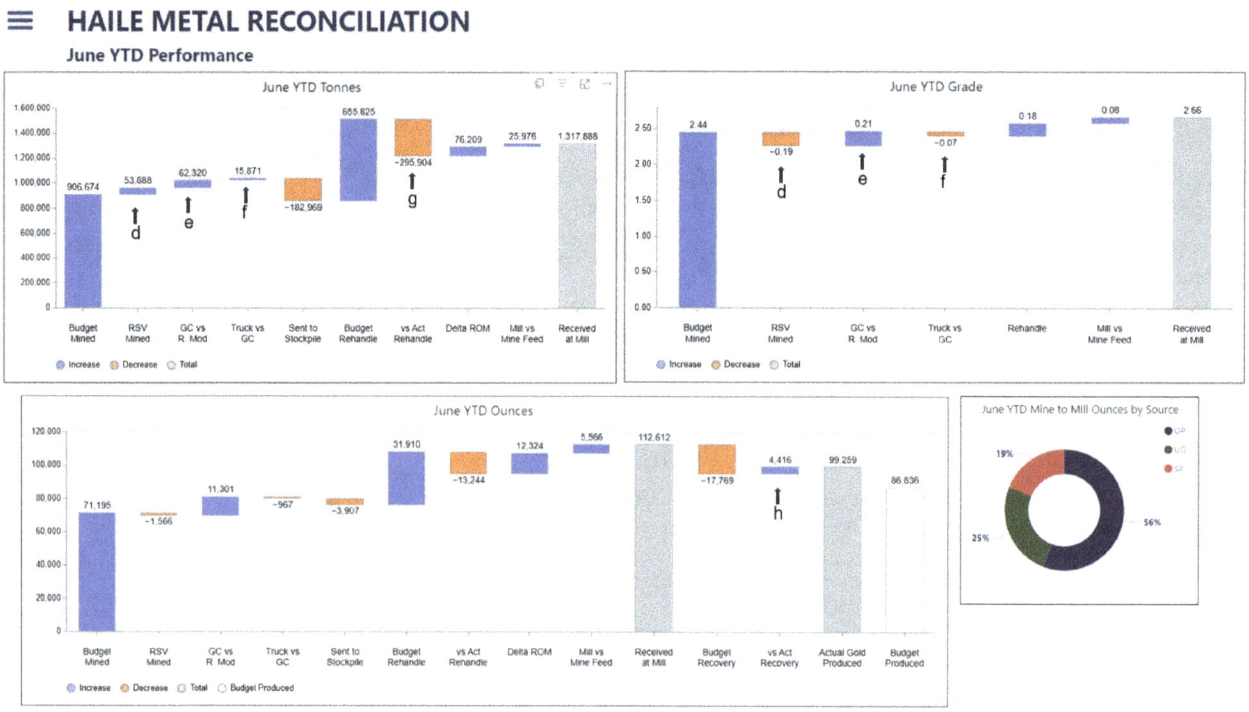

FIG 8 – Full chain reconciliation for YTD June 2025, tonnes, grade and ounces for Haile.

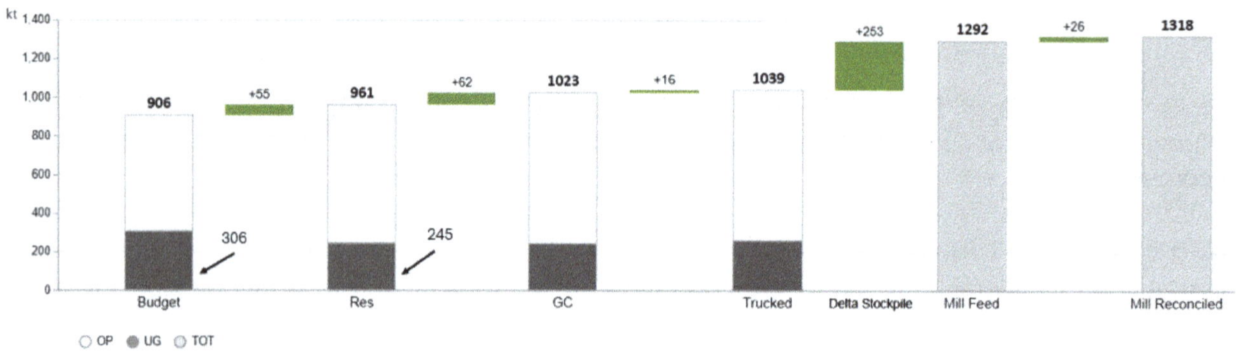

FIG 9 – Details full chain reconciliation for YTD June 2025, tonnes for Haile.

The F1 (GCD/RSV) reconciliation metric indicates that GCD outperforms the RSV model in tonnes and grade (point e). Note, the F2 (trucked/GCD) reconciliation metric indicates that mining dilution is low, with a very small impact on the grade relative to GCD (point f).

The waterfall charts also indicate the prioritisation of directly mined ore from the open pit ahead of budget (point g), resulting in the deferral of approximately 295 Kt of lower-grade stockpile feed.

Finally, the net effect of positive F1 and the prioritisation of higher-grade direct ex-pit ore, by deferring the low-grade stockpiles has resulted in higher than budgeted mill feed grade and positively influenced the overall mill recovery (point h).

This case study demonstrates that the MCR analysis approach can be used effectively to isolate links in a complex mining chain and identify positive and negative contributions to overall budget outcomes. The stockpile deferral was in response to limited mill throughput during maintenance periods, its deferral will need to be accommodated in future years.

These early signals, identified well before they manifest as budget variances, demonstrate the MCR system's value not only as a reconciliation and reporting tool, but as a forward-looking early warning system that equips geologists and planners to identify and communicate future risk.

At the time of completing the final edits for this paper, Haile had achieved its 2025 full-year guidance and Waihi had exceeded 2025 full-year guidance.

CONCLUSIONS

MCR reconciliation provides a structured framework for analysing variances along the mining chain, facilitating root-cause analysis rather over simplified explanations. By decomposing actual versus budget variances into compliance to plan, model performance (F1), extraction efficiency (F2), stockpile movements, Mine to Mill performance (F3), and processing recovery, the approach enables identification and isolation of root causes. This process requires teamwork across disciplines and business units.

Implementation of OceanaGold's MCR system has demonstrated significant operational value. The system improves decision-making by replacing assumptions with evidence-based insights, reducing time spent on misdirected investigations, and fostering transparency across geology, mining, and processing teams. Case studies from Waihi (2024) and Haile (2025) illustrate how the methodology isolates the sometimes elusive, true drivers of variance, whether mine plan slippage, unplanned dilution, stockpile deferrals, or material-flow misalignment. More importantly, it gives site personnel back some of their time to do what they do best, rather than repeatedly answering the same question: 'Where is the grade?'.

While reconciliation itself does not resolve performance issues, it functions as an essential diagnostic and governance tool and a focal point that guides targeted improvement efforts. The system has also driven cultural alignment across OceanaGold, establishing common terminology, shared understanding, and consistent reporting. In an industry where margins are tight and variability is inherent, such integrated reconciliation provides a critical foundation for reliably meeting production guidance and strengthening long-term resource and reserve confidence.

ACKNOWLEDGEMENTS

The authors would like to thank to OceanaGold Corporation, for the permission to publish this paper. Thanks also to Greg Hollett and the anonymous AusIMM reviewer for valuable feedback. Thank you to all OceanaGold's site champions for providing the best quality data every month.

REFERENCES

Dominy, S C, 2018. Geometallurgy—A route to more resilient mine operations, *Minerals, MDPI*, 8(12):560.

Fouet, T, Riske, R, Morley, C, Cook, A and Centofanti, J, 2009. Standardising the reconciliation factors required in governance reporting, in *Proceedings of the Seventh International Mining Geology Conference 2009*, pp 127–139 (The Australasian Institute of Mining and Metallurgy: Melbourne).

Hargraves, R, and Morley, C, 2014. Universal reconciliation – a multidisciplinary approach across the entire mining value chain to identify loss of vale and maximise operations performance, in *Proceedings Ninth International Mining Geology Conference 2014*, pp 411–419 (The Australasian Institute of Mining and Metallurgy: Melbourne).

Hunt, J A and Berry, R F, 2017. Geological contributions to geometallurgy: A review, *Geoscience Canada*, 44(3):103–118.

Journel, A G and Huijbregts, C J, 1978. *Mining Geostatistics* (Academic Press: London).

Macfarlane, A S, 2013. Reconciliation along the mining value chain, *The Journal of The Southern African Institute of Mining and Metallurgy*, 2015:115(8):679–685. https://doi.org/10.17159/2411-9717/2015/V115N8A3

Moore, J, Grant, M, Corley, D and Randa, W, 2023. Schrödinger's Kittens: Lifting the Lid on Resource Drill Hole Data after Mining, in *Proceedings of the Mineral Resource Estimation Conference 2023*, pp 141–150 (The Australasian Institute of Mining and Metallurgy: Melbourne).

Morley, C and Arvidson, S, 2017. Mine value chain reconciliation – demonstrating value through best practice, in *Proceedings of the Tenth International Mining Geology Conference 2017*, pp 279–292 (The Australasian Institute of Mining and Metallurgy: Melbourne).

Parker, H M, 2011. Reconciliation principles for the mining industry, in *Proceedings of the 35th APCOM Symposium 2011*, pp 47–67 (The Australasian Institute of Mining and Metallurgy: Melbourne).

Schofield, N, Moore, J and Carswell, J, 2012. Mine to Mill Reconciliation – Three case studies, in *Proceedings of the International Mine Management Conference 2012*, pp 161–172 (The Australasian Institute of Mining and Metallurgy: Melbourne).

Decoding geological contributions to coal loss and dilution through seam thickness simulation

R Saha[1], C Williams[2] and T Simpson[3]

1. Superintendent Geological Modelling, BMA, Brisbane Qld 4000. Email: ranjan.saha@bhp.com
2. Principal Resource Modelling, BMA, Brisbane Qld 4000. Email: craig.williams1@bhp.com
3. Principal Reconciliation, BMA, Brisbane Qld 4000. Email: thomas.simpson@bhp.com

ABSTRACT

Reconciliation of grade and tonnage is a critical component of mining operations, enabling the completion of the Plan-Do-Check-Act (PDCA) cycle. This process ensures that learnings from unexpected deviations—often arising from uncertainties in geological and reserve models are captured, ultimately improving resource recovery and mining efficiency. In coal mining, where individual seam packages vary stratigraphically in grade and thickness and are blended to form a single product, geological reconciliation presents unique challenges. The existing F-Series Reconciliations (Factor Series Reconciliations) provide a systematic and ongoing assessment of the reliability, accuracy, and validity of reserve model estimates by comparing them with actual production performance. These reconciliations support compliance with reporting requirements and are used to calibrate modifying factors within the reserve models. However, distinguishing coal loss and dilution due to geological factors (eg seam thickness uncertainty, model assumptions, and interpolations) from non-geological factors (eg blasting, plan compliance, wedge loss, floor dilution, and spillage) remains complex. These factors are often aggregated as singular under the F2 reconciliation process, which compares process plant feed with reserve model predictions.

To isolate and understand the geological contribution, particularly thickness uncertainty, this study employs conditional simulation on coal seam thickness. Simulations were conducted across various seams at appropriate mining block scales to quantify uncertainty and trace its volumetric contribution to overall loss and dilution within mining strips. This approach enhances understanding of coal uncovery (coal exposure after removal of overburden) and mining performance by identifying key geological drivers of deviation, such as *in situ* thickness variability and drill hole spacing. Integrating simulation-based validation of thickness variations into geological models strengthens the feedback loop to resources, which is an otherwise challenging task in coal mining.

INTRODUCTION

Reporting coal loss and dilution is an integral part of the overall reconciliation framework practised in open cut coal operations and is an expansion of the Factor (F-Series) series requirements. When only reporting standard F-Series metrics it becomes difficult to identify reasons for an area's underperformance. The need for more granular reporting and understanding becomes critical to isolate geological and non-geological factors.

Mine planning teams are responsible for applying assumptions (geological and non-geological) on loss and dilution in the reserve. The assumptions applied in the budget have a significant impact on plan stability and the data gathered by geoscience can add significant value to the accuracy of these assumptions and to the understanding of any deviations.

To isolate and understand the geological contribution to loss, a study has been undertaken on coal seam thickness and its inherent uncertainty, using conditional simulation, comparing uncertainty with actual mining losses. The aim here is to understand and isolate geological from non-geological contributing factors to coal losses.

GEOLOGICAL SETTING

BMA operates five open cut and one underground metallurgical coalmine in the Bowen Basin of Central Queensland. The coal asset's location is represented in Figure 1.

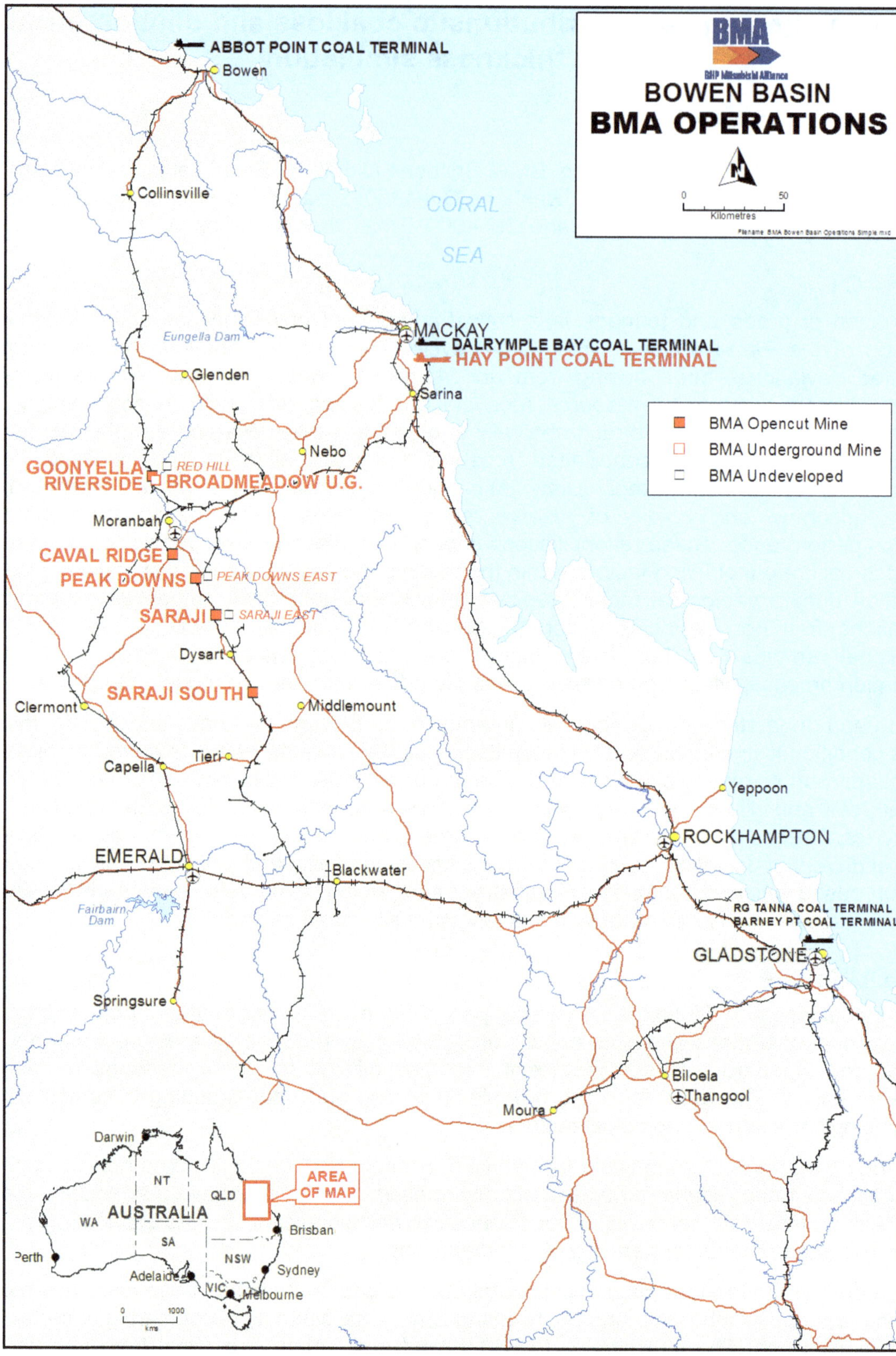

FIG 1 – Location map of BMA operations.

The Bowen Basin extends for more than 250 km north to south and up to 200 km east to west and is related to a group of Permo-Triassic basins in eastern Australia that includes the Sydney and Gunnedah Basins. A generalised stratigraphic column of the Bowen Basin is represented in Figure 2. The economic coal seams mined in the Bowen Basin primarily occur within the Rangal and Moranbah Coal Measures.

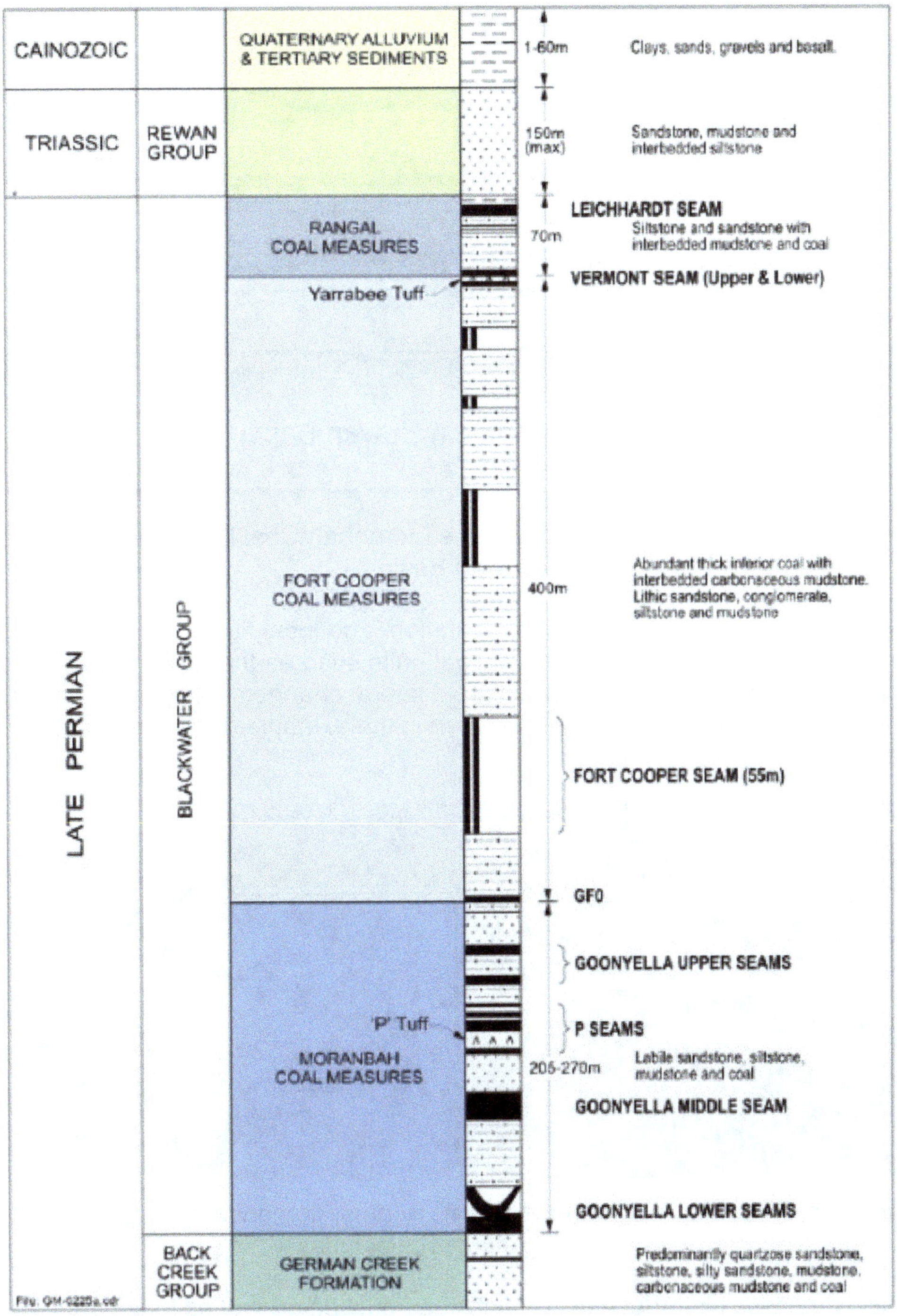

FIG 2 – Generalised stratigraphy of Bowen Basin.

Coal seams mined by BMA form part of the Moranbah Coal Measures (MCM). Within the MCM, different seams/seam groups exist which are distinctly heterogeneous in their inherent characteristics, like spatial continuity, thickness, rank, and related coal quality attributes. The heterogeneity is attributed to different depositional environments, provenance characteristics and post and syn tectonic structural regimes. Heterogeneity at individual seam/seam group level drives extraction strategies to meet volume and product commitments.

A schematic cross-section through the MCM (Figure 3) shows typical seam splitting and thickness variations in the basal Dysart seam package observed in our operations. This complexity is associated with higher rates of clastic sediment input at the base of the MCM. Increased levels of coal seam splitting contribute to coal seam thickness uncertainty in the geological model.

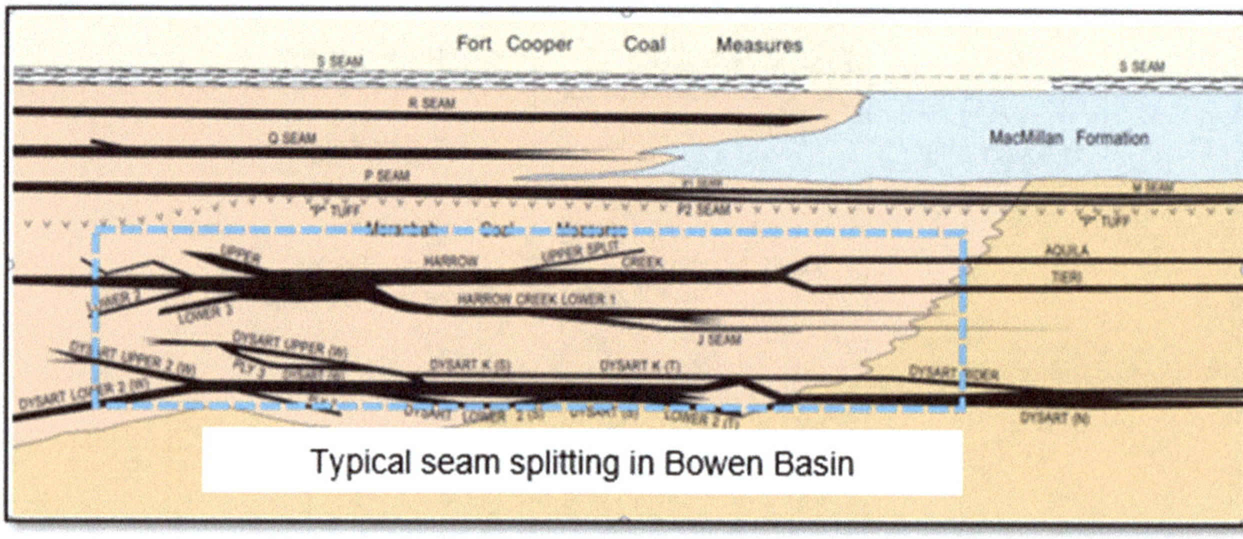

FIG 3 – Schematic cross-section through the Moranbah Coal Measures in the Northern Bowen Basin.

Seam splitting, subsidence rates, clastic sedimentation, and local fluvial environments contribute to significant heterogeneity in seam geometry. In coal units such as the Lower Dysart seam package, multiple generations of seam splitting produce rapid lateral changes in thickness, increasing the risk of mining unexpected partings or thin zones as seen in the example (Figure 4) of the Z-split between two Dysart seams.

FIG 4 – Example of Z-splitting between the Dysart Seams, demonstrating rapid lateral change in seam splitting relationships.

These geological processes introduce uncertainty that must be quantified to separate geological loss and dilution from operational inefficiencies.

EXISTING CHALLENGES IN RECONCILIATION

F-Series Reconciliations (also known as Factor Series Reconciliations, F1,2,3 Reconciliations or Geological Reconciliations) is the systematic, ongoing assessment of the reliability, accuracy and validity of the estimates made by the key geoscientific and mine planning models by comparing those estimates with actual production performance.

It is also the assessment of the effectiveness of extraction activities and each operation's ability to deliver the value outlined within the reserve model.

Many of the most important and influential inputs into the reserve model are derived from the resource model. In particular, the reserve model is dependent upon the validity of the geological interpretation and grade/quality estimation in the resource model. Therefore, even though the F-Series reconciliation calculations draw data from the reserve model, the process implicitly provides a test of the resource model prior to the application of modifying factors. It is one of the key sources of assurance complying to the JORC Code (2012) requirements. The F-Series Reconciliation form part of the 'Check' process in a 'Plan-Do-Check-Act' framework. Figure 5 gives a representation on the calculations involved in F Series reconciliation.

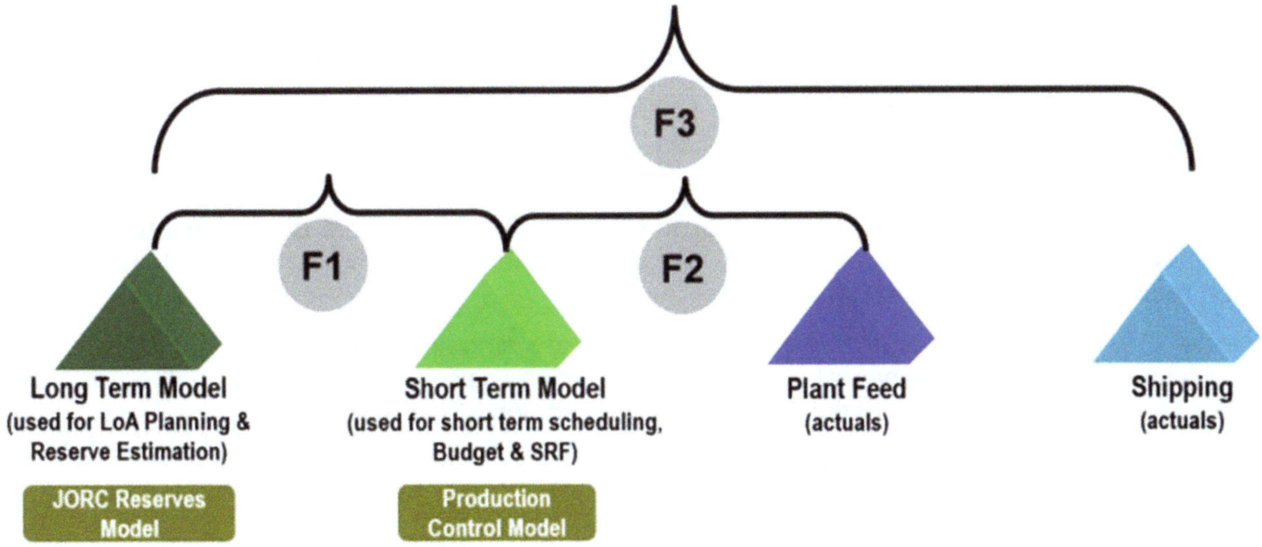

FIG 5 – F (Factor) Series Reconciliation Schematics.

The key component in the process is F2 reconciliation, which has the biggest impact in the mining value chain, when comparison is undertaken between the actual mine production and the short-term reserves model. Coal reconciliation is dominated by ROM-to-plant (F2) comparisons, with limited ability to trace discrepancies back to the *in situ* geological model (F1 reconciliation), due to:

- Conversion of *in situ* models to ROM models through the application of loss and dilution assumptions.
- Blending of multiple *in situ* seam source locations on the ROM pad prior to treatment through the plant.
- Lack of robust coal tracking systems from pit to port makes it difficult to align tonnage and quality data between models.
- Moisture variations and stockpile movements further complicate reconciliation, as these factors can mask true loss and dilution impacts.

As a result, geological drivers of coal loss and dilution (ACARP Project C13031; Holtham, Sanders and Scott, 2005), specifically seam thickness variability and seam splitting are aggregated with operational factors and cannot be isolated using traditional reconciliation workflows.

Key avenues of mining loss involve:

- Geology of seam roof, floor and corresponding seam thickness.
- Blasting efficiency and accuracy (eg blast damage/shunting/roof sheer).
- Size of overburden removal equipment and coal clean-up equipment.
- Presence of geological structures can displace or severely contort the coal, are usually left for separate scavenging operations. This coal is often left to mine on retreat from the area at a later (often not realised) time.
- Wet conditions.

So, the question remains unanswered, with multiple factors in play (ACARP Project C3017; The University of Queensland, 1995), how do we isolate the geological factors (eg seam thickness uncertainty, model assumptions, and interpolations) from non-geological factors (eg blasting, roof loss, plan compliance, wedge loss, floor dilution, and spillage) contributing to coal loss.

PRESENT STUDY

Conditional simulation workflows can be used to derive information about local geological uncertainty (Haase *et al*, 2019), model accuracy and/or possible model bias, thereby allowing for the identification of areas where:

- Model uncertainty is high, and additional drilling is required for improved resolution.
- Model uncertainty is within acceptable limits and coal losses outside of the budgeted reserve assumptions are due to operational issues.

Conditional simulation generates multiple equally probable realisations that reproduce both the global statistics and the spatial structure of the conditioning data. This ensemble represents a probabilistic description of *in situ* variability and can be used as a proxy for the most likely 'truth' when reconciling model predictions.

Journel (1974) demonstrated that estimation models are smoothed and therefore underestimate local geological variability, whereas conditional simulation realisations are not smoothed and are therefore useful for the quantification of geological uncertainty (Figure 6).

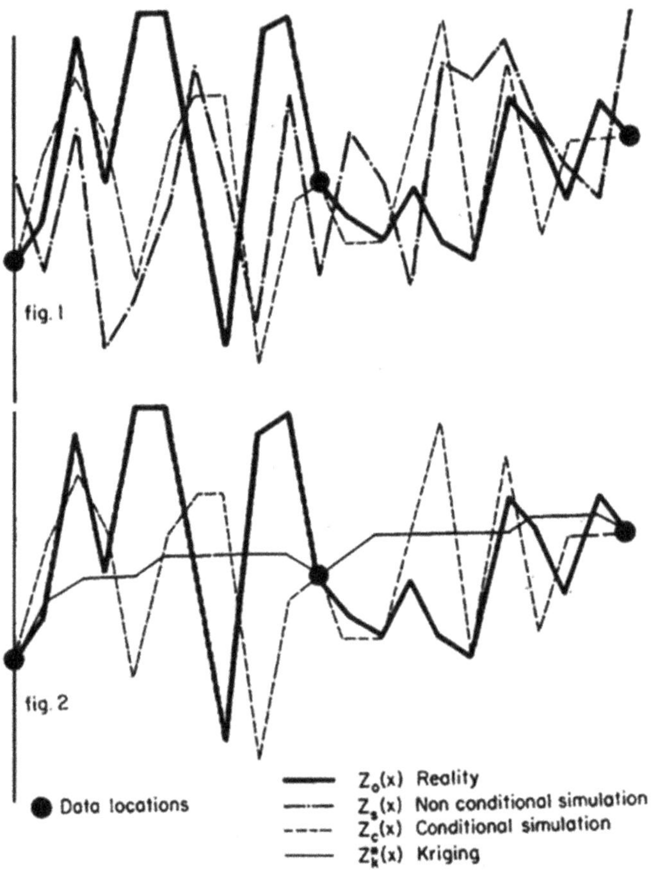

FIG. 1. Reality and simulations.
FIG. 2. Reality, simulation, and kriging.

FIG 6 – Conditional simulations compared to a smoothed (kriged) estimate reproduced from Journel (1974).

BHP has developed the Simulation Based Resource Evaluation (SBRE) Python package as a standardised platform for the design, testing and running of conditional simulation workflows for assets within the Group (Saha and Haase, 2022). SBRE incorporates a modular workflow approach, allowing for assets to use existing software platforms in the workflow, whilst also providing a standardised and validated simulation workflow tailored to commodity specific requirements.

This approach is particularly valuable in coal deposits where complex stratigraphy and variable drill hole spacing create uncertainty in seam thickness predictions. By reconciling deterministic model predictions against the distribution of simulated outcomes, deviations can be attributed to local geological uncertainty, model estimation error (variance) and any possible estimation bias.

The present study comprises of the following steps:

- Use existing true thickness sample points from drill holes.
- Construct variogram models (spherical) to understand spatial continuity.
- Generation of conditional simulations at sample support using Turning Bands method.
- Sampling of simulations at validation locations (synthetic drill holes) to build local conditional distributions.
- Sampling deterministic estimates at the same (synthetic) locations.
- Use of synthetic holes allows us to understand the spread of uncertainty for each seam/ thickness configuration. Simulated values at each synthetic hole location can be plotted as a boxplot and compared to the corresponding deterministic values (Figure 7).
- The above process gives an idea of the geological uncertainty at the synthetic drill hole locations. The information is used to understand the **residual uncertainty**.
- Once the base information for a seam is obtained, the next steps is to understand its effect on Loss% on a weekly mining block.
- Simulations are re-blocked to appropriate mining units emulating extraction volume/area per seam within weekly mining time frame.
- The standard deviation (±1 SD used in the study) of the simulated values can be used as a measure of the **irreducible error.**
- **The irreducible error** is the natural *in situ* variability that cannot be deterministically modelled.
- Loss obtained from weekly mining recovery report per seam basis is compared with the irreducible error.
- Any deviations outside the range of irreducible error, is then attributed to non-geological factors, which is then compared to observations in the pit reports for reconciliation.

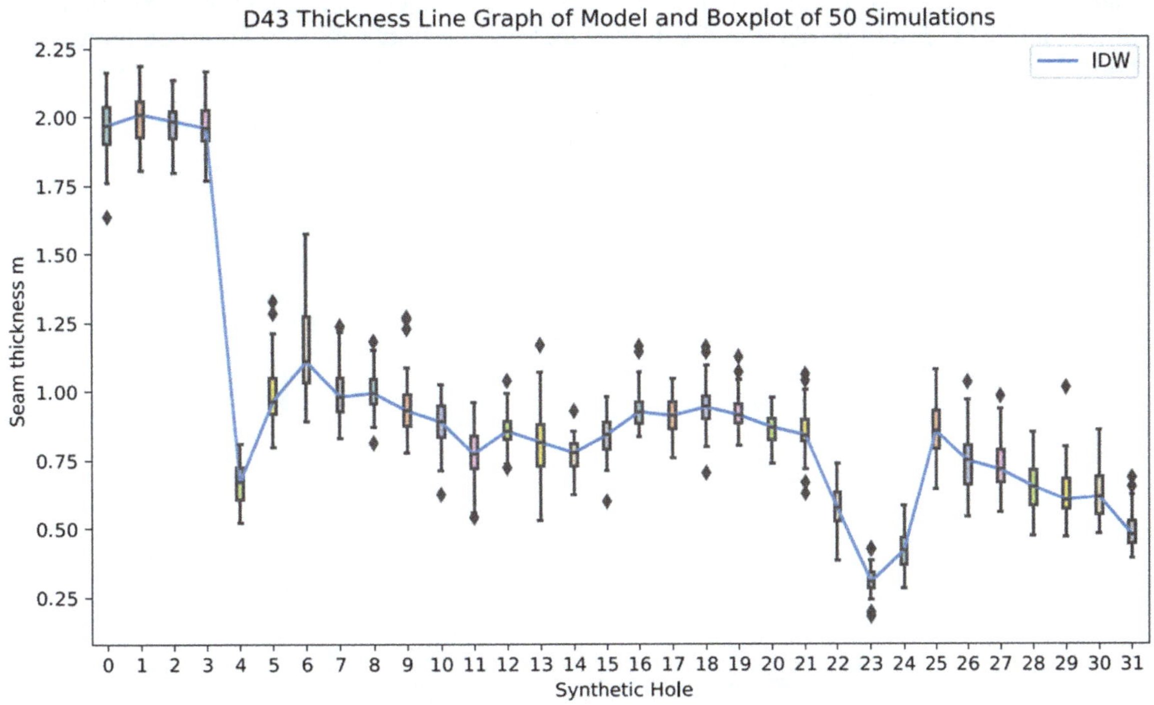

FIG 7 – Thickness boxplot for synthetic holes using conditional simulation compared to deterministic value (line plot).

OUTCOMES

Key objective of the present study is to understand the influence of *in situ* thickness uncertainty and compare the results with *as-mined* thickness and the pit recovery reports to determine geological and non-geological contributions to the loss percentage. The following simplistic cartoon in Figure 8 illustrates the scenario as described above.

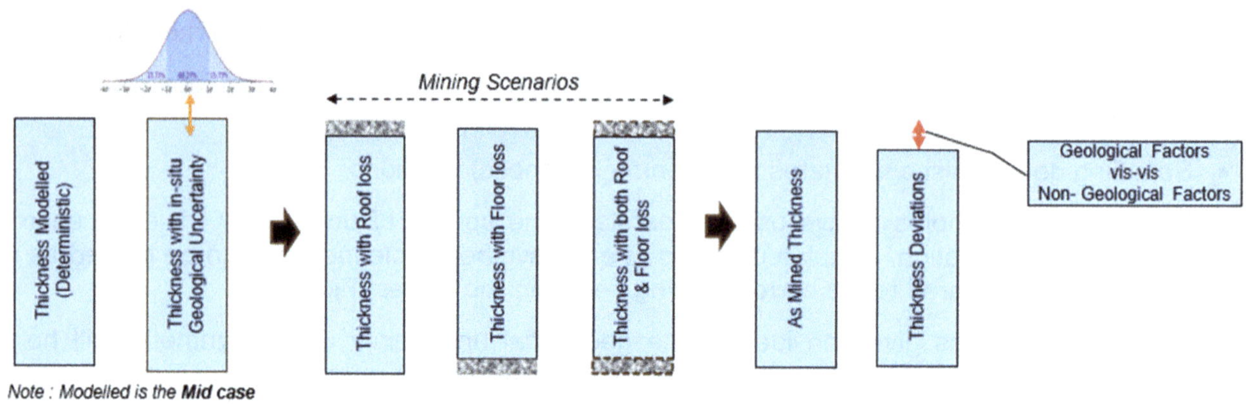

FIG 8 – Cartoon representation of the existing study.

Multiple seams varying in thickness from one operating mine site was selected for the study. The uncertainties associated with any measurement were assumed to be normally distributed. The normal distribution is described by two parameters: the mean and the spread or standard deviation as described in any statistics. Following are the key findings:

- Irreducible error decreases with thickness ie the greater the thickness of coal seam, the lower the error; hence geological factors have less contribution to the loss (Figure 9).
- Exceptions are noted for seams which are affected by complex seam splitting and faulting (h16).
- The loss associated with irreducible error remains within the expected threshold for mining thinner seams (≤ 2 m). While irreducible error can vary in both directions, in the context of mining, true loss is primarily attributed to net reduction in thickness. stemming from uncertainties and complexities in accurately delineating coal roof and floor positions (h00 and d46).

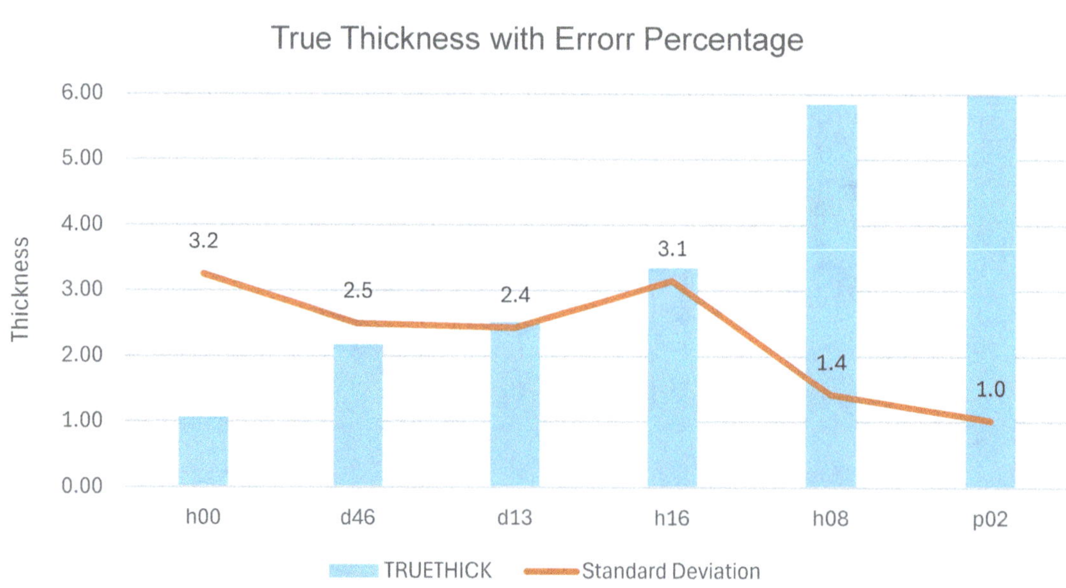

FIG 9 – Plot of seam thickness for different seams and their error percentage.

These observations are compared with the mining recovery reports to reconcile the study findings to actual seams within a specified recovery polygon. A key output of the mining recovery reports is a thickness comparison between the modelled seam and the mined seam. A cross plot of the seams/blocks used for the current study is shown in Figure 10. Coverage of overlapping survey roof and floor data is critical to achieve this, normally the comparison is displayed using a heatmap with a legend that clearly denotes any thickness difference. Photographs compiled during the pit inspection process are incorporated into the report with written context explaining what has occurred. Sections are also included to help visualise any variances between modelled versus actual roof/floor locations or thickness. Figure 11 shows the comparison between the modelled thickness with its tolerances based upon irreducible error and mined thickness. Figures 12–14 demonstrates key references to individual seams within mined out polygons and photography to reconcile and understand different contributing factors.

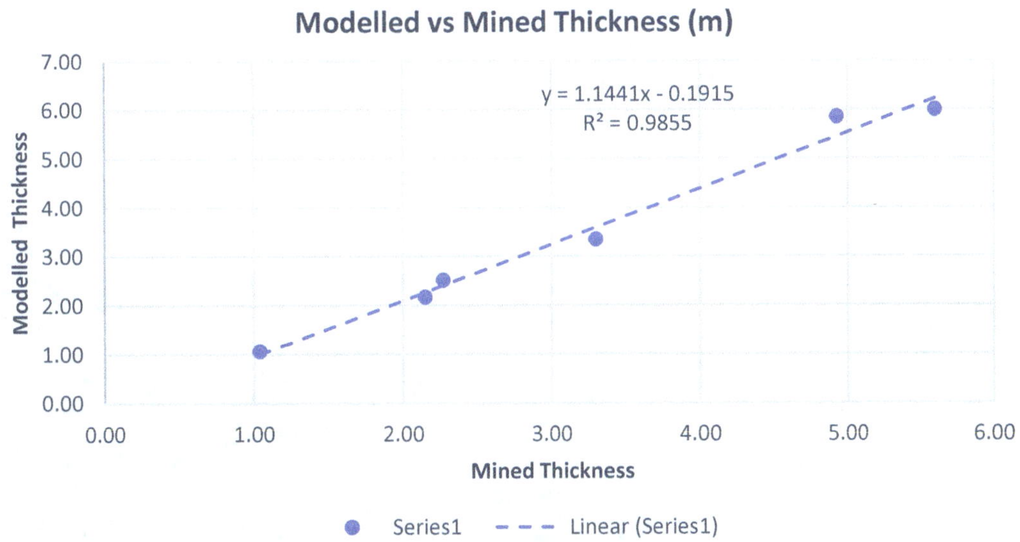

FIG 10 – Scatter plot modelled thickness vis-à-vis mine thickness.

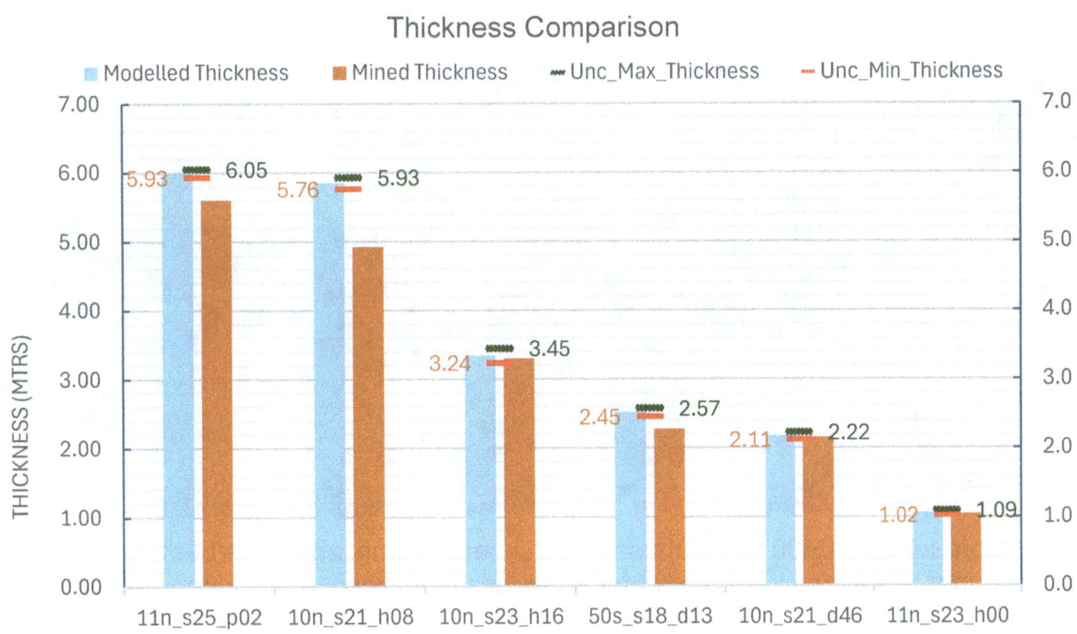

FIG 11 – Thickness comparison between modelled thickness (with error ranges) vis-à-vis realised mined thickness.

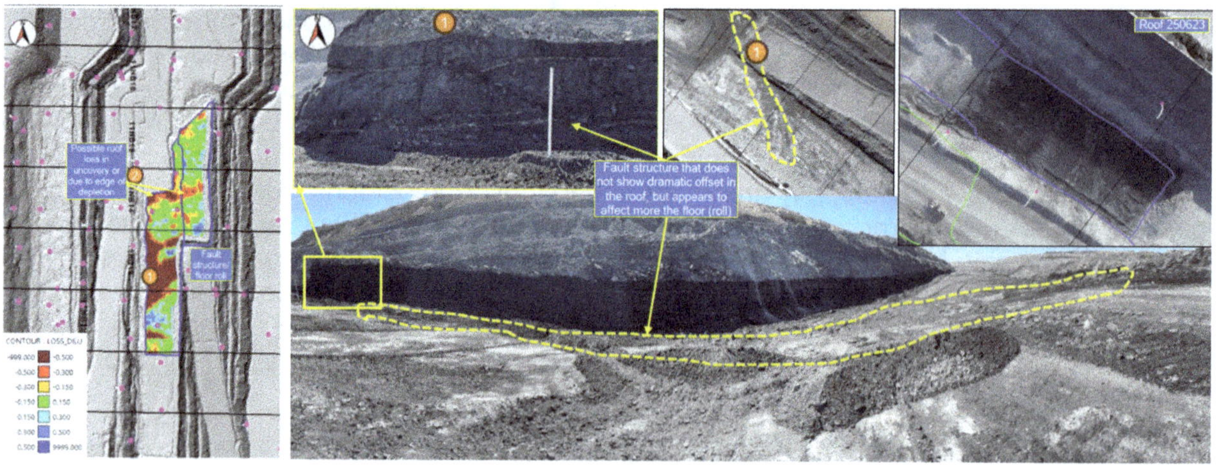

FIG 12 – Representative P02 (thick seam) shows the difference observed in thickness is due to a very localised structure (as indicated by reference 1 in the figure and red heat map area).

FIG 13 – Representative D46 (thin seam) shows very good compliance, but any differences with the mined thickness can be attributed to the irreducible error.

FIG 14 – Representative H00 (medium seam thickness) compliance has been good, but in the edges coal bunds are left which account for edge loss if not recovered, a non-geological factor.

CONCLUSIONS

Coal loss reconciliation is a complex process, and multiple factors, assumptions and ground conditions are at play. The present study has been aimed to understand the *in situ* irreducible error in thickness and its effect on loss. For thinner seams (less than 2 m) the irreducible error is within the bounds of coal roof/floor loss and dilution-hence all loss causes can be attributed to geological loss.

For seams greater than 2 m and above, isolation of geological factors from non-geological factors can effectively be understood using the study outcomes. This coupled with the mining reports serves as a reference to help explain losses beyond the irreducible error on thickness. Key contributing factors involve compliance to plan, equipment selectivity, drill and blast performance, mining hygiene etc.

As any improvement in coal loss has a direct relationship to the revenue. If we assume an average of 10 per cent of mineable coal is lost during open cut mining, a gain of even 1 per cent would equate to significant revenue increase and more saleable product.

It is also important to note for geological estimates that the best interpolator in the world cannot compensate for inadequate sampling data. Poorly sampled areas show up in the simulations as areas of high uncertainty and associated large irreducible error.

ACKNOWLEDGEMENTS

The authors would like to acknowledge BMA Geoscience and Mine Geology Teams for providing valuable inputs in framing the problem and working with the initial trials and looping back with the mining recovery reports to check experimental values to that of actual mining.

This project also would not have been a reality without the mentorship and technical stewardship of Ilnur Minniakhmetov, Principal Geoscientist BHP Coal, Centre of Excellence.

REFERENCES

Haase, B, Gillam, B, Rigby, A, Minniakhmetov, I, Jimenez, E, Joyce, L and Saha, R, 2019. Goldilocks: Value of information for geoscience in bulk mineral mining, in *Proceedings of the International Mining Geology Conference 2019*, pp 367–374 (The Australasian Institute of Mining and Metallurgy: Melbourne).

Holtham, P, Sanders, D and Scott, A, 2005. Improving the precision of tonnage reconciliation, ACARP report C13031, August.

JORC, 2012. Australasian Code for Reporting of Exploration Results, Mineral Resources and Ore Reserves (The JORC Code) [online]. Available from: <http://www.jorc.org> (The Joint Ore Reserves Committee of The Australasian Institute of Mining and Metallurgy, Australian Institute of Geoscientists and Minerals Council of Australia).

Journel, A G, 1974. Geostatistics for conditional simulation of ore bodies, *Economic Geology*, 69(5):673–687.

Saha, R and Haase, B, 2022. Unlocking resource confidence in coal through use of conditional simulation, in *Proceedings of the International Mining Geology Conference 2022*, pp 474–478 (The Australasian Institute of Mining and Metallurgy: Melbourne).

University of Queensland, The, 1995. The assessment and control of coal damage and loss, ACARP report C3017, September, Julius Kruttschnitt Mineral Research Centre (JKMRC).

Value creation

Plan to product

Mining the contradictions

B Byrne[1]

1. MAusIMM, Geology Manager, Northern Star Resources, Perth WA 6000.
   Email: bbyrne@nsrltd.com

ABSTRACT

As the gold price continues to rise, it has become increasingly viable to mine higher-risk, more complex ore systems closer to the economic cut-off. These opportunities carry narrower profit margins, making them more vulnerable to fluctuations in economic and mining cycles, and increasing the risk to favourable project economics. To manage this risk, operations often adopt accelerated mining strategies that minimises capital spend while maximising throughput. This is often in conflict with quality and delivery of optimal grade.

These competing relationships between cost, volume and grade are compounded by several contradictions. Robust geological understanding of uncertain ore systems requires more input data to capture complexity, yet time and cost constraints often limit data acquisition. Optimising grade from complex ore systems necessitates quality focused mining strategies, yet time and cost pressures favour larger, lower-cost mining fleets aimed at volume extraction. Furthermore, typical risk minimisation is challenged when short-term production targets are driven by locally complex and uncertain ore definitions, where small deviations in interpretation and execution can have disproportionate impacts on grade, dilution and loss.

To mitigate these conflicting priorities, Northern Star has developed several quantitative tools through collaborations between geology and mining teams. Central to this is the Geological and Mining Risk Matrix, a framework that supports strategic decision-making and communication across key stages of the mining cycle: planning, pre-blasting, post-blasting and extraction. The matrix enables teams to proactively identify and reduce uncertainty, dilution, and ore loss, ensuring complex orebodies are extracted optimally, and in alignment with Ore Reserve assumptions. Complementing this is the Revenue or Free Cash Flow Attribute – a clear, universally understood metric that communicates the economic significance of each ore block driving higher quality mining practices.

Together these tools help derisk the inherent geological uncertainty of complex ore systems, maximising margins in challenging, higher-risk projects.

INTRODUCTION

The sustained strength in the gold price has made it viable to develop a class of open pit deposits once set aside as too complex, too high-risk, or too close to the cut-off grade to be economic. These deposits are often characterised by irregular mineralisation and significant geological uncertainty at local scales which drives narrow profit margins. Under such conditions, operational performance becomes highly sensitive to short-term variations in grade, dilution, and ore–waste boundaries.

As these deposits operate within tight financial constraints, mining teams are often driven to adopt accelerated and high-volume mining strategies that minimise capital spend. This practice, centred on larger, lower-cost fleets, impacts the quality and control necessary to effectively recover grade from complex ore systems. At the same time, the geological understanding needed to support quality mining requires dense, high-quality data, yet drilling budgets and time frames are often reduced as mining accelerates. This creates a series of practical contradictions between cost-volume-grade that sit at the centre of complex ore extraction.

Economic extraction of complex orebodies requires explicit consideration of the geological risk at the local scale to minimise misallocation, dilution and loss. Operations that fail to account for this local complexity run the risk of grade underperformance, declining revenue, and a gradual drift away from the assumptions used to define the Ore Reserves.

This paper discusses these contradictions but more importantly outlines a framework developed at Northern Star to quantify local geological uncertainty, integrate mining capability, and strengthen decision-making through the mining cycle. The approach combines a Geological and Mining Risk

Matrix with an economically driven Revenue Attribute, ensuring that both risk and value are consistently recognised, communicated, and managed to support agile, fit-for-purpose mining practices.

The real opportunity lies not in the elevated gold price, but in building the capability to mine complex orebodies with intent, managing sensitive economic cut-off grades and protecting margins under elevated risk. This will unlock value from deposits once considered too difficult to mine.

GEOLOGICAL COMPLEXITY IN OPEN PIT GOLD SYSTEMS

Complex gold deposits differ from predictable orebodies not only in geometry but in the way grade behaves at the short range. They are often characterised by abrupt changes in mineralisation style, have structurally intricate geometries, mixed grade populations with narrow and discontinuous high-grade lenses, and variable weathering or lithological domains. These features reduce continuity and increase the local scale variability, making grade interpolation and ore boundary prediction more difficult. Furthermore, visual indicators are often subtle or entirely absent, increasing reliance on drilling, sampling, and detailed geological interpretation to define mineralisation with confidence.

Understanding these controls is critical because mining accuracy is governed by how well the geometry and the predictability of the mineralised lodes are resolved at the local scale. When complexity is understated or overlooked, mining teams are left trying to apply mining methods that may be fundamentally misaligned with the orebody's characteristics.

Table 1 outlines the main geological features of an orebody and how they can influence the risk when it comes to practising open pit mining strategies.

TABLE 1

Summary of key geological features contributing to local complexity and risk of open pit ore sources. Each feature is assessed independently.

	Low risk	Moderate risk	High risk
Dip*	Steep	Flat, and moderate to steep	Shallow
Lode width**	> 10 m	> 4 m	<4 m
Degree of weathering	Un-weathered	Moderately weathered	Highly weathered
Visual control	Clear	Subtle visibility	No visibility
Grade continuity	Continuous	Partly continuous	Discontinuous, short range
Structural complexity	None or simple deformation	Deformation that disrupts ore but not mining	Deformation that leads to major discontinuity
Grade distribution	Homogeneous	Partly homogeneous	Mixed, heterogeneous
Deleterious minerals or lithologies	Absent	Present and indirectly	Present
Geotechnical complications	Absent	Present an indirectly impacts ore	Present and directly impact ore or access to ore
Voids	No voids present	Indirect interactions with ore	Direct interaction with ore

* The risk associated with the dip is dependent on the digging technique between selective mining and conventional vertical mining of ore blocks.

** The risk associated with lode width will depend on the excavator bucket size. The numbers given are for large to shovel sized excavators with >4.5 m bucket width.

THE COST–VOLUME–GRADE CONTRADICTION

Like in all mineralised systems, operational performance relies on accurate ore–waste boundaries, maintaining grade, and minimising dilution and loss. Yet the narrow economic margins typical of high-risk and complex deposits, often push operations toward accelerated mining strategies and large, lower-cost mining equipment. In highly variable deposits, this becomes problematic particularly at the local scale: small deviations in blasting, digging, or load spotting can materially shift reconciled grade, and large equipment amplifies this effect by reducing practical selectivity. This is the central contradiction: high-throughput, low-cost mining operates at a scale that is fundamentally out of step with the geological scale of the mineralisation.

Grade control drilling is the primary tool for reducing this uncertainty, but when drill density is limited by time or budget, as is common in accelerated schedules, the uncertainty carried into production increases and directly affects extraction performance and revenue.

This mismatch flows into the broader cost–volume–grade contradiction. High-volume, accelerated mining reduces unit costs but erodes the precision needed to recover value from complex orebodies. Efforts to improve grade through increased selectivity, whether by tightening ore–waste boundaries or internal grade classification, require denser drilling, smaller excavators, or slower production. These measures are not favoured due to financial pressure or mill demand, meaning the required selectivity cannot be achieved at local scale. As uncertainty accumulates small operational misses compound, initiating a reinforcing cycle: declining grade drives higher mining rates, which further reduce selectivity and quality, which increases dilution and loss, and pushes grade even lower. Figure 1 outlines the cycle and its impact where the local risk is overlooked.

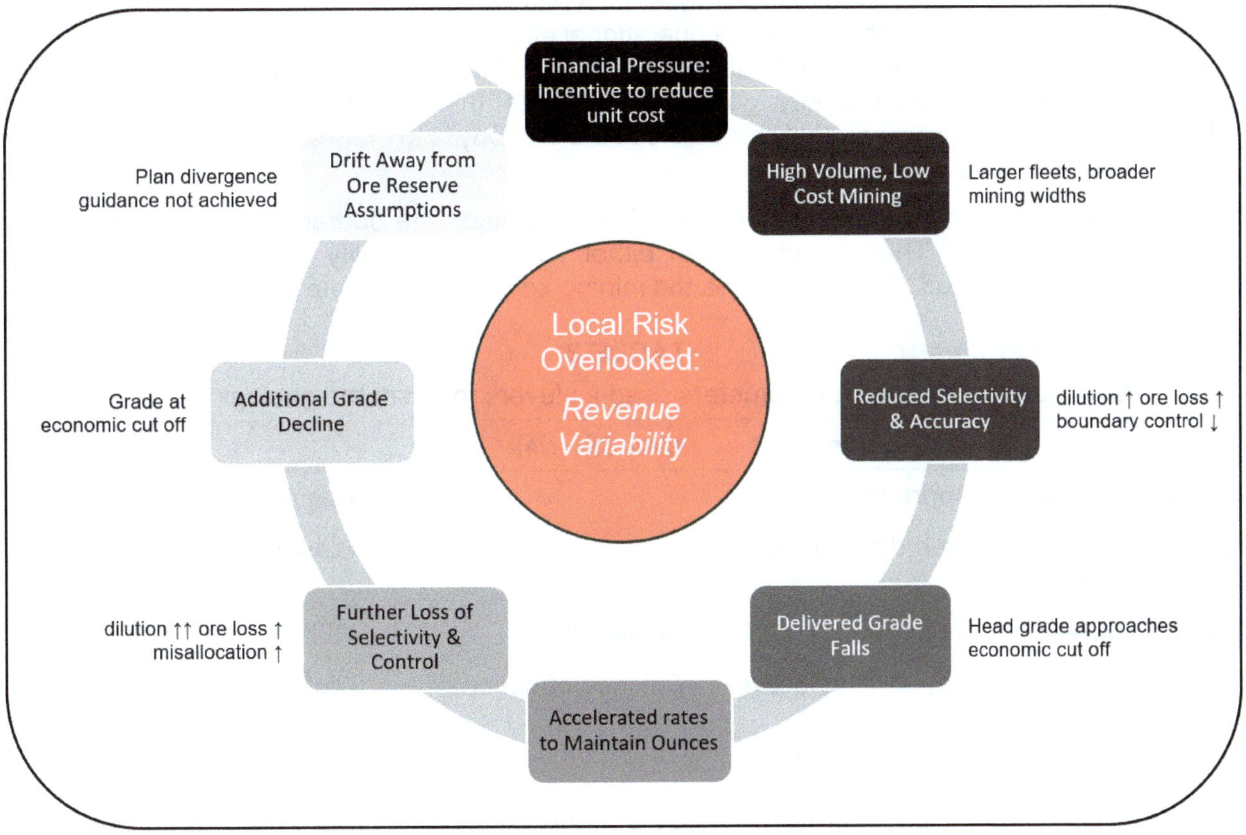

FIG 1 – The cost-volume-grade contradiction cycle.

If complexity is not explicitly recognised and built into planning, operations gradually drift away from the assumptions used to define Ore Reserves. Breaking this cycle, outlined in Figure 1, requires tools that link geological reality to mining method selection, drilling strategy, and production planning.

GEOLOGICAL AND MINING RISK MATRIX AND MITIGATION

To manage the challenges of complex ore systems, Northern Star developed a risk-based framework that quantifies geological complexity and the operational capacity to manage that complexity. The difference between the two produces the Total Risk Score (TRISK), indicating whether the mining strategy is aligned with the orebody conditions and whether risk is appropriately mitigated.

The system was collaboratively developed by geology, engineering, and operational mining teams and is designed to achieve three key objectives:

1. Communicate Geological Risk – clearly highlight local geological complexities and ensure all stakeholders understand their implications.
2. Define the Operational Plan – establish mining strategies that directly mitigate identified geological risks.
3. Enable Action – proactively adjust mine plans so that extraction strategies remain aligned with geological conditions to achieve optimised outcomes.

The application of the tool focuses on three primary outcomes:

1. Minimising dilution.
2. Minimising ore loss.
3. Maximising head grade commensurate with geological risk.

The framework evaluates two complementary sets of parameters. The first is the Geological Risk Parameters (GRISK), which describe the inherent characteristics of the orebody. These reflect natural complexity that cannot be altered by operational effort but must be understood and managed. Although not a natural feature, void interactions are included due to their strong influence on dilution and loss. Table 1 outlines the key geological features that define the GRISK parameters. Collectively, these parameters quantify how difficult the orebody is to mine selectively and how prone it is to dilution, loss, and misallocation.

The second set, the Mining Risk Parameters (MRISK), defines the operational levers available to mitigate geological risk. These controls can either reduce or amplify risk depending on how effectively they are applied. Table 2 outlines the mining controls that define the MRISK parameters.

TABLE 2

Summary of key mining parameters used as levers to offset the geological risk.

	Reduce risk	Maintain risk	Increase risk
Excavator size	Appropriate		Inappropriate
Material competency	Free dig material		Blasted material
Blasting options	Selective	Strike parallel	Oblique, centre lift
Mining options	Day shift, selective mining	Balance of options	Night shift, full flitch vertical mining
Marked up stocks	Sufficient		Insufficient
GPS control	GPS + free dig	GPS + blasted material	No GPS
Pit constraints	Space sufficient	Space confined	Space insufficient
Mining method	Along strike	End on, or up and down dip	Back select or top loading
Mining levels	Regular checks feasible		Regular checks not feasible

Minimum face length	Mined, available and used		Unavailable or available and not used
Shot/bench management	Dig lines and buffer strategies in place		Dig lines and buffers unavailable
Mine services	Washed + dozing and grading along strike	Mix of supervised and non-supervised	Not washed + dozing and grading cross strike
Sump locations	In waste	In ore	Across ore and waste
Ore spotting	Available		Unavailable
Supervision	Available and quality		Available and poor

GRISK and MRISK scoring

Prior to making either long-term or short-term decisions for mining, the GRISK and MRISK values are evaluated and calculated using either of the Risk Cards presented in Figure 2. The quantifying of the GRISK assumes that even for the most reliable and certain orebodies there is still some residual risk. Hence the lowest score that can be achieved is 12, with the highest risk rating at 52.

The higher the GRISK value the higher the geological risk.

The MRISK assumes that if no controls are in place to mitigate the risk the score is 0, and conversely if the controls are robust the maximum score is 52.

The higher the MRISK value the better the mitigation strategy.

OPEN PIT RISK CONTROL: LONG TERM RISK CARD

DATE: SHIFT: PIT: DIGGER: GEOLOGIST: ENGINEER/SUPERVISOR:

GEOLOGICAL RISK PARAMETRES		RISK RATING		
DIP		**full**	**sel**	**SCORE**
0-10 degrees		2	2	
10-30 degrees - optimal for selective		6	0	
30-65 degrees		4	4	
>65 degrees - optimal for full flitch		0	6	
LODE THICKNESS	***Shovel***			
>5m	>10m	2		
2.5m-5m	4m-10m	4		
<2.5m	<4m	6		
VOIDS (UG voids, Pit edges)				
No voids in shot/bench		0		
Indirect interaction with ore		4		
Direct interaction with ore		6		
DEGREE OF WEATHERING				
Un-weathered		2		
Moderately Weathered		4		
Highly weathered		6		
VISUAL CONTROL				
HW and FW clearly visible		2		
Moderately visible or one contact visible		4		
No visibility		6		
GRADE CONTINUITY and/or STRUCTURAL COMPLEXITY				
Continuity >30m along strike, down dip, & Low		2		
Continuity 10 - 30m along strike, down dip, & Medium		4		
Continuity <10m along strike, down dip, & High		6		
GRADE DISTRIBUTION				
Homogenous, similar DH and between holes		2		
Significant grades >2m from contacts, and/or clustering		4		
Significant grades within 2m of contacts, and/or heterogenous		8		
DELETERIOUS LITHOLOGIES, MINERALS				
Absent		0		
Present		4		
GEOTECHNICAL COMPLICATIONS				
Absent		0		
Present (Wall instability, Structural Complexity)		4		
TOTAL GEOLOGICAL RISK (GRISK) Higher the value, the greater the geological risk		12-52		

ACTION PLAN TO REDUCE TOTAL RISK AND IMPROVE ORE EXTRACTION

MINING RISK PARAMETRES	MITIGATE RATING		
EXCAVATOR SIZE			**SCORE**
Appropriate	4		
Inappropriate	1		
MATERIAL COMPETENCY			
Free Dig Material	4		
Blasted Material	1		
BLASTING OPTIONS			
Selective Waste & Ore	4		
Strike Parallel	2		
Oblique	1		
Centre Lift	-1		
MINING OPTIONS			
Full Flitch Mining OR	**DS**	**NS**	
1. Waste	2	2	
2. Ore	4	1	
OR			
Selective Flitch Mining	**DS**	**NS**	
1. Waste	4	-1	
2. Ore	1	-1	
MARKED UP STOCKS (optionality)			
Sufficient	4		
Insufficient	1		
GPS CONTROL			
GPS + Free Dig	4		
GPS + Blasted Material	1		
No GPS	0		
PIT CONSTRAINTS	**full**	**sel**	
Space sufficient	4	4	
Space confined	2	1	
Space insufficient	1	-1	
MINING METHOD			
Along Strike	4		
Up or down dip, or End on	2		
Back Select or Top Loading	1		
MINING LEVELS			
Regular level checks feasible	4		
Regular level checks NOT feasible	0		
MINIMUM FACE LENGTH			
Available and being used	4		
Unavailable or available and not being used	0		
SHOT/BENCH MANAGEMENT			
No Dig Lines (NDL's) and/or Buffer Strategies Used	4		
NDL's, Buffers Unavailable	0		
ORE SPOTTING & DESPATCH CONTROLLERS			
Geological supervision viable	4		
Geological supervision NOT viable	0		
SUPERVISION			
Available and Quality - as per standard	2		
Available and Poor - not per standard	0		
TOTAL MINING RISK (MRISK) Higher the value the better the Mitigation	0-52		

SIGNATURE GEOLOGY:	SIGNATURE ENGINEERING:
TOTAL RISK SCORING:	GRISK: Higher score the HIGHER the risk
HIGH RISK <15	GRISK: Lower score Better Geology
MODERATE RISK >=15 to <30	MRISK: Higher Score Better Mitigation
LOW RISK >=30	MRISK: Lower Score Poor Mining

TOTAL RISK = MRISK - GRISK

(a)

OPEN PIT RISK CONTROL: POST BLAST RISK CARD				
DATE: SHIFT: PIT:	DIGGER:	GEOLOGIST:	ENGINEER/SUPERVISOR:	

GEOLOGICAL RISK PARAMETRES		RISK RATING		
DIP		full	sel	SCORE
0-10 degrees		2	2	
10-30 degrees - optimal for selective		6	0	
30-65 degrees		4	4	
>65 degrees - optimal for full flitch		0	6	
LODE THICKNESS	***Shovel***			
>5m	>10m	2		
2.5m-5m	4m-10m	4		
<2.5m	<4m	6		
VOIDS (UG voids, Pit edges)				
No voids in shot/bench		0		
Indirect interaction with ore		4		
Direct interaction with ore (consider void fill and rilling also)		6		
DEGREE OF WEATHERING				
Un-weathered		2		
Moderately Weathered		4		
Highly weathered		6		
VISUAL CONTROL				
HW and FW clearly visible		2		
Moderately visible or one contact visible		4		
No visibility		6		
GRADE CONTINUITY and/or STRUCTURAL COMPLEXITY				
Continuity >30m along strike, down dip, & Low		2		
Continuity 10 - 30m along strike, down dip, & Medium		4		
Continuity <10m along strike, down dip, & High		6		
GRADE DISTRIBUTION				
Homogenous, similar DH and between holes		2		
Significant grades >2m from contacts, and/or clustering		4		
Significant grades within 2m of contacts, and/or heterogenous		8		
DELETERIOUS LITHOLOGIES, MINERALS				
Absent		0		
Present		4		
GEOTECHNICAL COMPLICATIONS				
Absent		0		
Present (Wall instability, Structural Complexity)		4		
TOTAL GEOLOGICAL RISK (GRISK) Higher the value, the greater the geological risk		12-52		

ACTION PLAN TO REDUCE TOTAL RISK AND IMPROVE ORE EXTRACTION

MINING RISK PARAMETRES			MITIGATE RATING
EXCAVATOR SIZE			SCORE
Appropriate	4		
Inappropriate	1		
MINING OPTIONS			
Full Flitch/Bench Mining	**DS**	**NS**	
1. Waste	2	2	
2. Ore	4	1	
OR			
Selective Flitch Mining	**DS**	**NS**	
1. Waste	4	-1	
2. Ore	1	-1	
MARKED UP STOCKS (optionality)			
Sufficient	4		
Insufficient	1		
GPS CONTROL			
GPS + Free Dig	4		
GPS + Blasted Material	1		
No GPS	0		
PIT CONSTRAINTS	**full**	**sel**	
Space sufficient	4	4	
Space confined	2	1	
Space insufficient	1	-1	
MINING METHOD			
Along Strike	4		
Up or down dip, or End on	2		
Back Select or Top Loading	-1		
MINING LEVELS			
Regular level checks feasible	4		
Regular level checks NOT feasible	0		
MINIMUM FACE LENGTH			
Available and being used	4		
Unavailable or available and not being used	0		
SHOT/BENCH MANAGEMENT			
No Dig Lines (NDL's) and/or Buffer Strategies Used	4		
NDL's, Buffers Unavailable	0		
MINE SERVICES			
Washed Down, Dozing/Grading along strike	4		
Washed Down, Dozing/Grading Clean up Supervised	2		
Not washed, Dozing/Grading Supervised	1		
Not washed, Dozing/Grading Unsupervised/cross strike	0		
SUMP LOCATIONS			
In WASTE	4		
In ORE	2		
Across Ore/Waste boundaries	0		
ORE SPOTTING & DESPATCH CONTROLLERS			
Geological Spotting and/or Ore Controllers available	4		
Geological Spotting and/or Ore Controllers NOT viable	0		
AVAILABLE SUPERVISION			
Available and Quality - as per standard	2		
Available and Poor - not per standard	0		
TOTAL MINING RISK (MRISK) Higher the value the better the Mitigation	0-52		

SIGNATURE GEOLOGY:	SIGNATURE ENGINEERING:	
TOTAL RISK SCORING: HIGH RISK <15 MODERATE RISK >=15 to <30 LOW RISK >=30	GRISK: Higher score the HIGHER the risk GRISK: Lower score Better Geology MRISK: Higher Score Better Mitigation MRISK: Lower Score Poor Mining	**TOTAL RISK = MRISK - GRISK**

(b)

FIG 2 – (a) Risk Card used for evaluating the GRISK and MRISK for long-term planning scenarios. (b) Risk card used for evaluating the GRISK and MRISK for short-term schedules.

The framework is designed for continuous use across the planning and execution cycle from long-term planning to short-term scheduling, and at key operational stages such as pre-blast, post-blast, and extraction. It is scalable and can be applied at different resolutions such as at the mine, bench, shot, lode, ore block scale, as shown in Figure 3.

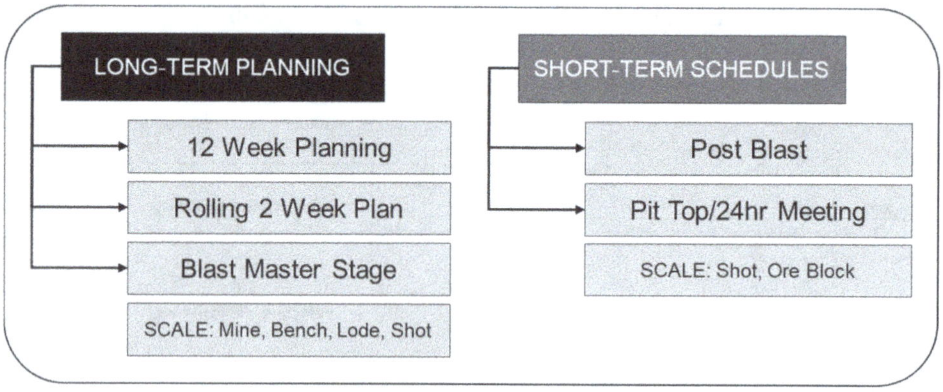

FIG 3 – The different scales and stages for the application of the GRISK and MRISK framework.

Operational testing demonstrated the need for two complementary Risk Cards:

1. Long-Term Risk Card, Figure 2a – supports strategic planning, design refinement, and pre-blast assessments and supports long-term planning.
2. Post-Blast Risk Card, Figure 2b – evaluates risk in the active mining phase and supports short-term scheduling.

The GRISK parameters remain consistent across both cards, while MRISK parameters are adjusted to reflect the operational requirements at each stage. Each card includes an agreed and signed Action Plan, which is captured in an action register. This ensures ongoing accountability and clarity of response.

The Risk Cards offer a standardised, auditable framework that is simple to use and embeds consistent risk management into daily operations.

The Total Risk Score is calculated by taking the GRISK score from the MRISK score, which indicates whether mining strategy and orebody characteristics are aligned, and risk mitigated. The range of the TRISK is from 40 to -52.

$$\text{TRISK} = \text{MRISK} - \text{GRISK}$$

TRISK scores are classified using a stoplight system informed by a calibrated lookup matrix as presented in Figure 4. These thresholds were refined through testing across multiple gold deposits of varying complexity to ensure alignment with geologists' qualitative assessments of geological risk and mining quality.

TRISK Classification:

- High Risk: < = 5.
- Moderate Risk: > 5 to < 25.
- Low Risk: ≥ 25.

The rating system is geared toward the recognition of geological risk and, as the TRISK lookup table demonstrates, an extremely complex orebody with a score of 52, even with a 'best practice' MRISK score of 52, will remain high risk with a TRISK score of 0.

Several examples can be followed in Figure 4. One example is where a 'best' mining practice of 47 offsets a deposit with a high geological risk of 38, to give a moderate TRISK of 9. Any slippage in the MRISK to an average quality pushes the TRISK to high, quickly.

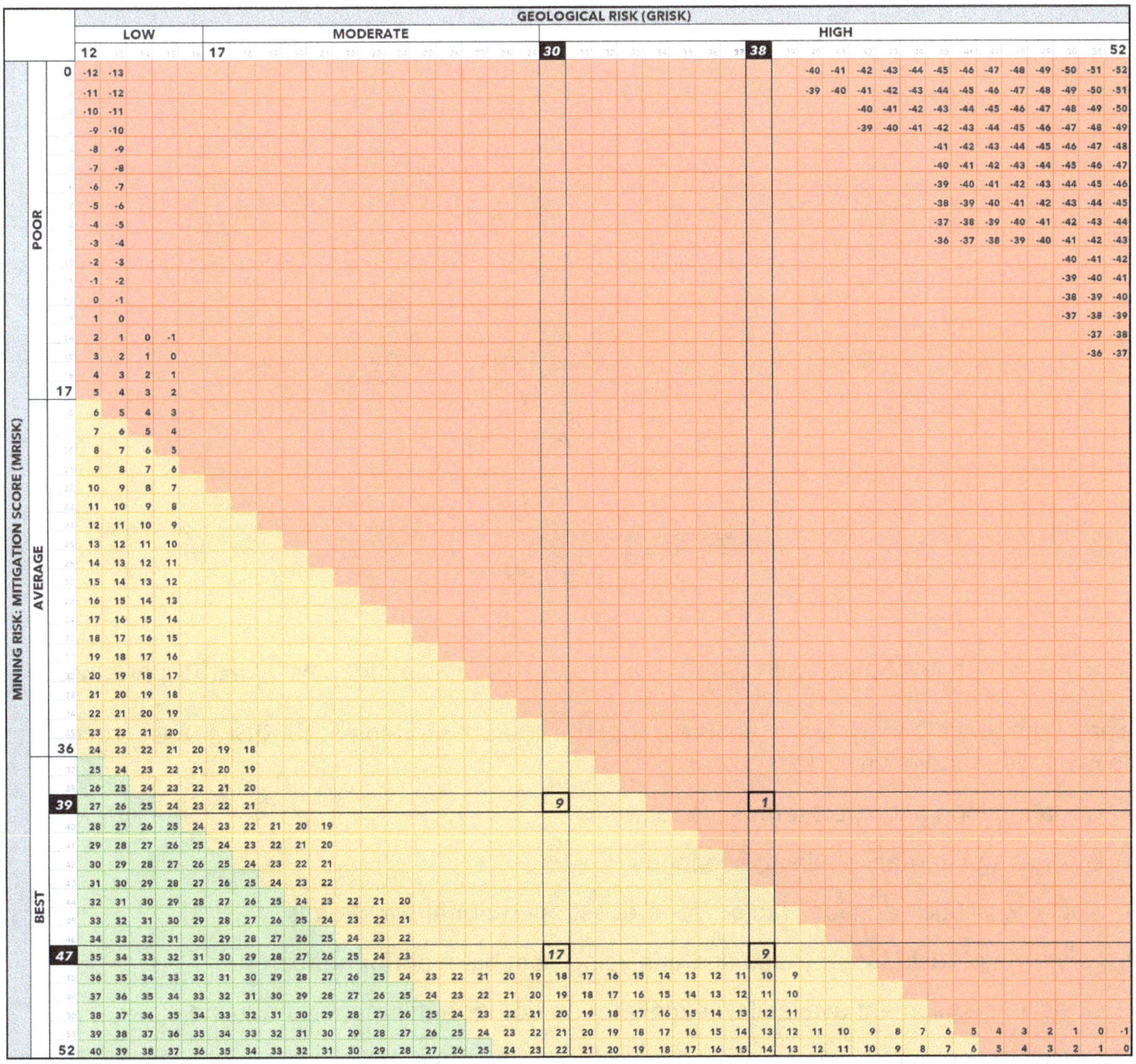

FIG 4 – The GRISK, MRISK and TRISK lookup table displaying worked examples, including an example that transfers from a high (1) to moderate (9) TRISK as the MRISK improves.

To maximise the accuracy and effectiveness of the risk assessment, prior preparation by the geology team is critical. Key preparatory steps include:

- Reviewing grade control drilling compliance.
- Validating drilling and sampling data (assays, logging accuracy).
- Updating domains and documenting interpretation uncertainty.
- Cross-checking drilling results with geological interpretation and models.
- Communicating structural or lithological features that may change risk.
- Presenting the risk spatially through bench-level risk plans, an example of which can be seen in Figure 5.

These steps ensure that geological inputs to the Risk Cards are evidence-based, consistent, and reflective of actual orebody conditions rather than assumptions.

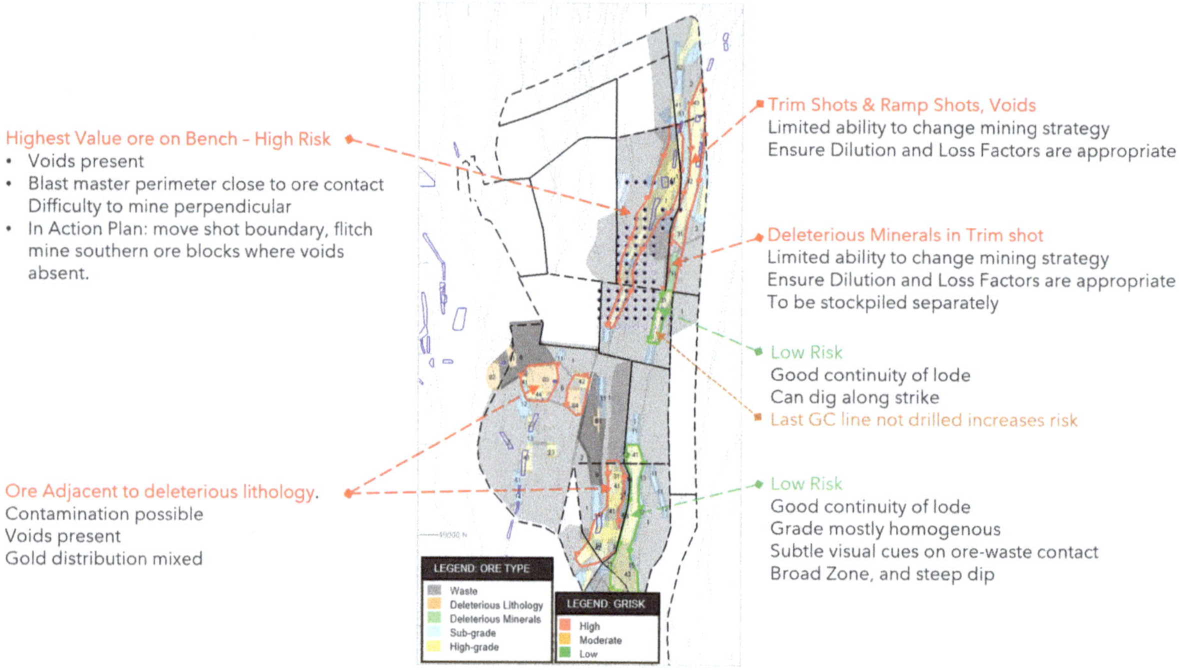

FIG 5 – Bench level plans with preparatory analysis of risk aligned with GRISK parameters.

Similarly, the engineering team ensures that mitigation strategies (MRISK) are practical and achievable. Preparation includes:

- Updating long and short-term schedules.
- Evaluating alternative mining scenarios.
- Applying dilution and loss factors to forecast operational constraints.
- Reviewing Blast Master Plans and analysing past blast performance.
- Checking equipment availability, excavator and other equipment scheduling.
- Identifying geotechnical complications that may affect extraction.

This preparation ensures operational decisions are aligned with geological risk and that mitigation strategies set out in the Risk Cards are realistic and executable.

The revenue attribute

In addition to the Geological and Mining Risk Matrix, Northern Star uses a complementary economic metric to communicate ore value: the Revenue Attribute (REV) and, where greater accuracy is needed, the Free Cash Flow Attribute (FCF). Unlike grade-only models, both attributes convert geological information into a dollar value per block by incorporating recovered ounces, dilution, loss, recovery, costs, and processing factors. This provides a clear and universally understood measure of economic importance, making value far more tangible than grade or ounces alone. This helps drive higher-quality mining practices and better short-term decisions.

The REV value is calculated for each ore block by:

$$\text{REV} = \text{ounces} \times (\text{dilution and loss factors}) \times \text{recovery} \times \text{reserve gold price}$$

This offers a simple block-by-block estimate of recovered revenue.

A more complete evaluation uses the FCF Attribute, defined as:

$$\text{FCF} = \text{Revenue} - \text{Costs}$$

Where Costs includes:

- mining costs (bench-tiered with depth)

- haulage costs
- processing costs
- royalties
- other operating or capital charges.

FCF provides a block-level view of economic margin and better reflects true mining cost variability with depth and haul distance. It is therefore the preferred measure for medium to long-term planning.

Presenting ore value in dollar terms, offers several advantages:

- Intuitive communication: Operators, engineers, and supervisors readily understand value expressed in revenue or cash flow.
- Visual clarity: REV/FCF can be displayed in 2D and 3D, making high-value ore blocks instantly recognisable as shown in Figure 6.
- Improved operational discipline: Teams can clearly see where careful, selective mining is essential, and where faster, bulk mining is acceptable.
- Value protection: Dollar values make the consequences of dilution, loss, or misallocation immediately visible, shifting decisions toward preserving value.

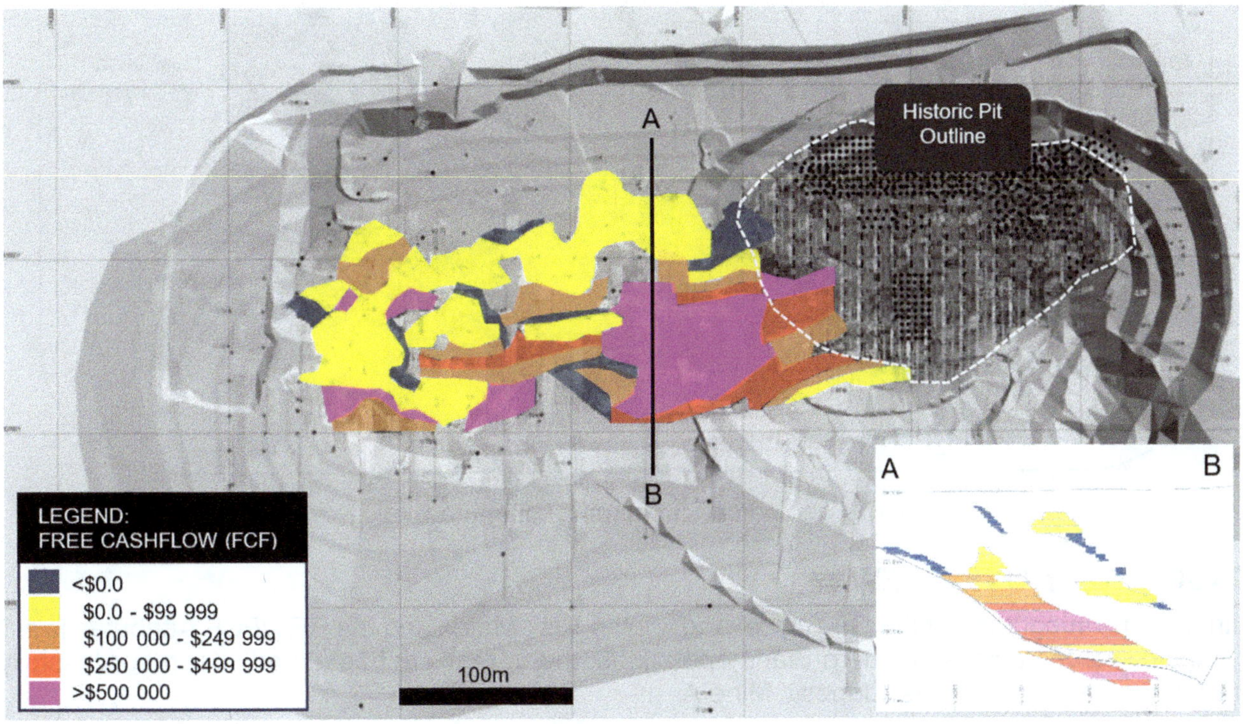

FIG 6 – Representative plan view and representative cross-section, (inset), displaying the FCF attribute.

Tool combination

When the FCF attribute is combined with GRISK/MRISK and TRISK, the result is a powerful, integrated decision-making framework. The framework ensures that operational effort, geological work, and mining controls are focused precisely where they yield the greatest financial return. It also reduces the likelihood that high-value material is lost through dilution, misclassification, or poorly aligned mining practices.

INTEGRATING RISK AND VALUE ACROSS THE MINING CYCLE

The combined use of GRISK, MRISK, TRISK, and the Revenue Attribute provides a structured way to manage geological uncertainty and optimise value. The framework ensures that risk is identified early, communicated clearly, and used to guide decisions at each stage of mining. It also supports reconciliation by highlighting where geological uncertainty is likely to cause deviation from plan.

In practice, these tools strengthen spatial compliance, reduce dilution, stabilise grade, and ensure short-term decisions remain consistent with operational expectations. When the inherent geological risk cannot be fully mitigated, the associated risks of misallocation, dilution, and loss must be recognised and incorporated into the planning frameworks.

By integrating geological knowledge, operational capability, and economic value, the framework helps navigate the inherent contradictions of complex ore systems and helps in protecting the value outlined in the original feasibility study. This cycle is outlined in Figure 7.

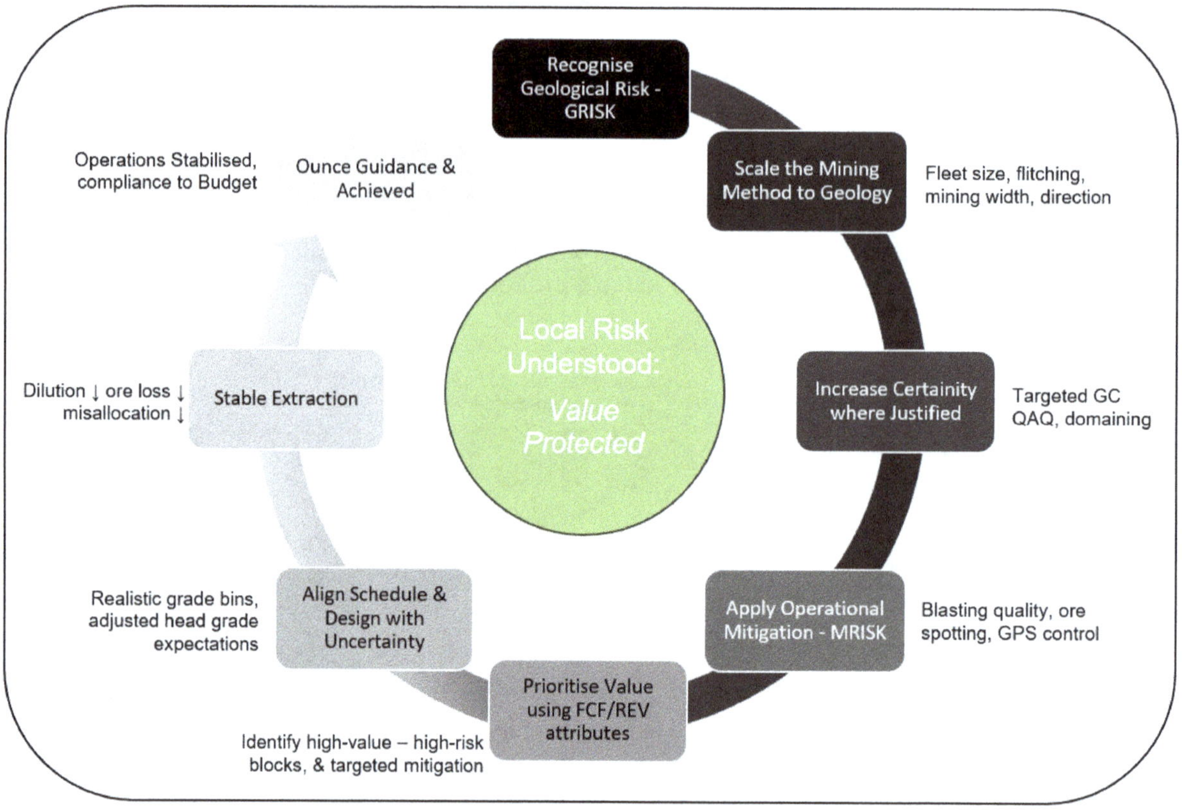

FIG 7 – The geological risk-mining mitigation-value cycle.

CASE STUDY OVERVIEW

Although the integration of the Geological and Mining Risk Matrix and the REV/FCF Attributes is still in its early stages across Northern Star's operations, the framework has already been successfully trialled at a bespoke high-risk deposit, Mine A.

This case study illustrates, in a practical setting, how geological complexity drives the cost–volume–grade contradiction, shapes mining decisions, and influences operational outcomes. It demonstrates the value of increased grade control drilling in reducing uncertainty, the effectiveness of the GRISK–MRISK approach in guiding fit-for-purpose mining strategies, and the power of the cash flow attribute in highlighting where value can be protected. Together, these elements show how an integrated risk and value framework can materially improve decision-making in complex, high-risk ore systems.

The economic gold at Mine A was confined to shallow (10–30°) south dipping, oxidised quart-sericite schists within an oxidised granite host, as shown in Figure 8a. It was characterised by thin (<10 cm) anastomosing and short-range quartz veins that plunged 30° to the SSE (Urquhart, 2017). As presented in Figures 8a and 9a, the main lode was variable in true width from 1 m to 10 m, with subsidiary hanging wall and footwall lodes <5 m in width.

FIG 8 – (a) shows the main mineralised shear zone at Mine A and its corresponding ore block mark up in the open pit with selective mining techniques. (b) is looking south and focuses on a ~10 cm sheared zone of quartz and gold at the hanging wall contact, with inset of dendritic gold found on oxidised shear planes. (b) and (c) highlights the visual variability in the hanging wall position of the main mineralised shear zone.

The main lode contained 85 per cent of the metal and value, was subtly visual, and the gold distribution was highly variable both downhole, (+70 per cent nugget effect), along strike and down dip. The gold mostly formed as visual blebs or dendritic forms <1 cm within the quartz veins, as seen in Figure 8b, however gold was also observed in the oxidised gangue and much of the quartz veining was also barren. As such the AU scatterplot comparing the original RC sample to the duplicate field sample was highly variable but also consistently random, as seen in Figure 9c (Bennison *et al*, 2019). The quality of sample collection was further challenged by excessive water within the deposit that impacted drilling.

The variability in geology and assays was evident at all scales: at the resource drill spacing 40 m × 40 m, and at the local grade control scale to 5 m × 4 m dice five pattern as presented in Figure 9. It was found that at a 5 m × 4 m grade control drill spacing, the gold distribution was mixed, and the ore-waste boundaries of the main lode remained spatially irregular. Figure 9a highlights these aspects of variability, with drill holes <1 m apart delivering 50 per cent variance in grade. Despite this variability the increased drill density highlighted improved grades in the hanging wall position which was coincidental with oxidised iron enrichment, as shown in Figure 8b.

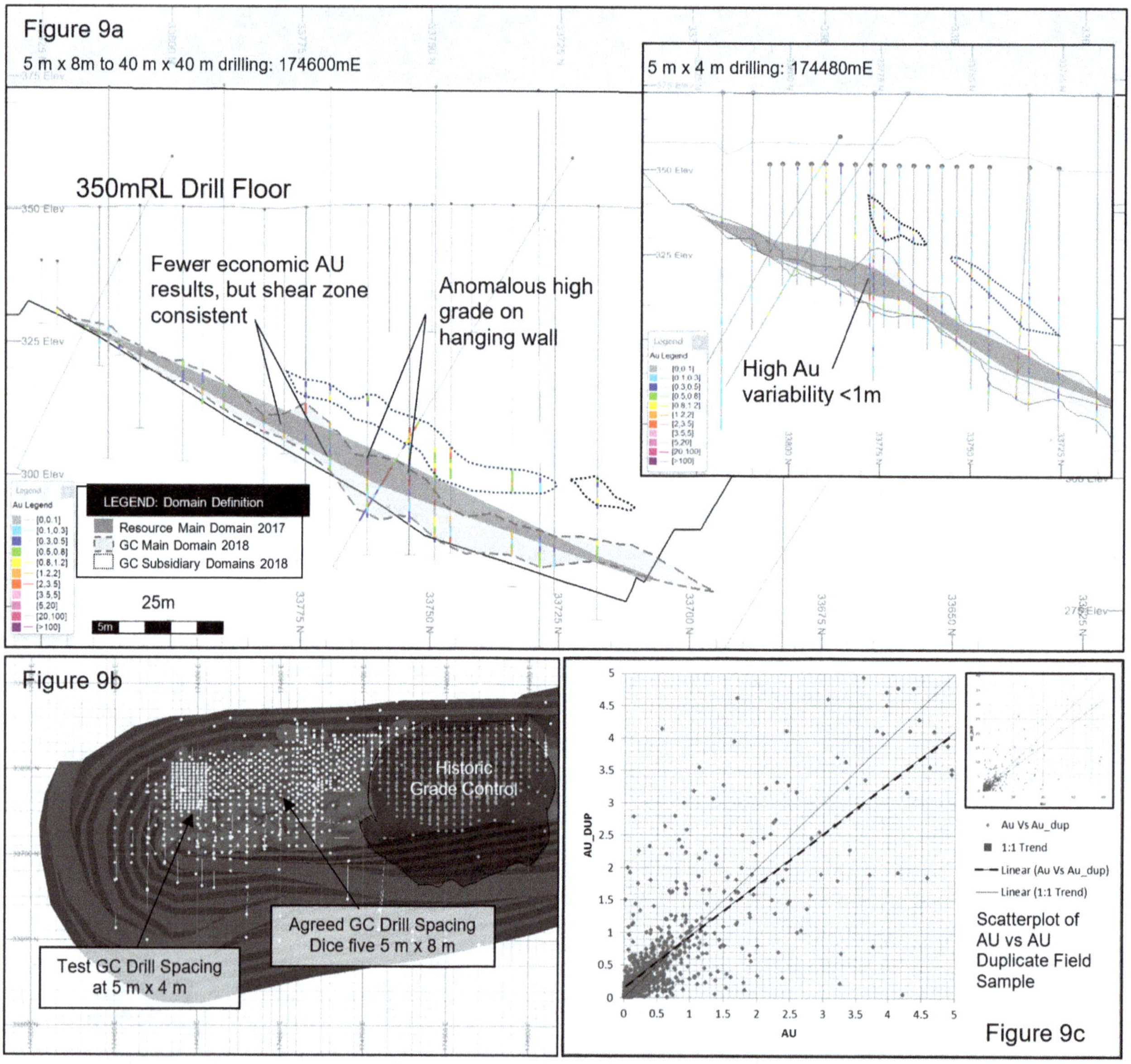

FIG 9 – (a) presents two representative cross-sections through Mine A at the various drill spacings and highlights the grade variability down drill holes, in adjacent drill drill holes and up and down dip. (b) is a plan view of the drilling spacings used and trialled at Mine A. (c) is a scatterplot of the original sampled AU compared to its field duplicate sample.

To ensure all the components of variability (field duplicate assay results, non-conformance of geology with gold occurrence, irregular ore-waste boundaries and mixed grade populations), were accounted for, a broader domaining policy was used.

The GRISK/MRISK and TRISK evaluation

The inherent geological risk was identified at the resource drilling stage, and the equivalent geological risk (GRISK) score was 38, placing it at the top end of the high-risk classification, (in Figure 10 the original score for each GRISK parameter is shaded). Consequently, time was given to appropriately grade control drill and test areas with closed space 5 m × 4 m grade control. In doing so, the geological risk score reduced to 30, with improved understanding of gold distribution and visual controls which supported the implementation and commitment to selective extraction of the main lode. However, with the inherent risk remaining high, mining strategies had to adjust to respect the geological uncertainty and ensure that dilution and ore loss was minimised. The score card for

Mine A can be seen in Figure 10 with GRISK and MRISK parameters scored, totalled and used to calculate the TRISK. Notably was the Action Plan that was successfully implemented.

OPEN PIT RISK CONTROL: LONG TERM RISK CARD

DATE: SHIFT: PIT: DIGGER: GEOLOGIST: ENGINEER/SUPERVISOR:

GEOLOGICAL RISK PARAMETRES		RISK RATING		
DIP		**full**	**sel**	**SCORE**
0-10 degrees		2	2	
10-30 degrees - optimal for selective		6	0	**0**
30-65 degrees		4	4	
>65 degrees - optimal for full flitch		0	6	
LODE THICKNESS	***Shovel***			
>5m	>10m	2		**2**
2.5m-5m	4m-10m	4		
<2.5m	<4m	6		
VOIDS (UG voids, Pit edges)				
No voids in shot/bench		0		**0**
Indirect interaction with ore		4		
Direct interaction with ore		6		
DEGREE OF WEATHERING				
Un-weathered		2		
Moderately Weathered		4		
Highly weathered		6		**6**
VISUAL CONTROL				
HW and FW clearly visible		2		
Moderately visible or one contact visible		4		**4**
No visibility		6		
GRADE CONTINUITY and/or STRUCTURAL COMPLEXITY				
Continuity >30m along strike, down dip, & Low		2		
Continuity 10 - 30m along strike, down dip, & Medium		4		
Continuity <10m along strike, down dip, & High		6		**6**
GRADE DISTRIBUTION				
Homogenous, similar DH and between holes		2		
Significant grades >2m from contacts, and/or clustering		4		
Significant grades within 2m of contacts, and/or heterogenous		8		**8**
DELETERIOUS LITHOLOGIES, MINERALS				
Absent		0		**0**
Present		4		
GEOTECHNICAL COMPLICATIONS				
Absent		0		
Present (Wall instability, Structural Complexity)		4		**4**
TOTAL GEOLOGICAL RISK (GRISK) Higher the value, the greater the geological risk		12-52		**30**

MINING RISK PARAMETRES	MITIGATE RATING		
EXCAVATOR SIZE			**SCORE**
Appropriate	4		**4**
Inappropriate	1		
MATERIAL COMPETENCY			
Free Dig Material	4		**4**
Blasted Material	1		
BLASTING OPTIONS			
Selective Waste & Ore	4		**4**
Strike Parallel	2		
Oblique	1		
Centre Lift	-1		
MINING OPTIONS			
Full Flitch Mining OR	**DS**	**NS**	
1. Waste	2	2	
2. Ore	4	1	
OR			
Selective Flitch Mining	**DS**	**NS**	
1. Waste	4	-1	**4**
2. Ore	1	-1	**1**
MARKED UP STOCKS (optionality)			
Sufficient	4		**4**
Insufficient	1		
GPS CONTROL			
GPS + Free Dig	4		**4**
GPS + Blasted Material	1		
No GPS	0		
PIT CONSTRAINTS	**full**	**sel**	
Space sufficient	4	4	**4**
Space confined	2	1	
Space insufficient	1	-1	
MINING METHOD			
Along Strike	4		**4**
Up or down dip, or End on	2		
Back Select or Top Loading	1		
MINING LEVELS			
Regular level checks feasible	4		
Regular level checks NOT feasible	0		**0**
MINIMUM FACE LENGTH			
Available and being used	4		**4**
Unavailable or available and not being used	0		
SHOT/BENCH MANAGEMENT			
No Dig Lines (NDL's) and/or Buffer Strategies Used	4		**4**
NDL's, Buffers Unavailable	0		
ORE SPOTTING & DESPATCH CONTROLLERS			
Geological supervision viable	4		**4**
Geological supervision NOT viable	0		
SUPERVISION			
Available and Quality - as per standard	2		**2**
Available and Poor - not per standard	0		
TOTAL MINING RISK (MRISK) Higher the value the better the Mitigation	0-52		**47**

ACTION PLAN TO REDUCE TOTAL RISK AND IMPROVE ORE EXTRACTION

AU mineralisation is highly variable, is shallow dipping with poor visibility along hanging wall and footwall contacts. Drilling indicates high grades at ore margins. ***Action Plan*** *: dig ore on day shift, main lode to be dug selectively with smaller excavator, stand off hanging wall and foot wall contacts an extra metre to ensure all metal is mined, dig along strike and dozing of floor is to be along strike with regular floor pick ups. All main lode ore must be spotted by a geologists. Because of the style of mineralisation, must mine all ore from hanging wall to footwall and not split into different grade parcels across strike.*

SIGNATURE GEOLOGY:	SIGNATURE ENGINEERING:		
TOTAL RISK SCORING: HIGH RISK <15 MODERATE RISK >=15 to <30 LOW RISK >=30	GRISK: Higher score the HIGHER the risk GRISK: Lower score Better Geology MRISK: Higher Score Better Mitigation MRISK: Lower Score Poor Mining	**TOTAL RISK = MRISK - GRISK**	**17**

FIG 10 – The calculation and record of the GRISK and MRISK score for Mine A using the Long-term risk card.

Sensitivity work with schedules and the financial balance sheet explored the trade-off between cost-volume-grade. The recognition of the high variability in gold distribution, the irregularity of the ore-waste boundary and elevated grades close to the hanging wall, and in parts the footwall, led to the successful mining strategy with a deliberate 1 m stand off on the upper and lower ore contacts. Financially it was more rewarding to accrue the metal than it was to reduce the dilution. That said, other mining strategies were put in place to ensure 'best practice' and that quality ore extraction was maintained. The smaller excavator was used to selectively mine the ore contacts along strike, over two flitches of a 5 m bench, with GPS control. Between the ore contacts, the ore was fully mined as broad ore blocks to account for the highly variable grade distribution and consequently minimise misallocation. And commitment was made to mine ore on day shift only when ore spotters could safely and effectively manage the extraction of the ore. For the non-blasted, free dig material the mitigation MRISK score of 47 successfully offset the GRISK of 30, to give a moderate TRISK score of 17. The risk could've been further mitigated if the pit conditions were drier, however water often impacted the quality of mining floors and levels. For blasted material the MRISK was slightly less effective at 39. These GRISK and mitigating MRISK scores, and the corresponding TRISK values are highlighted in Figure 4.

Case study outcomes

Despite drill spacing studies indicating that 5 m × 4 m was optimal for Mine A, the actual agreed and executed spacing was 5 m × 8 m dice five pattern. This was balanced against the cost of volume movements and schedule requirements. The risk of moving to a broader spacing was communicated clearly with all stakeholders, and it was deemed acceptable if there was no deviation from the 'best practice' strategy (Bennison *et al*, 2019).

At the life-of-mine and planning stages, understanding and acknowledging the risk of a Mine A, led to appropriate mining strategies, and a more realistic application of dilution and loss. Two factors that were afforded to the deposit, that was vital to getting this right, was TIME and GC Drill Density. Having both meant numerous sensitivity studies could be performed, where various estimates and techniques were compared (Byrne and Machukera, 2017), where full flitch mining versus selective mining was explored, where the impact of volume-grade-cost was calculated. It meant that target ounces were realistic, and it meant that dilution and loss values could be bespoke to the ore width and mining strategy and be applied on an ore block basis (Byrne, 2017).

On a bench to shot scale, to a week to day scale, the post blast risk card was used to refine the mining strategy, ensure compliance to the plan and operators were equipped with the knowledge and skill to carry out the work. This is where the FCF attribute played a critical role, as it identified the most valuable ore on the bench or in the shot and turned the grade value into a well understood dollar value. Figure 6 presents the FCF attribute highlighting the value of the ore blocks on the 305 mRL at Mine A. Not only did this allow high value ore blocks to be dug more carefully with a quality-focused strategy but also allowed subsidiary lodes with minimal value to be mined more quickly. This meant that the budget expectations of grade and volume were met (Bennison *et al,* 2019).

The application of the Geological and Mining Risk Matrix, combined with increased drill density and targeted mining controls, delivered several notable results (Bennison *et al*, 2019):

- Grade control drilling expanded the interpreted mineralised volume by ~49 per cent, costing an additional ~\$2M but generating an estimated \$40M increase in revenue.
- Actual cash flow exceeded budget by ~45 per cent, despite total operating costs increasing by ~30 per cent.
- Material movement costs were \$2/bcm higher than budgeted.
- Ore mining costs increased by approximately \$13/t relative to budget.
- Unplanned dilution averaged 13 per cent, with ore loss averaging 10 per cent across the deposit.
- Dilution varied strongly with mining method: blasted material mined non-selectively recorded ~21 per cent unplanned dilution, whereas free-dig selective mining achieved ~7 per cent unplanned dilution.

This case study reinforces the central theme of this paper: as higher gold prices unlock increasingly complex deposits, mining success depends on recognising geological uncertainty early and aligning mining strategies and controls. The learnings from Mine A demonstrate that when local risk is explicitly quantified and acted upon, operational outcomes improve, even in highly variable systems.

DISCUSSION AND RECOMMENDATIONS

Moving forward the standardised use of these tools and measuring the outcomes of embedding local geological risk into mine planning, will form the foundation for continuous improvement. The use of the tools and the side-by-side evaluations of risk will build understanding of geological risk, which is crucial to operational success. A more sophisticated understanding may lead to a step change in managing high-risk, complex orebodies, especially as increasingly marginal deposits come online in response to rising gold prices.

Figure 11 illustrates this trend clearly: as the gold price has increased, the geological risk of open pit deposits mined at Northern Star has also risen. This trend is unlikely to change, placing growing importance on understanding, quantifying, and managing geological uncertainty at the local scale, quickly, consistently, and with far greater operational precision.

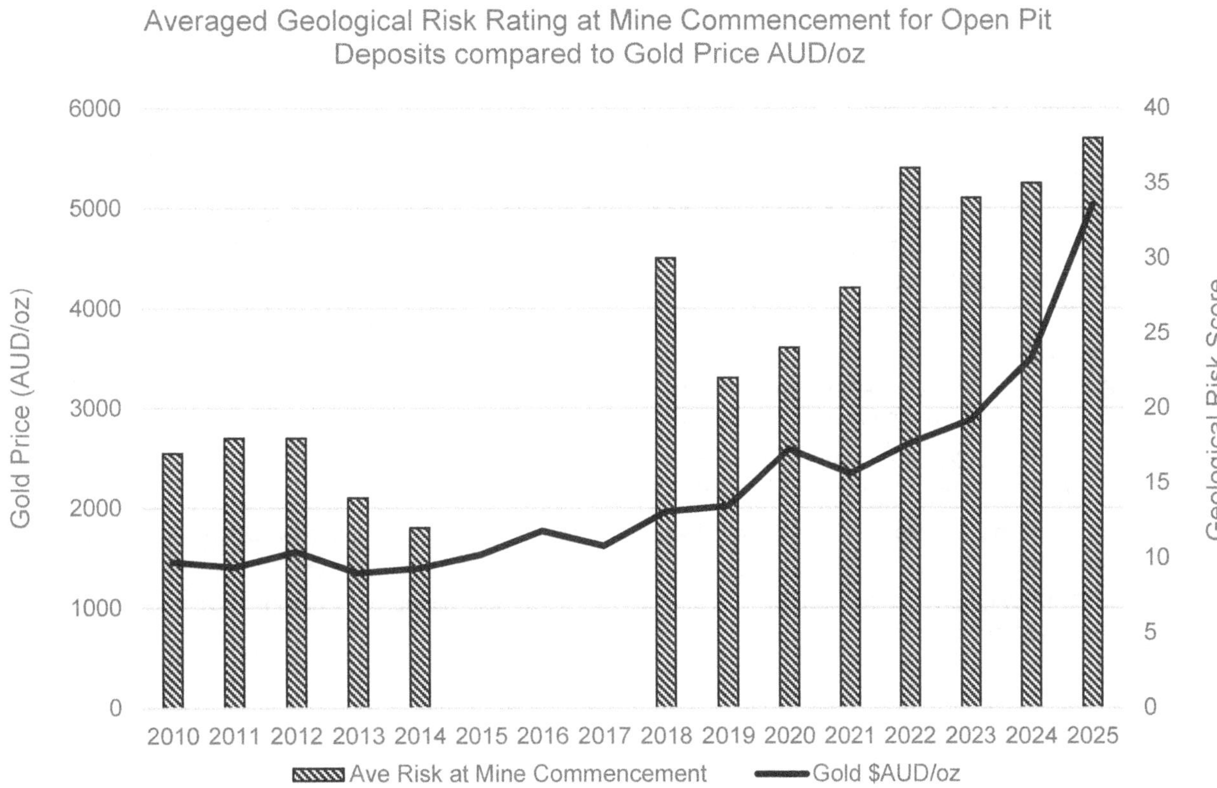

FIG 11 – Averaged geological risk rating evaluated at the commencement year for all gold open pit operations at Northern Star Resources.

Another equally important opportunity lies in how risk is communicated and shared across the organisation. While geological complexity is often discussed within technical teams, the operational consequences extend far beyond geology, affecting dilution, loss, cost, volume, grade, revenue, and ultimately business performance. As deposits become more variable and more complex with rising gold prices, the need for a clear, consistent, and organisation-wide understanding of risk becomes even more critical (Byrne, 2017).

Effective risk management therefore relies on shared visibility and common language across all functions: geology, mine operations, technical services, processing, finance, and senior leadership. A transparent and standardised platform for communicating risk, supported by tools such as TRISK and the FCF or Revenue Attributes, helps ensure that both long-term plans and short-term operational decisions around mining strategies, drill density and scheduling are grounded in a unified understanding of uncertainty, value, and consequence.

CONCLUSION

Complex, high-risk open pit gold systems present profound challenges when mined close to cut-off grade and under narrow profit margins. Geological variability, uncertain ore geometry, and limited visual controls increase the likelihood of dilution, ore loss, and deviation from plan. These risks are amplified by financial pressures that favour high-volume, low-cost mining methods that operate at a scale mismatched to the geology.

The Geological and Mining Risk Matrix, combined with the Revenue and Free Cash Flow Attributes, provides a practical and quantitative means to reconcile these contradictions. By explicitly recognising geological uncertainty and linking it to mining capability and economic value, the framework strengthens decision-making across the mining cycle. It improves spatial compliance, stabilises delivered grade, and preserves value in deposits where conventional approaches struggle.

As gold prices rise and increasingly complex deposits enter production, such integrated risk-and-value frameworks will be essential. They provide a structured way to navigate uncertainty, protect margins, and convert geological complexity from a constraint into an opportunity, unlocking value from orebodies that have long sat at the edges of economic viability.

ACKNOWLEDGEMENTS

The author would like to acknowledge the geology, engineering, and operational teams across Northern Star Resources whose expertise, collaboration, and willingness to share knowledge, challenge assumptions have been central to the development and trial of the Geological and Mining Risk Matrix and associated frameworks. The author would also like to acknowledge Northern Star Resources for their approval to publish this paper.

REFERENCES

Bennison, J, Burton, C, Byrne, B, Shelley, C and Smith, A, 2019. Mining Completion Report – Mine A Stage 1, Saracen Gold Mines, Thunderbox Operations 16 January 2019 report (internal).

Byrne, B and Machukera, J, 2017. Risk Analysis of the Mine A Estimation and Metal Content, Saracen Gold Mines Report, Thunderbox Operations 18 August 2017 report, (internal).

Byrne, B, 2017. The Mine A Project – Risk and Mitigation, Saracen Gold Mines, Thunderbox Operations 15 June 2017 report, (internal).

Urquhart, L, 2017. Exploration Blue Book of Mine A Deposit, Saracen Gold Mines, Thunderbox Operations 20 February 2017 report, (internal).

Engineering better outcomes – drill spacing as a tool for ore recovery, dilution control, and risk mitigation at South Kalgoorlie

J Mead[1]

1. Senior Mine Geologist, Northern Star Resources, Kalgoorlie WA 6430.
   Email: jmead@nsrltd.com

ABSTRACT

Drill spacing is often evaluated from a resource estimation and grade continuity perspective, with geologists advocating for tighter spacing to improve model accuracy. However, the cost of increased drilling should be considered with the tangible value addition, particularly the positive impact on revenue. At Northern Star Resources Ltd's South Kalgoorlie Operations, gold mineralisation is adjacent to a weak ultramafic unit in the footwall. This ultramafic is highly susceptible to overbreak, and in some cases has led to complete sterilisation of longhole stopes before ore is extracted. Understanding the ultramafics location is key to stope performance. This study assesses the impact of increasing grade control density from 20 m × 20 m to 10 m × 10 m. The aim is to improve orebody definition, refine the lithological model of the ultramafic, and enhance stope design, ultimately reducing dilution and offsetting the additional grade control drilling cost.

Current grade control drilling (20 m × 20 m) is conducted prior to the establishment of ore drives. Ore drives designed on this drill spacing aim to avoid intersecting the ultramafic contact while following the strike of the orebody. Although geological direction and mapping are used to refine drive and stope positioning, the contact is visually subtle and undulating resulting inconsistent mapping and subsequent re-development. Exposure of the ultramafic, either through development or longhole drilling of the above stopes, can lead to significant dilution, impacting the economics of mined stopes. While engineering controls have assisted, the most effective mitigation is avoiding exposure of the ultramafic altogether.

Tighter grade control spacing (10 m × 10 m) aims to better define the ultramafic contact while also increasing confidence in grade distribution. This enabled more precise drive and stope placement, optimised pillar design, and reduced geological risk during ore extraction. Where stoping followed the plan, observed dilution was reduced to 11 per cent from 26 per cent in comparably mined areas. This demonstrates that the additional cost of drilling is outweighed by improved mining execution and reduced dilution.

INTRODUCTION

Drill spacing in underground mining is traditionally evaluated from a resource estimation perspective, with spacing recommendations based on geological and grade continuity. However, the operational and economic implications of drill spacing extend beyond resource confidence. At the South Kalgoorlie Operations (SKO), key geological units present unique challenges for stope design and ore recovery.

Located in the Eastern Goldfields Superterrane, the Hampton Boulder Jubilee (HBJ) underground mine is located within the Boulder Lefroy shear zone approximately 35 km south of Kalgoorlie (part of the New Celebration mining district, Figure 1). The Northern ore zone (NOZ) orebody at HBJ is hosted on the hanging wall of an ultramafic unit that is highly susceptible to overbreak. Exposure of this ultramafic during development or stoping has historically resulted in significant footwall overbreak and, in extreme cases, complete sterilisation of stopes prior to ore extraction. While engineering controls have reduced some of the overbreak, the most effective strategy remains avoiding ultramafic exposure altogether.

Standard grade control practice for the NOZ orebody consists of underground diamond drilling completed at approximately 20 m × 20 m spacing prior to level development. Drive direction is completed under geology control, whereby the geologist provides direction cut by cut to optimise the lode position in the drive. Face sampling, backs mapping and sludge drilling is collected to refine the geological and grade control model for stoping. This paper presents a case study on the impact of tighter grade control drill spacing (10 m × 10 m). It evaluates improvements in geological modelling, stope design and planning, and quantifies the benefits in ore recovery, dilution control, and operational efficiency. Importantly, the analysis considers the economic relationship between drilling costs and mining performance.

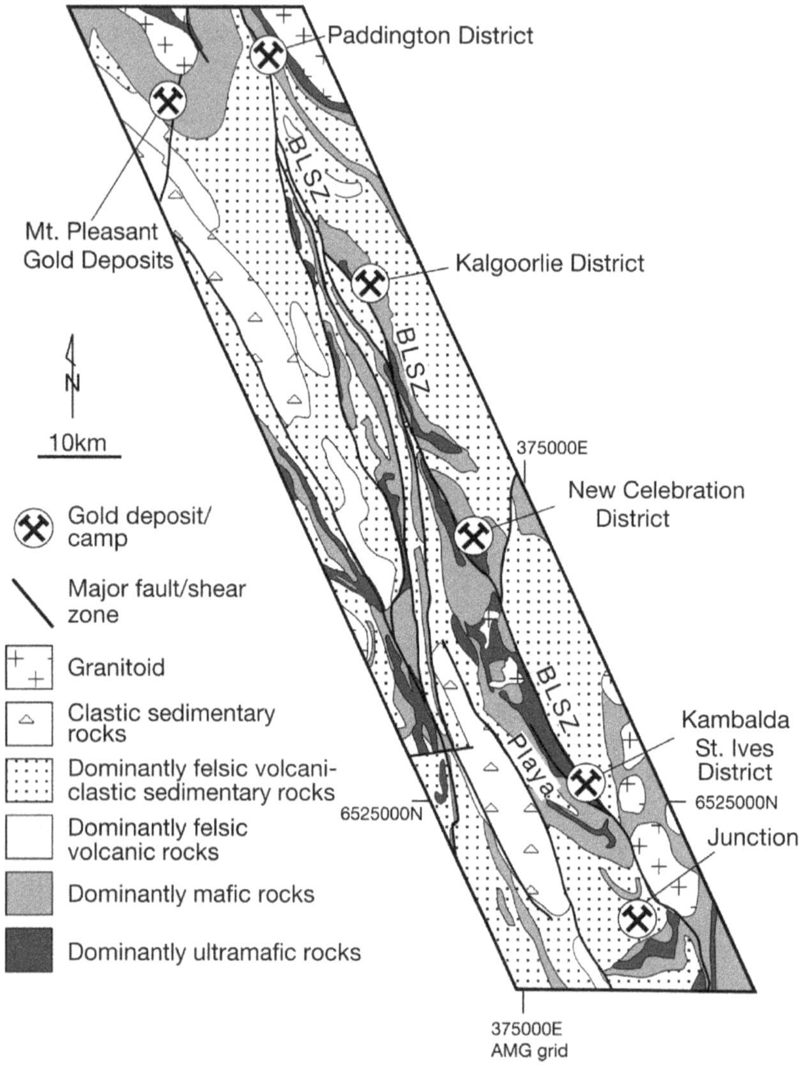

FIG 1 – Schematic map of the Boulder-Lefroy shear zone and surrounding geology and important gold camps (Weinberg *et al,* 2005).

The findings demonstrate that investment in geological data can deliver substantial operational and financial benefits, even in cost-constrained environments. This case study provides practical insights for geologists and engineers seeking to optimise and justify drill spacing for improved business outcomes and risk management.

GEOLOGY AND OPERATIONAL BACKGROUND

Regional geology

The New Celebration deposits encompassing the HBJ mine, are located within the Boulder Lefroy Shear Zone, a significant first-order structure in the Kalgoorlie terrane of the Eastern Goldfields Superterrane (Figure 2). The stratigraphy consists of mafic and ultramafic volcanic rocks overlain by felsic volcaniclastic, with later intrusion by dolerite and porphyry bodies. This sequence was shaped by multiple Neoarchean tectonic events, resulting in a complex structural architecture and greenschist facies metamorphism (Nichols and Hagemann, 2014).

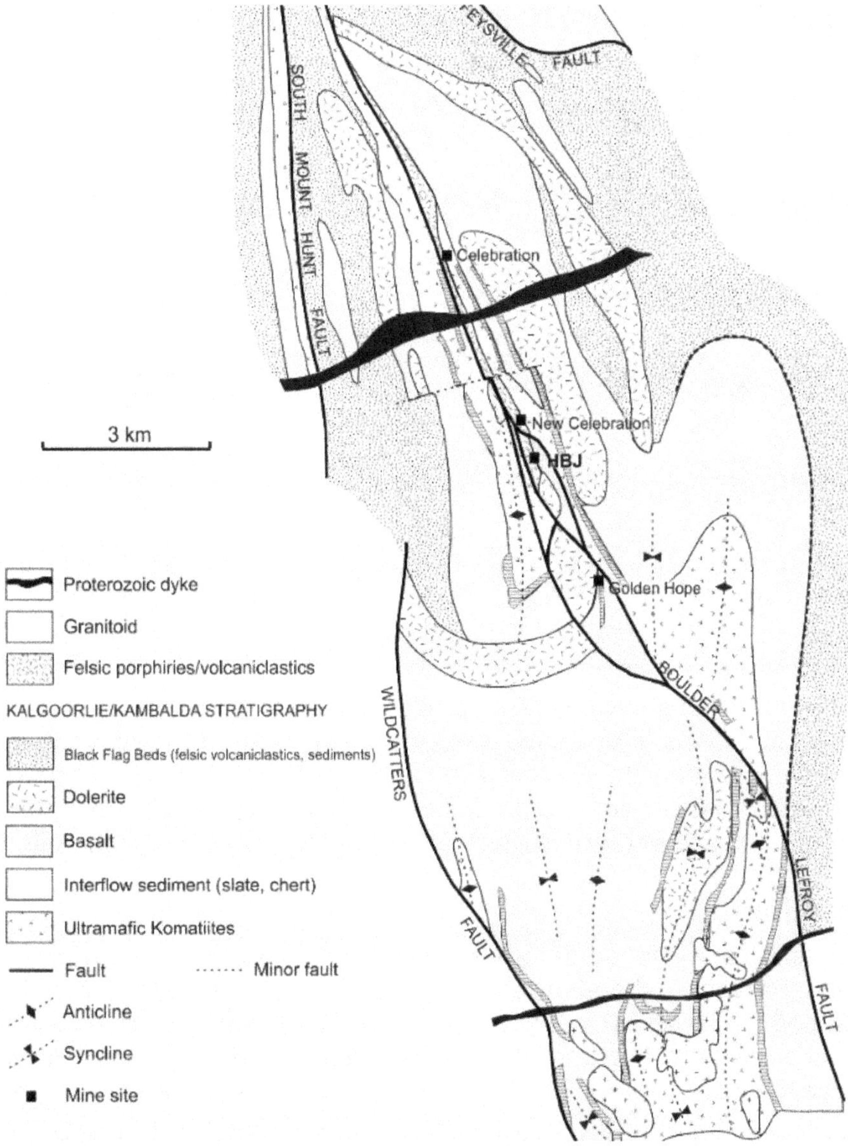

FIG 2 – Camp scale geology of the Kalgoorlie-Kambalda greenstone belt surrounding HBJ (Swager, 1989; as adapted in Mead, Sharrock and Ayris, 2024).

Local geology

The NOZ orebody at HBJ is situated on the hanging wall of the Celebration Fault, a splay of the Boulder Lefroy Shear Zone. The hanging wall consists of highly strained volcaniclastics that young to the west, bound by a mafic intrusive dolerite. The footwall consists of an ultramafic unit that young's eastward intruded by two distinct porphyry phases (Figure 3). A steeply west dipping mylonitic shear zone along the north striking Celebration Fault separates the two distinct zones. The entire hanging wall package is folded into an anticline plunging northward (Mead, Sharrock and Ayris, 2024).

Gold mineralisation at HBJ is primarily characterised by three styles. These include an altered mylonitic zone adjacent to the Celebration Fault, contact mineralisation associated with the porphyry intrusions and, late bucky quartz stockwork veins within the porphyry and dolerite intrusives. The mylonitic NOZ orebody has a strong plunge influenced by the shallow North dipping fold axis and a second subordinate southerly plunge linked to mineral lineations perpendicular to the folding. A series NNE trending faults influence the mineralisation continuity, causing boudinaging across the mylonitic and porphyry orebodies (Mead, Sharrock and Ayris, 2024). The NOZ mineralisation is visually indistinguishable from waste volcaniclastics surrounding it due to the intense alteration present.

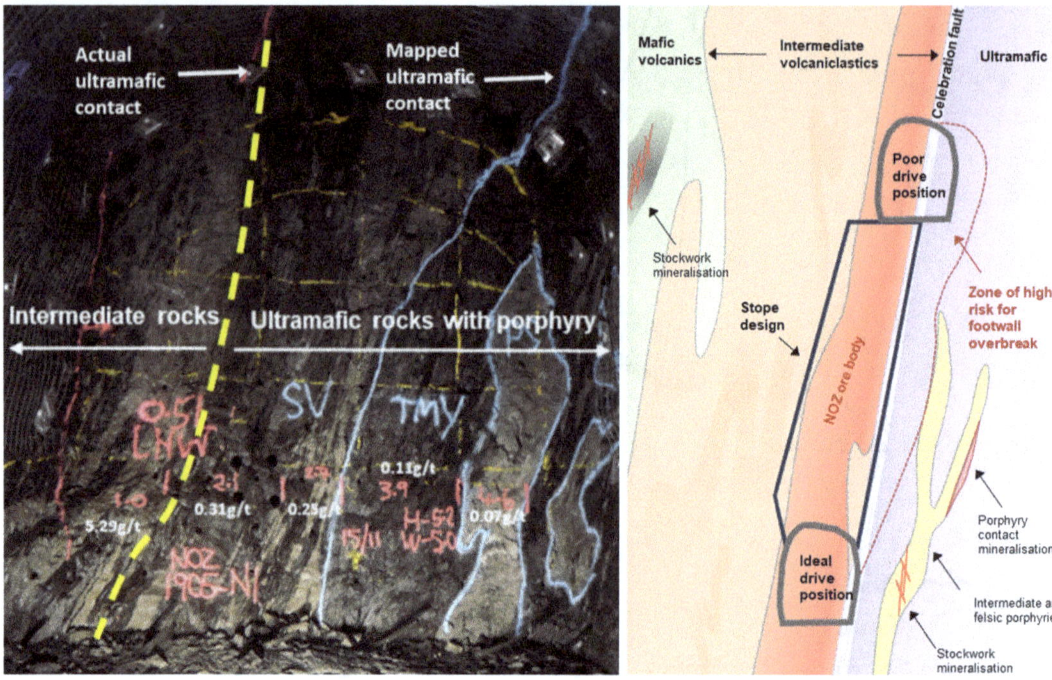

FIG 3 – (left) Face mapping showing mark up of volcaniclastics (SV) and mylonitic ultramafic (TMY) with porphyry intrusions (PS) versus assayed grades and XRF testing showing the actual boundary between the intermediate volcaniclastic and ultramafic package (adapted from Sinclair, 2021). (right) Schematic cross-section (not to scale) showing a type-example of the lithology, stope outlines, areas of grade and areas of low strength ultramafic contributing to dilution.

Implications to mining

The structural complexity and intense deformation make lithological boundaries difficult to identify visually underground. Pervasive alteration across the orebody makes different units appear similar, where intermediate porphyries and volcaniclastics being easily mistaken for one another (Figure 3). While tools such as hardness testing and portable XRF assist in distinguishing ultramafics from the volcaniclastics, these methods are typically applied in drill core rather than at the development face.

Accurate delineation of the ultramafic footwall ahead of time is critical for minimising stope dilution. Exposure of the ultramafic during development or longhole drilling can lead to significant overbreak, reducing stope grades and increasing waste tonnes. This risk underpins the need for robust geological modelling and input to mine planning.

The path to reduce dilution

Mineralisation at the HBJ mine extends over 4 km with a long history of surface and underground mining along its extents. The HBJ pit production was interrupted by major pit wall collapses of the ultramafic in the 2000s making access to high-grade open pit ore extremely difficult and effecting production for around 6 months (Dioro Exploration NL, 2009). Early underground mining at HBJ was completed with a combination of airleg and alimak stoping, using hydraulic sand fill. When longhole open stoping was introduced, large frequent rib pillars were required with sill pillars on every level (Brown, 2021). A number of stopes were unable to be surveyed and some development levels were abandoned before stoping due to ground failures and floor heave. During this period of mining, drives were completed on geology control with limited drilling and no lithology model. This meant limited consideration of the ultramafic during stope design with stopes focused on the ore position only.

In 2018, mining transitioned predominately into the NOZ orebody (Figure 4). Although the rock mass was more competent than previous mining areas, stope stability remained a challenge. In response, changes to mining practices were implemented from 2019. Increased emphasis was placed on drive positioning to minimise exposure of the ultramafic in development. Low density explosives were introduced for both development and stoping to reduce disruption of the ultramafic. Lithological modelling was further improved and actively incorporated into stope planning including drill and blast.

These measures resulted in a measurable reduction in dilution however, residual dilution risk associated with the ultramafic remained.

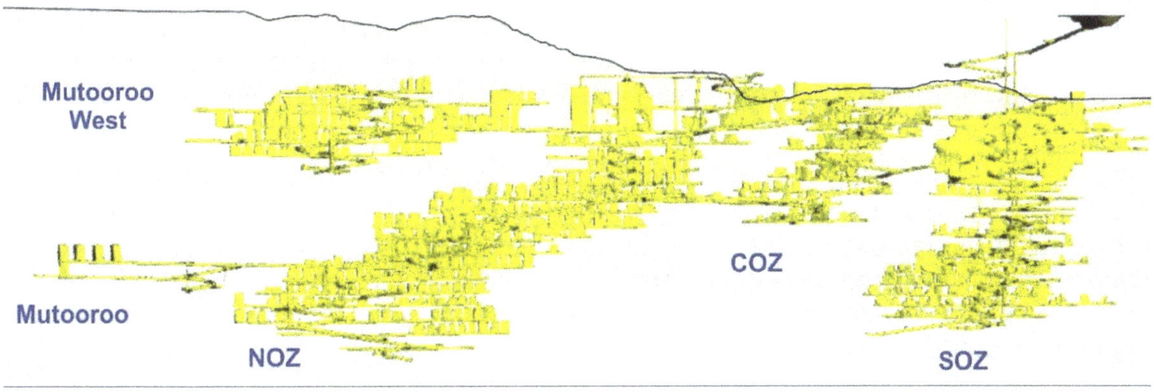

FIG 4 – Mining areas of the HBJ underground mine (Mead, Sharrock and Ayris, 2024).

PROBLEM STATEMENT AND MOTIVATION

Operational challenge

At the time increased grade control was proposed, recent levels on the NOZ orebody had experienced significant dilution. The 1905 level had seen an average of 26 per cent overbreak. Stoping was carried out using longhole open stoping, with minimum mining widths of less than 3 m, heights of 15–20 m, and strike lengths of approximately 20 m. Retreat stoping had commenced on the 1830 and 1855 level (Figure 5) towards a central access. Dilution was increasing with depth due to increased ground stress and time dependant failures becoming common. The removal of sill pillars to maximise ore extraction enabled significant failures from stopes above to propagate down into active stopes.

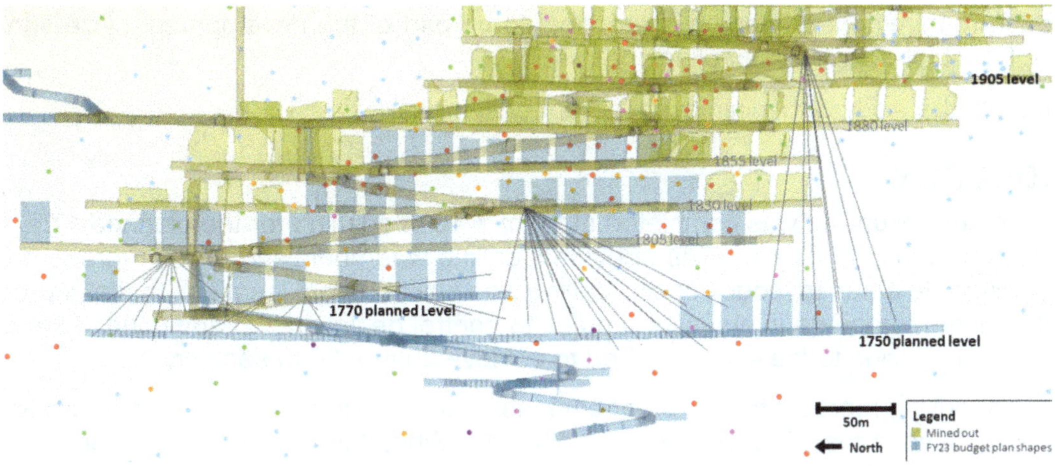

FIG 5 – Zoomed in view of the NOZ orebody mining front, showing planned 10 m × 10 m grade control in long section with heat map drill intercepts, current mine and planned FY23 development and stoping at the time of drill designing, May 2022.

The dilution of stopes in NOZ is predominantly observed on the footwall due to failure of the ultramafic unit. Its low rock strength (45–80 Mpa) makes it highly susceptible to overbreak. The challenge of avoiding the ultramafic in development drives and stopes is compounded by the subtle and undulating nature of the ultramafic contact, which is difficult to determine visually.

In FY23 the geology team motivated to select an area to tighten the drill spacing. The proposal aimed to deliver improved geological data, in particular define the local variability of the ultramafic, increase development rates through survey controlled development, improved stope dilution, ultimately delivering measurable operational and financial value.

Normal practice

The standard grade control practice for the NOZ orebody consists of diamond drilling at 20 m × 20 m spacing, providing adequate resource confidence but insufficient resolution of lithological boundaries for development to be completed under survey control. Geology drive direction is used for ore development with the objective to keep the ultramafic contact just outside the drive profile. Sludge drilling conducted by the development jumbo into each wall is also used to assist in identifying the position of the ultramafic, which can be visually distinct in drill returns. However, accurately following a subtle and undulating contact with geology direction had been unreliable.

Development drives under geology control can have several challenges: drives can be directed offline by geologists, increasing development costs (Rose, 2024); production inefficiencies when development is delayed pending geology input, increased risk of injury to geologists standing at the face to map and sample; and delays to incorporating data into grade control models, mine plan and stope shape updates.

Objective

The objective of this study was to evaluate the impact of tighter grade control drill spacing completed ahead of development. Specifically, aiming to:

- Assess improvements to geological confidence in defining the ultramafic footwall contact and orebody geometry allowing for survey control of ore drives.
- Enhance stope design, planning and execution to reduce dilution.
- Improve foresight for planning, enabling more accurate scheduling and resource allocation.
- Quantify economic relationships between additional drilling costs and the value gained through increased ore recovery and reduced dilution.
- Assess operational benefits, including increased development efficiencies from development cycle turnaround times and correct positioning (not redoing development).
- Reduce requirement on geology personnel to be part of the development cycle and remove them from high-risk face sampling.
- Complete the study with minimal spend and resources.

METHODOLOGY

Tighter grade control drilling was completed at 10 m × 10 m spacing for the planned 1750 level, on the main NOZ orebody (Figure 6). The additional drilling was prioritised for high-grade ore zones, excluding low-grade areas to reduce costs. Drilling was planned from locations that allowed for short hole lengths and appropriate intersection angles to control deviation. An additional 3 km of drilling was designed in addition to the 4 km required to achieve 20 m × 20 m spacing.

The additional drilling represented a 75 per cent increase in drilling and required a further 24 days of to complete beyond the 33 days originally planned. While this extended the model completion timeline, all drilling and modelling was completed ahead of level development and did not impede the mine plan.

To evaluate the value of grade control drilling to 10 m × 10 m spacing prior to development, the NOZ ore domains, Celebration Fault and ultramafic contact, were modelled and estimated twice with the tighter grade control campaign included and excluded. Extraction plans were then run on both models optimising rib pillar positions and stope sizes to maximise ore extraction. Mining was then complete on the level using the model with additional grade control data.

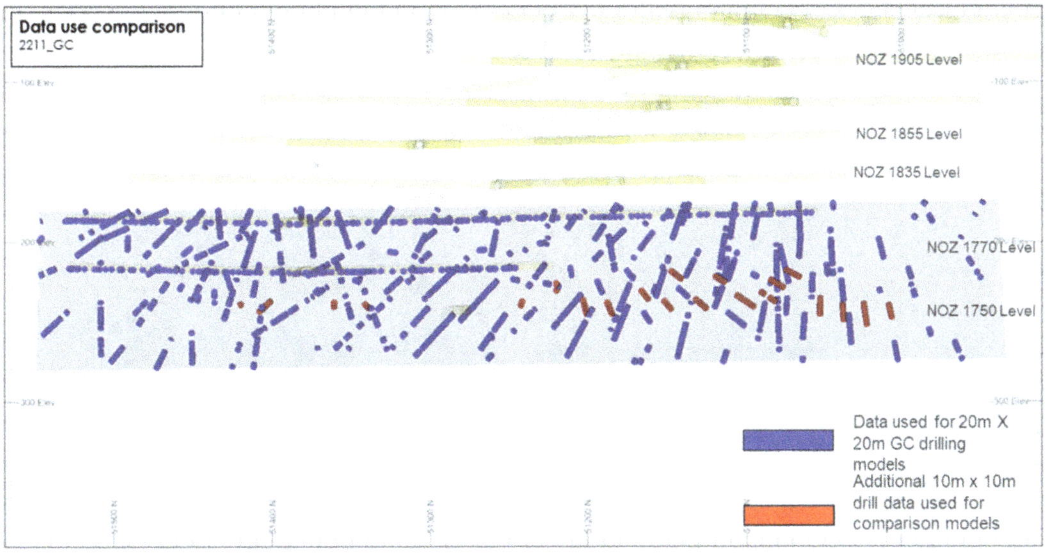

FIG 6 – Long-section view (looking east) showing data used for the comparison models.

RESULTS

To compare the outcomes of the 1750 Level results, the 1905 Level was chosen for comparison. The 1905 Level was the most recently completed mining level on the NOZ orebody and therefore provided the most relevant benchmark for assessing changes in planning and mining outcomes. The 1905 level is comparable to the 1750 ore drive in terms of northing, strike length (390 m on the 1905 Level and 440 m on the 1750 Level), mining method (both life-of-mine sill pillar levels) and geological setting.

Development

Tighter spaced grade control drilling allowed the 1750 South ore drive to be completed entirely under survey control. Increased geological confidence in the ultramafic and ore contacts allowed design decisions to be implemented at the level extraction planning stage that otherwise would not have been implemented. A comparison of the differing development designs is shown in Figure 7.

Exposure of the ultramafic contact in the 1750 South drive after development totalled 29 m, representing 11 per cent of the total development length. In comparison, the 1905 South drive experienced 105.5 m of ultramafic exposure, equivalent to 44 per cent of the development length. Maximum exposure of ultramafic in the 1750 South drive was 2.3 m in mapped faces, occurring where the ultramafic was locally dragged into the drive by a fault. Excluding this fault related exposure, the ultramafic contact did not extend more than 0.5 m into the drive. In comparison the maximum ultramafic exposure on the 1905 South drive reached 3.2 m in mapped faces, with an average exposure of 1.0 m.

The increased drilling data provided by the tighter grade control drilling reduced the requirement for intensive face sampling during development. On the 1750 South drive, 35 of the 77 cuts were sampled (45 per cent). In contrast, on the level above, the 1770 South development required 73 of 106 faces to be sampled (69 per cent), with the development frequently placed on hold over night shift to achieve a minimum sampling frequency of every second face. The 1905 South drive had 67 of 74 faces sampled (91 per cent) which reflects night shift geology coverage present at the time of developing the 1905. Night shift coverage was not present for the 1770 or 1750 development.

When development advance performance is compared, the first 77 cuts taken on the 1770 South level took 196 days to complete, compared with 146 days for the 77 cuts taken on the 1750 South drive to be completed. Assuming delays were primarily associated with geology placing the heading on hold to sample the following day, completion of tighter drilling ahead of time corresponded with an approximate increase in development rates of 30 per cent. The successful use of survey control resulted in no stripping, or redevelopment, in contrast to the 1905 level where stripping was required on multiple occasions due to the drive drifting off the ore and into the ultramafic. Reduced face sampling requirements also lowered exposure of geological personnel at the face by approximately 24 per cent on the 1750 South drive compared to the 1770 South drive.

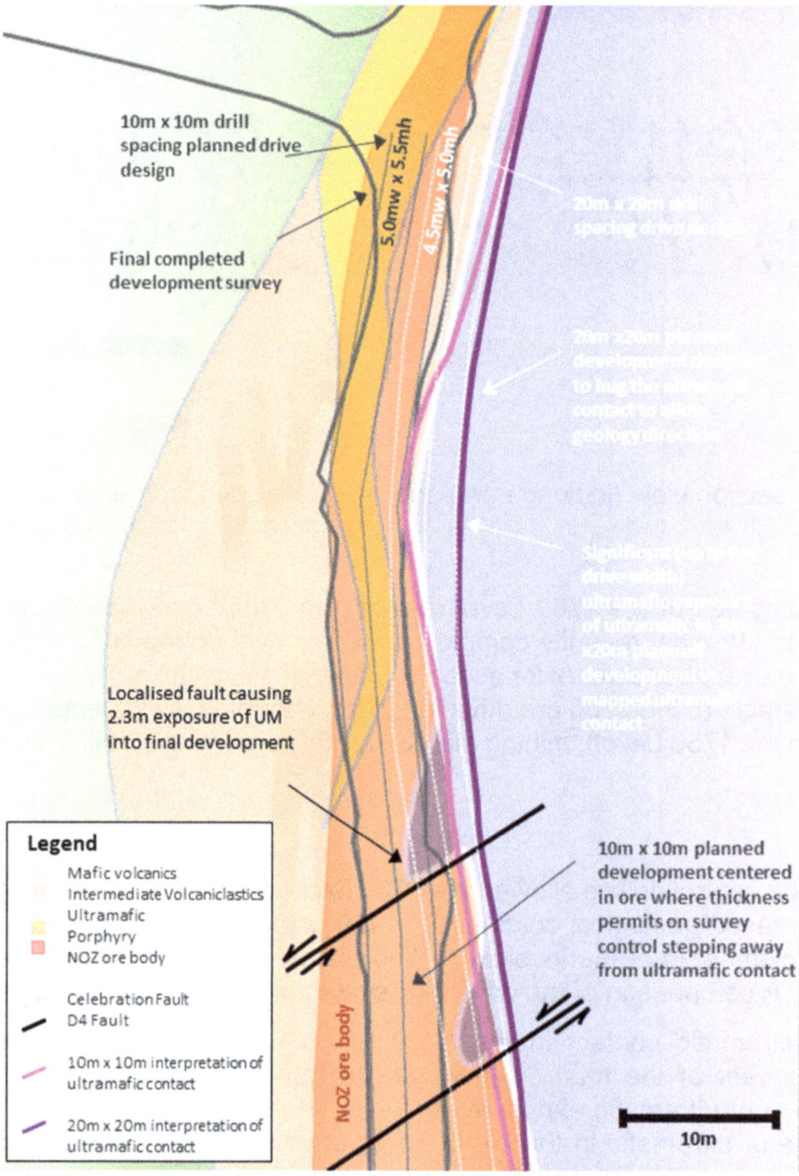

FIG 7 – Plan view of the 1750 South drive from access turnout to 70 m showing completed drive and mapping as well as planned development and ultramafic contacts based of the two comparison models.

Stopes

Mining of the 1750 level was completed in July 2024. Overall observed overbreak from stoping, where scans were achieved was 18 per cent. A reduction from the 26 per cent observed on the 1905 Level. Importantly, for stopes that were directly comparable to those on the 1905 level (followed the planned design, included adequate pillar support, and were not affected by other influences) overbreak was only 11 per cent (Figures 8 and 9).

Survey scans could not be completed for three of the stopes mined on the 1750 level. The first stope taken on the southern extent of 1750 South drive failed prior to full extraction of the ore. No dilution was mined. Two additional stopes mined to full height, excluding a sill pillar, suffered failure propagating down from the level above, choking the drawpoint and preventing scanning. The data for these stopes could not be included in the overbreak analysis. It is worth noting that one of the failed stopes yielded 57 per cent additional waste tonnes from failure of the hanging wall propagating to the level above prior to being abandoned.

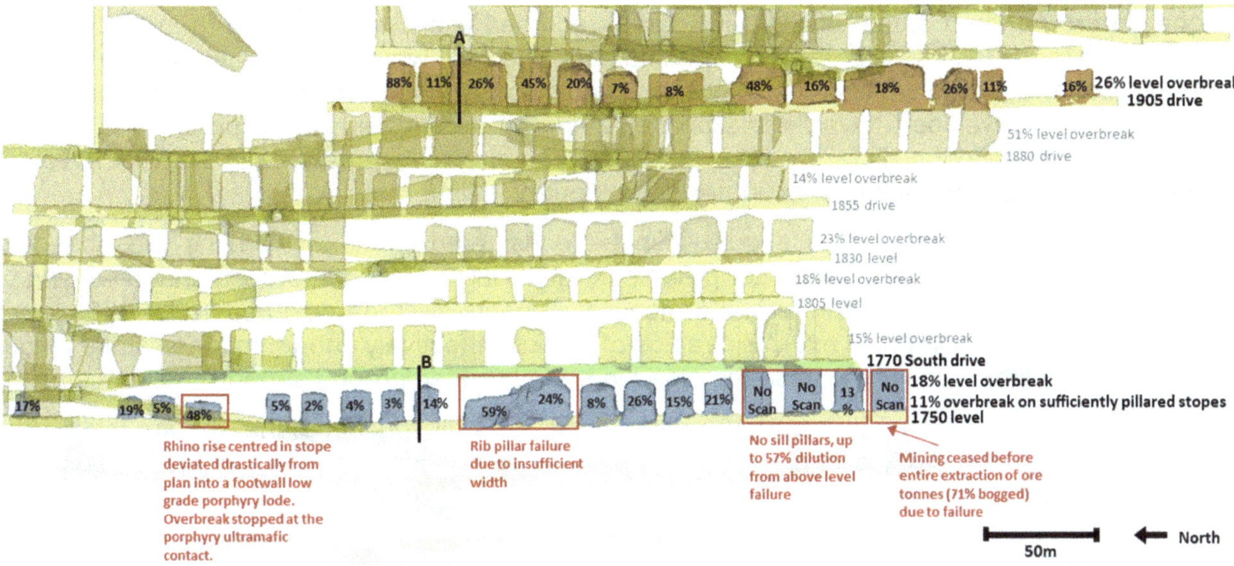

FIG 8 – Long-section view (looking east) showing data used for the stope overbreak based on designed stope versus scanned stope, two highlighted cross sections in Figure 9.

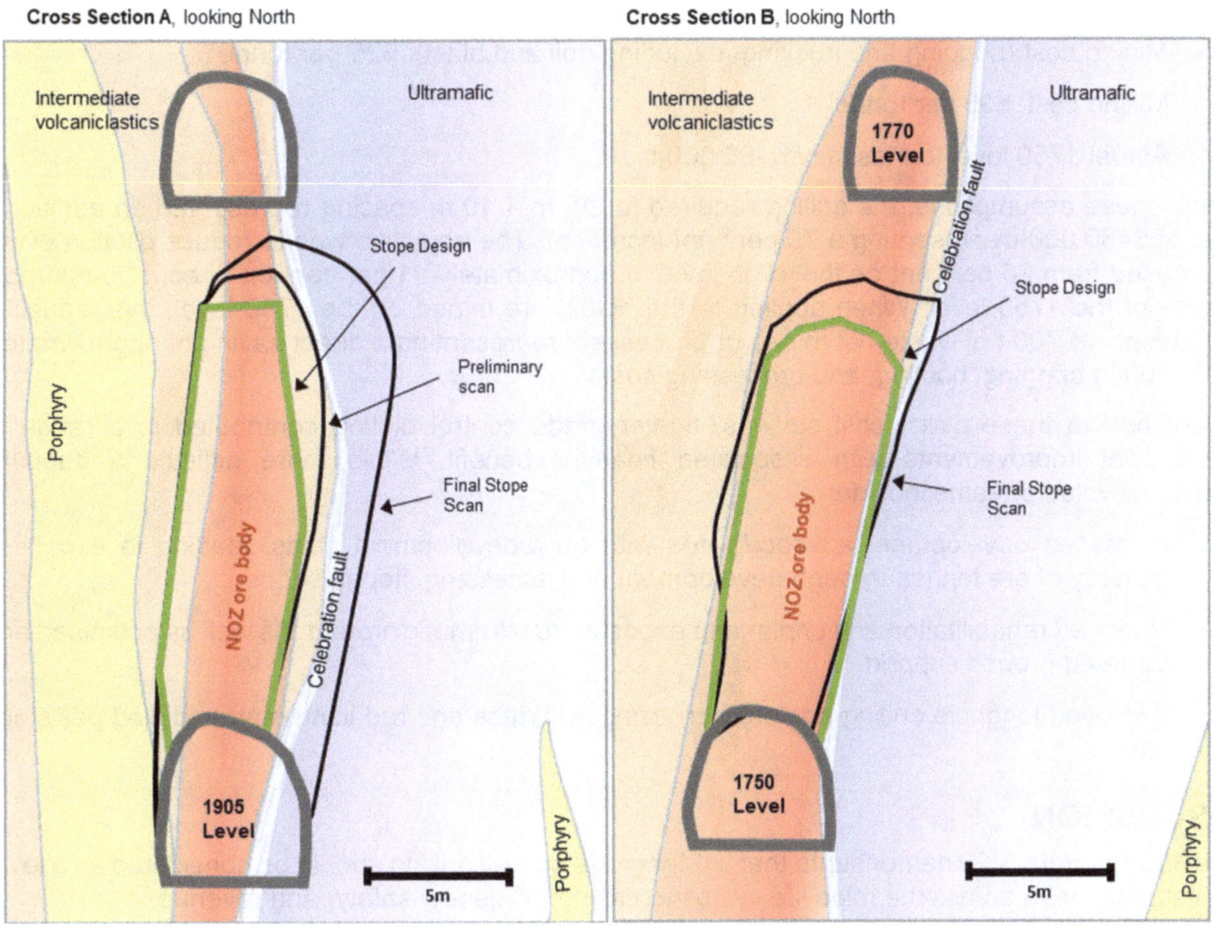

FIG 9 – Cross-sections A and B (highlighted in the long section Figure 8) of type stope design and survey scans on the 1905 (left) and 1750 (right).

One stope near the centre of the level experienced a significant rib pillar failure which was unrelated to the location of the ultramafic in the stope. Following the failure, the decision was made to continue extraction by mining the adjacent stopes sequentially, effectively combining them into a single enlarged stope.

A stope on the northern extent of the 1750 Level recorded a distinct deviation from the planned geometry due to issues associated with the rhino rise. The rise, which was intended to be centred within the stope, deviated significantly into a low-grade porphyry lode located in the footwall. As a result, overbreak developed towards this porphyry unit but terminated at the porphyry ultramafic contact. This stope and the rib failure was not considered directly comparable to the 1905 level for the purpose of assessing the effect of tighter grade control drilling

A number of geotechnical factors should also be considered when interpreting the observed outcomes on stope overbreak. The tighter grade control drilling provided improved understanding of the rock mass conditions, which in turn informed adjustments to stope design. On the 1750 level, this resulted in a reduction in the hydraulic radius of stopes, effectively shortening their strike length and height. The higher resolution data also aided in more targeted ground support.

These factors make it difficult to isolate the direct impact of drill spacing alone, though the overall improvement within the study scope suggests a positive correlation between increased geological certainty and reduced dilution.

Cost analysis

This analysis uses the listed assumptions to illustrate the relative revenue impact of tighter grade control drilling. These assumptions have been simplified and do not reflect actual SKO costs.

- Drilling cost: $150 per metre (incl assays).
- Mining cost (bogging and trucking, excluding drill and blast): $25 per tonne.
- Milling cost: $35 per tonne.
- Actual 1750 level tonnes mined: 98 000 t.

Under these assumptions, the drilling required for 10 m × 10 m spacing represented an additional cost of $450 000 (representing a 75 per cent increase). The objective was to reduce dilution which decreased from 26 per cent on the 1905 level to approximately 11 per cent on directly comparable stopes of the 1750 level. When applied to the actual ore mined on the 1750 level, this reduction equates to 14 700 t of waste not mined or processed, representing a direct saving of approximately $822 000 in bogging, hauling, and processing costs.

In addition to these direct cost savings, tighter grade control drilling contributed to a range of operational improvements with associated financial benefit. While more difficult to quantify, additional value streams include:

- Increased development advance rates with no redevelopment costs, leading to expedited delivery of ore tonnes through development and accessing stopes sooner.
- Reduced rehabilitation and unplanned exposure to ultramafic ground. As well as optimised and targeted ground support.
- Improved longhole drilling angles, increasing drill rates and reducing metres drilled per stope tonne.

DISCUSSION

The findings from SKO demonstrate that tighter grade control drilling should be considered as a lever for value creation across the mine life cycle increasing efficiency, safety, and revenue.

This study provides a framework for justifying drilling beyond estimation requirements. Through articulating the downstream benefits, such as reduced dilution, improved safety, planning foresight, accelerated mining and increased revenue geological data can be viewed as a value driver rather than a discretionary cost. While drill spacing for resource estimation remains important, the operational benefits observed at SKO supports the need to broaden the evaluation criteria for drill spacing and associated cost.

Observed dilution reductions, from 26 per cent on the 1905 level to 11 per cent on the 1750 level, demonstrates the operational impact of improved geological definition before mining commences.

While some dilution incurred was outside of this studies control, the improvement achieved within the study scope validates the approach. Importantly, this study reinforced that targeted geological spend can deliver substantial financial benefits.

When considered collectively, the operational improvements indicate that tighter grade control spacing delivered a material net positive outcome. When only the directly quantifiable savings from reduced dilution are considered, the additional drilling expenditure was fully offset. When the broader operational and geotechnical benefits are included, the value realised from tighter drilling is substantially greater than the incremental cost.

CONCLUSION

This study demonstrates how optimising drill spacing in relation to geological definition, not just resource estimation requirements, can play a role in operational performance and help justify the additional cost. At SKO, tighter grade control spacing improved important lithological models, reduced dilution, enabled better mining execution and economic outcomes.

ACKNOWLEDGEMENTS

The Author would like to acknowledge Northern Star Resources for their permission to use the data from South Kalgoorlie Operations in this publication.

REFERENCES

Brown, C, 2021. SKO Mining in Ultramafics, internal report, South Kalgoorlie Operations, Northern Star Resources, December 2021.

Dioro Exploration NL, 2009. Annual Report 2008–09, ASX announcement, published 27 November 2009. Available from: <https://announcements.asx.com.au/asxpdf/20091127/pdf/31md0yfjgbg4k1.pdf> [Accessed: December 2025].

Mead, V, Sharrock, B and Ayris, M, 2024. Directional change in underground drilling – a case study, in *Proceedings of the International Mining Geology Conference 2024*, pp 355–370 (The Australasian Institute of Mining and Metallurgy: Melbourne).

Nichols, S J and Hagemann, S G, 2014. Structural and hydrothermal alteration evidence for two gold mineralisation events at the New Celebration gold deposits in Western Australia, *Australian Journal of Earth Sciences*, 61(1):113–141.

Rose, S, 2024. The dilemma of geological control for underground ore development, *Coring Magazine*, 24 May 2024.

Sinclair, J, 2021. HBJ Mine — Northern Ore Zone (NOZ) Compilation, Geology Services Team, Northern Star Resources, February 2021, unpublished internal report.

Swager, C, 1989. Structure of Kalgoorlie greenstones – regional deformation history and implications for the structural setting of the Golden Mile gold deposits, *Western Australia Geological Survey Report*, 25:59–84.

Weinberg, R F, Van der Borgh, P, Bateman, R J and Groves, D I, 2005. Kinematic history of the Boulder–Lefroy shear zone system and controls on associated gold mineralization, Yilgarn Craton, Western Australia, *Economic Geology*, 100(7):1407–1426.

Implementing blast movement monitoring – a case study from Chatree Gold Mine

O Phongphanngam[1] and R Gray[2]

1. Senior Mine Geologist, Akara Resources, Phichit, Thailand, 66230.
   Email: owin@akararesources.com
2. Senior Geology Superintendent, Akara Resources, Phichit, Thailand, 66230.
   Email: rosemaryg@akararesources.com

ABSTRACT

Chatree Gold Mine in central Thailand is characterised by narrow, discontinuous epithermal quartz veins where ore control is highly sensitive to blast-induced movement. Following a six-year operational shutdown, the mine recommenced operations in 2024. During this restart phase, ore boundaries were marked using bamboo poles, providing only a guide to ore position. Mineralisation itself is not visually distinguishable in the field; however, dykes are identifiable and cross-cut mineralisation, providing visual reference that can be used to validate the accuracy of blast movement monitoring (BMM) offsets. BMM had been used prior to the mine's temporary closure and its 2025 reinstatement aimed to quantify movement and evaluate the technical and financial benefits of a measured approach.

Nine ore blasts were monitored with 148 sensors, achieving a 96 per cent detection rate. A campaign mill trial was conducted to assess geological performance under controlled conditions. Across the monitored blasts, use of BMM generated an estimated US$4.52 million in additional value, reaching payback in fewer than three blasts.

Financial year-to-date reconciliation shows that ounces milled are within 1 per cent of grade control (GC) predictions, confirming that BMM adjusted dig lines accurately represent ore boundaries. The tonne and grade variance approximately 10 per cent higher tonnes and 10 per cent, lower grade has been reviewed in detail. Of this variance, approximately 6 per cent represents claim over GC tonnes, likely reflecting spotting, digging and blasting related dilution. The remaining approximately 4 per cent is attributed to operational misallocation. Misallocation in this context refers to trucks tipping material to an incorrect destination. Material movements are currently tracked using spotters and truck sheets, and the additional tonnes were not recorded as being moved from the pit to the run-of-mine (ROM). A fleet management system will be introduced in 2026 to improve truck destination tracking and reduce misallocation risk. BMM has delivered measurable improvements in ore control performance. Next steps will focus on evaluating ore movement analysis tools, including Hexagon blast movement intelligence (BMI) and AI approaches, to determine whether their benefits outweigh their cost and workflow impact. Integration with fleet management is noted as supporting operational improvement.

INTRODUCTION

Chatree Gold Mine extracts gold from narrow epithermal quartz veins and stockwork zones that change width and continuity over short distances. In such environments, even modest blast related displacement can shift ore away from planned dig lines, reducing recovery and increasing dilution.

During the restart period, ore outlines were marked using bamboo poles as a temporary measure, but this provided minimal information on actual movement and made reconciliation difficult as production increased. To improve the reliability of ore control decisions, blast movement monitoring (BMM) was reintroduced in 2025 to provide measured three-dimensional movement data and support a more consistent workflow. The objective was to improve dig line accuracy, reduce dilution and strengthen confidence in ore delivery to the mill.

BACKGROUND

Narrow vein deposits are particularly prone to ore-waste boundary displacement during blasting. Without quantitative movement data, post blast ore position often results in ore loss and dilution.

BMM resolves this by determining post-blast sensor positions and generating displacement vectors for application to ore polygons prior to excavation, consistent with documented industry practice (Loeb and Thornton, 2014). Horizontal displacement guides dig line shifts, while vertical movement provides insight into bench scale behaviour and potential movement across flitch boundaries. Quarterly reviews have been completed with the drill and blast team to assess all blasts and measured movements, with findings used to refine blast design and implementation practices.

METHODOLOGY

Nine production blasts from A Pit were selected for assessment. A total of 148 BMM sensors were deployed, and post-blast detection was completed before material movement. A 96 per cent recovery rate of sensor locations post-blast was achieved.

Movement vectors were processed into horizontal and vertical components. Movement vectors were used to update dig lines. Vertical displacement was reviewed as part of quarterly assessments.

A campaign mill trial was completed to evaluate geological performance independent of blending. Ore from monitored blasts was processed as a discrete batch and compared with grade control (GC) estimates to validate ore boundary accuracy.

Evaluation included movement magnitude, dig compliance, delivered grade, reconciliation and financial return. Implementation factors such as workflow integration and training needs were also reviewed.

RESULTS

The displacement vectors showed both lateral and vertical movement, with vertical shifts locally reaching up to 4.68 m; horizontal reaching up to 7.38 m; and an average at 3.58 m.

A campaign mill trial was completed on ore mined between the 28th January and 19th February 2025, with material processed between the 3rd and 24th February. The GC *in situ* model for the trial areas predicted an average grade of 0.42 g/t Au. The processed material averaged 0.39 g/t Au. The mill grade was -7 per cent of the modelled grade, mined tonnes were 9 per cent higher than GC predictions, and the corresponding reduction in grade is consistent with dilution rather than errors in ore boundary interpretation. The mill trial therefore provided confidence that BMM adjusted dig lines accurately reflected post blast ore positions.

An economic evaluation was completed to quantify the financial benefit of applying BMM movement adjustments. This analysis compared BMM adjusted dig lines with unadjusted outlines (Figure 1). Across the nine monitored blasts, the combined effect of reduced dilution and improved ore recovery is estimated to have delivered approximately US$4.52 million in additional value, with payback achieved in fewer than three blasts. Table 1 highlights the movement discrepancy between bamboo poles (blast vector indicator – BVI) and BMMs in ore loss, dilution and misclassification.

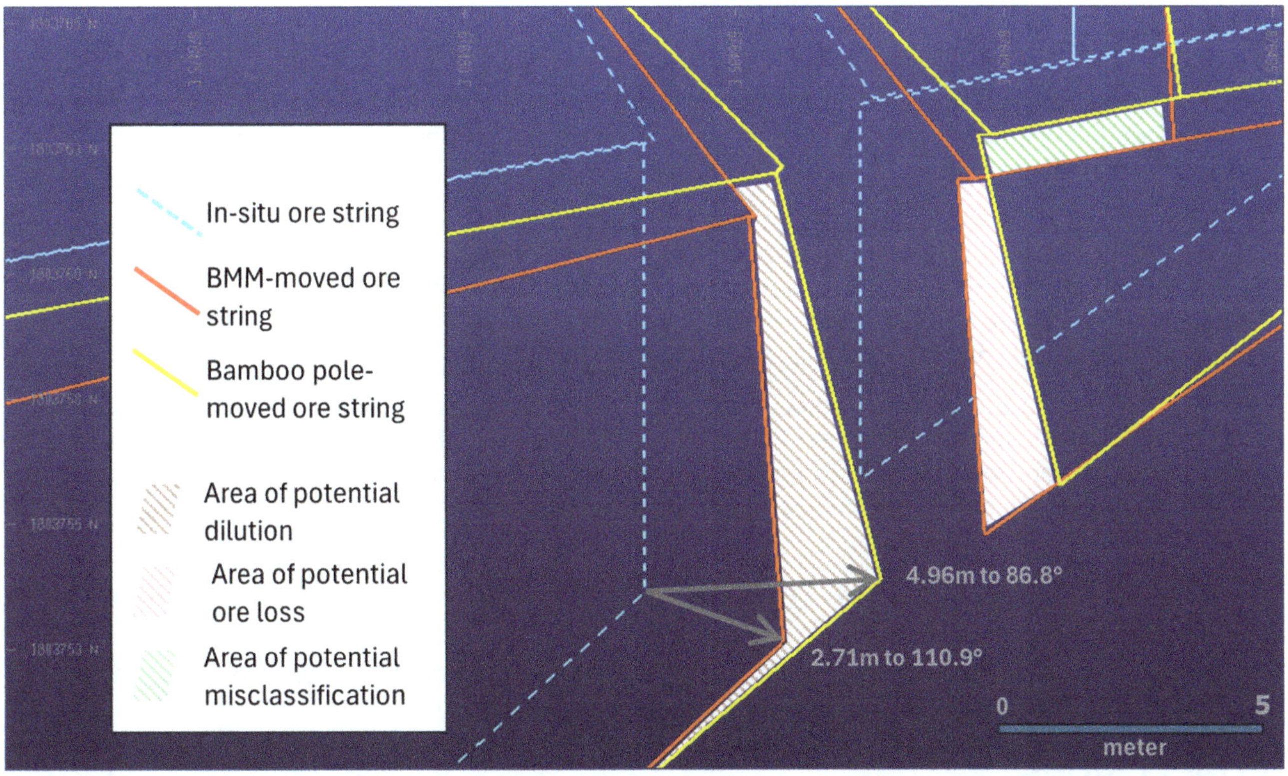

FIG 1 – Impact of blast movement technology in minimising ore loss, dilution and misclassification in block 96 in A West.

TABLE 1

Blast vector indicator (BVI) versus blast movement monitoring (BMM) economic value difference.

Blast	Ore Loss (USD)		Dilution (USD)		Misclassification (t)
29-098_100	$	355,209	$	180,085	2,920
29-098_096	$	108,738	$	78,050	3,742
29-118_119_120_064_133	$	68,233	$	83,094	6,942
47-088_090	$	181,014	$	99,401	1,204
29-097_099_133	$	137,898	$	84,083	5,539
20-102_104_106	$	265,966	$	135,975	9,553
20-108_111	$	265,966	$	116,126	8,239
Total	**$**	**1,383,024**	**$**	**776,814**	**38,139**

Financial year to date reconciliation further supports the technical outcomes of the trial and the ongoing use of BMM. Ounces milled are within 1 per cent of GC predictions, demonstrating that BMM adjusted boundaries are geologically consistent at production scale.

DISCUSSION

The results show that BMM provides the displacement accuracy required for reliable ore control at Chatree. The alignment between GC predicted and reconciled ounces indicates that adjusted dig lines accurately reflect post blast ore positions.

The controlled campaign mill trial reinforces these findings by removing operational variability. The displacement data also provides insight into vertical movement behaviour that cannot be captured with traditional methods.

The remaining tonnes grade variance is operational and arises from material routing inconsistencies. Addressing misallocation is therefore a related operational improvement but does not affect the geological accuracy achieved through BMM.

NEXT STEPS

BMM is now standard practice within grade control at Chatree. Next steps will focus on evaluating whether additional analytical tools provide sufficient geological or operational benefit to justify their cost and workflow impact. The 2025 BMM data set provides a consistent panel of measured blasts suitable for this assessment.

The first area of review will be Hexagon blast movement intelligence (BMI) which has the potential to provide a more detailed understanding of three-dimensional displacement behaviour. The evaluation will consider whether 3D modelling improves interpretation in complex zones, whether it leads to more accurate dig-line adjustments, and whether any improvement outweighs the additional processing time, training requirements and software cost.

The second focus area will be on other AI and machine learning tools, which will be tested using the 2025 measured movement data set. This work will assess whether predictive models can reliably estimate movement magnitude and direction, reduce manual interpretation time or support semi-automated dig-line updates. The assessment will determine whether the potential gains in efficiency or accuracy justify the effort.

A supporting operational initiative involves the implementation of a fleet management system to reduce misallocation.

Collectively, these evaluations aim to determine whether supplementary tools provide a net benefit beyond the current BMM workflow.

CONCLUSIONS

The 2025 BMM trial at Chatree confirms that measured blast movement significantly improves ore control in narrow epithermal vein systems. BMM reduced dilution, improved dig line accuracy and generated strong financial returns. Future work will focus on evaluating advanced analysis tools to determine whether their benefits outweigh their cost and implementation effort.

ACKNOWLEDGEMENTS

The authors acknowledge the contributions of the geology, survey and mining teams at Chatree and the support of Kingsgate Consolidated Limited for approval to publish the findings.

REFERENCES

Loeb, J and Thornton, D, 2014. A Cost Benefit Analysis to Explore the Optimal Number of Blast Movement Monitoring Locations, in *Proceedings Ninth International Mining Geology Conference 2014,* pp 441–449 (The Australia Institute of Mining and Metallurgy: Melbourne).

The integration of geological information into the mine plan

C Rowett[1], M Hamilton[2] and D O'Rielly[3]

1. Technical Services Manager and Chief Geologist, Hillgrove Resources, Kanmantoo SA 5252. Email: caitlin.rowett@hillgroveresources.com.au
2. Principal Engineer, Strategic Mine Engineering Pty Ltd, Aldgate SA 5154. Email: mark.hamilton@strategicmineeng.com.au
3. Geology Superintendent, Hillgrove Resources, Kanmantoo SA 5252. Email: daniel.orielly@hillgroveresources.com.au

ABSTRACT

The Kanmantoo Copper Mine (KCM) ceased open pit operations in 2019 and commenced the development of an Underground Operation in 2023. This switch in operation type created a new challenge in the way geological information including grade estimations are integrated into the development of a Mine Plan suitable for a selective underground Mining Method.

The Kanmantoo Cu-Au-Bi-Ag deposit is hosted within Cambrian age metamorphosed sediments with copper distribution controlled by sulfide dominant veins and vein stockworks located within a shear system. This mechanism of mineralisation provides unique challenges such as discontinuous mineralised veins, varying veining abundance, changing plunge and dip of mineralisation across the lode systems, varied intensity of alteration and deposit scale boudinaging. These geological considerations need to be a central to the development of a mine plan in order to achieve maximum economic return from the mining sequence, while also giving consideration to the geological uncertainty when determining mining sequence.

To achieve the most robust mine plan the information requirements for the end users of the geological interpretations needs to be considered in the determination type of estimations that are created. This allows the information collected from geological observation and drill hole information to be integrated into mine plans and appropriate geological confidence levels can be accounted for in the Mine Planning process.

As a result the development of the geological estimation methodology at Kanmantoo has developed significantly to facilitate the end goal of a robust mine plan that honours the geological variability and identifies variability in information confidence while honouring the geology of the deposit.

INTRODUCTION

To achieve the most robust and executable mine plan a strong understanding of the mineral system geology must be clearly communicated between the geology and engineering teams. As a part of this communication the users and the subsequent workflows that these geological interpretations will be used for needs to be considered when determining the type and form of estimations created. The information collected and interpreted from geological observation and geochemistry results need to be characterised in a format that can be used to communicate geological risk and be subsequently scheduled into a detailed mine plan that accounts for the geological risk and opportunity.

The Kanmantoo Mineral System is complex in nature and the best mining outcomes come from the application of geology into the mining process.

DEPOSIT GEOLOGY

The Kanmantoo Copper Gold Deposit is situated 55 km south-east of Adelaide with the area having a long history of copper mining. Mining first commenced at the Kanmantoo Deposit in the 1800s as a series of small high-grade copper and copper-gold underground mines producing over 24 000 t of ore at around 8.5 per cent copper (Rolley and Wright, 2017). Between 1970–1976 (Verwoerd and Cleghorn, 1975) a single open pit was developed on the Kavanagh lode system and approximately 4.1 Mt at 0.87 per cent Cu, 0.06 g/t Au was extracted from the open pit, in 1976 operations ceased due to low copper prices. From 2011 to 2019 Hillgrove Resources (HGO) mined three open pits at

Kanmantoo for approximately 26.9 Mt at 0.58 per cent Cu, 0.1 g/t Au. In 2023 Hillgrove Resources commenced underground mining on the Kavanagh Cu-Au lode system which has continued to grow in output and expanded to encapsulate the Nugent area in 2025.

The Kanmantoo Trough of the Cambro-Ordovician Kanmantoo Province of south-east South Australia is host to numerous precious metal deposits including Cu, Pb, Zn, Ag, Au, As, Bi over its 300 km span of which the Kanmantoo Cu-Au deposit is the largest known. Approximately 95 per cent of the Kanmantoo Province is under the cover of the Tertiary-Quaternary Murray Basin. The Kanmantoo Trough rift basin formed between 521–511 Ma and the Cu-Au mineralisation at Kanmantoo is hosted within the Tapanappa Formation of the Kanmantoo Group. The Tapanappa Formation is described as a sequence of turbidites, siltstones and pelites with minor Fe sulfide members (Rolley and Wright, 2017).

The Delamerian Orogen has impacted the Kanmantoo Group with three periods of deformation between 514 and 490 Ma (Foden *et al,* 2006). Within the Kanmantoo deposit, D2 is the dominant deformation event observed with upright north–south folds with shallow southerly plunges and a dominant axial-plane parallel S2 fabric (Offler and Flemming, 1968).

Regional Buchan-style metamorphism observed to upper amphibolite facies has affected the Kanmantoo Province during the Delamerian Orogen. Within the Kanmantoo group peak metamorphism is observed as coinciding with peak deformation during D2. The Kanmantoo deposit is within the andalusite-staurolite metamorphic zone (Offler and Flemming, 1968).

Magmatism within the Kanmantoo Province correlates with each of the three Delamerian age deformation events, S-type intrusives dominant with D1, S- and I-type intrusives with D2, and A-type intrusives post deformation (Foden *et al,* 2020). The Geological Survey of South Australia (GSSA) indicates that post deformation magmatism continues through to 465 Ma and possibly 390 Ma. Minor I- and S- felsic magmatism is seen at the Kanmantoo deposit and local districts that is pre and syn peak deformation.

The main copper orebody at Kanmantoo (Kavanagh) is discordant both to bedding and to all mineral lineations. Kavanagh is partly sympathetic with the axial planar schistosity of the D2 Kanmantoo Syncline which averages ~65° -> 084°, whereas Nugent is rotated further to the north-east. The rotation is hypothesised as resulting from the ongoing deformation that occurred during emplacement of the mineralised fluids. Figure 1 shows the broad distribution of the copper zones at Kanmantoo and their general copper geometry relative to the S2 axial planar schistosity.

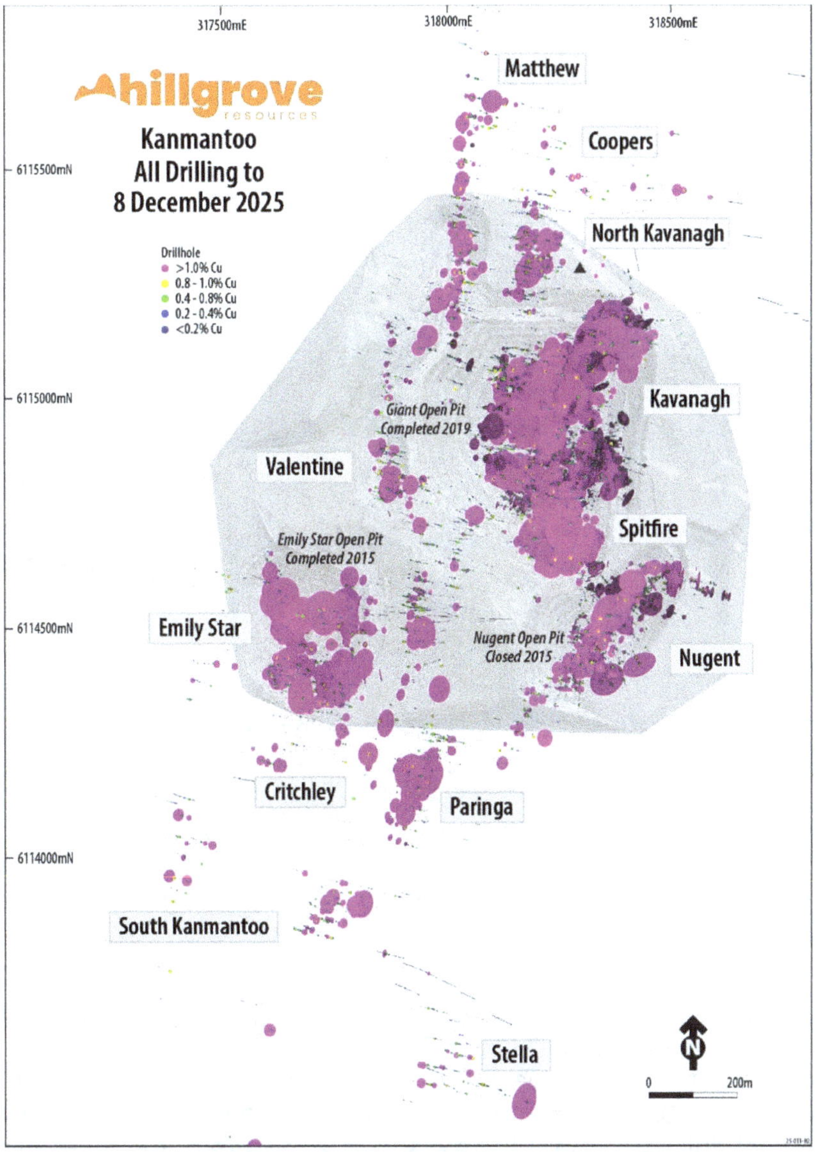

FIG 1 – Plan view of all drilling and open pit blasthole assays greater than 0.8 per cent Cu illustrating the various mineralised lodes.

The Kanmantoo mineralisation consists of quartz poor sulfide veins and veinlets dominated by pyrrhotite and chalcopyrite with minor magnetite, bismuthinite, and pyrite. The sulfide mineralisation overprints all metamorphic assemblages and overprints/infills all peak deformation structures. Age dating suggests mineralisation occurs at 490±3 Ma (Lyon, 2012). Copper and silver are strongly correlated, as are gold and bismuth with the later pair often overprinting chalcopyrite and pyrrhotite. Alteration mineralogy is dominantly garnet, biotite, chlorite with minor potassium feldspar. The mineralisation to date has been drilled to 800 m below surface as a continuous structural system. Mineralisation at the Kanmantoo deposit is interpreted to be epigenetic (Oliver *et al,* 1998) and the result of basinal metamorphic fluids mixing with granite derived fluids channelled along pre-existing structures by magmatic driven geothermal fronts during syn- to post-peak-deformation.

At a mine scale the geometry of the alteration is repeatable and predictable with a halo as evidenced by the presence or absence of andalusite, garnet and chlorite and sulfur grades of ~ 0.6 per cent S. Within the alteration corridor (Figure 2) the geometry and scale of the structurally controlled Cu-Au mineralisation is highly variable, with the variability of the geometry and scale of the Cu-Au zones increasing with increasing copper grade (Rowett, Rolley and O'Rielly, 2023). This is a result of the natural copper cut-off grades (COG) which correspond the three major types of mineralisation observed in the system. These are:

- S2 controlled copper (0.2 to 0.4 per cent Cu).

- epigenetic sulfide dominant veins (0.4 to 0.8 per cent Cu).
- the brecciated sulfide vein stockworks (>0.8 per cent Cu).

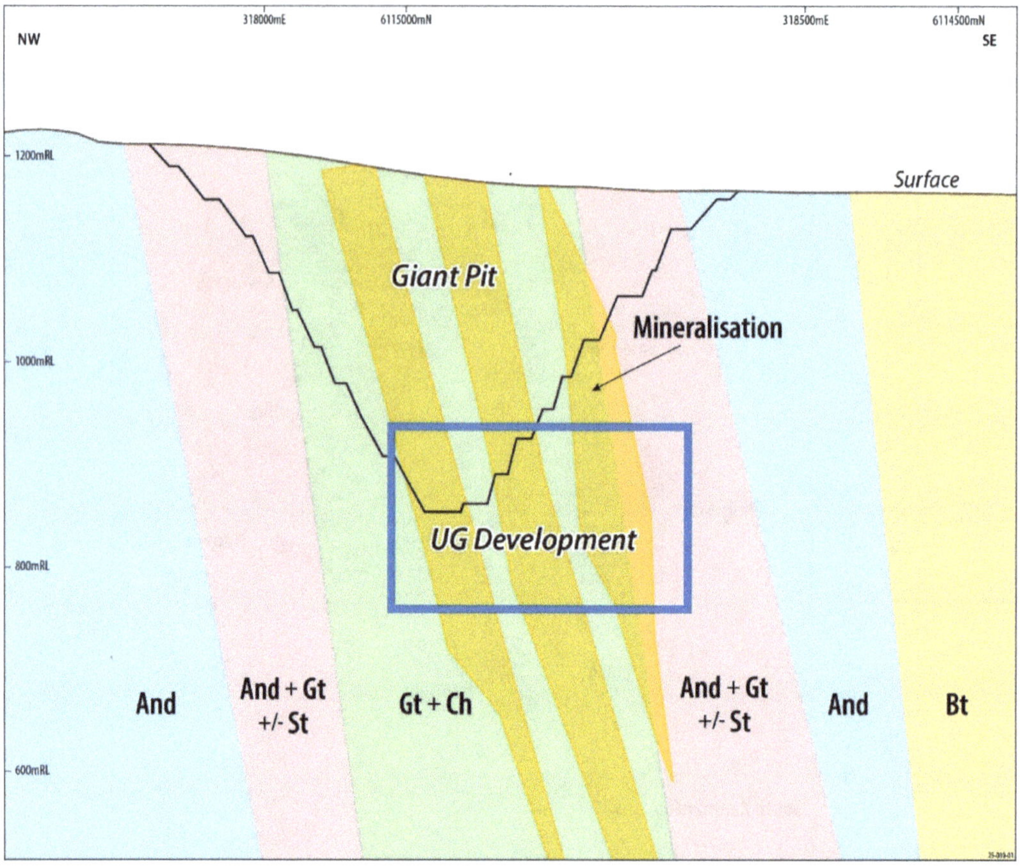

FIG 2 – Relationship of alteration and mineralisation within the Kavanagh Mineral System.

The highest grades are associated with the brecciated stockworks where massive chalcopyrite is observed as seen in Figure 3. The geometry and extent of these zones is the hardest to predict based on the spacing of drill hole data. Figure 4 shows a mixture of the three mineralisation styles occurring within one core tray.

FIG 3 – Example of high-grade Cu-Au mineralisation from NQ core hole 23KVUG064 through the East Kavanagh Lode from 51.39 m to 55.40 m downhole.

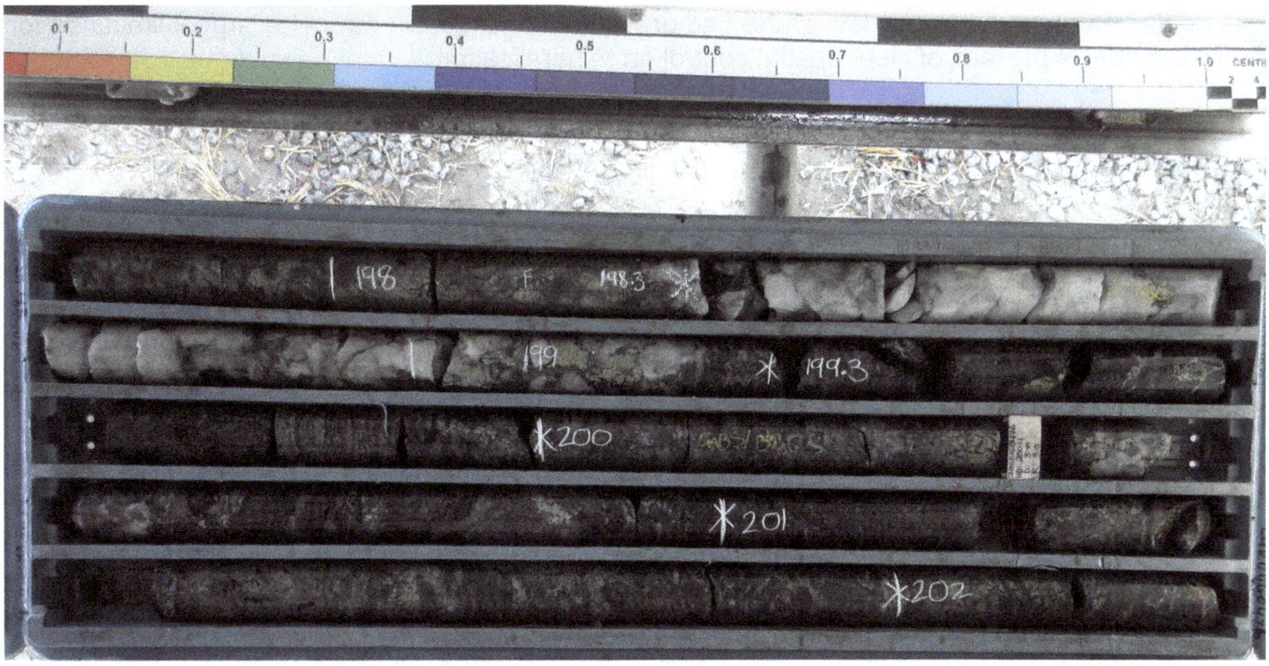

FIG 4 – Example of high-grade Cu-Au mineralisation from NQ core hole 24KVUG476 through the East Kavanagh Lode from 197.82 m to 202.27 m downhole.

MINING METHOD SELECTION

Based on the deposit geology the operation has elected to use a modified version of longhole open stoping for bulk production extraction. These stopes are accessed via strike drive development with additional of cross-cuts developed where the mineralisation width increases to a point requiring additional void for the rise firings. The production drill design is executed by Simba drill rigs before and hauled from the stope using a combination of manual and remote loaders for haulage by an underground trucking fleet to surface. The stope design guidelines used at Kanmantoo has been developed to ensure that highest grade stockwork systems are encapsulated within the stope boundaries. This is done by designing the development extents based on a 0.8 per cent Cu COG and designing the stope geometries on a 0.6 per cent Cu COG. The use of a lower COG for stoping ensures that all +0.8 per cent Cu material is encapsulated within the stope boundary despite being harder to predict spatially.

MODELLING METHODOLOGY

Based on the variability observed in the spatial location and geometry of the highest grade zones an estimation system that provides the required information to produce a robust mine plan needed to be created. This estimation system needed to be specific to underground and needed to convey the geological interpretation of both grade and spatial location to the mining engineering team to facilitate mine planning work.

Historic mineral resource estimation systems

During open pit mining the Kanmantoo Mineral Resource Estimate was approached in a number of different ways. Initially it was an Ordinary Kriged (OK) estimation within tight geological domains with hard boundaries between the ore and waste domains. These estimations used all drilling completed on-site including historical drill holes and were validated by numerous external reviews. Unfortunately, these estimations resulted in a significant over estimation in the grade of the material predicted for mining in the open pit.

In 2015 a change in the estimation methodology was completed transitioning to a Multiple Indicator Kriged (MIK) model. MIK is often used to estimate recoverable resources and uses large scale panels to report the proportion of various COG contained within the panel. This recoverable estimation technique worked well in the open pit where free selection of ore was available, blasthole sampling facilitated a 3 m × 3 m sample density on average and where all material from the open pit was hauled as ore to either the low-grade pad or Run-of-Mine (ROM) pad. The largest limitation for the MIK model is the lack of clear spatial control on mineralisation.

Historic mineral resource estimation systems limitations for underground mine planning

Due to the success of the open pit estimations from 2015 onwards a MIK estimation was used for early underground mineral resource estimations. The 2022 MRE for Kavanagh was an MIK estimation which utilised a panel size of 4 m W × 20 m L × 25 m H (Hillgrove Resources Pty Ltd, 2022). This MIK estimation was used in all early mine planning evaluation works and excelled in communicating the low geological confidence and overall geological risk however was severely limited in providing mineralisation spatial guidance. To improve the spatial location information within the estimation opportunities for a localised MIK estimation were investigated.

With early mine production coming online in 2023 reconciliation of the production versus the mineral resource estimation highlighted the need for a new estimation system with stopes reconciling on average 30 per cent higher than the MIK model predicted.

Key to detailing the mine planning application of MIK is understanding that each 4 × 20 × 25 m model cell (6200 t) also contains tonnes and grade estimates at differing copper cut-offs. For example, the full cell may contain 6200 t @ 0.55 per cent, at a cut-off of 0.6 per cent it contains 4000 t at 0.85 per cent and at 0.8 per cent cut-off it contains 2500 t at 1.06 per cent. Each cell contains tonnes and grade estimates over the same range of cut-offs.

To utilise the MIK model in underground mine planning a system to manage the large scale panels had to be developed. This process known internally as the 'Hybrid Partial' method evaluated the use of full panels and partial panels and when each were used to facilitate the mining inventory available for the mine plan. A 'full cell' is where the full 6200 t (from the 4 × 20 × 25 m panel at a 3.1 density) is used, at the overall average cell grade. Partial cell is where the tonnes and grade for a nominated cut-off is utilised, for example 0.6 per cent. This implies that it is feasible to selectively mine part of the cell and that the mined portion is spatially aligned to the adjacent full cell panel. 'Hybrid partial' is part way between full and partial. This means utilising the midpoint of rock tonnes and metal tonnes between the full and partial values. Again, this implies that a selective mining outcome is practical for the cell, although not as aggressive as partial.

The detailed process as was developed in 2022 to generate life-of-mine plans is as follows, with the numbered points outlining the key steps, with illustrations following that attempt to improve clarity on a numbered point were necessary. Note, Deswik CAD was used as the graphical and data interface.

1. Input: 2022 MRE MIK model.
2. Import model cells with Full cell Cu >0.6 per cent. Display solid.
3. Second pass, import model cells with >25 per cent tonnes and >0.6 per cent Cu. Display transparent.
4. Cells from steps 2 and 3 displayed together. Working from bottom-up within a production panel, outlines are drawn around cells which preliminary delineate stopes.
5. Many rules drive the delineation of outline strings (outcome shown in Figure 5):
    a. Minimum stope width: one full panel (4 m east–west).*
    b. Maximum stope width: six panels (24 m).**
    c. If >six adjacent panels are above cut-off, a pillar must be left between stopes.***
    d. Minimum pillar width is two panels (8 m). Can be partials.
    e. Where pillars are necessary, checks vertically below are critical to ensure a pendant pillar is not created. In the instance that it is, the pillar or lower stope is modified to remove pendant pillar, while honouring all other rules.
    f. Pillars cannot start or stop mid-stope along strike length. If this is required, the stope stops and a new stope is commenced.

g. Stopes typically limited to 60 m along strike (some allowed to 80 m on occasion).

h. Stopes can gain or loose width along strike, but only allowed to change by a maximum two cell widths for each 20 m panel strike length. This is to avoid an unrealistic hanging or footwall that is likely to fail in mining.

6. The strings produced in step 5 are used to flag model cells that fall within stopes. These strings are used to stamp a unique naming attribute for each stope.

7. Cells flagged from step 6 are displayed graphically in Deswik.

8. Individual cells are flagged as full, partial or hybrid partial based on the following logic (as per Figure 6):

   a. Full is allocated if a given cell has an adjacent cell in the transverse direction (east west), strike direction (north south) or vertically above. Additionally, if a stope is only one cell wide (4 m) with adjacent cells along strike, it is flagged for full extraction to avoid planning for <4 m wide stopes.

   b. Partial is allocated to cells that have no adjacency in the transverse direction only. If an identified stope is only two cells wide, it is permissible to allocate two adjacent cells as partial.****

   c. Hybrid partial is allocated to cells that would otherwise have been allocated as Full but has either no adjacent cell directly above or along strike.

9. Cells of common stopes are merged in Deswik to create solids. Additionally, separate solids are generated for 20 m strike length sections within stopes. These are used as backfill solids in the mine schedule.

10. Excel is utilised to manage calculation of tonnes and grade contained within each stope, honouring the flagging of full, partial or hybrid.

11. Ore Drive development is generated in Deswik based on final stope set (detailed later in the mining section). Development solids are generated, and trimmed to stope boundaries, keeping only the sections of development that fall within stopes. Solids are interrogated against the Full Cell within the MIK_Mar22 model. Using excel, the development tonnes and grade within each individual stope is depleted from the stope tonnes and grade. This avoids double counting tonnes, whilst allowing development ore to be processed in the period it is mined.

12. Final stope shapes within the Deswik project have development depleted tonnes and grade imported and stamped on the solid as an attribute. This is subsequently utilised in the development of the mine schedule.

* Where step 3 identifies only a single width stope, the full cell inventory is utilised except where it is flagged as a hybrid-partial. See point 8 for flagging.

** Given partial cells are applied on the stope sides, the actual maximum stope width is usually more like 20 m.

*** This version of mine plan does not consider a chequer-board sequence with primary and secondary stopes, enabled by fully consolidated material against the exposed wall. Future versions will consider this and it will be adopted if it adds value.

**** The exception to this is where two adjacent partial cells have a cumulative tonnage of >6000 t. If this is the case a manual process is undertaken where either one cell is flagged as full, or the stope is removed from the set. This is to avoid planning for <4 m wide stopes.

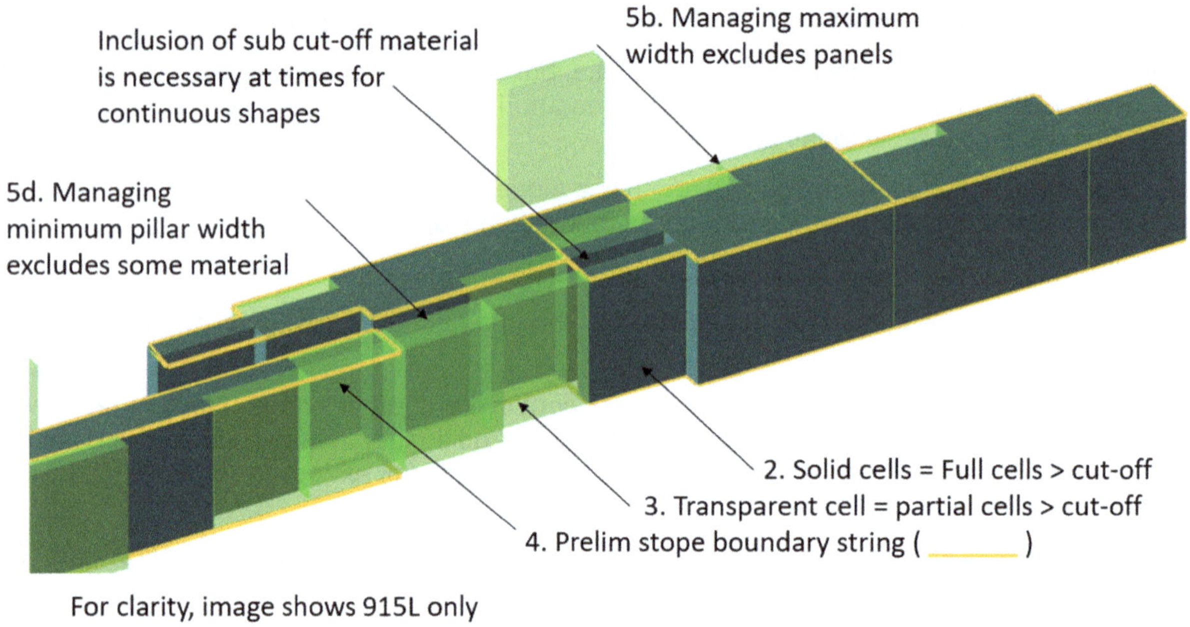

(a)

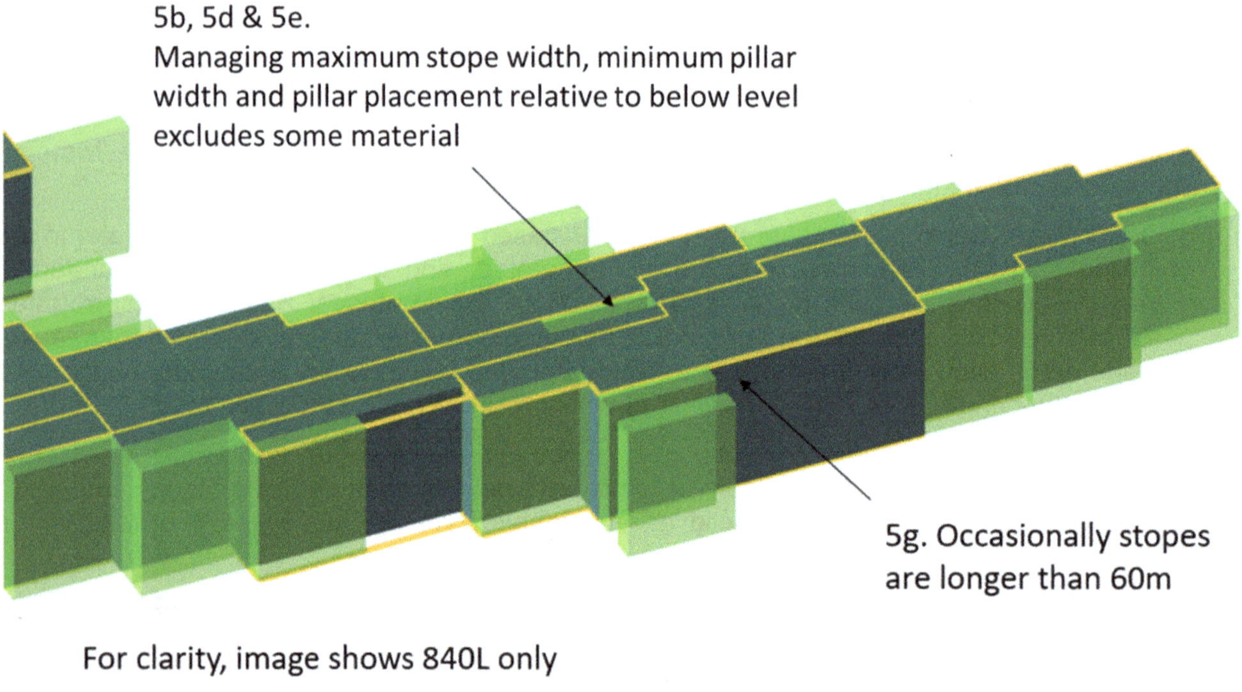

(b)

FIG 5 – (a) Isometric view illustrating 915L, highlighting application of stope flagging rules (steps 2–5). (b) Isometric view illustrating 840L, highlighting application of stope flagging rules (steps 2–5).

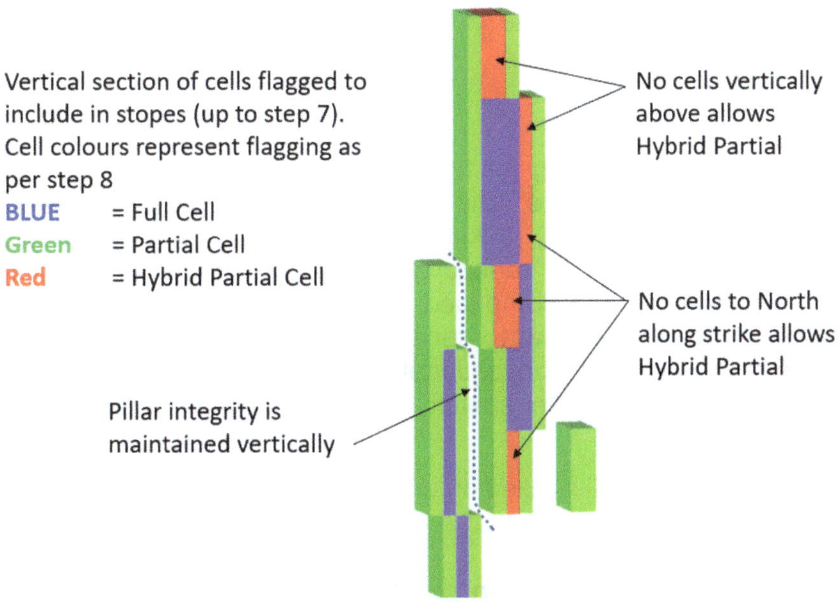

(a)

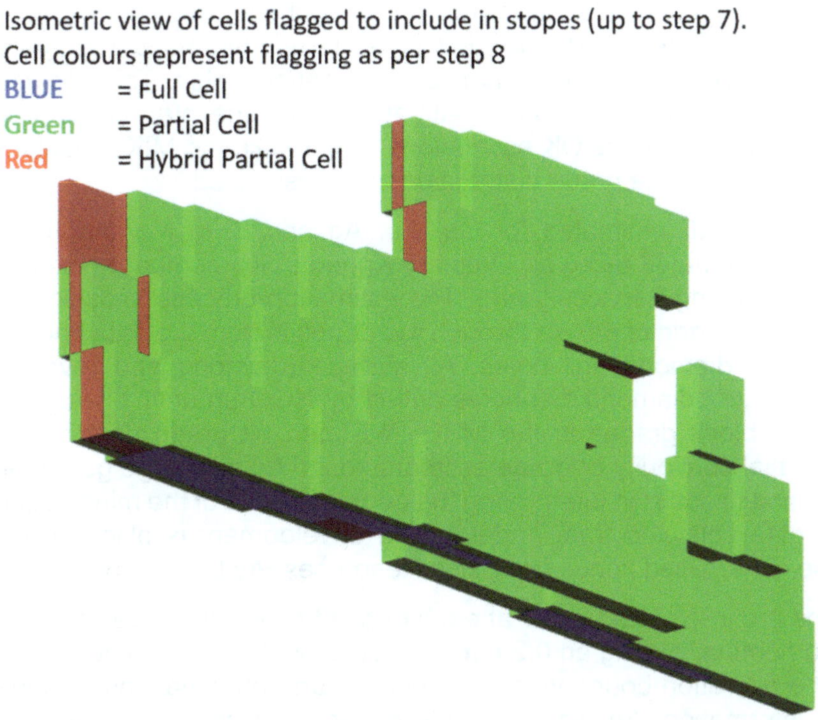

(b)

FIG 6 – (a) Cell flagging for model recovery, per step 8. (b) Outcome of cell flagging as per step 8.

The developed process was capable of flowing through to a fully functioning mine schedule with all necessary scheduling tasks plus output all required revenue and cost drivers.

However, the key disadvantages to using the MIK model for executable mine planning follow:

- Very manual and slow process required to develop mine plans. More than a week to turnaround a full mining scenario (this includes each cut-off grade under analysis).
- When planning a partial cell, there is a complete lack of spatial understanding of the inventory above cut-off. Not feasible to guide executable operational where small selective stopes are utilised.

Ultimately it was determined that this style of model was unable to be used to support selective underground mining.

LMIK modelling

The next iteration of resource modelling for the Kanmantoo Underground mine was a Localised Multiple Indicator Kriged (L-MIK) estimate. This built upon the MIK estimates, but estimated internal sub blocks at 2 × 5 × 5 m model cells by ranking the individual subblocks and assigning the grade to each ranked subblock based on the ranked histogram which was binned for each subblocks tonnage The model processing was completed using multiple software packages; one to the underlying MIK estimate and then another to generate the ranked sub-blocks. Once produced, the LMIK model was able to be utilised via conventional mine planning processes with full cell estimates used in interrogation of solids and the spatial distribution of grade able to be visualised for development planning.

The limitation of this method was that it relied on uncommon software systems and was very slow to process large data sets.

Current mineral resource estimation systems

To improve the spatial accuracy of the Mineral Resource Estimate (MRE) and allow for systematic mine planning processes a new estimation system was sought. This new estimation system was to build on the knowledge of previous iterations of MREs and the success of each individual technique. As a result the 2025 Kanmantoo MRE was developed to use large scale 0.2 per cent Cu domains as per the earlier MIK estimations but with an OK estimation within. This allowed 0.2 per cent Cu COG for the estimation domains allowed the estimation to see the internal low-grade and prevented the overestimation seen in previous OK estimations. And unlike the MIK models the OK estimation allowed for spatial location of the grade distribution to be visualised.

The 2025 Kanmantoo MRE estimated for Cu, Au, Ag and Bi in Kavanagh, Nugent and North Kavanagh have been estimated using an Ordinary Kriged (OK) method by experts from ERM (part of the ERM Technical Mining Services team). The estimation process to assign the Cu, Au and Ag grades to a the defined 3D grid of panels through the deposits from the data collected from sampling of RC percussion and diamond drill holes. All surface (diamond and reverse circulation) and underground grade control diamond drill holes drilled by HGO up until 11 August 2025 have been used to estimate the block grades in the MRE. No open pit grade control data is used in the estimation of the spatial continuity or grade estimates but the knowledge gained from mining of the open pit has been used to assist in interpreting the general trends of the mineralised zones. Mapping and Spectral (IR) data collected from underground development is also used to assist with the interpretation of the mineralised zones (Hillgrove Resources Pty Ltd, 2025).

Geological domains are interpreted in 3D at each deposit by Hillgrove geologists based on the drill hole mineralisation nominally using an 0.2 per cent Cu COG to identify each mineralisation zone in conjunction with the alteration boundaries and underground observations, including mapping from and spectral data information. Analysis of grade distributions for Cu, Au, Ag, Bi, S and Fe was undertaken, with reference to the various combinations of geological domains coded onto the sample files. Statistical analysis via cumulative distribution frequency plots, together with spatial assessment through contact plots and visual inspection of desurveyed drill holes coloured with grade values, were utilised to determine which variables and geological domains could reasonably be grouped together to create estimation domains for grade interpolation. A similar process was undertaken for the variable density. Following this variogram models are created for the grade variables Cu, Au, Ag, Bi, S and Fe for the generated domains.

To capture high-grade mineralisation that is discontinuous and therefore hard to constrain to a single domain, a Categorical Indicator Kriging (CIK) block modelling approach has been applied. Indicator coding of assayed intervals falling outside of the wireframed mineralised zones have been used to construct a probability-based block model to define the 'un-wireframed' Cu mineralised zones. These areas are defined as low confidence and classified accordingly in accordance with guidelines contained in the JORC Code. The volume of additional Cu mineralisation defined through this process is geologically reasonable for the deposit.

Additional detail on the specifics of the estimation process can be found in the ASX release (Hillgrove Resources Pty Ltd, 2025).

Use of the mineral resource estimation in mine planning

The generation of the Mining Inventory used in the Ore Reserve Estimation (ORE), Life-of-mine (LOM) plan and rolling 13 Week Plan is based on the Mineral Resource Estimation and subsequent internal grade control block model estimation updates provided by the geology team.

Stope shapes within the plans are guided by Stope Optimiser shapes created in Deswik. As a first pass, shapes generated on CuEq 0.8 per cent cut-off are used to determine ore drive layouts and lateral development extents. Subsequently a CuEq 0.6 per cent set of shapes is used to guide final stope shapes. To ensure the mineability and stability, each shape has then been individually designed taking into account geometries required for stable stope walls and regionally appropriate pillars. These designs are in line with the existing stope design parameters used on-site. Development designs have been optimised for the ORE stope set including capital infrastructure such as ventilation, dewatering, power and mine services allocated to facilitate this mine plan. All the development design work has been based on the existing design guidelines utilised in current operations.

Consequently, all geological interpretation needs to be accounted for in the underlying model as this is how all geological information is conveyed into the mine plan.

The current Mineral Resource Estimation process of an OK estimation into the modelled mineralisation domains provides the spatial location of the known mineralisation and the geological confidence. This facilitates the mine plan allowing for the timing of lower confidence material and the balance of grade material on a monthly interval.

FUTURE WORK

To build on the existing estimation process further work needs to be completed to add more economic parameters to the Resource Estimation utilised in mine planning. This will allow for a Nett Smelter Return (NSR) to be evaluated per block which will allow for a detailed economic assessment to be completed on all scales of the mine plan from stope to LOM.

This continues to ensure that geological information is converted into a direct economic outcome and is validated in each level of mine planning.

To assist in the speed of providing geological information and a simple way to convey geological risk and uncertainty into the mine plan the implementation of conditional simulation into geological modelling workflows will be assessed.

CONCLUSIONS

Following the journey from project to production it has become clear that the purpose of an estimation model and the subsequent work it will inform must be at the forefront when determining the estimation methodology. It is also exceptionally clear that a strong working relationship between the geologists creating the estimation and the mining engineers using the model must be created and maintained to ensure that the estimation provides the required information and will facilitate a clear mine plan.

ACKNOWLEDGEMENTS

The authors wish to acknowledge the hard work and continuing dedication of the Hillgrove Resources Team from both the Open Pit and Underground Operations. Their willingness to engage and optimise new and innovative processes for the dissemination of geology knowledge underpins the success of our system. Without the teams courage to elicit changes to modelling processes and to work as a cohesive technical services team across disciplines to achieve the best technical outcomes based on the strong geological and mining engineering principles suited to the Kanmantoo Mineral System. In particular we would like to thank Peter Rolley for his unwavering dedication to teaching all of us that 'data is king' and the importance of evidence based solutions.

We also thank Hillgrove Resources for permission to publish this paper.

REFERENCES

Foden, J D, Elberg, M A, Dougherty-Page, J and Burtt, A, 2006. The timing and duration of the Delamerian Orogeny: Correlation with the Ross Orogen and Implications for Gondwana assembly. *The Journal of Geology*, 156:10–33.

Foden, J, Elburg, M, Turner, S, Clark, C, Blades, M L, Cox, G, Collins, A S, Wolff, K and George, C, 2020. Cambro-Ordovician magmatism in the Delamerian orogeny: Implications for tectonic development of the southern Gondwanan margin, *Gondwana Research*, 81:490–521.

Hillgrove Resources Pty Ltd, 2022. Updated Kavanagh Underground Mineral Resource Estimate, ASX Release. Available from: <https://investors.hillgroveresources.com.au/announcements/4150856>

Hillgrove Resources Pty Ltd, 2025. Hillgrove Resources delivers ore reserve extension at 1.0% CuEq*, ASX Release. Available from: <https://investors.hillgroveresources.com.au/announcements/7235533>

Lyon, N, 2012. Evidence for magmatic hydrothermal mineralisation at Kanmantoo Copper deposit, South Australia, Hons Thesis, The University of Adelaide.

Offler, R and Flemming, P D, 1968. A Synthesis of Folding and metamorphism in the Mount Lofty Ranges, South Australia. *American Journal of Science*, 12(2):245–266.

Oliver, N H S, Dipple, G M, Cartwright, I and Schiller, J, 1998. Fluid flow and metasomatism in the genesis of the amphibolite-facies, pelite-hosted Kanmantoo copper deposit, South Australia, *Journal of the Geological Society of Australia*, pp 181–218.

Rolley, P and Wright, M, 2017. Kanmantoo Copper Deposits, *Australian Ore Deposits,* Monograph 32, pp 667–670 (The Australasian Institute of Mining and Metallurgy: Melbourne).

Rowett, C, Rolley, P and O'Rielly, D, 2023. Opening the Geologists Tool Box: Building a Grade Control System, in *Proceedings of the International Mine Geology Conference 2023*, pp 231–242 (The Australasian Institute of Mining and Metallurgy: Melbourne).

Verwoerd, P and Cleghorn, J H, 1975. Kanmantoo copper orebody, in *Economic Geology of Australia and Papua New Guinea* (ed: C Knight), pp 560–565 (The Australasian Institute of Mining and Metallurgy: Melbourne).

From drill core to decision-making – a framework for early resource characterisation

A Vatandoost[1], S Badock[2], M Saba[3], A Tan[4] and G Farcas[5]

1. Technical Manager, Resource Characterisation, Fortescue, Perth 6000. Email: adel.vatandoost@fortescue.com
2. Senior Manager Exploration, Exploration, Fortescue, Perth 6000. Email: stuart.badock@fortescue.com
3. Project Metallurgist, Resource Characterisation, Fortescue, Perth 6000. Email: moji.saba@fortescue.com
4. Senior Metallurgist, Resource Characterisation, Fortescue, Perth 6000. Email: aden.tan@fortescue.com
5. Senior Metallurgist, Resource Characterisation, Fortescue, Perth 6000. Email: gerald.farcas@fortescue.com

ABSTRACT

Each orebody is inherently unique, exhibiting specific processing behaviours that can significantly impact project economics. Early-stage ore characterisation is therefore critical to maximising the long-term value of orebodies, processing plants and associated infrastructure.

This paper presents a resource characterisation framework implemented across Fortescue's Iron Ore Exploration projects, highlighting how early geometallurgical insights are embedded in practice. The process begins with strategic planning of diamond drilling and sampling to ensure materials are representative of orebody variability and suitable for downstream testing.

Drill core samples undergo high-resolution digital scanning to capture core imagery, mineralogical, geotechnical and bulk density data sets. Hyperspectral analysis plays a vital role in detecting potential processing risks, such as the presence of swelling clays and minerals not familiar to the naked eye.

Geological and mineralogical data sets inform the selection of metallurgical composites, each tailored to reflect specific ore types and variability. Metallurgical test work is conducted at independent metallurgical laboratories and forms the backbone of the characterisation workflow. These tests evaluate physical properties, beneficiation potential and processing response, particularly scrubbing, screening, magnetic and gravity separation methods.

All data generated throughout the program from initial logging to test work outcomes are captured in acQuire™ database platform. This enables traceability and security, supports data integrity and facilitates efficient analysis and data integration to build geometallurgical models. These models support mine planning decisions, enhance value optimisation and ensure a smooth transition from Exploration to studies and operations.

By embedding this workflow early in the exploration phase, we can better inform strategic decisions, improve confidence in resource development and align orebody knowledge with plant design and capabilities. The goal is to deliver a fit-for-purpose characterisation approach that drives value from drill core to decision-making.

INTRODUCTION

The mining industry is operating in an environment of increasing geological complexity, tightening economic margins and heightened expectations for predictable production outcomes. Despite this, exploration, resource definition, mine planning and processing often function as independent disciplines resulting in fragmented data sets, inconsistent sampling, test work planning and practices and limited visibility of downstream processing opportunities and constraints during the early stages of project evaluation.

Compounding these challenges is the evolving nature of iron ore mineral discoveries. A large proportion of new discoveries tend to be either low-tonnage satellite deposits, lower-grade resources or orebodies with elevated deleterious elements (eg alumina, silica, phosphorus). Some discoveries are also constrained by environmental or heritage sensitivities and obligations, limiting their development potential. In many regions, these deposits also sit far from existing infrastructure or require significant capital for development making their viability highly sensitive to early uncertainty. These realities emphasise the need for robust resource characterisation to better understand physical and chemical properties, processing constraints and value drivers of an orebody well before a project reaches development decision points. Improved geometallurgical understanding not only reduces technical uncertainty but also supports broader industry objectives around energy efficiency and lower emission operations by enabling more predictable processing pathways and minimising unnecessary rehandling of waste material.

Integrated geometallurgical workflows and predictive behavioural modelling have been widely explored in literature (Lamberg, 2011). The framework presented here builds on similar principles while being customised for iron ore mineralisation and organisational requirements. This work represents the first implementation of an integrated, multi-disciplinary resource characterisation workflow at Fortescue, including the organisation's first systematic migration of metallurgical data sets into an integrated database system.

This paper presents a unified framework for geometallurgical data integration aimed at bridging the gap between exploration, development and operational geometallurgical decision-making. The emphasis is on building a continuous data value chain from drill core to processing plant using systematic test work, consistent data structures and cross-disciplinary collaboration. The framework emphasises standardised data acquisition, structured test work, process behaviour-based geological interpretation and unified platforms that support both modelling and operational decision-making. The aim is to present a practical framework for systematic geometallurgical data integration and demonstrate how early-stage characterisation improves domain definition and risk management.

FRAMEWORK OVERVIEW

Modern geometallurgical frameworks increasingly adopt integrated data structures and multi-disciplinary modelling approaches to improve predictability and orebody knowledge (Parian *et al*, 2021; Lind, Lamberg and Oliveira, 2018). The resource characterisation framework specifically implemented for Fortescue's hematite iron ore projects in the Pilbara is underpinned by four pillars: representivity, integration, traceability, and decision-focus. These pillars establish the strategic intent of the workflow, ensuring that all data collected during exploration contributes meaningfully to understanding orebody variability and guiding downstream planning decisions. Representivity ensures that sampling captures the true geological, geochemical and mineralogical spectrum of variability. Integration links geological logging, digital core scanning, mineralogy, geotechnical properties, and metallurgical testing into a coherent interpretive system. Traceability provides an auditable pathway from drill core to test work outputs, preserving data integrity and enabling consistent QA/QC. Decision-focus ensures that insights generated at each stage support study progression, product evaluation and operational readiness.

Operationally, the framework is delivered through five interconnected stages that form a continuous pathway from diamond drill coring to geometallurgical interpretation. These stages include strategic drilling and sampling, high-resolution core characterisation, metallurgical composite selection, metallurgical testing and data migration and management. Together, these stages ensure that early drilling programs generate more than geological observations, they create actionable geometallurgical understanding. The deliberate integration of these pillars and stages allows the framework to support rapid early insights, progressive refinement as additional drilling becomes available, and long-term continuity of technical knowledge across the project life cycle.

STRATEGIC DRILLING AND SAMPLING

Within this framework, strategic drilling forms the foundation for all subsequent characterisation stages. During project development, decisions relating to life-of-mine, production rate, product strategy and capital expenditure depend heavily on understanding the processing behaviour of the ore. Viability of a resource project is significantly enhanced when metallurgical characteristics can be defined early using representative samples obtained through carefully designed drilling programs.

Within the framework a multi-purpose PQ (~85 mm diameter) diamond drill program is adopted across Pilbara exploration and development projects to obtain *in situ* materials for ore characterisation studies. The selection of project areas for drilling is primarily driven by the absence of geometallurgical knowledge and their importance for downstream decision-making. Drill programs in the targeted project area support geological understanding of the project and also enable selection and preparation of samples for metallurgical, geotechnical, environmental and hydrogeological testing.

Early drilling campaigns were designed to access fresh, representative material depending on the study stage of project area. Sampling plans are aligned with lithology and stratigraphic domains as defined in the Inferred Resource model. Drill designs are informed by existing information from reverse circulation (RC) drill holes and aim to target key mineralisation and stratigraphical units for variability testing.

Sampling representation is an important and critical part to ensure metallurgical test work results reflect the true processing variability of the orebody. It is assessed both spatially and when comparing assays of stratigraphy units with characteristics of their population; that is, the resource model.

HIGH RESOLUTION CORE CHARACTERISATION

High resolution spectral scanning formed a central component of the framework providing consistent, objective mineralogical and structural information that enhances early orebody understanding and supports the integration of geological and metallurgical data sets within the geometallurgical framework. Hyperspectral scanning (Haest *et al,* 2011) provided mineralogical information that cannot be reliably identified through visual logging alone. This was particularly important for detecting swelling clays, differentiating goethite-hematite assemblages and identifying gibbsite with direct implications for beneficiation performance and product quality. The early visibility of these minerals played a key role in defining geometallurgical domains and informing metallurgical composites design.

The digital scanning outputs provided a consistent, high-resolution data set (core imagery and mineral logs) that bridged the gap between geology, mineralogy and metallurgical response. The integration of these data sets significantly improved early domaining and formed the foundation for targeted metallurgical testing.

METALLURGICAL COMPOSITE SELECTION

The material obtained through diamond drilling campaigns is classified into unique geometallurgical domains. Core recovery throughout the drilling program is critical since higher core loss within a selected composite interval will not be representative. Intervals from each domain are generally around 8–10 m long, representing mining bench height in most cases and are used for metallurgical evaluation and development of upgrade regression models for product quality prediction. Composited intervals are constrained by a series of criteria, mainly stratigraphical units, textural, physical attributes (hardness) and mineralogical observations from hyperspectral analysis while ensuring sufficient mass is obtained for physical testing. As ore hardness and mineralogy influences particle size distribution of the final product, composite intervals are characterised by stratigraphic units and hardness ie hard, medium and friable. Presence of internal waste also needs to be monitored when selecting intervals for testing since the presence of a significant volume of internal waste can inadvertently bias metallurgical performance both in terms of upgrade and recovery.

METALLURGICAL TESTING

The metallurgical test workflow sheet adopted for the exploration hematite projects was designed for ore characterisation, in particular, lump and fines mass and grade splits and to assess amenability of material for upgrading through either dry or wet processing.

Baseline testing focused on simple, high value techniques suitable for early exploration stages, including particle size distribution, ore friability measurement and deleterious elements deportment analysis. Other beneficiation test work such as gravity and magnetic separation is considered in the test workflow sheet subject to review of baseline flow sheet test results. The impact analysis of preliminary upgrade regression models on mining inventory and product strategy is critical to investigate more detailed processing scenarios (Vatandoost *et al,* 2013). With declining ore grade and increasing water constraints, it is important to establish whether a wet processing plant may be required at any stages of the mine life and whether the proposed processing scenario delivers the expected product strategy.

Ore comminution characterisation was performed selectively for domains where physical strength characteristics indicated relevance to mine planning and processing. The outcomes of this early stage test work were not intended to replicate feasibility level design parameters but to identify processing behaviours that influence domain definition, metallurgical risk and product strategy decision. By linking metallurgical response back to geological and mineralogical attributes, the program enabled rapid identification of lump potential, clay-related processing challenges, and limitations in upgrade performance.

DATA MIGRATION AND MANAGEMENT

The critical component of the resource characterisation workflow was the establishment of a unified data management structure within the acQuire™ database, serving as one source of truth. Historically geoscientific data sets have been well supported by enterprise systems, whereas metallurgical data has been more difficult to store, standardise and integrate. Metallurgical data sets are inherently more complex than conventional geological data, comprising multi-stage process outputs, mass balance calculations, derived recovery metrics and composite-based results that often do not align naturally with interval-based drill hole database structures. Historically, this complexity has resulted in metallurgical data being stored in isolated spreadsheets and laboratory reports, limiting traceability and cross-disciplinary integration. Through a coordinated data migration initiative, all program data sets including logging, assays, hyperspectral outputs, mineralogy, physical properties and for the first time metallurgical test work were imported into a single structured environment. This involved streamlined data receival from vendors into an in-house system.

The metallurgical data migration project involved building digital log sheets in collaboration with vendor labs and creating a user-friendly calculation template that can be used by the Exploration data management team and incorporated into acQuire™. Achieving metallurgical data integration represented a significant milestone overcoming a long-standing challenge and enabling full life cycle visibility of each sample. This centralised system now supports rigorous QA/QC, fast and efficient cross-disciplinary analysis and a robust foundation for developing geometallurgical models and domain interpretation. A live dashboard, GEMEX (Exploration Geometallurgy) was developed using Microsoft Power BI to accelerate data analytics, visualisation and decision-making. The outcome is traceable and an auditable data pipeline that preserves the value of early characterisation and improves continuity between exploration, studies and operations.

KEY INSIGHTS

Early application of the characterisation framework generated several meaningful insights across exploration projects. Targeted metallurgical testing identified domains with lump product potential, enabling early evaluation of product pathways and associated reductions in crushing requirements and CO_2 emissions.

Hyperspectral data identified the presence of smectite ie a group of swelling clay minerals that were not visible through traditional geological and textural logging, enabling early recognition of potential processing and material handling risks. Since swelling clays can pose various risks to mining operations, including material handling, ground stability issues, processing challenges and tailings management, a detailed mapping of their occurrences in each drilled hole provided a means to incorporate the risk profile in the resource model and quantify their distribution across the deposit.

In several studies, the integrated data set demonstrated limited or no upgrade benefit via beneficiation, allowing future studies to optimise sampling and downstream test work and focus resources more effectively on areas with higher uncertainty or greater processing risk. For example, ore physical property assessment in some of the studies highlighted zones of high-grade and friable material which, despite favourable chemistry, pose challenges for mining, processing and material handling.

A significant milestone was the successful migration of metallurgical data into the acQuire™ database achieving full traceability from core interval through to processing response, something that has historically been difficult to implement. Collectively these insights demonstrate how early integrated characterisation can reduce technical uncertainty, refine study assumptions and create tangible value well ahead of feasibility-level study.

CONCLUSIONS

Early-stage orebody characterisation provides a critical foundation for understanding processing behaviour, identifying technical risks and unlocking value opportunities well before a project enters formal study phases. The framework presented in this paper demonstrates a practical and scalable approach for integrating geological, mineralogical, geotechnical and metallurgical data from the first drill core onwards. By combining strategic sampling, high resolution digital scanning, targeted composite selection and metallurgical testing the workflow delivers rapid and meaningful geometallurgical insights at a point in the project life cycle where they have the greatest influence.

Overall, the approach demonstrates that geometallurgical thinking does not need to begin at feasibility or operations; it can and should start once drill core is obtained. A fit-for-purpose early-stage characterisation framework enables organisations to make more informed decisions, reduce technical surprises and maximise value from their iron ore assets from the first drill hole through to long-term operational performance. Looking forward, the continued application of this framework and expansion of the underlying geometallurgical database will provide a foundation for advanced data analytics. As larger, more diverse data sets are accumulated, there is substantial potential to apply machine learning and pattern recognition techniques to identify proxies for ore behaviour in sparsely sampled or unsampled projects. This represents a significant opportunity to improve predictability, reduce uncertainty, and further enhance the value of integrated resource characterisation. The transition toward predictive geometallurgical intelligence will assist in lowering operational energy intensity and aligns directly with Fortescue's long-term decarbonisation strategy.

ACKNOWLEDGEMENTS

The authors wish to thank Fortescue Ltd for permission to publish this paper. We would also like to thank and acknowledge former and present Fortescue staff for their contributions to the Resource Characterisation program, in particular Sophie Sperring, Vanessa Squires and Agustin Bejerman who were involved in the design and implementation of the data migration project. Bureau Veritas is also acknowledged, in particular Ryan Carrington, for collaborative engagement and technical input during the implementation of the data migration.

REFERENCES

Haest, M, Cudahy, T, Laukamp, C, Ramanaidou, E, Gregory, S, Stark, J C and Podmore, D, 2011. Characterisation of Bedded and Channel Iron Ore Deposits Using CSIRO's Hylogging Systems, in *Proceedings of the Iron Ore Conference*, pp 249–255 (The Australasian Institute of Mining and Metallurgy: Melbourne).

Lamberg, P, 2011. Geometallurgy – Towards integrated models for mine planning, in *Proceedings of the 14th International Conference of the International Association for Mathematical Geosciences (IAMG)*, pp 1–6.

Lind, G, Lamberg, P and Oliveira, D, 2018. Geometallurgy – A route to more resilient mine operations, *Minerals*, 8(12):560.

Parian, M, Lamberg, P and Rosenkranz, J, 2021. Process simulations in mineralogy-based geometallurgy of iron ores, *Mineral Processing and Extractive Metallurgy: Transactions of the Institutions of Mining and Metallurgy,* 130(1):25–30. doi:10.1080/25726641.2018.1507072

Vatandoost, A, Beven, J, Campbell-Hardwick, S and Young, J, 2013. A Geometallurgical Approach for Iron Ore Product Evaluation, in *Proceedings of the The Second AusIMM International Geometallurgy Conference*, pp 259–266 (The Australasian Institute of Mining and Metallurgy: Melbourne).

Author index

www.ingramcontent.com/pod-product-compliance
Lightning Source LLC
LaVergne TN
LVHW060210210926
848481LV00008B/2687
* 9 7 8 1 9 2 2 3 9 5 5 5 9 *